Quantitative Process Control Theory

AUTOMATION AND CONTROL ENGINEERING

A Series of Reference Books and Textbooks

Series Editors

FRANK L. LEWIS, Ph.D.,
Fellow IEEE, Fellow IFAC
Professor
Automation and Robotics Research Institute
The University of Texas at Arlington

SHUZHI SAM GE, Ph.D.,
Fellow IEEE
Professor
Interactive Digital Media Institute
The National University of Singapore

Intelligent Diagnosis and Prognosis of Industrial Networked Systems, *Chee Khiang Pang, Frank L Lewis, Tong Heng Lee, Zhao Yang Dong*

Classical Feedback Control: With MATLAB® and Simulink®, Second Edition, *Boris J. Lurie and Paul J. Enright*

Synchronization and Control of Multiagent Systems, *Dong Sun*

Subspace Learning of Neural Networks, *Jian Cheng Lv, Zhang Yi, and Jiliu Zhou*

Reliable Control and Filtering of Linear Systems with Adaptive Mechanisms, *Guang-Hong Yang and Dan Ye*

Reinforcement Learning and Dynamic Programming Using Function Approximators, *Lucian Buşoniu, Robert Babuška, Bart De Schutter, and Damien Ernst*

Modeling and Control of Vibration in Mechanical Systems, *Chunling Du and Lihua Xie*

Analysis and Synthesis of Fuzzy Control Systems: A Model-Based Approach, *Gang Feng*

Lyapunov-Based Control of Robotic Systems, *Aman Behal, Warren Dixon, Darren M. Dawson, and Bin Xian*

System Modeling and Control with Resource-Oriented Petri Nets, *Naiqi Wu and MengChu Zhou*

Sliding Mode Control in Electro-Mechanical Systems, Second Edition, *Vadim Utkin, Jürgen Guldner, and Jingxin Shi*

Optimal Control: Weakly Coupled Systems and Applications, *Zoran Gajić, Myo-Taeg Lim, Dobrila Skatarić, Wu-Chung Su, and Vojislav Kecman*

Intelligent Systems: Modeling, Optimization, and Control, *Yung C. Shin and Chengying Xu*

Optimal and Robust Estimation: With an Introduction to Stochastic Control Theory, Second Edition, *Frank L. Lewis, Lihua Xie, and Dan Popa*

Feedback Control of Dynamic Bipedal Robot Locomotion, *Eric R. Westervelt, Jessy W. Grizzle, Christine Chevallereau, Jun Ho Choi, and Benjamin Morris*

Intelligent Freight Transportation, *edited by Petros A. Ioannou*

Modeling and Control of Complex Systems, *edited by Petros A. Ioannou and Andreas Pitsillides*

Wireless Ad Hoc and Sensor Networks: Protocols, Performance, and Control, *Jagannathan Sarangapani*

Stochastic Hybrid Systems, *edited by Christos G. Cassandras and John Lygeros*

Hard Disk Drive: Mechatronics and Control, *Abdullah Al Mamun, Guo Xiao Guo, and Chao Bi*

Automation and Control Engineering Series

Quantitative Process Control Theory

Weidong Zhang
Shanghai Jiaotong University, Shanghai, People's Republic of China

CRC Press
Taylor & Francis Group
Boca Raton London New York

CRC Press is an imprint of the
Taylor & Francis Group, an **informa** business

CRC Press
Taylor & Francis Group
6000 Broken Sound Parkway NW, Suite 300
Boca Raton, FL 33487-2742

First issued in paperback 2017

Version Date: 20111020

ISBN 13: 978-1-138-07753-9 (pbk)
ISBN 13: 978-1-4398-5557-7 (hbk)

Visit the Taylor & Francis Web site at
http://www.taylorandfrancis.com

and the CRC Press Web site at
http://www.crcpress.com

Dedicated
to My Parents

Contents

List of Figures

List of Tables

Symbol Description

Abbreviations

DMC	Dynamic matrix control
DOF	Degree-of-freedom
IAE	Integral absolute error
IMC	Internal model control
ISE	Integral squared error
ITAE	Integral of time multiplied by absolute error
LFT	Linear fractional transformation
LHP	Left half-plane
LMI	Linear matrix inequality
LQ	Linear quadratic
LQG	Linear quadratic Gaussian
MAC	Matrix algorithmic control
MIMO	Multi-input/multi-output
MP	Minimum phase
MPC	Model predictive control
NMP	Non-minimum phase
PID	Proportional-integral-derivative
QPCT	Quantitative process control theory
RGA	Relative gain array
RHP	Right half-plane
SISO	Single-input/single-output
SSV	Structured singular value
SVD	Singular value decomposition

Symbols

$C(s)$ ($\boldsymbol{C}(s)$)	Unity feedback loop controller
$C_1(s)$ ($\boldsymbol{C}_1(s)$)	Controller of the reference loop
$C_2(s)$ ($\boldsymbol{C}_2(s)$)	Controller of the disturbance loop
$d(s)$ ($\boldsymbol{d}(s)$)	Disturbance at the plant output
$d'(s)$ ($\boldsymbol{d}'(s)$)	Disturbance at the plant input
$e(s)$ ($\boldsymbol{e}(s)$)	Tracking error
$G(s)$ ($\boldsymbol{G}(s)$)	Nominal plant (plant model)
$\tilde{G}(s)$ ($\tilde{\boldsymbol{G}}(s)$)	Real plant

$G_A(s)$ $(\boldsymbol{G_A}(s))$	All-pass part of $G(s)$ $(\boldsymbol{G}(s))$
$G_D(s)$ $(\boldsymbol{G_D}(s))$	Time delay part of $G(s)$ $(\boldsymbol{G}(s))$
$G_{MP}(s)$ $(\boldsymbol{G_{MP}}(s))$	MP part of $G(s)$ $(\boldsymbol{G}(s))$
$\boldsymbol{G_N}(s)$	All-pass part of $\boldsymbol{G_O}(s)$
$G_O(s)$ $(\boldsymbol{G_O}(s))$	Rational part of $G(s)$ $(\boldsymbol{G}(s))$
$H(s)$ $(\boldsymbol{H}(s))$	Matrix for internal stability verification
H_2	Set of all stable strictly proper functions without poles on the imaginary axis
H_∞	Set of all stable proper functions without poles on the imaginary axis
$J(s)$ $(\boldsymbol{J}(s))$	Filter
K	Gain of a plant
K_C	Gain of a PID controller
k_j	Multiplicity of a RHP zero z_j
k_{ij}	Largest multiplicity of z_j in the ith column of $\boldsymbol{G}^{-1}(s)$
K_u	Ultimate gain
l_j	Multiplicity of a RHP pole p_j
l_{ij}	Largest multiplicity of p_j in the ith row of $\boldsymbol{G}(s)$
$L(s)$ $(\boldsymbol{L}(s))$	Open-loop transfer function
$\boldsymbol{M}(s)$	Matrix for robustness verification
p_j	jth pole
$Q(s)$ $(\boldsymbol{Q}(s))$	IMC controller
$Q_{opt}(s)$ $(\boldsymbol{Q_{opt}}(s))$	Optimal IMC controller
$r(s)$ $(\boldsymbol{r}(s))$	Reference
$R(s)(\boldsymbol{R}(s))$	Controller of the Smith predictor
r_p	Number of RHP poles of a plant
r_z	Number of RHP zeros of a plant
$S(s)$ $(\boldsymbol{S}(s))$	Nominal sensitivity transfer function
$\tilde{S}(s)$	Real sensitivity transfer function
$T(s)$ $(\boldsymbol{T}(s))$	Complementary sensitivity transfer function (closed-loop transfer function)
T_D	Derivative constant of a PID controller
T_F	Filtering constant of a PID controller
T_I	Integral constant of a PID controller
T_p	Resonance peak
t_r	Rise time
T_u	Ultimate period
$u(s)$ $(\boldsymbol{u}(s))$	Controller output
$\hat{u}(t)$	Constrained controller output
$\boldsymbol{v_j}$, $\boldsymbol{v_{jk}}$	Direction of the zero z_j
$W(s)$ $(\boldsymbol{W_{p1}}(s), \boldsymbol{W_{p2}}(s))$	Performance weighting functions
$\boldsymbol{W_1}(s), \boldsymbol{W_2}(s)$	Uncertainty weighting functions
$y(s)$ $(\boldsymbol{y}(s))$	Plant output
z_j	jth zero

Greek Characters

α_i	Smallest relative degree of all elements in the ith column of $\boldsymbol{Q_{opt}}(s)$
γ	Closed contour contained in Ω
$\delta_m(s)$ ($\boldsymbol{\delta_m}(s)$)	Uncertainty
$\Delta(s)$ ($\boldsymbol{\Delta}(s)$)	Normalized uncertainty
$\Delta_m(s)$ ($\boldsymbol{\Delta_m}(s)$)	Uncertainty profile
$\boldsymbol{\Delta_p}(s)$	Performance block in $\boldsymbol{\Delta}(s)$
$\boldsymbol{\Delta_u}(s)$	Uncertainty block in $\boldsymbol{\Delta}(s)$
θ	Time delay of a plant
θ_{ij}	Time delay of the ijth element of $\boldsymbol{G}(s)$
θ^{ij}	Prediction of the ijth element of $\boldsymbol{G}^{-1}(s)$
θ_{li}	Largest prediction of the ith column of $\boldsymbol{G}^{-1}(s)$
θ_{si}	Smallest time delay of the ith row of $\boldsymbol{G}(s)$
λ, λ_i	Performance degree
λ_{ij}	ijth relative gain
$\lambda_{ei}[\boldsymbol{T}(j\omega)]$	ith eigenvalue of $\boldsymbol{T}(j\omega)$
$\mu[\boldsymbol{M}(j\omega)]$	Structured singular value of $\boldsymbol{M}(j\omega)$
$\rho[\boldsymbol{T}(j\omega)]$	Spectral radius of $\boldsymbol{T}(j\omega)$
σ	Overshoot
$\sigma_i(\boldsymbol{T}(j\omega))$	ith singular value of $\boldsymbol{T}(j\omega)$
$\bar{\sigma}[\boldsymbol{T}(j\omega)]$	Maximum singular value of $\boldsymbol{T}(j\omega)$
$\underline{\sigma}[\boldsymbol{T}(j\omega)]$	Minimum singular value of $\boldsymbol{T}(j\omega)$
τ	Time constant of a plant
Ω	Simply connected open subset of the complex plane

Special Notation

$\|\cdot\|_1$	1-norm
$\|\cdot\|_2$	2-norm
$\|\cdot\|_\infty$	∞-norm
$:=$	Is defined as
$\forall$	For all
$\in$	Belong to
$\otimes$	Element-by-element product
$\bar{z}_r(\bar{\boldsymbol{A}})$	Complex conjugate
adj	Adjoint
deg	Degree of a polynomial
det	Determinant of a matrix
diag	Diagonal matrix
$\boldsymbol{T}^T(s)$	Transpose of a matrix or vector
$\boldsymbol{T}^H(j\omega)$	Complex conjugate transpose of a matrix: $\boldsymbol{T}^H(j\omega) = \bar{\boldsymbol{T}}^T(j\omega)$
$\boldsymbol{T}^*(s)$	Conjugate transpose of a system: $\boldsymbol{T}^*(s) = \boldsymbol{T}^T(-s)$

Im	Imaginary part of a complex number
$N_+(s)$	Polynomials with roots in the closed RHP
$N_-(s)$	Polynomials with roots in the open LHP
Re	Real part of a complex number
sup	Supremum
Trace	Trace of a matrix

Preface

Since the Industrial Revolution, control systems have played important roles in improving product quality, saving energy, reducing emissions, and relieving the drudgery of routine repetitive manual operations. In the past hundred years, many theories have been proposed for control system design. However, there are three main problems when some of these advanced control theories are applied to industrial systems:

1. These theories depend on empirical methods or trial-and-error methods in choosing weighting functions.
2. Both the design procedures and results are complicated for understanding and using.
3. The controllers cannot be designed or tuned for quantitative engineering performance indices (such as overshoot or stability margin).

In this book, an improved theory called the Quantitative Process Control Theory is introduced to solve these problems. This new theory has three features:

1. When using the theory, the designer is not required to choose a weighting function.
2. The design is suboptimal and analytical. It is easy to understand and use.
3. The controller can be designed or tuned for quantitative engineering performance indices.

These features enable the controller to be designed efficiently and quickly.

Mathematical proofs are provided in this book for almost all results, especially when they contribute to the understanding of the subjects presented. This will, I believe, enhance the educational value of this book. As few concepts as possible are introduced and as few mathematical tools as possible are employed, so as to make the book accessible. Examples are presented at strategic points to help readers understand the subjects discussed. Chapter summaries are included to highlight the main problems and results. At the end of each chapter, exercises are provided to test the reader's ability to apply the theory he/she has studied. They are an integral part of the book. There is no doubt that a serious attempt to solve these exercises will greatly improve one's understanding.

The methods developed here are not confined to process control. They are equally applicable to aeronautical, mechanical, and electrical engineering. To stress this point, examples with different backgrounds are adopted. With a few exceptions, these examples are based on real plants, including

- Paper-making machine
- Heat exchanger
- Hot strip mill
- Maglev
- Nuclear reactor
- Distillation column/Heavy oil fractionator
- Jacket-cooled reactor
- Missile
- Helicopter/Plane
- Anesthesia

The book is divided into 14 chapters. Important topics that are covered include

1. Introduction and review of classical analysis methods (Chapter 2)
2. Essentials of the robust control theory (Chapter 3)
3. H_∞ and H_2 proportional-integral-derivative controllers for stable plants with time delay (Chapters 4 and 5)
4. Quasi-H_∞ and H_2 controllers for stable plants with time delay (Chapter 6)
5. Quasi-H_∞ and H_2 controllers for integrating plants with time delay (Chapter 7)
6. Quasi-H_∞ and H_2 controllers for unstable plants with time delay (Chapter 8)
7. Complex control strategies, including two degrees-of-freedom control, cascade control, anti-windup control, and feedforward control (Chapter 9)
8. Analysis of multi-input/multi-output control systems (Chapter 10)
9. Classical multi-input/multi-output system design, including decentralized control and decoupling control (Chapter 11)
10. Quasi-H_∞ decoupling control for plants with time delay (Chapter 12)

11. H_2 optimal decoupling control for plants with time delay (Chapter 13)
12. Multivariable H_2 optimal control (Chapter 14)

This book is intended for a wide variety of readers. It is appropriate for higher level undergraduates and graduates in engineering, beginners in the research area of robust control, and engineers who want to learn new design techniques. It is assumed that readers have had an undergraduate course in classical control theory. A prior course on optimal control or process control would be helpful but is not a requirement.

This book has grown out of 15 years of research. The procedure is always much harder than anyone anticipates. I received financial support from the National Science Foundation of China, the Alexander von Humboldt Foundation, Germany, and the National Science Fund for Distinguished Young Scholars, China, which enabled me to pursue the research. I am vastly indebted to many people who have helped and inspired me to start, continue, and complete this book.

My first thanks goes to Professor Shengxun Zhang and Professor Youxian Sun, Zhejiang University. They brought me into the area of process control. I am grateful for the continuing help and support from Professor Xiaoming Xu, Professor Yugeng Xi, Professor Songjiao Shi, Professor Zuohua Tian, and Professor Xinping Guan at Shanghai Jiaotong University. I am also greatly indebted to Professor F. Allgöwer and Professor C.A. Floudas, who hosted me at the University of Stuttgart and Princeton University, respectively, as a visiting professor during the writing of this book.

The first six chapters of this book have been classroom tested for several years at Shanghai Jiaotong University. Many students have contributed their time to the book. I would like to thank my PhD students F. S. Alcántara Cano, Danying Gu, Daxiao Wang, and Mingming Ji for particularly helpful suggestions.

The book makes limited use of the material from several books. In particular, I want to express my sincere appreciation to Morari and Zafiriou (1989), Doyle et al. (1992), and Dorf and Bishop (2001).

Family members are a source of special encouragement in a job of this magnitude, and I send love and thanks to my parents and my son in this regard.

Lastly, I thank my wife, Chen Lin. She read the manuscripts of different versions and made corrections in her spare time. She gave hundreds of suggestions on editing, grammar, and technical problems. This book would not be the same without her enormous care and patience.

Weidong Zhang

About the Author

Weidong Zhang received his BS, MS, and PhD degrees from Zhejiang University, China, in 1990, 1993, and 1996, respectively, and then worked as a postdoctoral fellow at Shanghai Jiaotong University. He joined Shanghai Jiaotong University in 1998 as an associate professor and has been a full professor since 1999. From 2003 to 2004 he worked at the University of Stuttgart, Germany, as an Alexander von Humboldt Fellow. From 2007 to 2008 he held a visiting position at Princeton University. In 2011 he was appointed chair professor at Shanghai Jiaotong University.

Dr. Zhang's research interests include control theory and its applications, embedded systems, and wireless sensor networks. He is probably the earliest researcher on the automotive reversing ultrasonic radar in China. He has many years of industry experience and was a control engineering consultant at Atmel Corporation (Shanghai R&D Center) in 2005.

Dr. Zhang is the author of more than 200 refereed papers and holds 15 patents. He is a recipient of National Science Fund for Distinguished Young Scholars of China.

Correspondence address:
Prof. Weidong Zhang
Department of Automation
Shanghai Jiaotong University
Shanghai 200240, P. R. China
Email: wdzhang@sjtu.edu.cn
Web: automation.sjtu.edu.cn/wdzhang (in English)
automation.sjtu.edu.cn/ipac (in Chinese)

1

Introduction

CONTENTS

A control system is an interconnection of components that provides the desired system output for a given input. The object to control is called the plant, while the device to generate the input to the plant is called the controller. The control system is most often based on the principle of feedback, whereby the controller adjusts the input to the plant so as to keep the deviation between the desired value and the actual value of system output as small as possible.

The process of constructing a basic feedback control system generally involves two steps: developing the plant model and designing the controller. The goal of this book is to present a quantitative design theory that captures the essential issues, solves practical problems, and provides interested readers with new materials for further study.

1.1 A Brief History of Control Theory

By looking back on the development of control theory, this section reviews the main trend and important developments in the area.

Although automatic control devices of various sorts date back to antiquity, it was J.C. Maxwell who provided the first rigorous mathematical analysis for feedback control systems in 1868 (Figure 1.1.1). His work on the stability analysis of a centrifugal governor is generally taken as the starting point of control theory development. The pioneering work of Bode and Nyquist prior to World War II paved the way for the development of control theory. As the core of classical control theory, the frequency domain technique they presented not only has evident engineering and physical meanings, but also makes it possible to give acceptable solutions to practical problems. Even today, the technique is an indispensable means for analyzing and designing control systems.

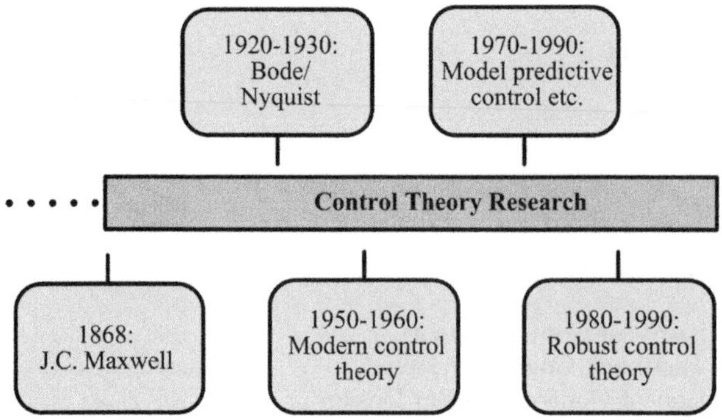

FIGURE 1.1.1
Selected historical developments of control systems.

Since the frequency domain technique originated from the impetus of practice, rather than rigorous systematic theories, the formulation is far from mathematization. The frequency domain technique provides tools for control system design, yet the design procedure remains very much an art, and normally results in non-unique feedback systems. There are some problems that need to be clarified for in-depth and thorough studies. For example:

1. What is the mathematical objective for control system design?
2. How is the control system optimized in the design procedure?

In view of these problems, the modern control theory was proposed in the early 1960s. New theoretical tools were introduced and some important problems, such as optimality, controllability and observability, were considered. Modern control theory provides the unique optimal solution for the design of control systems and makes it possible to solve multivariable control problems in a unified framework.

Since the appearance of modern control theory, there has been a strong desire to apply it to industrial systems. Unfortunately, the results were much less than expected in many such cases.

A number of possible reasons for this failure can be identified. For example, modern control theory adopts the state space method. The problem studied in this method is in fact a mathematical problem. There, engineering intuition has very little effect. For engineers familiar with the frequency response of physical systems, it is difficult to use the sophisticated mathematical theory in solving practical control problems. More serious is that the theory does not address the model uncertainty problem, which is of practical importance. Nevertheless, this problem can be treated with the classical control theory by using notions like gain margin and phase margin.

Regardless of the design technique used, the controller is always designed based on the information as to the dynamic behavior of plant. It is almost impossible to exactly model a real physical plant. There is always uncertainty. Therefore, it is desirable that the controller be insensitive to the model uncertainty; that is, the controller should be robust. Since the late 1970s, robustness has become a major objective of control study and related achievements have formed robust control theory.

The robust control theory based on the state space technique provided elegant solutions to both optimal and robust design problems and thus seemed to hold high promise for applications. However, even though robust control theory has developed for several decades, its effect in industrial practice is still not obvious nowadays. Both the design procedure and result of the new theory are too complicated for engineers to use. The level of abstraction makes it accessible only by the researchers in this special area and the selection of weighting functions still depends on trial and error. In addition, practical requirements in the design of control systems are usually quantitatively specified in terms of time domain response (such as overshoot, amplitudes of coupled responses, and so on) or frequency domain response (such as resonance peak, stability margin, and so on). For example, the design specification might be that the worst-case overshoot is 5% when uncertainty exists. This specification is not easy to reach with those developed methods.

In parallel to the development of the state space method, a class of new algorithms exemplified by model algorithmic control (MAC) and dynamic matrix control (DMC) were invented and successfully applied to industrial systems. These algorithms were internally related to some classical methods, such as the Dahlin algorithm and Smith predictor. Some modern robust control characteristics had been incorporated in them in an ad hoc fashion. These algorithms are now generally known by the generic term model predictive control (MPC). Another theory developed at the same time, which can be regarded as a frequency domain version of MPC, is the internal model control (IMC). IMC explains some important problems of control system design in a simple and direct framework and thus has a profound influence on feedback control theory. Since these related methods are based on a firm footing and provide simple design means, they greatly boost the application of advanced control theories in industry.

In the past two decades, the most active direction in the control area is the linear matrix inequality (LMI). The main attraction of LMIs is that they are very flexible, so that a variety of problems can be expressed as LMIs. Nevertheless, although the LMI method provides a powerful tool to solve control problems, it is much more complicated than other methods. The complexity of the design procedure and result is a main obstacle for the application of the LMI method.

Almost all real plants involve nonlinearity. One may think that almost all methods introduced above are developed for linear systems and thus are not applicable to nonlinear systems. This is somewhat misleading. The control engineering practice in the past hundred years shows that most nonlinear plants can be controlled well by controllers developed for linear systems. In many cases, a well-tuned PID controller is enough.

1.2 Design of Feedback Control Systems

The most elementary feedback control system has three components: a plant, a sensor to measure the system output, and a controller. Usually, the actuator is lumped in with the plant. The input of the feedback control system, which is called the reference (or set-point), is the desired output of the system. The controller is normally an equation or an algorithm. What the controller does is to compare the system output with the reference and, if an error exists, to manipulate the plant input so that the error is driven toward zero.

In this book, only linear systems are considered. A system is linear if the principle of superposition applies. The principle of superposition states that the response produced by the simultaneous applications of two different forcing functions is the sum of the two individual responses.

Figure 1.2.1 shows the feedback control system of a paper-making process, of which the goal is to produce paper with constant basis weight. The basis weight, denoting the thickness of paper, is the weight in grams of a single sheet of paper with the area of 1 m^2. In the system, the plant is the paper-making machine, the input of the paper-making machine, referred to as the control variable (or manipulated variable), is the flow rate of stock with certain consistency, and the output of the paper-making machine, referred to as the controlled variable, is the basis weight of paper. The higher the flow rate of stock, the heavier the basis weight of paper; contrarily, the lower the flow rate of stock, the lighter the basis weight of paper. The controller is normally implemented with a computer. The actuator is the valve adjusting the flow rate of stock. The sensor is the basis weight gauge. The reference is the desired basis weight. The computer compares the actual basis weight from the basis weight gauge with the desired basis weight (that is, the reference), determines the deviation (that is, the error), and sends out a signal to the valve to draw the basis weight to the desired value.

The design problems of most control systems are similar to that of the paper-making system. The procedure generally involves two steps:

1. Analyzing—What dynamic behavior does the plant have? What is the control objective?
2. Design—How to design the controller to satisfy the requirement?

The procedure may require judgments and iterations (Figure 1.2.2).

Models

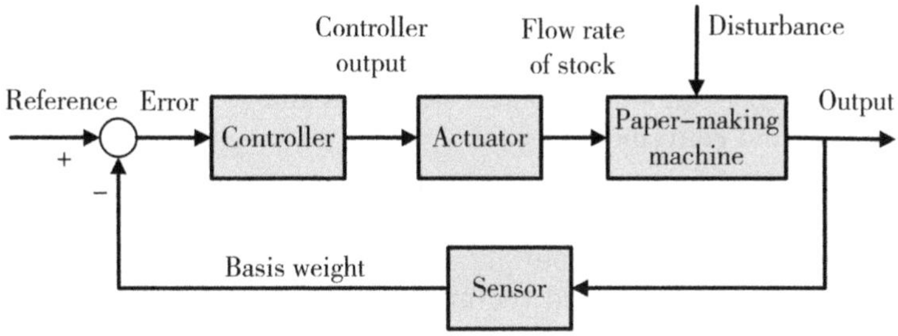

FIGURE 1.2.1
Paper-making process control.

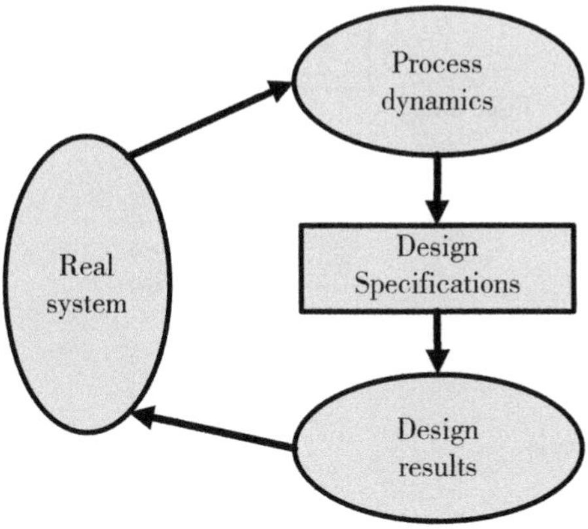

FIGURE 1.2.2
Design procedure of a feedback control system.

To design a control system, the designer needs to know how the plant output is quantitatively influenced by its input over time. In other words, a mathematical model that describes the dynamic behavior of the plant must be obtained. What should a dynamic model provide? It should capture the main dynamic behavior of a physical plant and predict the input-output response. Models are the basis of control system design. What is more, with the help of models the designer can adjust control strategies and controller parameters in an economical and convenient way.

The model used for control system design is referred to as the nominal plant. It can be built with a mechanism-based method (that is, build the model by applying the laws of physics, chemistry, and so on) or with an identification method (that is, build the model from the measured input-output data) (Figure 1.2.3). In the procedure, there are always dynamics that cannot be incorporated into the model. The difference between the nominal plant and the real plant is the uncertainty. All of the uncertain plants form a family, in which the nominal plant can be regarded as its "center." Details of the uncertainty might be unknown, but a bound of it, in some cases, can be estimated.

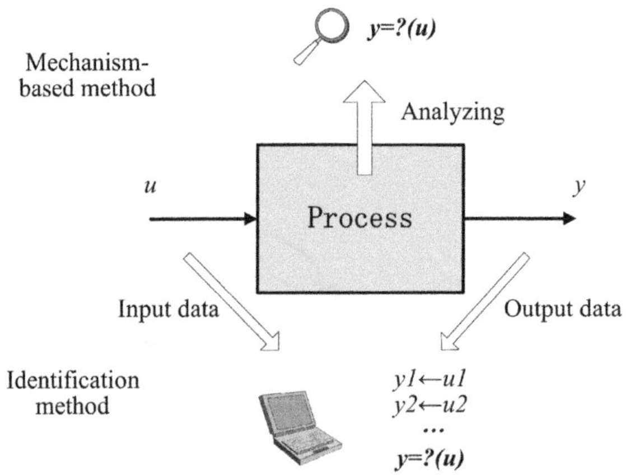

FIGURE 1.2.3
Two modeling methods.

Compared with control systems in other areas, the uncertainty problem is more prominent in industrial systems. This is due to not only technical reasons, but also economical reasons. To make sure the design result can be successfully applied to practical systems, enough importance must be attached to the uncertainty problem.

For some plants, if a change takes place at the input, its effect can be instantly observed from the behavior of the output. This is not true for many industrial plants. For example, in the paper-making process, when the flow rate of stock changes, its effect on basis weight cannot be observed instantaneously because of the transport time required for the fluid to flow through piping. The transport time is the so-called time delay (or dead time). In industry, plants with time delay are commonly encountered.

Objectives

Generally speaking, the objective of a control system is to make the output behave in a desired way by manipulating the input. In terms of operating features of the system, there are two kinds of common objectives:

1. Regulator problem—To keep the system output close to some equilibrium point.
2. Servo problem—To keep the difference small enough between the output of the system and the given reference.

In most cases, the design for the regulator problem is identical to that for the servo problem.

The main problem investigated in process control is the regulator problem. Such systems are required to have good disturbance rejection capability; that is, the system output is kept close to the reference in the presence of disturbance, with certain precision. Nevertheless, this does not mean that the reference never changes; instead it does not change often. As a matter of fact, the product during the change of the reference is usually useless. For instance, in the paper-making process, the basis weight is 70g at one phase and 90g at another phase. During the period that the reference is changed from 70g to 90g, all of the produced paper is waste.

Design

Loosely speaking, there are two sorts of methods to design a controller. One is the traditional empirical method. The designer chooses a controller according to the plant dynamics. Normally the controller is a proportional-integral-derivative (PID) controller. The controller parameters are adjusted by rules of thumb after the controller has been installed. The other is the model-based method. In this method, both the structure and parameters of the controller are derived based on models. Since experience is still necessary, both art and science are involved in designing such a controller. Almost all advanced design methods are based on models. The most important merit of such a method lies in its rigor. Even in engineering disciplines, rigor can lead to clarity and methodical solutions to problems.

In industrial systems, a controller with parameters that cannot be adjusted is seldom used because of several reasons:

1. It is a challenge to obtain the exact information about the plant uncertainty.
2. The operating point of control system may offset or change.

3. The design requirements of control systems may be changed after the system comes into operation.

The controller commonly used (for example, the PID controller) has a fixed structure and adjustable parameters. The static and dynamic requirements of the closed-loop system are met by selecting proper values for the controller parameters in the field. This field adjusting procedure is the so-called controller tuning.

The achievable performance of the whole control system relates to the plant. In some cases, obvious improvement of performance can only be obtained by modifying the plant itself rather than the controller.

The design procedure of a control system can be complicated. Fortunately, the computer technology developed rapidly in the past several decades; some excellent software packages (such as MATLAB®) are available. These software packages provide powerful tools for control system design. With them the designer can simulate uncertain systems, adjust and compare the performances of different control systems, and understand how the fundamental control constraints affect the system.

1.3 Consideration of Control System Design

In the past hundred years, many theories have been proposed for control system design. Among them the linear quadratic (LQ) optimal control and the H_∞ optimal control are the most important two. However, there exist three problems when some of these advanced control theories are applied to industrial systems:

1. These theories depend on empirical methods or trial-and-error methods in choosing weighting functions.
2. Both the design procedures and the results are complicated for understanding and using, in particular, when the plant involves a time delay.
3. The controller cannot be designed or tuned for quantitative engineering performance indices (such as overshoot or stability margin).

To solve these problems, the Quantitative Process Control Theory (QPCT) is proposed in this book by extending the ideas and methods of classical control theory, optimal control theory, and robust control theory. In QPCT, the following problems are considered:

1. How can the choosing of weighting functions be simplified?

   Most of the advanced design methods are optimization-based methods. The first step of these methods is choosing weighting functions. Unfortunately, even today there are no clear rules for choosing weighting functions. Only empirical methods or trial-and-error methods are available. This implies that different designers will obtain different controllers even if the same method is used. The designers do not know whether the best controller is obtained. In QPCT, a fixed weighting function is adopted. The designer is not required to choose weighting functions. The role of weighting functions is substituted by introducing a filter to the optimal controller.

2. Can the control system with time delay be analytically designed with optimal methods?

   The design methods adopted in practice are usually empirical methods. The controller cannot be analytically designed with optimal methods even if the plant is rational. Many optimal design methods have been proposed. However, most of them are numerical methods, which involve intricate design procedures. The merit of the analytical design is that the designer can use formulas to design a controller. In this way, the design task is significantly simplified. In QPCT, the analytical optimal controller is obtained by employing the controller parameterization and the plant factorization techniques.

3. How is the order of the optimal controller related to the plant order?

   The controller designed with the optimal design methods is usually of high order. For high-order controllers there exist many problems in realization and application. It is desirable to know the relationship between the order of the optimal controller and the plant order, so as to understand how a low-order controller can be obtained. For many methods, the relationship is not known. In QPCT, the result is analytical. Hence, there is a direct relationship between the controller order and the plant order.

4. How is the new design theory related to classical performance indices?

   The main advantage of modern control theory is that the controller can be designed with the optimal method. However, the theory was not widely adopted in industrial systems. To some extent, this is because the performance index of modern control theory has little relationship with engineering design requirements. For example, in practice, the performance index is normally given in terms of overshoot, stability margin, amplitudes of coupled responses, etc., which are difficult to describe in modern control theory. In QPCT, the performance degree is defined, the quantitative relationship between the optimal performance indices and engineering performance indices is built based on the performance degree, and the quantitative design method is presented.

5. Can the performance and the robustness be easily tuned?

   Designers may find it difficult for classical design methods to make a clear and reasonable tradeoff between conflicting performance indices, for example, nominal performance and robustness. Some new methods can make the tradeoff if the uncertainty profile is exactly known. However, the profile is usually difficult to obtain in practice owing to technical or economical reasons. In addition, design requirements may be changed and the uncertainty may be offset. To reduce the maintaining cost, instead of redesigning the control system, it is desirable that the design requirements be met by tuning in these cases. In QPCT, a simple and effective tuning method is provided to solve this problem.

6. Is the design method applicable to different input signals?

   In practice, the most frequently encountered signals are steps and pulses (a pulse can be obtained by combining two opposite steps). Therefore, many design methods have a default assumption; that is, the input signal is a step. Nevertheless, other signals, like ramps, may be encountered. In this case, it is desirable that the developed design method should still work. QPCT provides a design method that is applicable to different input signals.

To sum up, the goal of QPCT is to simplify the design procedure on the premise of ensuring good performance, and design controllers for quantitative performance requirements. In this theory, good performance is ensured with the help of optimal design procedures, simplicity is achieved by avoiding choosing the weighting function and designing the controller analytically, and quantitative design is realized by analyzing the relationship between the closed-loop response and controller parameters (Figure 1.3.1). The most important feature of QPCT is that it is "no-weight," "analytical," and "quantitative." With this feature, the theory provides an easy way to design a controller efficiently and quickly.

As we know, auto-tuning control is an important method for enhancing the automation level of a control system. Auto-tuning control involves two steps. The first step is identifying the model. The controller conducts its own process behavior test. The second step is parameter computing. The controller parameters are computed accordingly based on the obtained model. As all work is finished in the field computer, design formulas are necessary for the parameter computation. The LQ control and H_∞ control do not work here. They require choosing weighting functions and carrying out numerical computation. The design method in this book is particularly suitable for the requirement of auto-tuning control.

This book focuses exclusively on the frequency domain method. This is because, on one hand, the frequency domain method is easy to understand; on the other hand, the resulting controllers are easy to implement and use.

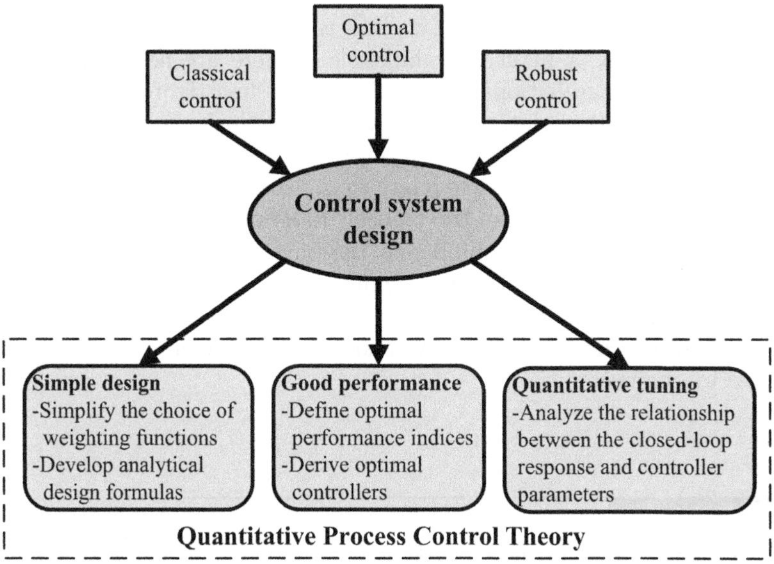

FIGURE 1.3.1
Philosophy of QPCT.

Some problems encountered frequently in the state space method can also be dealt with within this framework:

1. Stability analysis. In the analysis of stability, it was already known that the internal stability could be tested only with input-output information and all stabilizing controllers could be parameterized.
2. Optimal control. It will be shown that the optimal solution can be achieved with only input-output information.
3. State control. If a state variable needs to be controlled, the control problem can be formulated in such a way that the variable is chosen as an output.
4. State observer. If it is necessary, a state feedback system with a state observer can be converted into an output feedback system with a compensator.
5. Design constraints. Some design constraints (for example, the constraint on the control variable) can directly be considered in the new framework.

For beginners and engineers, it might be easier to understand those basic concepts of control theory in the frequency domain method than in the state space method. For example, in a single-input/single-output (SISO) system, controllability and observability of a plant is related to the zero-pole cancellations in the plant; a plant is stabilizable if there are no right half-plane (RHP) zero-pole cancellations in the plant. In a less rigorous way, one can simply understand the relationship between the transfer function model and the state space model as follows: the input and output relate to the whole transfer function, while the states relate to the coefficients of the terms in the numerator and denominator of the transfer function.

It is important to bear in mind that the purpose of adopting new control strategies in industry is not for the techniques themselves, but is to strive for more profit. Hence, control theory study should satisfy practical requirements. The results should have theoretical warranty, and also can be understood and accepted by engineers.

1.4 What This Book Is about

A good design theory should provide engineers with a framework to cast their control problems and to deal with fundamental tradeoffs and constraints in a systematic and rigorous way. The design procedure should be sufficiently easy and effective and applicable to a larger number of similar problems. This is a recurring theme throughout the book.

This book mainly considers the plant with time delay. Although the control problem of systems with time delay emerges in industry, the methods developed here are not confined to this area. As the rational system is a special case of the system with time delay, the study on the system with time delay is theoretically of primary importance. Most of the results given in this book can be directly used for rational systems.

This book emphasizes more on the design method than on the analysis method. The main content is sketched in Figure 1.4.1. The general layout is: Chapters 2 and 3 are preliminary knowledge; Chapters 4 to 8 are devoted to the design of SISO control systems; Chapter 9 discusses complex control strategies; Chapters 10 to 14 deal with the design of multi-input/multi-output (MIMO) control systems. In principle, the SISO design for stable plants is a special case of that for unstable plants, and the SISO material can be regarded as a special case of the MIMO material. For tutorial reasons they are treated separately, so that the study difficulty increases gradually. The design procedure is first introduced for special SISO cases, and then for the general SISO case, the decoupling case, and the general MIMO case. The difficulty in studying the MIMO system is larger than that in studying the SISO system. Chapter 14 is the most challenging part of this book. The analysis of stability and performance is involved in different chapters.

Mathematical proofs are provided in this book for almost all results, especially when they contribute to the understanding of the subjects presented.

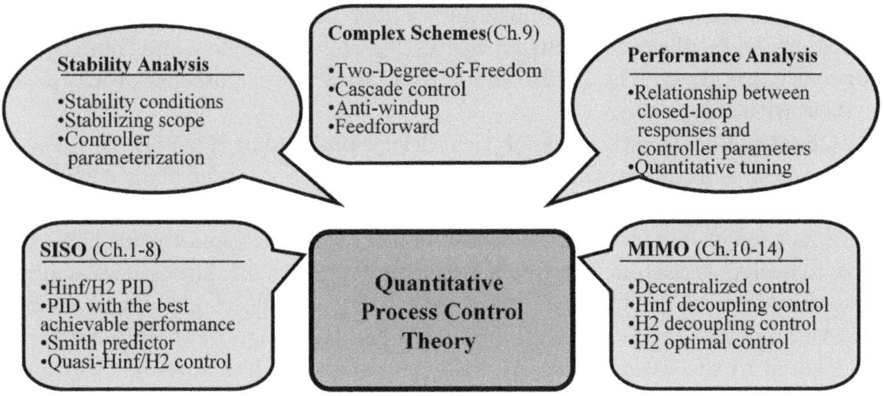

FIGURE 1.4.1
Main content of this book.

To help readers master those important ideas and methods, they are repeated with different backgrounds in different sections. The main results of this book are a series of formulas for controller design. They are introduced in the following way: the design formulas are derived by utilizing the proved theorems; design examples are presented at strategic points to help readers understand the use of the corresponding design formula. With a few exceptions, these examples are based on real plants. With the help of MATLAB, one can conveniently repeat the result in these examples, because almost all controllers are given in the analytical form, even for MIMO systems. The analytical controllers, as well as the plant models, are boxed off.

To make the book accessible, as few concepts as possible are introduced and as few mathematical tools as possible are employed, while keeping the mathematics reasonably rigorous. Chapter summaries are included to highlight the main problems and results. Exercises are provided at the end of each chapter to test the reader's ability to apply the theory he/she studies. Some exercises are straightforward, while others are much more challenging. The exercise with a star means that the knowledge about the state space method is needed, which is provided only to the readers with this background.

Chapter 2 introduces the classification of dynamic systems and the analysis methods of classical control theory. Time domain and frequency domain analysis methods of classical control theory have been widely used in industrial systems. Not only can they provide engineering and physical insights, but they are the important fundamentals for developing new theories and understanding the related topics. In Chapter 2, some principles on comparing different controllers are also discussed.

Chapter 3 introduces the basic concepts of robust control theory, including the definitions of norm, system gain, closed-loop specification, and controller parameterization. Robust stability and robust performance are defined, and necessary and sufficient conditions for testing robust stability and robust performance are given. The final topic is to analyze the robustness of a typical system with time delay.

Chapter 4 deals with the analytical design problem of H_∞ PID controllers for the first-order plant with time delay and the second-order plant with time delay. The basic idea is to approximate time delay with a rational approximation, then design the controller based on H_∞ optimization. In this chapter, how to design a controller for quantitative performance and robustness is illustrated in detail. The stabilizing scope of a PID controller is also investigated.

An analog of the H_∞ design method is the H_2 design method. Chapter 5 is devoted to the design of an H_2 PID controller. Analytical design methods are developed. The performance limit that is achievable by a PID controller is studied by utilizing rational transfer functions to approximate irrational functions involving a time delay. At the end of this chapter, the filter design problem is discussed.

In Chapter 6, the design problem of the general stable plant with time delay is discussed. With a rigorous treatment on the time delay, a quasi-H_∞ controller and an H_2 controller are analytically derived. It is shown that there exists close relationship between the two controllers and many other well-known control strategies, such as the Smith predictor, IMC, Dahlin algorithm, deadbeat control, inferential control, predictive control, and PID controller.

Chapters 2 to 6 constitute a basic treatment for control system design. Chapters 7 to 9 consider several special control problems, which are seldom discussed in current textbooks and monographs.

For stable plants, many control methods have been developed, for example, the PID controller and the Smith predictor. However, the control method for integrating plants is not so popular. Chapter 7 deals with the control problem of integrating plants. Analytical design procedures are developed, the closed-loop performance is analyzed, and the performance limit is discussed as well.

Chapter 8 concentrates on the control problem of unstable plants. A complete treatment on the optimal design problem of a linear system with time delay is given. This is accomplished by developing a simple yet effective parameterization of all stabilizing controllers, which allows us to compute the optimal controller for both stable and unstable plants with time delay. An analytical design procedure is presented and how to obtain quantitative performance and robustness is discussed.

Until now, only basic control strategies are considered for SISO feedback control systems. Chapter 9 studies two degrees-of-freedom (2 DOF) control and several other complex control strategies such as cascade control, anti-windup control, and feedforward control. The controller design problems for the optimal rejection of input disturbance and for the plant with multiple time delays are also discussed.

The control configurations considered in the preceding chapters are confined to the plant with a single output, requiring a single manipulated input.

Such a SISO system is relatively simple. The practical plant may have two or more outputs, requiring two or more manipulated inputs. Chapter 10 focuses on the analysis of MIMO systems, including the definition of zero and pole, design specification, uncertainty description, and robustness test criterion.

The design problem of a MIMO system is very challenging as compared to that of a SISO system, because there is performance tradeoff among different outputs, as well as control action tradeoff among different inputs. In Chapter 11, the classical analysis and design methods for MIMO systems are reviewed, including the pairing problem, the decentralized control, and the decoupler design.

Chapter 12 introduces the quasi-H_∞ decoupling control for MIMO systems with time delay. The main problem of the classical decoupler design is that the method cannot be applied to non-minimum-phase (NMP) plants and unstable plants. It is also difficult to analyze the effect of the decoupler on the closed-loop performance and robustness. These problems can be solved by using the quasi-H_∞ decoupling control. In the method, the design of the decoupler and the controller is finished in one step. The result is analytical and can be applied to unstable NMP plants. The performance and robustness can be quantitatively tuned.

Chapter 13 focuses on the H_2 decoupling control for MIMO systems with time delay. H_∞ control and H_2 control are two prevailing design methods. When the system performance is specified in terms of ∞-norm, simple and easy-to-understand results can be obtained for robustness analysis. For controller design, however, the H_2 control can provide more elegant results because of the orthogonality of 2-norm. In this chapter, the H_2 optimal controller is analytically derived for decoupling control. The resulting controller can be regarded as a natural extension of the SISO controller.

The last chapter of this book is Chapter 14, of which the subject is the multivariable H_2 optimal control. The decoupling control is very important in practice, since the tuning can be significantly simplified in a decoupled system. Nevertheless, the optimal control in a general sense (whose response may be non-decoupled) is more important in theoretical research. Compared with the decoupling optimality, the general optimality implies that the minimum error is reached. In this chapter, the optimal controller is analytically derived based on the proposed plant factorization and the controller parameterization.

This book makes use of a lot of materials from published papers and books. A detailed description is given in Notes and References.

1.5 Summary

The development of feedback control theory in the past several decades makes the subject more rigorous and more applicable. This necessitates a new description for the discipline. The goal of this book is to introduce an advanced design theory that is as compendious as possible and can be applied to a wide range of practical control problems.

In this book, the SISO design, the decoupling system design, and the MIMO optimal design are dealt with within a unified framework, in which the SISO design is a special case of MIMO design. This framework has the following notable features:

- The design problem of control system is endowed with a clear theory. The controller is derived based on the optimal control theory. The optimal design procedure guarantees good performance.

- The choosing of weighting functions is simplified. This is a big obstacle in practice when some advanced design methods are used. In the design procedure introduced in this book, the designer is not required to choose a weighting function.

- The optimal controller is derived with an analytical method. The controller order is directly related to the plant order. The designer can obtain the controller by using design formulas directly. Compared with numerical methods, the analytical design significantly reduces the design workload.

- The relationship between the new theory and the classical performance indices is established. The designer can design or tune the controller for quantitative performance and robustness. Field tuning is usually necessary in practice. On one hand, the exact uncertainty cannot be estimated through a few tests. On the other hand, the working condition or the design requirement may be changed after a control system comes into operation.

- Only input and output information is used. The information about the state variable is not required. The use of state variables implies that more sensors are needed. However, to reduce the cost, it is desirable to use as few sensors as possible in practice. If instead an observer is used to estimate state variables, both the structure and the implementation will become complicated.

This book focuses on the frequency domain method; nevertheless, the author strongly believes that the design methods, as well as the key ideas presented here, have special reference to other design methods, because there are certain relationships among different methods. What makes the frequency domain method attractive is that it is intuitive, which makes designers more easily master the essential of design problems.

Exercises

1. This exercise is used to illustrate a possible situation where the design method presented in this book can be applied.

   The head section of a paper-making machine consists of several tanks as depicted in Figure E1.1. The thick stock is mixed with the recycled water (called the white water) in mixing tanks. The head-box delivers the diluted suspension of fibers to a fine mesh screen called the wire. The amount of stock flowing onto the wire can be controlled by the flow rate of thick stock. Such a plant can be described by a model in the form of the first-order plant with time delay:

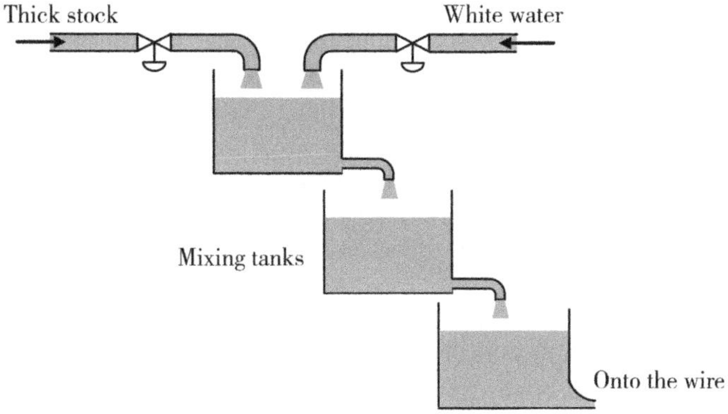

FIGURE E1.1
Head section of a paper machine.

$$G(s) = \frac{Ke^{-\theta s}}{\tau s + 1}.$$

where K=0.26, τ=2.5, and θ=0.5. As the real plant is very complex, the model parameters are uncertain. There exists a 20% error on K and θ respectively and a 30% error on τ. The design requirements are that the overshoot is less than 10% and the rise time is as short as possible.

   (a) Assume that there is no uncertainty. Design a controller to satisfy the requirements.

   (b) How to tune the controller to satisfy the design requirements for all uncertain plants?

(c) Now, due to the change of the working point, the error of θ increases to 30%. Furthermore, it is found that 5% overshoot is a more reasonable performance specification. The operator is not able to redesign the controller. Is it possible to achieve the requirements for all uncertain plants only by tuning the controller?

With regard to the design methods that you are familiar with, give some suggestions on the way to satisfy the above design requirements.

Notes and References

Mayr (1970) reviewed the history of automatic control in detail.

There is a lot of literature about the control of paper-making machines. Sun (1993), Zhang and Sun (1995), and Zhang (1996) discussed the control of low-speed paper-making machines.

Section 1.3 follows the discussion in Zhang (1998). Some ideas in this book are inspired by the work in Morari and Zafiriou (1989) and Doyle et al. (1992). Readers may regard this book as a sequel of the famous IMC method (Morari and Zafiriou, 1989).

For auto-tuning, please refer to Astrom and Hagglund (2005).

The following books provide excellent summaries of current thoughts: Zhou et al. (1996), Goodwin et al. (2001), Brosilow and Joseph (2002), Skogestad and Postlethwaite (2005), Qiu and Zhou (2009), and Wu et al. (2010). For the introduction to LMIs, readers can refer to Boyd et al. (1994).

The plant in Exercise 1 is from Sun (1993, p. 36).

2

Classical Analysis Methods

CONTENTS

This chapter and the next are most fundamental to the design methods in this book.

Before learning how to design a control system, the dynamic behavior of the plant, or equivalently the mathematical model of the plant, should be investigated. In the first part of this chapter, dynamic models of typical plants are introduced and the rational approximations of time delay are discussed. In the remainder of this chapter, the basic methods for time domain analysis and frequency domain analysis in classical control theory are reviewed; further discussion on design requirements and performance comparison are given. The analysis methods of classic control theory have been widely used in practical control systems for a long time. These methods are the basis for developing the new theory in this book and for understanding the related topics.

2.1 Process Dynamic Responses

A control system is generally designed based on the mathematical model of a plant. The model provides a functional relationship between the input and the output of the plant. Practical plants can be different equipments or units, for example, distillation column, paper-making machine, disk, maglev, etc. Although the physical and chemical phenomena taking place in these plants are different, their models are essentially similar from the viewpoint of control theory. These plants are usually described by a linear time-invariant causal model $G(t)$, where t is the continuous time variable. Causality means that $G(t) = 0$ for $t < 0$. In such a system, the output depends only on the current and previous inputs.

Let $G(s)$ denote the transfer function of $G(t)$. Then $G(s)$ is in the form of a proper transfer function with real coefficients and time delay. A transfer function $G(s)$ is proper if $G(s)|_{s=j\infty}$ is finite, is strictly proper if $G(s)|_{s=j\infty} = 0$, and is bi-proper if $G(s)|_{s=j\infty}$ is a nonzero constant. All transfer functions that are not proper are improper. In particular, for a rational transfer function $G(s)$, it is proper if its degree of denominator is greater than or equals its degree of numerator, is strictly proper if its degree of denominator is greater than its degree of numerator, and is bi-proper if its degree of denominator equals its degree of numerator.

The linear models of industrial plants described by transfer functions normally fall into three categories:

Stable plant with time delay For some plants, when the original mass or energy equilibrium is upset by a change at the input, the output will eventually reach a new equilibrium. Such plants do not have closed RHP poles. They are called stable plants. Stable plants are usually described by the following model:

$$G(s) = \frac{K}{(\tau_1 s + 1)(\tau_2 s + 1)...(\tau_n s + 1)} e^{-\theta s}, \tag{2.1.1}$$

where K is a real constant denoting the static gain, θ is a positive real constant denoting the time delay, and $\tau_i (i = 1, 2, ..., n)$ have positive real parts and denote the time constants. The following first-order model is frequently used in practice:

$$G(s) = \frac{K}{\tau s + 1} e^{-\theta s}. \tag{2.1.2}$$

The model can be well illustrated by utilizing a shower (Figure 2.1.1). Assume that the temperature of warm water, the flow rate of hot water, and the flow rate of cold water are constants. At the time t, turn up the valve of hot water by a small increment Δq. This will make the temperature of warm water gradually increase until a new steady value is reached. Let the final increment of the temperature be Δc. At the time t_1 the increment is $0.632\Delta c$, and at the time t_2 the increment reaches Δc and does not change anymore. The warmer water flows along the pipe and appears at the outlet at the time t_3. Then the gain of the process is $K = \Delta c/\Delta q$. The time constant τ denotes the speed of temperature change. More precisely, $\tau = t_1 - t$. The time delay θ is the time the warmer water goes through the pipe: $\theta = t_3 - t_2$.

Unstable plant with time delay When the original mass or energy equilibrium in some plants is upset by a change at the input, the output will

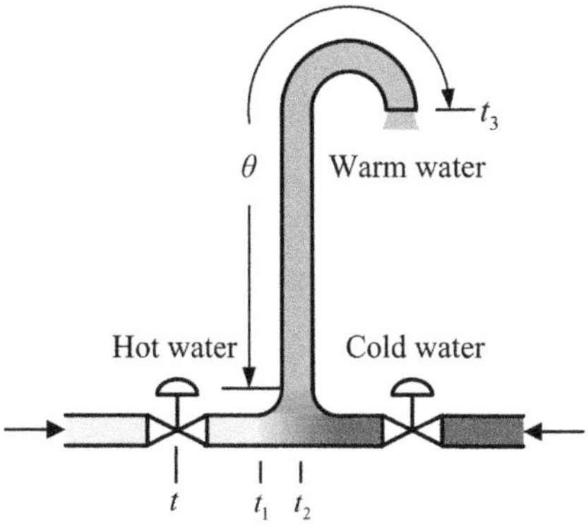

FIGURE 2.1.1
A shower.

increase or decrease faster and faster until the physical limit is reached. Such plants have RHP poles and thus are called unstable plants. Unstable plants can be described by

$$G(s) = \frac{K}{(-\tau_1 s + 1)(-\tau_2 s + 1)...(-\tau_m s + 1)\cdot (\tau_{m+1} s + 1)...(\tau_n s + 1)} e^{-\theta s}, \tag{2.1.3}$$

where K is a real constant denoting the static gain, θ is a positive real constant denoting the time delay, and $\tau_i (i = 1, 2, ..., m, ..., n)$ have positive real parts and denote the time constants. The first-order unstable plant can be written as

$$G(s) = \frac{K}{\tau s - 1} e^{-\theta s}. \tag{2.1.4}$$

Integrating plant with time delay For some other plants, when the original mass or energy equilibrium is upset by a change at the input, the output will increase or decrease with a fixed speed until the physical limit is reached. Such plants are called integrating plants. Integrating plants have one or more poles at the origin. In this book, it is assumed that they do not have open RHP poles. Those with poles both at the origin and in the open RHP are included in the unstable plants. The integrating plants are critical plants between stable ones and unstable ones. In general, they are regarded as the special cases of unstable plants. The integrating plants are usually modeled as

$$G(s) = \frac{K}{s^m (\tau_1 s + 1)(\tau_2 s + 1)...(\tau_n s + 1)} e^{-\theta s}, \tag{2.1.5}$$

where K is a real constant denoting the static gain, θ is a positive real constant denoting the time delay, $\tau_i (i = 1, 2, ..., n)$ have positive real parts and denote the time constants, and m is a positive integer. The first-order integrating model is

$$G(s) = \frac{K}{s} e^{-\theta s}. \tag{2.1.6}$$

The step responses of the three plants are shown in Figure 2.1.2.

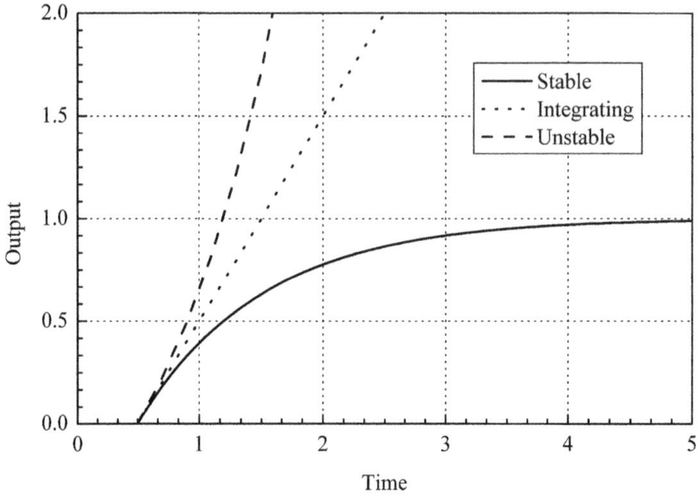

FIGURE 2.1.2
Step responses of three different plants.

Stable plants, integrating plants, and unstable plants are distinguished by their pole positions. According to their zero positions, plants can be classified as minimum-phase (MP) plants and NMP plants. A plant is NMP if its transfer function contains zeros in the closed RHP or contains a time delay. Otherwise, the plant is MP. Sometimes, to avoid the difficulty of treating the time delay in plants, the plant with time delay is modeled as an NMP rational plant.

Most practical plants can be covered by the three categories. To get general results in theoretical study, the following plant model may sometimes be used:

$$G(s) = \frac{KN_+(s)N_-(s)}{M_+(s)M_-(s)} e^{-\theta s}, \tag{2.1.7}$$

where K is a real constant denoting the static gain, θ is a positive real constant denoting the time delay. The subscript "+" denotes that all roots are in the closed RHP, and "−" denotes that all roots are in the open left half-plane (LHP); that is, $N_+(s)$ and $M_+(s)$ are the polynomials with roots in the closed RHP, and $N_-(s)$ and $M_-(s)$ are the polynomials with roots in the open LHP. It is assumed that $N_+(0) = N_-(0) = M_+(0) = M_-(0) = 1$, and $\deg\{N_+\} + \deg\{N_-\} \leq \deg\{M_-\} + \deg\{M_+\}$. Here $\deg\{\cdot\}$ denotes the degree of a polynomial. The first assumption implies that the constant terms of these polynomials are normalized as 1, and is made solely to simplify the statement. The second assumption is a normal one, with which the plant is proper. An improper plant does not exist physically. The plant with finite imaginary axis poles can be regarded as a special case of (2.1.7). In design problems, it is normally assumed that the plant does not have any finite imaginary axis zeros. Practical plants seldom have finite zeros on the imaginary axis. In case this happens, a slight perturbation can be introduced to the zeros in order to use the design method in this book. For example, substitute $(s+0.01)/(s+1)$ for $s/(s+1)$.

There may exist some cases that are more complicated. For example, there are multiple time delays in the plant:

$$G(s) = \frac{1}{\tau_1 s + 1} e^{-\theta_1 s} + \frac{1}{\tau_2 s - 1} e^{-\theta_2 s}. \tag{2.1.8}$$

It is a challenge to deal with such a model. Normally, this case should be avoided in modeling.

2.2 Rational Approximations of Time Delay

In control system design, a rigorous treatment on the time delay involved in the plant is very difficult. This is because time delay is an irrational function. It is of infinite dimension. Most design methods that have been developed so far are based on rational functions. They can only be applied to plants of finite dimension. A widely adopted method to overcome this problem is to approximate time delay by employing rational functions.

As we know, time delay can be expressed as the limit of a large number of first-order lags in series:

$$e^{-\theta s} = \lim_{n\to\infty} \left(\frac{1}{1 + \theta s/n} \right)^n. \tag{2.2.1}$$

This implies that time delay has an infinite number of poles, which makes the analysis and design of control systems challenging. A natural idea is to approximate time delay with a finite number of lag elements, for example let $n = 1$.

Another method is to approximate time delay with the Taylor series expansion:

$$e^{-\theta s} = \lim_{n\to\infty} 1 - \theta s + \theta^2 s^2/2! + ... + (-1)^n \theta^n s^n/n!. \tag{2.2.2}$$

In both mathematics and applications, the Taylor series expansion is frequently utilized to approximate a function and analyze its property.

Rational fraction expressions can be used as the tool of approximation and provide better results. A typical method is the Pade approximation. The basic idea is to make the power series expansion of a rational function match a given power series expansion with as many terms as possible.

Assume that the formal power series expansion of a function $F(s)$ is given as

$$F(s) = c_0 + c_1 s + c_2 s^2 + ... \tag{2.2.3}$$

Let m and n be nonnegative integers. The Pade approximant of $F(s)$ is a rational fraction given by

$$\frac{V_{mn}(s)}{P_{mn}(s)} = \frac{a_m s^m + a_{m-1} s^{m-1} + ... + a_0}{b_n s^n + b_{n-1} s^{n-1} + ... + b_0}. \tag{2.2.4}$$

To obtain a unique solution, one has to set a scale. It might as well take $b_0 = 1$. This leaves $m + n + 1$ unknowns in the fraction. Let the Taylor series expansion of the Pade approximant match the first $m + n + 1$ terms of the given power series expansion of $F(s)$, that is,

$$\begin{aligned} & (b_n s^n + b_{n-1} s^{n-1} + ... + b_0)(c_0 + c_1 s + c_2 s^2 + ... + c_{m+n} s^{m+n}) \\ = \ & a_m s^m + a_{m-1} s^{m-1} + ... + a_0. \end{aligned} \tag{2.2.5}$$

Compare the coefficients of $1, s, ..., s^{m+n}$ in the two sides of the equation. One obtains the following equations:

$$\begin{bmatrix} a_0 \\ a_1 \\ ... \\ a_m \end{bmatrix} = \begin{bmatrix} c_0 & 0 & 0 & ... & 0 \\ c_1 & c_0 & 0 & ... & 0 \\ ... & ... & ... & ... & ... \\ c_m & c_{m-1} & c_{m-2} & ... & c_{m-n} \end{bmatrix} \begin{bmatrix} b_0 \\ b_1 \\ ... \\ b_n \end{bmatrix}, \tag{2.2.6}$$

$$\begin{bmatrix} c_{m+1} & c_m & ... & c_{m-n+1} \\ c_{m+2} & c_{m+1} & ... & c_{m-n+2} \\ ... & ... & ... & ... \\ c_{m+n} & c_{m+n-1} & ... & c_m \end{bmatrix} \begin{bmatrix} b_0 \\ b_1 \\ ... \\ b_n \end{bmatrix} = \begin{bmatrix} 0 \\ 0 \\ ... \\ 0 \end{bmatrix}. \tag{2.2.7}$$

Here $c_i = 0$ when $i < 0$. Since $b_0 = 1$, the second set of equations can be rewritten as

$$\begin{bmatrix} c_m & c_{m-1} & ... & c_{m-n+1} \\ c_{m+1} & c_m & ... & c_{m-n+2} \\ ... & ... & ... & ... \\ c_{m+n-1} & c_{m+n-2} & ... & c_m \end{bmatrix} \begin{bmatrix} b_1 \\ b_2 \\ ... \\ b_n \end{bmatrix} = - \begin{bmatrix} c_{m+1} \\ c_{m+2} \\ ... \\ c_{m+n} \end{bmatrix}. \tag{2.2.8}$$

If this set of equations has a solution, the coefficients of $P_{mn}(s)$ can be obtained from (2.2.8) and the coefficients of $V_{mn}(s)$ can be obtained from (2.2.6).

For exponential functions, the Pade approximant has a more clear expression. The m/n Pade approximant of time delay can be written as

$$e^{-\theta s} \approx \frac{V_{mn}(\theta s)}{P_{mn}(\theta s)}, \tag{2.2.9}$$

where

$$\begin{aligned} V_{mn}(\theta s) &= \sum_{j=0}^{m} \frac{(m+n-j)!m!}{(m+n)!j!(m-j)!}(-\theta s)^j, \\ P_{mn}(\theta s) &= \sum_{j=0}^{n} \frac{(m+n-j)!n!}{(m+n)!j!(n-j)!}(\theta s)^j. \end{aligned}$$

It can be verified that $P_{mn}(\theta s) = V_{nm}(-\theta s)$. When $m = n$, the all-pass Pade approximant is obtained. For SISO systems, a transfer function is all-pass if its magnitude equals 1 at all points on the imaginary axis. The terminology comes from the fact that a filter in the form of an all-pass transfer function passes the input sinusoids of all frequencies without attenuation. All zeros of the all-pass Pade approximant are in the open RHP and all poles of the all-pass Pade approximant are in the open LHP. The zeros and the poles are the mirror-images of each other.

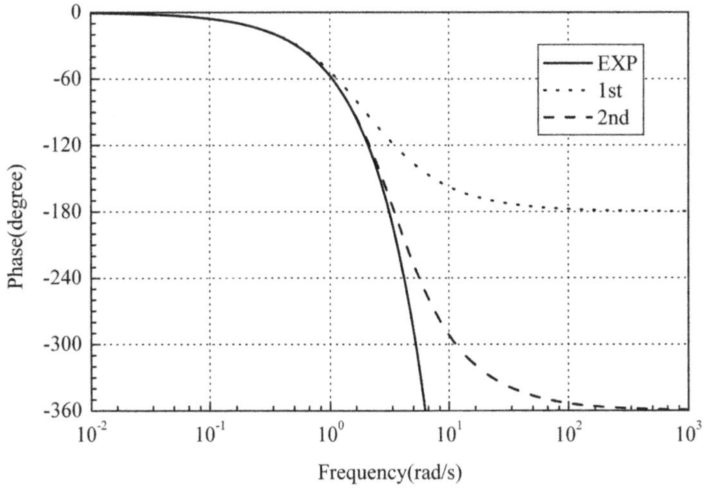

FIGURE 2.2.1
Phases of the all-pass Pade approximant and time delay.

The all-pass Pade approximant is frequently used in practice. Compared with the Taylor series expansion, the all-pass Pade approximant has two features:

1. The all-pass Pade approximant provides better precision than the Taylor series expansion of the same order.
2. In the all-pass Pade approximant, the magnitude characteristic of time delay is preserved; the only difference is the phase (Figure 2.2.1).

A closed-loop system is stable if its characteristic equation has no roots in the closed RHP. Although the rational approximation can arbitrarily approximate time delay, they were seldom used in classical control theory to analyze the stability of the closed-loop system. The main reason is that sometimes the rational approximation cannot guarantee the correctness of the result even though a high-order rational approximation is used.

Example 2.2.1. *To see how the designer may be misled, consider the simple system shown in Figure 2.2.2. The characteristic equation of the closed-loop system is*

$$1 + \frac{1}{s}e^{-s} = 0.$$

or

$$1 + se^{s} = 0.$$

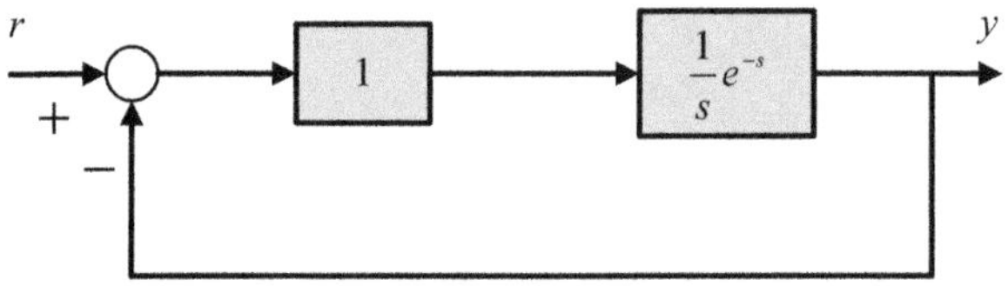

FIGURE 2.2.2
A simple control system.

It is easy to verify that this closed-loop system is stable. With the Taylor series expansion of different orders, different roots for the characteristic equation can be obtained:

$1 + s = 0$	*The root is* $s = -1$
$1 + s + s^2 = 0$	*The roots are* $s = -0.5 \pm j0.8660$
$1 + s + s^2 + s^3/2 = 0$	*The roots are* $s = -1.5437, -0.2282 \pm j1.1151$

From these results, it can be concluded that the closed-loop system is stable. However, a higher order Taylor series expansion gives the following equation:

$$1 + s + s^2 + \frac{s^3}{2} + \frac{s^4}{3!} + \frac{s^5}{4!} + \frac{s^6}{5!} = 0,$$

which has two roots in the RHP:

$$s = 0.1041 \pm j3.0815.$$

Although the rational approximation did not provide the desired result in the above example, this does not imply that it definitely cannot be used for the analysis and design of a control system. It will be shown in Chapters 4 and 5 that, with appropriate methods, the rational approximation can be used to analyze and design control systems, and satisfactory responses can be obtained.

2.3 Time Domain Performance Indices

The most elementary feedback control system has two components: a plant to be controlled and a controller to generate the input to the plant. Consider the unity feedback control loop shown in Figure 2.3.1, where $C(s)$ denotes the controller, $G(s)$ denotes the plant, $r(s)$ is the reference, $e(s)$ is the error, $u(s)$ is the controller output, $y(s)$ is the system output, $d'(s)$ is the disturbance at the plant input, and $d(s)$ is the disturbance at the plant output. Measurement noise is very small and thus can be neglected.

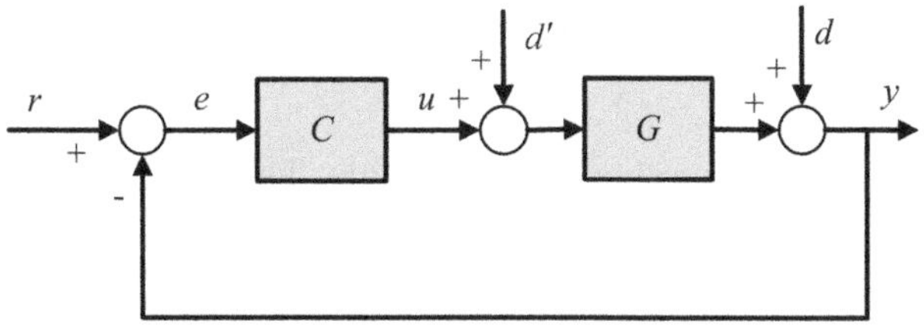

FIGURE 2.3.1
Elementary feedback control loop.

The unity feedback control loop has played a vital role in the study of control theory. On one hand, it is the most widely used structure. On the other hand, most of the non-unity feedback control loops can easily be converted into the unity feedback control loop. The unity feedback control loop is the main control structure discussed in this book. The response of the system shown in Figure 2.3.1 depends on not only the model, but also the input and the initial condition. Without loss of generality, it is a common practice to use the standard initial condition; that is, initially the system is at rest with its output and all time derivatives thereof being zero.

In the analysis and design of control systems, performances of various systems should be compared on the same basis. This may be achieved by specifying a particular test signal for the input and then comparing the responses of different systems to the test signal.

Test signals are abstracted from practical signals. They should be chosen as follows:

1. The response characteristic of the represented signal is reflected as much as possible.
2. The form of the test signal is as simple as possible so that the system response can be analyzed easily.

Four types of time domain test signals are listed here. They are frequently used in the design of control systems.

Impulse The impulse signal is usually denoted by $\delta(t)$:

$$\delta(t) = \begin{cases} 0 & t \neq 0 \\ \infty & t = 0 \end{cases},$$
$$\int_0^{\infty} \delta(t)\,dt = 1.$$

No ideal impulse exists in the real world. In practice, the impulse can be approximated by a rectangular pulse:

$$r(t) = \begin{cases} A & 0 \leq t \leq 1/A \\ 0 & t < 0, t > 1/A \end{cases}.$$

Here A is a constant denoting the magnitude. The Laplace transform of the impulse is 1.

Step The step signal is defined by

$$r(t) = \begin{cases} 0 & t < 0 \\ A & t \geq 0 \end{cases},$$

where A is a constant denoting the magnitude. The step signal with $A = 1$ is called the unit step. The Laplace transform of the unit step is $1/s$.

Ramp Mathematically, a ramp signal is defined as follows:

$$r(t) = \begin{cases} 0 & t < 0 \\ At & t \geq 0 \end{cases},$$

Here A is a constant. A ramp signal with $A = 1$ is called the unit ramp. The Laplace transform of the unit ramp is $1/s^2$.

Sinusoidal The sinusoidal signal with the amplitude A and the frequency ω can be expressed as

$$r(t) = \begin{cases} 0 & t < 0 \\ A \sin \omega t & t \geq 0 \end{cases} .$$

Its Laplace transform is $A\omega/(s^2 + \omega^2)$.

Test signals play a fundamental role in the design of control systems. Many design indices are based on test signals. The use of test signals is justified because of the relationship existing between the response of a system to a test signal and the capability of the system to cope with the real signal. It is impossible to design a controller that works well for all types of inputs. The designer has to decide which inputs are the most important and the most frequent, and then to design a controller for the corresponding test signal.

There are two types of inputs in the feedback control system shown in Figure 2.3.1: the reference and the disturbance. In most cases, they can be approximated as pulses, steps, or ramps. The step is the most frequently encountered signal. The pulse can be obtained by combining two opposite step signals. The ramp is seldom encountered. Normally, practical signals can be viewed as the result obtained by impulsing a linear stable system with a step or a pulse. This is why the controller is often designed for a step input. In a step signal, there are abundant sinusoidal signals whose amplitudes decrease with frequencies. Therefore, the step signal is a reasonable and sufficiently strict choice for both theoretical research and application.

Looking back at the history of control theory, it can be seen that the development of control theory is closely related with the description of the response of a control system and the corresponding performance indices.

In classical control theory, the performance of a control system is usually characterized in terms of transient response and steady-state response in the time domain. The transient response is the response that goes from the initial state to the steady state. The steady-state response refers to the manner in which the system output behaves as the time approaches infinity. People usually pay more attention to the transient response. On one hand, the magnitude and the time of the transient response are usually limited in practice. As a result, it is difficult to obtain the required transient response. On the other hand, there are always disturbances in a real system. The system almost always goes from one state to another.

Performance indices of the transient response are generally defined by utilizing the unit step response (Figure 2.3.2).

Overshoot σ The overshoot is the maximum value of the unit step response, measured from unity. It is common to use the percentage overshoot.

Rise time t_r In overdamped systems, the rise time is the time required for the unit step response to rise from 10% to 90% of its steady-state value. For

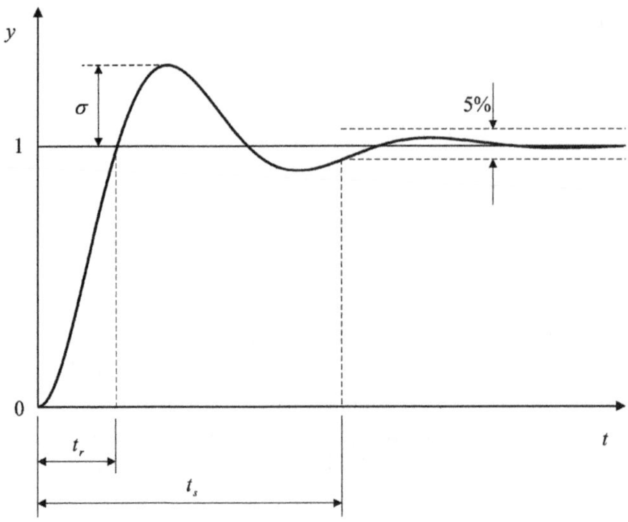

FIGURE 2.3.2
Step response curve for time domain performance indices.

underdamped systems, the rise time is specified as the time when the unit step response reaches the steady-state value for the first time.

Settling time t_s The settling time is the time required for the unit step response to reach and stay within a given error band (usually 5% of the steady-state value).

One major objective of control systems is to reject the effect of disturbances and keep the system output as close as possible to the reference. Although there exists an internal relationship between the reference response and the disturbance response, the reference response cannot give a quantitative estimation. To provide this information, two transient performance indices are defined here for the unit step disturbance at the plant input (also referred to as the load). They are shown graphically in Figure 2.3.3.

Perturbation peak ρ The perturbation peak is the maximum value of the disturbance response, measured from the steady-state value.

Recovery time t_{rs} The recovery time is the time required for the disturbance response to reach and stay within a given error band (usually 5% of the steady-state value).

Now consider the steady-state performance. The steady-state performance is measured with the steady-state error. The steady-state error of a system is the difference between the desired output and the real output of the system as

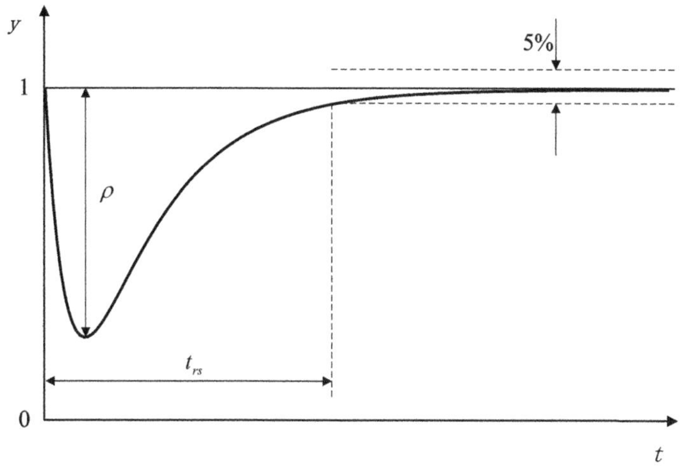

FIGURE 2.3.3
Disturbance response for time domain performance indices.

time goes to infinity (that is, the transient response reaches the steady state). It reflects how accurately the output can be controlled. It is desirable that the steady-state error should gradually vanish.

Consider the system shown in Figure 2.3.1. Let $L(s) = G(s)C(s)$ be the open-loop transfer function. From the Final Value Theorem it is known that the steady-state error of a stable system is

$$\lim_{t\to\infty} e(t) = \lim_{s\to 0} \frac{sr(s)}{1 + L(s)}. \tag{2.3.1}$$

It can be seen that the steady-state error depends on not only the input but also the open-loop transfer function of the system. Whether a given system exhibits a steady-state error for a given type of input is determined by the system type. The definition of the system type is given as follows.

Definition 2.3.1. *The system is said to be of Type m if $L(s)$ has m poles at the origin.*

Now we study the effect of the input and the system type on steady-state error. Only the step input and the ramp input are considered here, since it is rather rare to design a system of Type 3 or higher in practice.

From (2.3.1) the steady-state error for a step reference (that is, $r(s) = 1/s$) is

$$\begin{aligned}\lim_{t\to\infty} e(t) &= \lim_{s\to 0} \frac{1}{1 + L(s)} \\ &= \frac{1}{1 + \lim_{s\to 0} L(s)}.\end{aligned} \tag{2.3.2}$$

It is a nonzero constant for a Type 0 system. If the zero steady-state error is required, the system type is at least one.

The steady-state error for a ramp reference (that is, $r(s) = 1/s^2$) is given by

$$\begin{aligned}\lim_{t\to\infty} e(t) &= \lim_{s\to 0}\frac{1}{s+sL(s)}\\ &= \frac{1}{\lim_{s\to 0}[sL(s)]}.\end{aligned} \tag{2.3.3}$$

In this case, a Type 0 system is incapable of following a ramp at the steady state. The steady-state error is infinite. A Type 1 system can follow a ramp with a nonzero steady-state error. A Type 2 or higher system can follow a ramp with zero steady-state error.

The classical performance indices, such as overshoot and rise time, are intuitive and have been widely used in practice. However, it is difficult to describe them mathematically. This prevents people from studying the design problem in a rigorous way. To optimize the system performance with mathematical methods, integral performance indices were presented. These indices were defined in different forms.

Integral Absolute Error (IAE):

$$\mathrm{IAE} = \int_0^\infty |e(t)|dt.$$

Integral Square Error (ISE):

$$\mathrm{ISE} = \int_0^\infty e^2(t)dt.$$

Integral of Time Multiplied by Absolute Error (ITAE):

$$\mathrm{ITAE} = \int_0^\infty t|e(t)|dt.$$

Among these performance indices, the ISE index is the only one that is widely applied. This is because it is easier to perform the optimization procedure with the ISE index than with other indices, while other indices cannot provide superior performance to the ISE index. Based on the ISE index, the LQ optimal control and linear quadratic Gaussian (LQG) optimal control were developed. They are the most important parts that compose the modern control theory.

Another ISE-based index is the H_2 index. In H_2 optimal control, the system input is assumed to be the impulse, or equivalently, the white noise with zero mean and unit variance. The basic design objective is to search for a controller such that the ISE of the system is minimized for this particular input.

One can also minimize the worst ISE resulting from all energy-bounded inputs. In this case, the performance index can be written as

$$\sup_{r(t)} \int_0^{\infty} e^2(t)dt.$$

The index motivates the appearance of H_∞ optimal control.

One major shortcoming of the H_2 and H_∞ performance indices is that they do not correlate with the classical performance indices. The designer does not know whether the response specified in terms of classical performance indices (such as overshoot, stability margin, the amplitude of coupled responses, and so on) can be reached by an optimal/suboptimal controller. As a direct result, there is a gap between the LQ/H_2/H_∞ control theory and the classical techniques. Many engineers, who usually depend on engineering and physical insights, are reluctant to accept these new theories.

2.4 Frequency Response Analysis

When a linear system is subject to a sinusoidal input, its ultimate response is also a sustained sinusoidal wave. This characteristic constitutes the basis of frequency response analysis, the most conventional method used by engineers for the analysis and design of control systems. In the frequency response analysis, the frequency of the input is varied over the range of interest; the responses resulting from the input are studied.

Consider the unity feedback control system, of which the open-loop transfer function is $L(s)$. The closed-loop transfer function $T(s)$ can be written as

$$T(s) = \frac{L(s)}{1 + L(s)}. \tag{2.4.1}$$

The frequency response of the system is

$$T(j\omega) = \frac{L(j\omega)}{1 + L(j\omega)},$$

which can be expressed with magnitude and phase as

$$T(j\omega) = |T(j\omega)| \angle T(j\omega). \tag{2.4.2}$$

For a stable MP system, the magnitude has a one to one relationship with the phase. If the magnitude curve of a system is specified, then the phase curve is uniquely determined, and vice versa. This, however, does not hold for an NMP system.

Similar to the time domain method, in the frequency response analysis, some points with special characteristics are selected to define the performance indices (Figure 2.4.1).

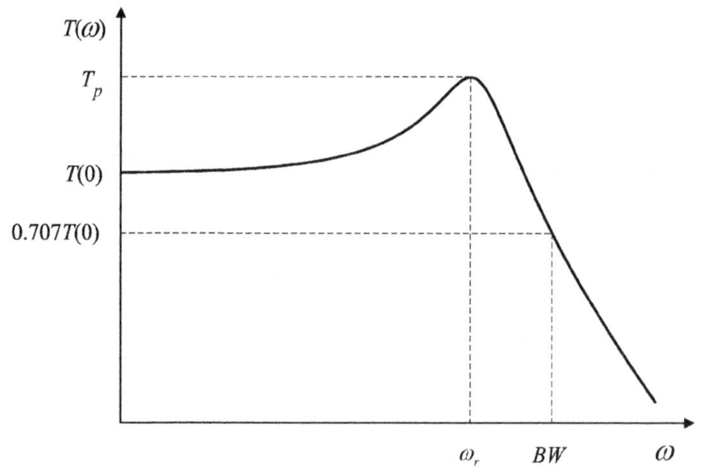

FIGURE 2.4.1
Magnitude curve for frequency response analysis.

Resonance peak T_p The resonance peak is the maximum value of the magnitude of the closed-loop frequency response.

Resonance frequency ω_r The resonance frequency is the frequency at which the resonance peak occurs.

Bandwidth BW The bandwidth is the frequency range beyond which the signal magnitude drops down by more than 3dB.

The resonance peak is indicative of the relative stability of a system. A large resonance peak corresponds to a large overshoot. This implies that the system response is not very steady. The bandwidth reflects the speed of the transient response. If the bandwidth is large, high frequency signals can pass through the system. As a result, the system response is fast.

A distinct feature of the frequency response analysis is that the stability, as well as the performance, of the closed-loop system can be determined from the characteristics of the open-loop response. At the early stage of control theory, this feature reduced the level of difficulty in control system design.

The stability of the closed-loop system can be tested by means of the Nyquist plot of the open-loop transfer function. Let us begin with the Nyquist path: it starts at the origin, goes up the positive imaginary axis, turns into the RHP following a semicircle of infinite radius, and comes up the negative imaginary axis to the origin again. As a point $s = j\omega$ makes one circuit around this curve, the point $L(j\omega)$ traces out a curve called the Nyquist plot of the transfer function $L(s)$ (Figure 2.4.2).

The Nyquist path must not pass through zeros or poles of $L(s)$. If $L(s)$

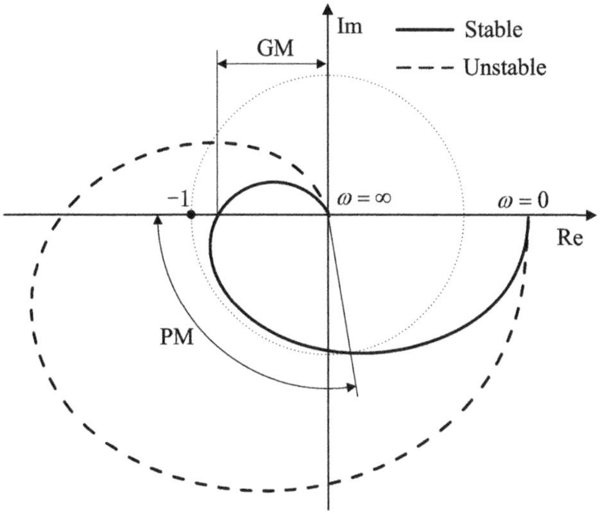

FIGURE 2.4.2
Nyquist plot of a stable open-loop system.

has finite zeros or poles on the imaginary axis, the path must be modified by an infinitesimal detour. The path goes around these zeros and poles so that the zeros and poles on the imaginary axis are regarded as they are in the open LHP, while all of the zeros and poles in the closed RHP are enclosed by this path.

Theorem 2.4.1. (Nyquist Stability Criterion) *Let n denote the total number of poles of $L(s)$ in the RHP. The closed-loop system is stable if and only if the Nyquist plot of $L(s)$ does not pass through the point $(-1, 0)$ and encircles it n times counterclockwise.*

The stability of the closed-loop system can also be tested with the Bode plot of $L(s)$. A Bode plot consists of two graphs: one is the logarithm graph of the magnitude $|L(j\omega)|$; the other is the graph of the phase angle $\angle L(j\omega)$. Both are plotted versus the frequency (Figure 2.4.3).

Theorem 2.4.2. (Bode Stability Criterion) *A closed-loop system is unstable if the frequency response of $L(s)$ has a magnitude greater than or equal to 1 at the frequency where the phase angle is $-180°$.*

Compared with the algebra stability criterion, one advantage of the Nyquist stability criterion and the Bode stability criterion is that they can cope with systems having a time delay. Write the open-loop transfer function as $L(s) = L_o(s)e^{-\theta s}$, where $L_o(s)$ is the delay-free part and θ is the time delay. The magnitude of the open-loop transfer function remains the same as

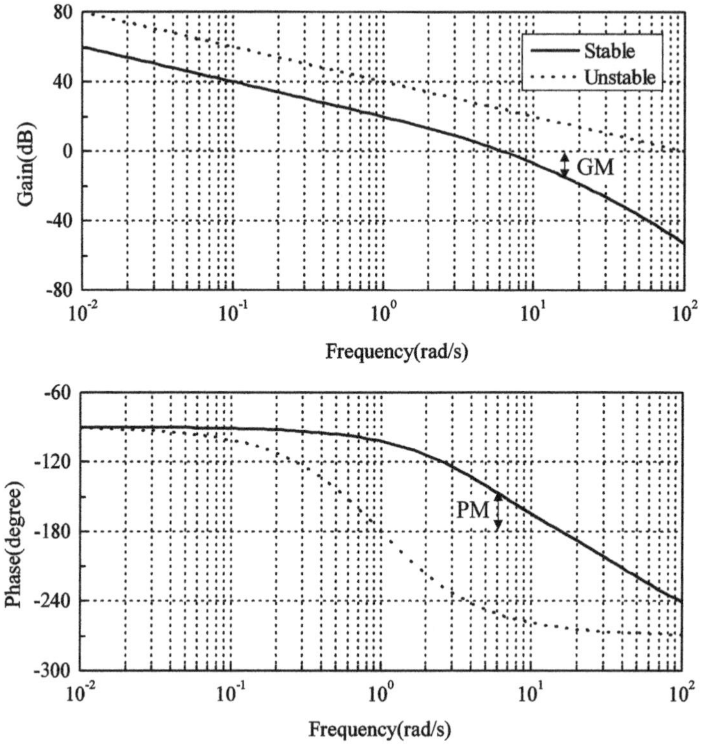

FIGURE 2.4.3
Bode plot of $L(j\omega)$.

that of its delay-free part, while the phase angle becomes

$$\angle L(j\omega) = \angle L_o(j\omega) - \theta\omega. \tag{2.4.3}$$

The Bode plot can be used not only to test the stability of the closed-loop system, but also to describe the relative stability or the stability margin.

Gain Margin (GM) Let ω_c be the frequency where the phase angle is $-180°$(namely phase crossover frequency). The gain margin is the reciprocal of the magnitude at the frequency.

$$\text{GM} = |L(j\omega_c)|^{-1}, \angle L(j\omega_c) = -180°. \tag{2.4.4}$$

Phase Margin (PM) Let ω_g be the frequency where the magnitude is 1 (namely gain crossover frequency). The phase margin is the additional phase lag needed to distabilize the system.

$$\text{PM} = 180° + \angle L(j\omega_g), |L(j\omega_g)| = 1. \tag{2.4.5}$$

The gain margin and phase margin measure the distance from the critical point to the Nyquist plot in a certain direction (Figure 2.4.2). In the classic control theory, they are important metrics of stability and frequently used as the performance index for control system design. To determine the relative stability, both the gain margin and the phase margin have to be given.

2.5 Transformation of Two Commonly Used Models

The so-called modeling is the process of constructing the mathematical, physical, or conceptual simulation of a real-world object or phenomenon. People usually use experimental methods or mechanism-based methods to get a model. In this section, only experimental methods are considered.

Almost all design methods are model-based despite the different forms of model. There are two typical experimental methods in process control: the step response method and the ultimate cycle method. The step response method is based on the information from the open-loop step response. The obtained model is usually a transfer function in the form of the first-order plant with time delay. The ultimate cycle method utilizes a closed-loop procedure in which only a proportional controller is used. By tuning the controller, a sustained oscillation can be built, which gives the information as to the ultimate gain and the ultimate period.

With regard to the two types of models, there are two types of design methods. One is based on the step response model, for example, the Cohen-Coon (C-C) method. The other is based on the ultimate cycle model, for example, the Ziegler-Nichols (Z-N) method.

Historically, the two models and the two corresponding design methods are developed independently. Since the plants they describe are the same one, they must be internally equivalent. Here, the quantitative relationship between the two methods will be derived.

First, assume that the step response model is expressed as follows:

$$G(s) = \frac{K}{\tau s + 1} e^{-\theta s}, \tag{2.5.1}$$

and it is desirable to compute the three parameters from the obtained ultimate cycle model. The ultimate cycle method goes through the following steps: First, the controller is taken to be a proportional controller K_C and K_C is set to be a small value. Then, K_C is increased from small to large by small steps until a sustained oscillation with constant amplitude occurs for a reference change. The value of K_C that results in the oscillation is referred to as the ultimate gain K_u. The period of the oscillation is recorded as the ultimate period T_u. The ultimate frequency is $\omega_u = 2\pi/T_u$ (Figure 2.5.1).

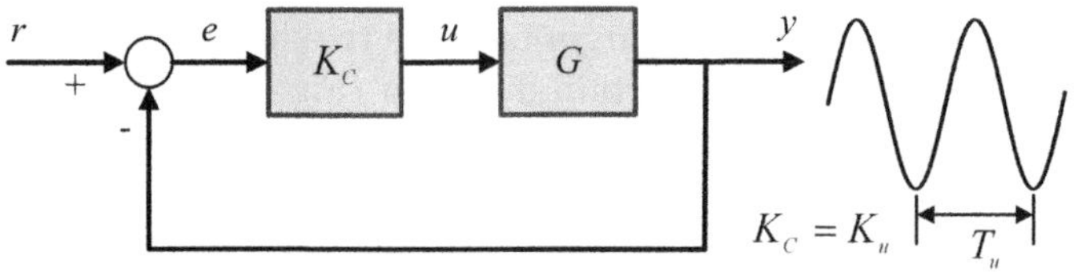

FIGURE 2.5.1
Ultimate cycle method.

The Nyquist plot and the Nyquist stability criterion can be used to explain the ultimate cycle method. An increase of the controller K_C will make all points on the Nyquist plot radially move outward from the origin. When K_C increases to the extent that the Nyquist plot passes through the point $(-1, 0)$, the stability limit is reached and a sustained oscillation occurs. Accordingly, we have

$$\angle K_u G(j\omega_u) = -\pi, \tag{2.5.2}$$
$$|K_u G(j\omega_u)| = 1. \tag{2.5.3}$$

Since

$$K_u G(j\omega_u) = \frac{K_u K e^{-j\theta\omega_u}}{j\tau\omega_u + 1}, \tag{2.5.4}$$

the relationship between the step response model and the ultimate cycle model can be expressed as

$$-\theta\omega_u - \arctan(\tau\omega_u) = -\pi, \tag{2.5.5}$$

$$\frac{K_u K}{\sqrt{(\tau\omega_u)^2+1}} = 1. \tag{2.5.6}$$

When the plant gain K is known (if not, one can carry out the procedure twice), the time constant of the step response model is

$$\begin{aligned}\tau &= \frac{\sqrt{(K_u K)^2-1}}{\omega_u} \\ &= \frac{T_u\sqrt{(K_u K)^2-1}}{2\pi},\end{aligned} \tag{2.5.7}$$

and the time delay of the step response model is

$$\begin{aligned}\theta &= \frac{\pi - \arctan(\sqrt{(K_u K)^2-1})}{\omega_u} \\ &= \frac{T_u}{2} - \frac{T_u}{2\pi}\arctan(\sqrt{(K_u K)^2-1}).\end{aligned} \tag{2.5.8}$$

Hence, if the ultimate cycle model is obtained, the step response model can be exactly calculated by employing K_u and ω_u.

One may desire to calculate the ultimate cycle model K_u and T_u when K, θ, and τ are known in the first order plus time delay model. It is a challenge to obtain analytical expressions by solving the above equations. Here they are derived by the time domain analysis. Interested readers can compare the precision and complexity of the two methods.

Without loss of generality, assume that a unit step reference is used in the ultimate cycle method. The system output can be written as

$$y(s) = \frac{1}{s}\frac{K_C K e^{-\theta s}}{\tau s + 1 + K_C K e^{-\theta s}}. \tag{2.5.9}$$

Let

$$f(s) = \frac{K_C K e^{-\theta s}}{\tau s + 1 + K_C K e^{-\theta s}}. \tag{2.5.10}$$

Then

$$y(s) = \frac{f(s)}{s}, \tag{2.5.11}$$

where $f(s)$ is an irrational function. To derive the ultimate cycle model, $f(s)$ is expanded with a rational approximation. The Pade approximation is employed.

Assume that the Maclaurin series expansion (that is, the Taylor series expansion of a function about 0) of $f(s)$ is

$$f(s) = f(0) + f'(0)s + \frac{f''(0)}{2!}s^2 + \frac{f^{(3)}(0)}{3!}s^3 + \frac{f^{(4)}(0)}{4!}s^4 + ..., \tag{2.5.12}$$

where

$$
\begin{aligned}
f(0) &= \frac{K_C K}{1+K_C K}, \\
f'(0) &= -\frac{K_C K(\theta+\tau)}{(1+K_C K)^2}, \\
f''(0) &= -\frac{K_C K(K_C K\theta^2 - \theta^2 - 2\theta\tau - 2\tau^2 + 2K_C K\theta\tau)}{(1+K_C K)^3}, \\
f^{(3)}(0) &= -\frac{\begin{array}{c} K_C K(\theta^3 + 3K_c^2K^2\theta^2\tau - 12K_C K\theta^2\tau + 6\tau^3 \\ -12K_C K\theta\tau^2 + 6\theta\tau^2 + K_c^2K^2\theta^3 + 3\theta^2\tau - 4K_C K\theta^3) \end{array}}{(1+K_C K)^4}, \\
f^{(4)}(0) &= -\frac{\begin{array}{c} K_C K(K_c^3K^3\theta^4 + 4K_c^3K^3\theta^3\tau - 11K_c^2K^2\theta^4 \\ -44K_c^2K^2\theta^3\tau - 48K_c^2K^2\theta^2\tau^2 + 11K_C K\theta^4 \\ +44K_C K\theta^3\tau + 84K_C K\theta^2\tau^2 + 72K_C K\theta\tau^3 \\ -\theta^4 - 4\theta^3\tau - 12\theta^2\tau^2 - 24\theta\tau^3 - 24\tau^4) \end{array}}{(1+K_C K)^5}.
\end{aligned}
$$

Let the Pade approximation of $f(s)$ be

$$
f(s) = \frac{a_2 s^2 + a_1 s + a_0}{b_2 s^2 + b_1 s + 1}. \tag{2.5.13}
$$

Then

$$
\begin{bmatrix} a_0 \\ a_1 \\ a_2 \end{bmatrix} = \begin{bmatrix} f(0) & 0 & 0 \\ f'(0) & f(0) & 0 \\ f''(0)/2! & f'(0) & f(0) \end{bmatrix} \begin{bmatrix} 1 \\ b_1 \\ b_2 \end{bmatrix} \tag{2.5.14}
$$

$$
\begin{bmatrix} f''(0)/2! & f'(0) \\ f^{(3)}(0)/3! & f''(0)/2! \end{bmatrix} \begin{bmatrix} b_1 \\ b_2 \end{bmatrix} = - \begin{bmatrix} f^{(3)}(0)/3! \\ f^{(4)}(0)/4! \end{bmatrix}. \tag{2.5.15}
$$

This leads to

$$
\begin{aligned}
a_0 &= f(0), \\
a_1 &= b_1 f(0) + f'(0), \\
a_2 &= b_2 f(0) + b_1 f'(0) + f''(0)/2!, \\
b_1 &= -\frac{f''(0)f^{(3)}(0)/12 - f'(0)f^{(4)}(0)/24}{f''(0)^2/4 - f'(0)f^{(3)}(0)/6}, \\
b_2 &= -\frac{-f^{(3)}(0)^2/36 + f''(0)f^{(4)}(0)/48}{f''(0)^2/4 - f'(0)f^{(3)}(0)/6}.
\end{aligned}
$$

The system output is

$$
y(s) = \frac{a_2 s^2 + a_1 s + a_0}{s(b_2 s^2 + b_1 s + 1)}
$$

$$= \frac{a_2(s^2 + p_1 s + p_0)}{b_2 s[(s+q)^2 + \omega^2]}, \tag{2.5.16}$$

where

$$p_0 = \frac{a_0}{a_2}, \; p_1 = \frac{a_1}{a_2}, \; q = \frac{b_1}{2b_2}, \; \omega^2 = \frac{1}{b_2} - \frac{b_1^2}{4b_2^2}.$$

Evidently, to obtain a sustained oscillation q should be zero, or equivalently, $b_1 = 0$. This implies that the ultimate frequency is

$$\omega_u = \frac{1}{b_2}. \tag{2.5.17}$$

Then, the ultimate period is

$$T_u = 2\pi \sqrt{\frac{\theta(KK_C\theta^3 + 6KK_C\theta^2\tau + 12KK_C\theta\tau^2 + \theta^3 + 6\theta^2\tau + 18\theta\tau^2 + 24\tau^3)}{12(KK_C\theta^2 + 4KK_C\theta\tau + 6KK_C\tau^2 + \theta^2 + 4\theta\tau + 6\tau^2)}}.$$

Since

$$b_1 = -\frac{KK_C\theta^3 + 5KK_C\theta^2\tau + 8KK_C\theta\tau^2 - \theta^3 - 5\theta^2\tau - 12\theta\tau^2 - 12\tau^3}{2(KK_C\theta^2 + 4KK_C\theta\tau + 6KK_C\tau^2 + \theta^2 + 4\theta\tau + 6\tau^2)},$$

$b_1 = 0$ (the K_C at this moment is K_u) gives

$$KK_u = \frac{\theta^3 + 5\theta^2\tau + 12\theta\tau^2 + 12\tau^3}{\theta^3 + 5\theta^2\tau + 8\theta\tau^2}.$$

The ultimate gain is

$$K_u = \frac{\theta^3 + 5\theta^2\tau + 12\theta\tau^2 + 12\tau^3}{K(\theta^3 + 5\theta^2\tau + 8\theta\tau^2)}. \tag{2.5.18}$$

Thus far, the analytical relationship between the step response model and the ultimate cycle model has been built. It can be seen that the two models can be directly converted into each other. One may use a controller based on the step response model when the ultimate cycle model is obtained, and vice versa.

2.6 Design Requirements and Controller Comparison

This section discusses several problems about design requirements and controller comparison.

An evident feature of control theory is that it is closely related to practice. Let us see what requirements a practical control system should satisfy.

If the plant is stable, an overshoot larger than 50% is rarely accepted. How large the overshoot should be depends on the plant. Some plants have a strict limitation on overshoot. For example, in some chemical processes the controlled variable is temperature or pressure. If the temperature or pressure exceeds the limit of the equipment, an accident may happen. For some other plants, like the paper-making process, there is no strict limitation on overshoot. However, an excessive overshoot implies a large error, which may cause the saturation problem.

The rise time is used to characterize the system response speed and reflects the tracking ability of a system from one angle. It is desirable that the rise time be fast. However, in the system with an overshoot, the rise time and the overshoot usually conflict with each other. In general, the faster the rise time, the larger the overshoot (Figure 2.6.1). To obtain a proper overshoot, the response speed may have to be sacrificed.

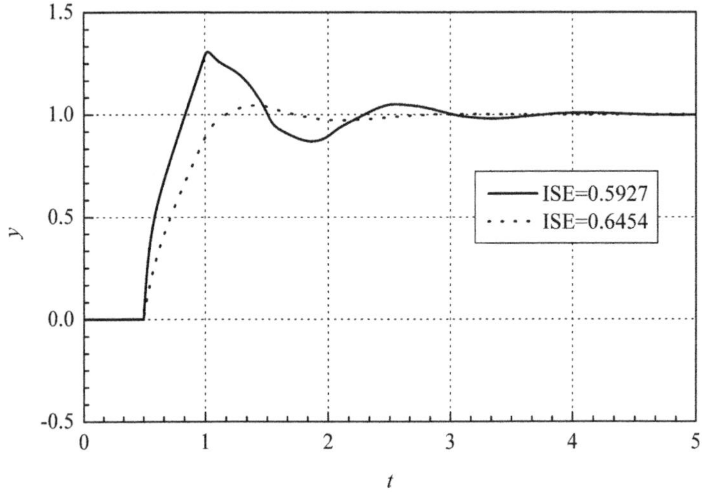

FIGURE 2.6.1
Systems with different overshoots.

The settling time is related to the duration of transient response. It describes the tracking ability of a system from another angle. Sometimes the system performance is measured with integral criteria, such as ISE. With the same overshoot, the faster the settling time, the smaller the ISE, and the better the performance. Thus, it is desirable that the settling time be as short as possible. The settling time relates to overshoot and rise time. Normally, in a system with an overshoot, the larger the overshoot, the faster the rise time, and the longer the settling time; in a system without overshoot, the faster the rise time, the shorter the settling time.

There always exist disturbances in a real system. To guarantee the product quality, it is desirable that the system be insensitive to disturbances. The disturbance response and the reference response relate to each other. In many cases, for the system with an overshoot, the smaller the perturbation peak, the larger the overshoot.

The relationship among overshoot, rise time, settling time, integral indices, reference response, and disturbance response implies that they are not independent. Once the requirement on one index is given, a restriction is imposed on other indices at the same time. A desired controller should provide fast and steady response. With respect to this objective, one can take the overshoot of 5% to 15% if no other requirements are given. Empirically, such an overshoot can, in many cases, provide a good tradeoff between performance and robustness.

Any new design method has to be compared with typical design methods to illustrate its advantage. As there is no unified standard, the comparison among different design methods could be difficult and sometimes unfair.

If the order of a plant is high or a plant contains a time delay, the order of the obtained controller will be high. One may have to use model reduction techniques. The order of the controller is interrelated with the closed-loop response. It is certainly easier to reach good performance for a higher-order controller than for a lower-order controller. In this case, comparison of controllers with different orders may be unfair.

In controller comparing, a plausible viewpoint is that the controller with good performance should have a small overshoot. As a matter of fact, the controller with good performance, in many cases, has a large overshoot (Figure 2.6.1). The reason is that the controller cannot exactly cancel the zeros and poles of the plant when the plant is of high order or has a time delay. Another related point of view is that the performance is improved if the overshoot is reduced. However, a reduced overshoot can usually be obtained by sacrificing the performance. For a fair comparison, a common basis is necessary. For example, when the overshoots of different controllers are compared, they should have the same rise time.

In robust control theory, the system performance is classified as nominal performance and robust performance. The nominal performance is the performance of the system with an exact model, while the robust performance is the performance of the uncertain system. Robust performance in fact involves the requirement on both nominal performance and robust stability. The two aspects compete. Improved robust stability can be obtained in two ways:

1. Sacrifice the nominal performance.
2. Do not sacrifice the nominal performance.

When comparing different controllers, one may have to consider whether the nominal performances is sacrificed for robust stability. Ideally, it is desirable that the robust stability be obtained without sacrificing the nominal performance.

In addition to the performance comparison, an important aspect in evaluating a design method is the practicability, that is, whether the design method is easy to use and whether the design procedure and result are simple enough. In many applications, the problem of practicability is even more important than the performance problem. For example, in an ultra supercritical power plant there are more than 100 loops. The time for configuration and tuning is very limited. In this case, it is impossible to use those complicated methods.

The best way to evaluate the design methods in this book may be applying other design methods to the examples given in this book and answering the following questions:

1. Which method provides better performance (for example, ISE)?
2. Which method is easier to understand and use?
3. Which method can reach practical performance specifications more easily?

2.7 Summary

In this chapter, plants are categorized into three types according to their features. The simplest models are

$$\text{Stable plant:} \quad \frac{K}{\tau s+1}e^{-\theta s},$$
$$\text{Integrating plant:} \quad \frac{K}{s}e^{-\theta s},$$
$$\text{Unstable plant:} \quad \frac{K}{\tau s-1}e^{-\theta s}.$$

Practical performance requirements in control system design are frequently expressed in terms of time domain response index, such as overshoot, rise time, and settling time. Besides these, this chapter defines two new indices about disturbance rejection: perturbation peak and recovery time. The advantage of these time domain indices is that they are very intuitive. However, it is difficult to use these indices to design the optimal controller. The optimal controller can be designed based on the H_2 index and H_∞ index given in this chapter.

The frequency domain analysis is reviewed. The method can be used for both system analysis and design. The Nyquist stability criterion provided by the method has been an important theoretical tool all along.

There are two typical modeling methods in process control: the step response method and the ultimate cycle method. The relationship between the two models is studied, and analytical formulas describing the relationship are provided in this chapter.

This chapter ends with a discussion about design requirements and method comparison. Basic design requirements are formulated and several rules are proposed for controller comparison.

Exercises

1. Based on the frequency response technique, an alternative rational approximation of time delay is obtained. The second order approximation can be written as

$$e^{-\theta s} = \frac{[\theta s - (2.57 + 1.97j)][\theta s - (2.57 - 1.97j)]}{[\theta s + (2.57 + 1.97j)][\theta s + (2.57 - 1.97j)]}.$$

   Compare the precision of the 2/2 Pade approximant with that of the new approximation by plotting their phase curves for $\theta = 1$.

2. The open-loop transfer function of the unity feedback control loop is given by

$$L(s) = \frac{1}{s(s+1)}.$$

   Compute the overshoot and the rise time.

3. In the unity feedback loop, the measurement of the output $y(s)$ may be corrupted by a measurement noise signal $n(s)$(Figure E2.1). Find the transfer function from $n(s)$ to $y(s)$ and compare it with the closed-loop transfer function.

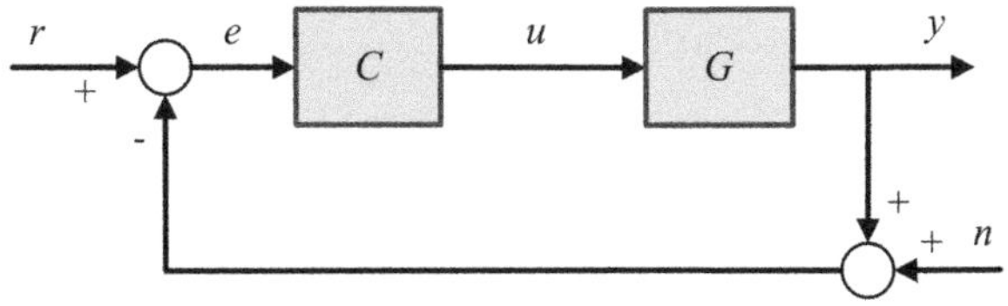

FIGURE E2.1
Unity feedback loop with measurement noise.

4. Consider the following plant

$$G(s) = \frac{K}{\tau s + 1} e^{-\theta s}.$$

It is known that the gain of $G(s)$ is $K = 1$. With the ultimate cycle method, it is obtained that the ultimate gain is $K_u = 2.26$ and the ultimate period is $T_u = 3.09$. Compute τ and θ in $G(s)$.

6*. Assume that there is an improper transfer function, for example, $s + 1$.

(a) Can it be described by the state space method?

(b) Give the state space realization of $(s + 1)/(\lambda s + 1)$. Here λ is a very small positive real number.

Notes and References

The MP system is restricted to be stable in some textbooks. This book adopts a different definition, which is identical to that in Morari and Zafiriou (1989).

The rational approximations in Section 2.2 can be found in many books, for example, Saff and Varga (1977). The rational approximation has been adopted in many different design methods, for example, Stahl and Hippe (1987) and Rivera et al. (1986).

Example 2.2.1 is from Marshall (1979).

The performance indices about disturbance rejection in Section 2.3 can be found in Zhang et al. (1999). In process control, the load response is far more important than the reference response (Shinskey, 2002).

Figure 2.4.2 and Figure 2.4.3 are respectively drawn based on Figure 1.2 and Figure 1.3 in Maciejowski (1989) with different models.

Section 2.5 closely follows the paper by Gu et al. (2005).

There are many books about the classical control theory, for example, Dorf and Bishop (2001), Kuo (2003), Ogata (2002), and D'Azzo and Houpis (1988).

Exercise 1 is adapted from Stahl and Hippe (1987).

3

Essentials of the Robust Control Theory

CONTENTS

There always exists uncertainty in practice. Uncertainty means that we cannot exactly predict the output of a system even if the input is known. Hence, it is most desirable that the controller be robust; that is, the controller is insensitive to uncertainty. This chapter begins with some basic concepts of robust control, such as norm, system gain, design specification, and controller parameterization. Then robust stability and robust performance are studied based on these concepts; the necessary and sufficient conditions for robust stability and robust performance are given. The final topic is to discuss the robustness of the first-order system with time delay. This chapter sets the scene for further study in the subsequent chapters by formulating important concepts and results in robust control theory.

3.1 Norms and System Gains

In this section, the norms for signals and systems, as well as the system gains, are introduced.

The objective of a control system is to make the output behave in a desired way by manipulating the input. The most widely used objective is to keep the plant output close to the reference. For example, in a paper-making process the basis weight and the moisture content of the paper should be kept close to the required values. To decribe the performance of such a system, the "size" of certain signals should be defined. The "size" of a signal can be defined by introducing norms. "The signal is small" means "the norm of the signal is small."

Consider a signal $r(t)$ in time domain. $r(t) = 0$ for $t < 0$. A norm is a nonnegative real number, denoted by $\|r(t)\|$, that possesses the following properties:

1. $\|r(t)\| = 0$ if and only if $r(t) = 0$, $\forall t$.
2. $\|\alpha r(t)\| = |\alpha|\|r(t)\|$, α is any real number.
3. $\|r_1(t) + r_2(t)\| \leq \|r_1(t)\| + \|r_2(t)\|$.

Several frequently used signal norms are listed as follows:

1-norm. The 1-norm of a signal $r(t)$ is the integral of its absolute value:

$$\|r(t)\|_1 := \int_{-\infty}^{\infty} |r(t)|\, dt. \tag{3.1.1}$$

2-norm. The 2-norm of $r(t)$ is

$$\|r(t)\|_2 := \left[\int_{-\infty}^{\infty} r^2(t)\, dt\right]^{1/2}. \tag{3.1.2}$$

Suppose that $r(t)$ is the current through a 1Ω resistor. The instantaneous power equals $r(t)^2$ and the energy equals $\|r(t)\|_2^2$.

∞-norm. The ∞-norm of $r(t)$ is the least upper bound of its absolute value:

$$\|r(t)\|_\infty := \sup_t |r(t)|. \tag{3.1.3}$$

Consider a linear time-invariant and causal system $T(t)$, of which the input is $r(t)$ and the output is $y(t)$. In time domain, the input-output model of such a system can be expressed in the form of a convolution equation:

$$\begin{aligned} y(t) &= T(t) * r(t) \\ &= \int_{-\infty}^{\infty} T(t-\tau) r(\tau)\, d\tau. \end{aligned}$$

Let $T(s)$ denote the transfer function of $T(t)$. Norms can also be defined for the system $T(s)$:

2-norm.

$$\|T(s)\|_2 := \left[\frac{1}{2\pi}\int_{-\infty}^{\infty} |T(j\omega)|^2\, d\omega\right]^{1/2}. \tag{3.1.4}$$

∞-norm.

$$\|T(s)\|_\infty := \sup_\omega |T(j\omega)|. \tag{3.1.5}$$

The following theorem tells us when the two norms are finite.

Theorem 3.1.1. *The 2-norm of $T(s)$ is finite if and only if $T(s)$ is strictly proper and has no poles on the imaginary axis. The ∞-norm of $T(s)$ is finite if and only if $T(s)$ is proper and has no poles on the imaginary axis.*

Proof. Assume that $T(s)$ is strictly proper and has no poles on the imaginary axis. Then the Bode magnitude plot attenuates at high frequencies. We can always find a low-pass transfer function whose Bode magnitude plot is higher than that of $T(s)$. For example, for sufficiently large positive c and sufficiently small positive τ,

$$|c/(\tau j\omega + 1)| \geq |T(j\omega)|, \forall \omega.$$

It is easy to verify that the 2-norm of $c/(\tau s+1)$ equals $c/\sqrt{2\tau}$. Hence $T(s)$ has a finite 2-norm.

When $T(s)$ is proper and has no poles on the imaginary axis, its Bode magnitude plot is still finite. Therefore, $T(s)$ has a finite ∞-norm. □

Now consider how to compute the two norms.

Suppose that $T(s)$ is strictly proper and has no poles on the imaginary axis. We have

$$\begin{aligned} \|T(s)\|_2^2 &= \frac{1}{2\pi}\int_{-\infty}^{\infty} |T(j\omega)|^2 \, d\omega \\ &= \frac{1}{2\pi j}\int_{-j\infty}^{j\infty} T(-s)T(s)\, ds \\ &= \frac{1}{2\pi j}\oint T(-s)T(s)\, ds. \end{aligned}$$

To compute the contour integral, a theorem is needed. Assume that Ω is a simply connected open subset of the complex plane, $a_1, ..., a_n$ are finitely many points enclosed by the closed contour γ, and the interior region of γ is completely contained in Ω.

Theorem 3.1.2 (Residue Theorem). *If $F(s)$ is analytic in Ω except for at $a_1, ..., a_n$, then*

$$\oint_\gamma F(s)ds = 2\pi j \sum_{i=1}^{n} \text{Res}F(a_i).$$

Let Ω be equal to the LHP and γ be the contour consisting of the imaginary axis and an infinite-radius semicircle in the LHP. By Theorem 3.1.2, $\|T(s)\|_2^2$ equals the sum of the residues of $T(-s)T(s)$ at its poles in the LHP.

$\|T(s)\|_\infty$ can be obtained by choosing a series of frequency points for $T(s)$:

$$\{\omega_1, ..., \omega_n\},$$

and then searching for

$$\max_{1\leq k\leq n} |T(j\omega_k)|.$$

If $T(s)$ is stable, Parseval's theorem gives a property of the 2-norm:

$$\|T(s)\|_2 = \|T(t)\|_2. \tag{3.1.6}$$

The following lemma can be viewed as a special case of Theorem 3.1.2.

Lemma 3.1.3 (Cauchy Integral Theorem). *If $F(s)$ is analytic in Ω, then*

$$\oint_\gamma F(s)ds = 0.$$

Denote the complex conjugate of a complex number $c = \mathrm{a}+\mathrm{b}i$ as $\bar{c} = \mathrm{a}-\mathrm{b}i$. The following theorem gives another property of the 2-norm. This property is referred to as the orthogonality in some literature.

Theorem 3.1.4. *If $T_1(s)$ has no poles in Re $s > 0$ while $T_2(s)$ has no poles in Re $s < 0$, then*

$$\|T_1(s) + T_2(s)\|_2^2 = \|T_1(s)\|_2^2 + \|T_2(s)\|_2^2. \tag{3.1.7}$$

Proof.

$$\begin{aligned} & \|T_1(s) + T_2(s)\|_2^2 \\ = \; & \frac{1}{2\pi}\int |T_1(j\omega) + T_2(j\omega)|^2 d\omega \\ = \; & \|T_1(s)\|_2^2 + \|T_2(s)\|_2^2 + 2\mathrm{Re}\left[\frac{1}{2\pi}\int \overline{T_1(j\omega)}T_2(j\omega)d\omega\right]. \end{aligned}$$

Now, it suffices to show that the last integral equals zero. Convert it into a contour integral by closing the imaginary axis with an infinite-radius semicircle in the LHP:

$$\frac{1}{2\pi}\int \overline{T_1(j\omega)}T_2(j\omega)d\omega = \frac{1}{2\pi j}\oint T_1(-s)T_2(s)ds.$$

A simple application of Lemma 3.1.3 shows that the right-hand side of the above equation equals zero. □

The ∞-norm of $T(s)$ equals the distance from the origin to the farthest point on the Nyquist plot of $T(s)$. It also appears as the peak on the Bode magnitude plot of $T(s)$. An important property of the ∞-norm is that it is sub-multiplicative:

$$\|T_1(s)T_2(s)\|_\infty \leq \|T_1(s)\|_\infty \|T_2(s)\|_\infty. \tag{3.1.8}$$

Now the question of interest is that, when it is known how large the input is, how large is the output going to be? For example, the input is a signal with its 2-norm less than or equal to 1. What is the least upper bound of the 2-norm of the output? The answer to this question correlates with an important concept called the system gain.

Consider a linear system with the input $r(t)$, the output $y(t)$, and the transfer function $T(s)$. It is meaningless to consider an unstable system. Thus, $T(s)$ should be stable. Also $T(s)$ is strictly proper. The system gains are shown in Table 3.1.1.

TABLE 3.1.1
System gains for SISO systems.

	$r(t)=\delta(t)$	$\|r(t)\|_2$
$\|y(t)\|_2$	$\|T(s)\|_2$	$\|T(s)\|_\infty$
$\|y(t)\|_\infty$	$\|T(t)\|_\infty$	$\|T(s)\|_2$

Entry (1, 1). *Here (1, 1) denotes the first row and the first column of Table 3.1.1. The denotation applies to the rest. The meaning of this entry is that the energy of the output is the square of the 2-norm of the system transfer function when the input is the impulse.*

Proof. If $r(t)=\delta(t)$, then

$$\begin{aligned} y(t) &= \int_{-\infty}^{\infty} T(t-\tau)\delta(\tau)\,d\tau \\ &= T(t). \end{aligned}$$

Therefore, $\|y(t)\|_2 = \|T(t)\|_2$. As $T(s)$ is stable, $\|T(s)\|_2 = \|T(t)\|_2$. □

Entry (2, 1). *The meaning of this entry is that the amplitude of the output is the ∞-norm of the system impulse response function when the input is the impulse.*

Proof. Since $y(t)=T(t)$, the result is evident. □

Entry (1, 2). *The meaning of this entry is that the energy of the output is bounded by the square of the ∞-norm of the system transfer function when the input is a signal of which the energy is bounded by unity.*

Proof. $\|T(s)\|_\infty$ is an upper bound on the system gain:

$$\begin{aligned} \|y(t)\|_2^2 &= \|y(s)\|_2^2 \\ &= \frac{1}{2\pi}\int_{-\infty}^{\infty} |T(j\omega)|^2 |r(j\omega)|^2\,d\omega \\ &\le \|T(s)\|_\infty^2 \frac{1}{2\pi}\int_{-\infty}^{\infty} |r(j\omega)|^2\,d\omega \end{aligned}$$

$$= \|T(s)\|_\infty^2 \|r(s)\|_2^2.$$

Since $\|r(t)\|_2$ is bounded, $\|r(t)\|_2 = \|r(s)\|_2$ and

$$\|y(t)\|_2^2 \le \|T(s)\|_\infty^2 \|r(t)\|_2^2.$$

To show that $\|T(s)\|_\infty$ is the least upper bound, it is enough to prove that the identity holds for at least one input. Choose a frequency ω_o where

$$|T(j\omega_o)| = \|T(s)\|_\infty.$$

Similar to the time domain impulse, construct a frequency domain impulse signal $\delta_f(j\omega)$:

$$|\delta_f(j\omega)| = \begin{cases} \infty & \omega = \omega_o \\ 0 & \omega \ne \omega_o \end{cases},$$

with

$$\frac{1}{2\pi}\int_{-\infty}^{\infty} |\delta_f(j\omega)|^2 \, d\omega = 1.$$

For this specific input we have

$$\begin{aligned} \|y(t)\|_2^2 &= \frac{1}{2\pi}\int_{-\infty}^{\infty} |T(j\omega)|^2 |\delta_f(j\omega)|^2 \, d\omega \\ &= |T(j\omega_o)|^2 \\ &= \|T(s)\|_\infty^2. \end{aligned}$$

□

There exists no ideal frequency domain impulse in a real system. However, the impulse can be approximated by the following band-pass function:

$$|\delta_f(j\omega)| = \begin{cases} \sqrt{\pi/2\epsilon} & |\omega - \omega_o| < \epsilon \text{ or } |\omega + \omega_o| < \epsilon \\ 0 & \text{else} \end{cases},$$

where ϵ is a small positive number.

Entry (2, 2). *The meaning of this entry is that the amplitude of the output is bounded by the 2-norm of the system transfer function when the input is a signal of which the energy is bounded by unity.*

Proof. With the Cauchy-Schwarz inequality, we have

$$\begin{aligned} |y(t)| &= \left| \int_{-\infty}^{\infty} T(t-\tau) r(\tau) d\tau \right| \\ &\le \left(\int_{-\infty}^{\infty} T^2(t-\tau)\, d\tau \right)^{1/2} \left(\int_{-\infty}^{\infty} r^2(\tau)\, d\tau \right)^{1/2} \\ &= \|T(t)\|_2 \|r(t)\|_2. \end{aligned}$$

$T(s)$ is stable. Hence $\|T(t)\|_2 = \|T(s)\|_2$ and

$$\|y(t)\|_\infty \leq \|T(s)\|_2.$$

To show that $\|T(s)\|_2$ is the least upper bound, it is enough to prove that the identity holds for at least one input. Choose the following input:

$$r(t) = \frac{T(-t)}{\|T(t)\|_2}.$$

It is easy to verify that $\|r(t)\|_2 = 1$ and

$$\begin{aligned} |y(0)| &= \left| \int_{-\infty}^{\infty} T(-\tau) \frac{T(-\tau)}{\|T(t)\|_2} d\tau \right| \\ &= \frac{1}{\|T(t)\|_2} \left| \int_{-\infty}^{\infty} T^2(-\tau) d\tau \right| \\ &= \|T(t)\|_2. \end{aligned}$$

□

In the literature about robust control theory, several important concepts are frequently used. For convenience of reading and understanding, these concepts are explained here.

L_2 space

The set of strictly proper rational functions without poles on the imaginary axis.

L_∞ space

The set of proper rational functions without poles on the imaginary axis.

Hardy space

The set of rational functions without poles in the open RHP.

H_2 space

The intersection of the Hardy space and the L_2 space. RH_2 space denotes the overall of real functions in the H_2 space.

H_∞ space

The intersection of the Hardy space and the L_∞ space. RH_∞ space denotes the overall of real functions in the H_∞ space.

When we say a rational function belongs to L_2 space, it means that the function is strictly proper and has no poles on the imaginary axis. Similarly, a rational function that belongs to the H_∞ space implies that the function is stable proper and has no poles on the imaginary axis. The rest is similar.

3.2 Internal Stability and Performance

The concept of stability describes the ability of a system to return to equilibrium after a small disturbance. In a control system, it cannot be tolerated that a small disturbance at one location leads to unbounded signals at some other locations. We hope to guarantee bounded internal signals for all bounded external signals. It is not enough to look only at the stability of the closed-loop transfer function. Even though the closed-loop transfer function is stable, internal signals could be unbounded. To guarantee bounded internal signals, the closed-loop system must be internally stable.

Definition 3.2.1. *A linear time-invariant control system is internally stable if the transfer functions between any two points of the system are stable.*

In a control system, any two points can be selected for testing internal stability, but some choices are equivalent. In the system shown in Figure 2.3.1, there are only two independent outputs and two independent inputs. One can choose $r(s)$ and $d'(s)$ as inputs and $y(s)$ and $u(s)$ as outputs. The closed-loop system is internally stable if and only if all elements in the transfer function matrix $\boldsymbol{H}(s)$ from $r(s)$ and $d'(s)$ to $y(s)$ and $u(s)$ are stable:

$$\begin{bmatrix} y(s) \\ u(s) \end{bmatrix} = \boldsymbol{H}(s) \begin{bmatrix} r(s) \\ d'(s) \end{bmatrix}, \tag{3.2.1}$$

where

$$\boldsymbol{H}(s) = \begin{bmatrix} \dfrac{G(s)C(s)}{1+G(s)C(s)} & \dfrac{G(s)}{1+G(s)C(s)} \\ \dfrac{C(s)}{1+G(s)C(s)} & \dfrac{-G(s)C(s)}{1+G(s)C(s)} \end{bmatrix}.$$

Recall the concept of stability. A system is stable if and only if all poles of its closed-loop transfer function are in the open LHP (or equivalently, the closed-loop transfer function has no poles in Re $s \geq 0$). Compared with the concept of internal stability, the concept of stability is not a complete one, because the RHP zero-pole cancellation in the feedback loop is not considered.

The zero-pole cancellation means that there is a zero and a pole at the same point. It may occur in a component (for example, in the controller) or between two separated components (for example, between the controller and the plant). Some cancellations are removable, like $(s-1)/(s-1) = 1$. Such a transfer function is unstable before the cancellation is removed. Some cancellations are not removable, like $(e^{-s}-1)/s$. In general, such a case happens in systems with time delay.

Example 3.2.1. *This example is used to illustrate the difference between the stability and the internal stability.*

In Figure 2.3.1, take

$$C(s) = \frac{s-1}{s+1}, G(s) = \frac{1}{(s-1)(s+1)}.$$

It is easy to verify that the transfer function from $r(s)$ to $y(s)$ is stable, but the one from $d'(s)$ to $y(s)$ is not. Therefore, the system is internally unstable.

Before discussing the performance, let us first define the sensitivity function:

$$S(s) := \frac{1}{1+G(s)C(s)}, \tag{3.2.2}$$

which describes the effect of the disturbance $d(s)$ on the output $y(s)$, or the effect of the reference $r(s)$ on the error $e(s)$:

$$S(s) = \frac{y(s)}{d(s)} = \frac{e(s)}{r(s)}. \tag{3.2.3}$$

Let $T(s)$ denote the transfer function from the reference $r(s)$ to $y(s)$, namely the closed-loop transfer function:

$$T(s) = \frac{y(s)}{r(s)} = \frac{G(s)C(s)}{1+G(s)C(s)}. \tag{3.2.4}$$

As

$$T(s) = 1 - S(s), \tag{3.2.5}$$

$T(s)$ is also called the complementary sensitivity function.

The sensitivity function $S(s)$ quantifies how sensitive $T(s)$ is to the variation in $G(s)$; that is, the limit of the ratio of a relative perturbation in $T(s)$, $\Delta T(s)/T(s)$, to a relative perturbation in $G(s)$, $\Delta G(s)/G(s)$:

$$\lim_{\Delta G(s)\to 0} \frac{\Delta T(s)/T(s)}{\Delta G(s)/G(s)} = \frac{dT(s)}{dG(s)}\frac{G(s)}{T(s)} = S(s).$$

Now let us see the relationship among the closed-loop response, $S(s)$, and $T(s)$. The exact model is commonly referred to as the nominal plant. In practice, one can take the "center" of all uncertain plants as the nominal plant. Under the nominal condition, it is desirable to make $|S(j\omega)|$ as small as possible (Figure 3.2.1). A smaller $|S(j\omega)|$ implies that the changes of the reference and disturbance have less effect on the system output. Such a system has a higher ability on disturbance rejection. Algebraically, $|S(j\omega)|$ can be made equal to 0 for all frequencies, which is referred to as the perfect control. However, this is impossible in practice. In a real system, $G(s)$ is usually strictly proper. Then

$$\lim_{s\to\infty} S(s) = \lim_{s\to\infty} \frac{1}{1+G(s)C(s)} = 1. \tag{3.2.6}$$

Generally speaking, $|T(j\omega)|$ should be made as close as possible to unity. A $|T(j\omega)|$ close to unity means that the system has large bandwidth and thus good tracking ability. However, due to the limitation $T(s)+S(s)=1$, $|T(j\omega)|$ can be made equal to unity only within a finite frequency range (Figure 3.2.1).

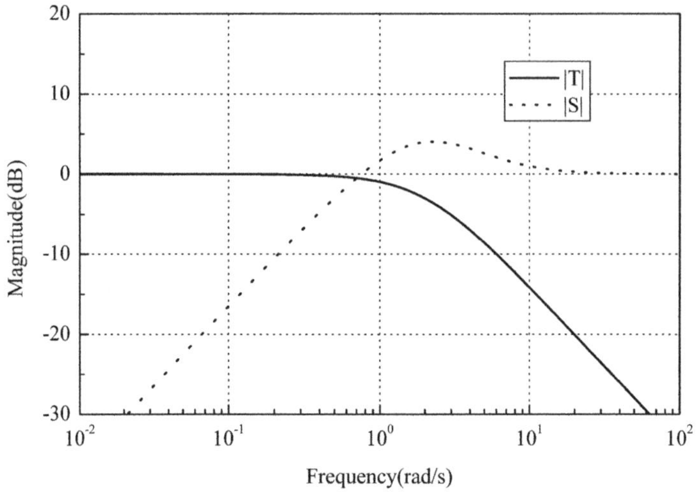

FIGURE 3.2.1
Sensitivity function and complementary sensitivity function.

Consider the performance problem. The most basic objective of a feedback controller is to keep the error between the plant output $y(t)$ and the reference $r(t)$ small when the overall system is affected by the external disturbance and the plant uncertainty. In order to quantify performance, an index of "smallness" for the error has to be defined. A widely used performance index is the H_2 index:

$$\min \int_0^{\infty} e^2(t)dt = \min\|e(t)\|_2^2. \tag{3.2.7}$$

The controller structure and parameters can be determined from the solution to the above optimization problem. If the reference $r(s)$ is known, a weighting function $W(s) = r(s)$ is introduced to normalize the reference so that the system input $r'(s)$ is the impulse (Figure 3.2.2). Use Table 3.1.1: if the input is the impulse, the energy of the output $e(t)$ is the square of the 2-norm of the system transfer function:

$$\|e(t)\|_2 = \|W(s)S(s)\|_2, \quad \frac{r(s)}{W(s)} = 1.$$

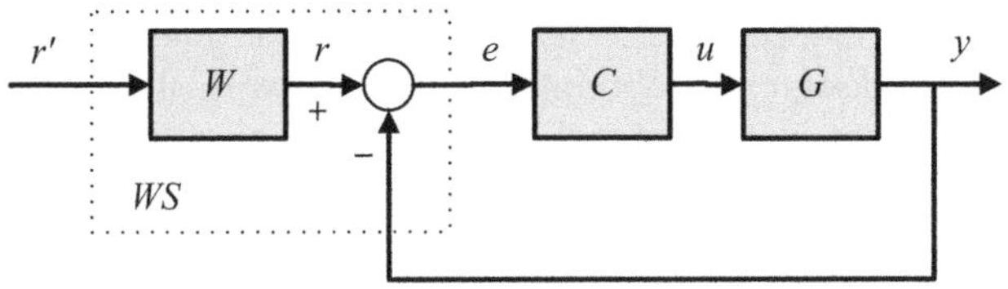

FIGURE 3.2.2
Normalization of the system input.

Hence, the H_2 performance index in the Laplace domain can be written as

$$\min\|W(s)S(s)\|_2. \tag{3.2.8}$$

Assume that $\min\|W(s)S(s)\|_2 \to \epsilon$. Include the effect of the constant ϵ into the weighting function $W(s)$; the index can be expressed in the form of

$$\|W(s)S(s)\|_2 < 1. \tag{3.2.9}$$

Another widely used performance index is the H_∞ index:

$$\min_{r(t)}\sup \int_0^\infty e^2(t)dt = \min_{r(t)}\sup\|e(t)\|_2^2. \tag{3.2.10}$$

Again, a weighting function $W(s)$ is introduced to normalize the system input. Utilize Table 3.1.1: if the input is a signal of which the energy is bounded by unity, the energy of the output is bounded by the square of the ∞-norm of the system transfer function:

$$\sup_{r(t)}\|e(t)\|_2 = \|W(s)S(s)\|_\infty, \quad \left\|\frac{r(s)}{W(s)}\right\|_2 \leq 1.$$

The design index in the frequency domain can be written as

$$\min\|W(s)S(s)\|_\infty \quad \text{or} \quad \|W(s)S(s)\|_\infty < 1. \tag{3.2.11}$$

The two indices try to minimize their respective norms of the weighted sensitivity function. In some literature, the problem is referred to as the weighted sensitivity problem.

It is noticed that, to treat the design problem in a unified mathematical form, a weighting function is introduced to the performance index. It is assumed that the reference $r(s)$ is generated by a normalized input $r'(s)$ passing through a transfer function block $W(s)$. Then the system gain in Table 3.1.1 can be applied. $W(s)$ is called the performance weighting function in some literature, because the obtained performance depends on the choice of $W(s)$.

It should be emphasized that the choice of norm is not crucial. Normally, the choice of performance index is not crucial, either. The essential objective for control system design is to obtain the controller with the desired response. The tradeoff inherent in the control system means that although the norms or performance indices may differ to a large extent, the obtained responses are not very different. The design requirement can be achieved by applying different norms and performance indices. Comparatively, it might be more important to choose a norm or a performance index that is mathematically convenient.

In most industrial systems, the designer mainly concerns the regulator problem. With such an objective, the obtained system usually has a good ability for disturbance rejection. At the same time, the objective involves a requirement on good tracking ability as well. In most cases, the regulator problem and the servomechanism problem have the same objective, because the effect of the disturbance $d(s)$ on the output $y(s)$ is the same as that of the reference $r(s)$ on the tracking error $e(s)$. The conclusion can also be explained by means of the sensitivity function and the complementary sensitivity function. To obtain a controller with good ability of disturbance rejection, it is desirable to minimize $|S(j\omega)|$. For all control systems there exists the constraint $S(s) = 1 - T(s)$. To minimize $|S(j\omega)|$ means to make $|T(j\omega)|$ close to 1, namely, to design a controller with good tracking ability.

In the last chapter, the relationship between the asymptotic tracking property and the system type was studied. Now it is shown how the asymptotic tracking property relates to the sensitivity function.

Theorem 3.2.1. *Assume that the closed-loop system is internally stable and is of Type m. Then the sensitivity function satisfies*

$$\lim_{s\to 0} \frac{S(s)}{s^k} = 0, \quad k = 0, 1, ..., m-1.$$

As $t \to \infty$ the closed-loop system perfectly tracks references of the form $\sum_{k=0}^{m} a_k s^{-k}$, where a_k are real constants.

Proof. With $S(s) = 1/[1 + L(s)]$, the result follows directly from the Final Value Theorem. □

In particular, for step and ramp inputs we have the following corollary.

Corollary 3.2.2. *Assume that the closed-loop system is internally stable.*

1. *If the reference is a step, then the tracking error tends to be zero as $t \to \infty$ if and only if $S(s)$ has at least one zero at the origin.*
2. *If the reference is a ramp, then the tracking error tends to be zero as $t \to \infty$ if and only if $S(s)$ has at least two zeros at the origin.*

Note that the number of the zeros of $S(s)$ at $s = 0$ is the same as that of the poles of $L(s) = G(s)C(s)$ at $s = 0$.

3.3 Controller Parameterization

This section discusses how to characterize all controllers that stabilize the closed-loop system. Assume that the plant $G(s)$ in the unity feedback loop is stable. Introduce $Q(s)$, which denotes the transfer function from $r(s)$ to $u(s)$:

$$Q(s) = \frac{C(s)}{1 + G(s)C(s)}. \tag{3.3.1}$$

The controller of the unity feedback control system can be obtained with the inverse relationship:

$$C(s) = \frac{Q(s)}{1 - G(s)Q(s)}. \tag{3.3.2}$$

Then the matrix $\boldsymbol{H}(s)$ defined in the last section can be rewritten as

$$\boldsymbol{H}(s) = \begin{bmatrix} G(s)Q(s) & [1 - G(s)Q(s)]\,G(s) \\ Q(s) & -G(s)Q(s) \end{bmatrix}. \tag{3.3.3}$$

Evidently, the system is internally stable if and only if $Q(s)$ is stable.

In fact, we have proved the following theorem.

Theorem 3.3.1. *Assume that $G(s)$ is stable. All of the stabilizing controllers of the unity feedback loop can be expressed as*

$$C(s) = \frac{Q(s)}{1 - G(s)Q(s)},$$

where $Q(s)$ is any stable transfer function.

What the theorem gives is the well-known Youla parameterization. Let us look at a simple explanation of the result in terms of the IMC structure.

Throughout this book, $G(s)$ is used to denote the plant. If it is necessary to distinguish between the nominal plant and the real plant, $G(s)$ is used to denote the nominal plant and $\widetilde{G}(s)$ is used to denote the real plant.

By equivalently transforming the unity feedback control loop in Figure 2.3.1, a new feedback configuration can be obtained (Figure 3.3.1). Regard the structure contained in the dashed box as $Q(s)$. The new configuration is the so-called IMC structure. $Q(s)$ is referred to as the IMC controller. If $G(s)$ is viewed as a reference model, the model reference adaptive control can be embedded in the structure. One can also regard $G(s)$ as a predictive model. Then, the structure interprets the basic idea of MPC algorithms.

Assume that the model is exact, that is, $G(s) = \widetilde{G}(s)$. Figure 3.3.1 shows that the stability of the closed-loop system is determined only by $Q(s)$. This is exactly the conclusion given by Youla parameterization.

The Youla parameterization is a simple yet perfect result. What is the significance of Youla parameterization?

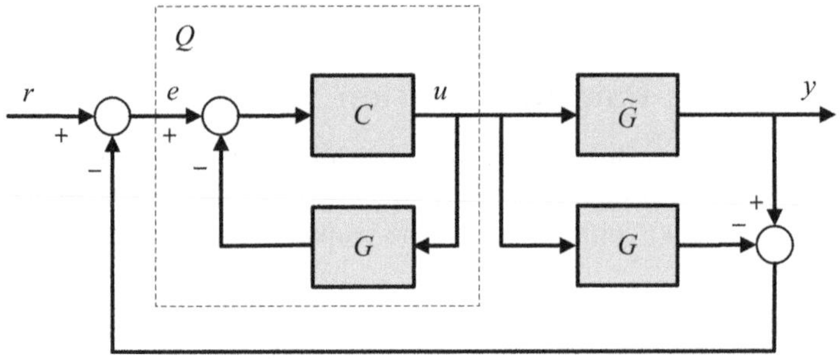

FIGURE 3.3.1
Explanation of Youla parameterization.

1. As we know, most design problems can be formulated in this way: given a plant $G(s)$, design a controller $C(s)$ such that the feedback system is internally stable and the output $y(s)$ asymptotically tracks a step reference $r(s)$. The procedure is greatly complicated when the stability and performance are treated simultaneously. With Youla parameterization, one can consider the stability and the performance separately. First, describe all stabilizing $C(s)$s in terms of a stable transfer function $Q(s)$. Then, search the optimal $C(s)$ among the stabilizing controllers (Figure 3.3.2). This search is much easier.

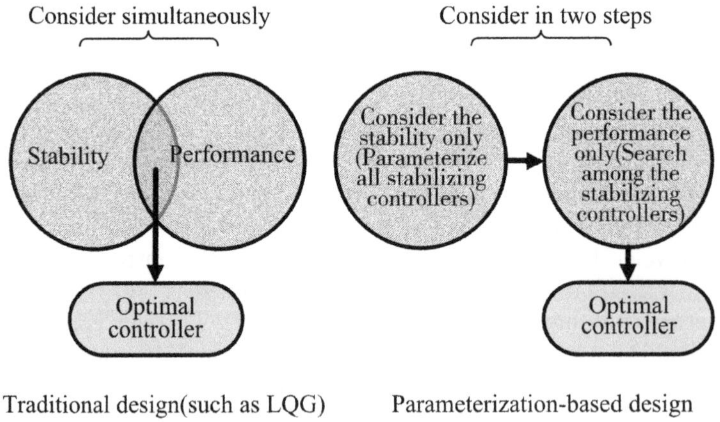

FIGURE 3.3.2
Two different design procedures.

2. The effect of the unity feedback controller $C(s)$ on the sensitiv-

ity function and complementary sensitivity function is complicated. Nevertheless, $Q(s)$ relates to the sensitivity function and complementary sensitivity function in a "linear" manner, which can significantly simplify the design task for optimal controllers.

$$\begin{aligned} S(s) &= \frac{1}{1+G(s)C(s)} \\ &= 1 - G(s)Q(s), \end{aligned} \tag{3.3.4}$$

$$\begin{aligned} T(s) &= \frac{G(s)C(s)}{1+G(s)C(s)} \\ &= G(s)Q(s). \end{aligned} \tag{3.3.5}$$

It is evident that $Q(s)$ should be proper, since an improper transfer function is not realizable physically. However, it would be convenient for presentation to temporarily relax the requirement of properness in controller design. When an improper $Q(s)$ is used, the mathematically precise readers may interpret the improper transfer function as a shorthand notation of its proper approximation. For example, a proper approximation of $s+1$ is $(s+1)/(\lambda s+1)$, where λ is a very small positive real number. Such a problem will be encountered in the next chapter.

Sometimes, the plant is not stable. To find all stabilizing controllers, one might try as follows. Write $G(s)$ in the form of its coprime factorization:

$$G(s) = \frac{V(s)}{U(s)}, \tag{3.3.6}$$

where $V(s)$ and $U(s)$ are stable, proper, real, and rational. There exist two stable proper real rational functions $X(s)$ and $Y(s)$ satisfying the equation

$$V(s)X(s) + U(s)Y(s) = 1. \tag{3.3.7}$$

By checking $\boldsymbol{H}(s)$, it is known that $C(s) = X(s)/Y(s)$ is a stabilizing controller:

$$\begin{aligned} \boldsymbol{H}(s) &= \begin{bmatrix} \dfrac{G(s)C(s)}{1+G(s)C(s)} & \dfrac{G(s)}{1+G(s)C(s)} \\ \dfrac{C(s)}{1+G(s)C(s)} & \dfrac{-G(s)C(s)}{1+G(s)C(s)} \end{bmatrix} \\ &= \begin{bmatrix} V(s)X(s) & V(s)Y(s) \\ X(s)U(s) & -V(s)X(s) \end{bmatrix}. \end{aligned}$$

Notice that

$$V(s)\underbrace{[X(s)+U(s)Q(s)]}_{Stable} + U(s)\underbrace{[Y(s)-V(s)Q(s)]}_{Stable} = 1$$

for any stable $Q(s)$. It turns out that all of the controllers with which the feedback system is internally stable can be expressed as

$$C(s) = \frac{X(s) + U(s)Q(s)}{Y(s) - V(s)Q(s)}, \quad (3.3.8)$$

where $Q(s)$ is any stable transfer function. It is easy to verify that the result reduces to Theorem 3.3.1 when $G(s)$ is stable. In this case,

$$V(s) = G(s),\ U(s) = 1,\ X(s) = 0,\ Y(s) = 1.$$

3.4 Robust Stability and Robust Performance

Almost all controller design methods are based on the model in one form or another. Since the model may be inaccurate, it is most desirable that the designed controller should perform well even when the uncertainty occurs. With respect to this objective, it is natural to propose the following questions: how does the uncertainty affect the stability, and what limitation is imposed on the achievable performance? This introduces two important concepts: robust stability and robust performance.

The general notion of robust stability is that the internal stability holds for all plants in the uncertain model family, while robust performance implies that the internal stability and nominal performance hold for all plants in the family of uncertain models. In order to analyze robust stability and robust performance, not only a nominal model should be provided, but also some description about the uncertainty should be given.

Uncertainty can be described in many different ways. Unfortunately, only a few can be adopted when the mathematical convenience is considered.

A description of uncertainty is the structured uncertainty. A frequently used structured uncertainty is the parameter uncertainty. Consider the following plant:

$$G(s) = \frac{a}{s^2 + 2s + 1}. \quad (3.4.1)$$

The constant a in the plant is known to be in some interval. The parameter uncertainty can be described by $a_{\min} \le a \le a_{max}$.

For rational systems and some specific parameter uncertainty there is a famous test for robust stability.

Consider the following characteristic equation:

$$s^n + a_{n-1}s^{n-1} + ... + a_1 s + a_0 = 0, \quad (3.4.2)$$

where n is a nonnegative integer. The uncertain parameters vary between an upper bound and a lower bound:

$$\alpha_i \leq a_i \leq \beta_i,\ i = 0, 1, ..., n-1. \tag{3.4.3}$$

To ascertain the stability of this system, one might have to investigate all possible combinations of parameters. Fortunately, it is possible to investigate a limited number of worst case polynomials. As a matter of fact, it is sufficient to analyze only four polynomials.

Theorem 3.4.1 (Kharitonov's Theorem). *The closed-loop system (3.4.2) is robust stable if and only if the following four polynomials are stable:*

$$\begin{aligned}
q_1(s) &= \alpha_0 + \alpha_1 s + \beta_2 s^2 + \beta_3 s^3 + \alpha_4 s^4 + \alpha_5 s^5 + ...,\\
q_2(s) &= \alpha_0 + \beta_1 s + \beta_2 s^2 + \alpha_3 s^3 + \alpha_4 s^4 + \beta_5 s^5 + ...,\\
q_3(s) &= \beta_0 + \beta_1 s + \alpha_2 s^2 + \alpha_3 s^3 + \beta_4 s^4 + \beta_5 s^5 + ...,\\
q_4(s) &= \beta_0 + \alpha_1 s + \alpha_2 s^2 + \beta_3 s^3 + \beta_4 s^4 + \alpha_5 s^5 + ...
\end{aligned}$$

The proof of this theorem is beyond the core scope of this book and thus is omitted.

Another uncertainty description is the unstructured uncertainty. Assume that the nominal plant $G(s)$ and the uncertain plant $\widetilde{G}(s)$ have the same number of RHP poles. One can understand the assumption as follows: $G(s)$ and $\widetilde{G}(s)$ are similar. The unstructured uncertainty can be written in the form of a multiplicative uncertainty description:

$$\widetilde{G}(s) = G(s)\left[1 + \delta_m(s)\right],\ \delta_m(s) = \Delta(s)\Delta_m(s). \tag{3.4.4}$$

Here $\Delta_m(s)$ is a fixed stable transfer function. $\Delta(s)$ is a variable stable transfer function satisfying $\|\Delta(s)\|_\infty \leq 1$. It can be viewed as a normalized uncertainty. For each frequency ω,

$$|\delta_m(j\omega)| \leq |\Delta_m(j\omega)|.$$

$|\Delta_m(j\omega)|$ provides the uncertainty profile.

The unstructured uncertainty can be described as a disk in the complex plane: at each frequency ω, the point $\widetilde{G}(j\omega)$ lies in the disk with the center $G(j\omega)$ and the radius $|\Delta_m(j\omega)|$, as shown in Figure 3.4.1. Hence, $\widetilde{G}(s)$ is described by a family of models, rather than a single model.

The unstructured uncertainty is important for two reasons:

1. All models used in controller design should include unstructured uncertainty to cover unmodeled dynamics (for example, nonlinearity).
2. With the unstructured uncertainty, simple and general results can be obtained for not only robust stability but also robust performance. The price is that the description might be conservative.

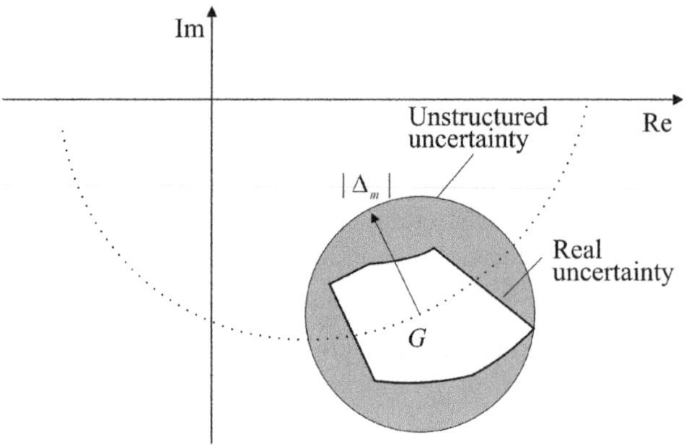

FIGURE 3.4.1
Disk for describing the unstructured uncertainty.

> As an example, the region with an irregular shape in Figure 3.4.1 represents the parameter uncertainty. The uncertainty is usually described by a disk, which is easy to treat mathematically.

The following theorem provides the condition for testing robust stability.

Theorem 3.4.2. *Assume that the nominal closed-loop system is internally stable. $C(s)$ provides robust stability if and only if*

$$\|\Delta_m(s)T(s)\|_\infty < 1.$$

Proof. By assumption, the nominal feedback system is internally stable. By the Nyquist criterion, it is known that for every frequency ω the Nyquist plot of $L(j\omega) = G(j\omega)C(j\omega)$ does not pass through the point $(-1, 0)$, and its number of counterclockwise encirclement equals the sum of the number of poles of $G(s)$ in the RHP and the number of poles of $C(s)$ in the RHP. It is also known that $G(s)$ and $\widetilde{G}(s)$ have the same number of RHP poles. Thus, $C(s)$ provides robust stability if and only if the Nyquist band of $\widetilde{L}(j\omega) = \widetilde{G}(j\omega)C(j\omega)$ does not include the point $(-1, 0)$ for every frequency ω.

Consider the simple geometric argument in Figure 3.4.2. For a frequency ω, $\widetilde{L}(j\omega)$ is a disk in the complex plane. The disk has the center $L(j\omega)$ and the radius $|\Delta_m(j\omega)G(j\omega)C(j\omega)|$. The radius denotes the maximum perturbed scope of $L(j\omega)$. The distance from $L(j\omega)$ to the point $(-1, 0)$ is $|1 + G(j\omega)C(j\omega)|$. It is clear that $\widetilde{L}(j\omega)$ does not include the point $(-1, 0)$ for every frequency ω if and only if

$$|1 + G(j\omega)C(j\omega)| > |\Delta_m(j\omega)G(j\omega)C(j\omega)|.$$

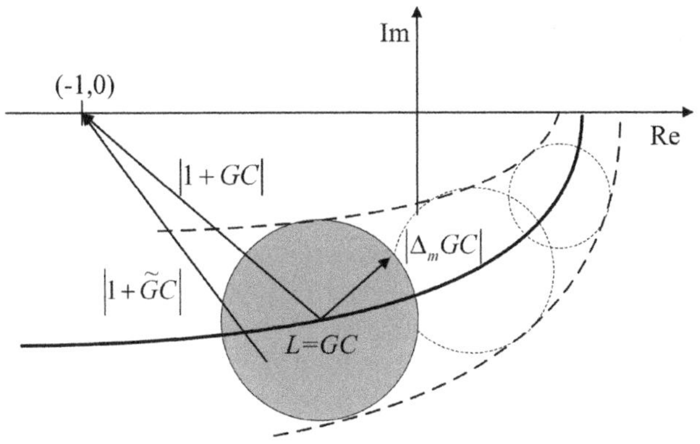

FIGURE 3.4.2
Graphical interpretation for robust stability.

As

$$T(j\omega) = \frac{G(j\omega)C(j\omega)}{1 + G(j\omega)C(j\omega)},$$

the above inequality is equivalent to

$$|\Delta_m(j\omega)T(j\omega)| < 1, \forall\omega.$$

This completes the proof. □

It can be observed from the test condition for robust stability that good robustness can be obtained by decreasing the effect of $T(s)$.

Now consider the robust performance. Recall that when the nominal feedback system is internally stable, the performance requirement is $\|W(s)S(s)\|_\infty < 1$, where

$$S(s) = \frac{1}{1 + G(s)C(s)}.$$

When there exists uncertainty, $S(s)$ is perturbed to

$$\begin{aligned}\widetilde{S}(s) &= \frac{1}{1 + [1 + \delta_m(s)]\, G(s)C(s)} \\ &= \frac{S(s)}{1 + \delta_m(s)T(s)}.\end{aligned}$$

The robust performance condition should be

$$\|W(s)\widetilde{S}(s)\|_\infty < 1.$$

Theorem 3.4.3. *Assume that the nominal closed-loop system is internally stable. A necessary and sufficient condition for robust performance is*

$$|W(j\omega)S(j\omega)| + |\Delta_m(j\omega)T(j\omega)| < 1, \forall\omega. \tag{3.4.5}$$

Proof. In view of Figure 3.4.2, the distance from the point $(-1, 0)$ to a point in the disk is $|1+\widetilde{G}(j\omega)C(j\omega)|$ and the smallest distance form the point $(-1, 0)$ to the point in the disk is $|1 + G(j\omega)C(j\omega)| - |\Delta_m(j\omega)G(j\omega)C(j\omega)|$ for every frequency ω. Then

$$|1+\widetilde{G}(j\omega)C(j\omega)| \geq |1+G(j\omega)C(j\omega)| - |\Delta_m(j\omega)G(j\omega)C(j\omega)|.$$

By taking the inverse of both sides of the above inequality, it is easy to obtain that

$$|\widetilde{S}(j\omega)| \leq \frac{|S(j\omega)|}{1-|\Delta_m(j\omega)T(j\omega)|}.$$

The robust performance requires

$$\|W(s)\widetilde{S}(s)\|_\infty < 1.$$

The condition holds if and only if

$$|W(j\omega)S(j\omega)| + |\Delta_m(j\omega)T(j\omega)| < 1, \forall\omega.$$

□

The test condition for robust performance also has a nice graphical interpretation. When both sides of the test condition are multipled by $|1 + G(j\omega)C(j\omega)|$, two disks are constructed for every frequency ω: one with the center $(-1, 0)$ and the radius $|W(j\omega)|$; the other with the center $G(j\omega)C(j\omega)$ and the radius $|\Delta_m(j\omega)G(j\omega)C(j\omega)|$. The test condition holds if and only if these two disks disjoint (Figure 3.4.3).

Notice that robust performance implies both robust stability and nominal performance. It is desirable to make $\|W(s)S(s)\|_\infty$ small for good nominal performance and at the same time make $\|\Delta_m(s)T(s)\|_\infty$ small for good robust stability. Unfortunately, the interdependence of $S(s) + T(s) = 1$ makes the objective a challenge. Improving the nominal performance worsens the robust stability and pushes the system close to instability. Conversely, good robust stability may be obtained by sacrificing the nominal performance. This explains why the system tuned improperly sometimes has a good adaptive capacity. How to reach a good tradeoff between nominal performance and robust stability is the main problem in control system design.

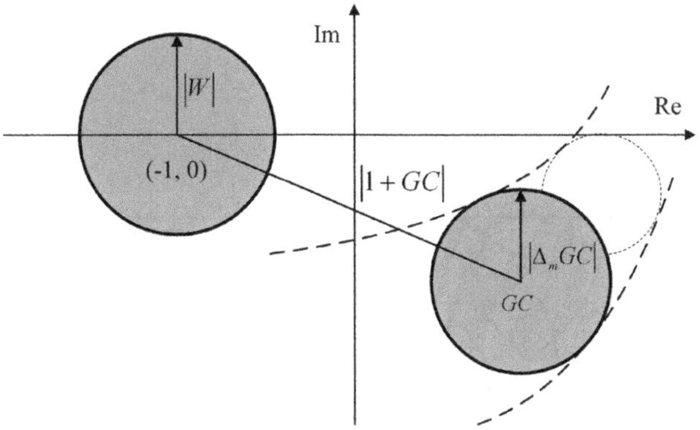

FIGURE 3.4.3
Graphical interpretation for robust performance.

3.5 Robustness of the System with Time Delay

In control system design, the model is usually of low order. The uncertainty of the low-order model is often described in the form of a parameter uncertainty description. This section discusses how the parameter uncertainty of the first-order plant with time delay affects the stability and performance of the closed-loop system. We choose the first-order plant with time delay because

1. This model is widely used in practice, especially in industry.
2. It is easy to deal with the first-order plant and the result provides good understanding for complex systems.

Assume that the following model has been obtained from the experimental data:

$$\widetilde{G}(s) = \frac{\widetilde{K}\exp(-\widetilde{\theta}s)}{\widetilde{\tau}s + 1}, \tag{3.5.1}$$

where $\widetilde{K}$ is the gain, $\widetilde{\tau}$ is the time constant, $\widetilde{\theta}$ is the time delay, and

$$\begin{aligned}\widetilde{K} &\in [\widetilde{K}_{min}, \widetilde{K}_{max}],\\ \widetilde{\tau} &\in [\widetilde{\tau}_{min}, \widetilde{\tau}_{max}],\\ \widetilde{\theta} &\in [\widetilde{\theta}_{min}, \widetilde{\theta}_{max}].\end{aligned}$$

The uncertain model family for this plant is shown in Figure 3.5.1.

According to design requirements, design methods can roughly be classified into two kinds:

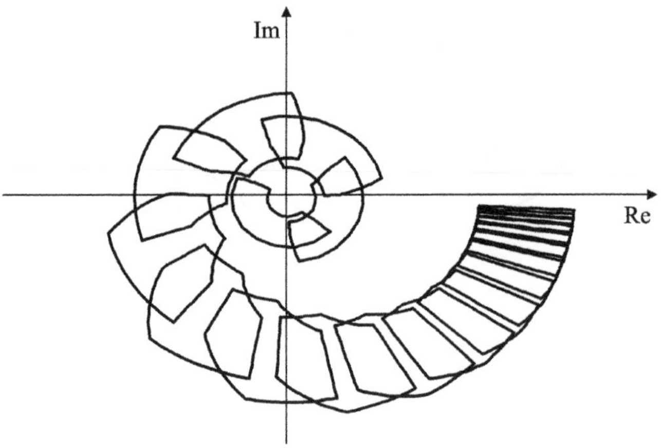

FIGURE 3.5.1
Uncertain model family for the first-order plant with time delay.

1. The design requirements involving the specification on robustness are imposed on the nominal system. The controller is designed for the nominal plant, and then used for the real plant.
2. The design requirements are proposed for the uncertain system. The controller is designed based on the nominal plant and the associated uncertainty, and then used for the real plant.

In both of the two methods, a nominal plant is needed. In this example, the "center" of the uncertain plant is chosen as the nominal plant:

$$G(s) = \frac{Ke^{-\theta s}}{\tau s + 1}, \tag{3.5.2}$$

where

$$\begin{aligned} K &= \frac{\widetilde{K}_{min} + \widetilde{K}_{max}}{2}, \\ \tau &= \frac{\widetilde{\tau}_{min} + \widetilde{\tau}_{max}}{2}, \\ \theta &= \frac{\widetilde{\theta}_{min} + \widetilde{\theta}_{max}}{2}. \end{aligned}$$

The parameter uncertainties are

$$\begin{aligned} |\delta K| \leq \Delta K &= |\widetilde{K}_{max} - K| < |K|, \\ |\delta \tau| \leq \Delta \tau &= |\widetilde{\tau}_{max} - \tau| < |\tau|, \\ |\delta \theta| \leq \Delta \theta &= |\widetilde{\theta}_{max} - \theta| < |\theta|. \end{aligned}$$

Then the uncertain model family can be written as:

$$\widetilde{G}(s) = \frac{(K + \delta K)e^{-(\theta+\delta\theta)s}}{(\tau + \delta\tau)s + 1}. \tag{3.5.3}$$

Now assume that there is ±20% or ±40% perturbation for the gain, time constant, and time delay, respectively. Let us see how the changes of the three parameters affect the closed-loop response. Typical responses of the closed-loop system are shown in Figures 3.5.2–3.5.4.

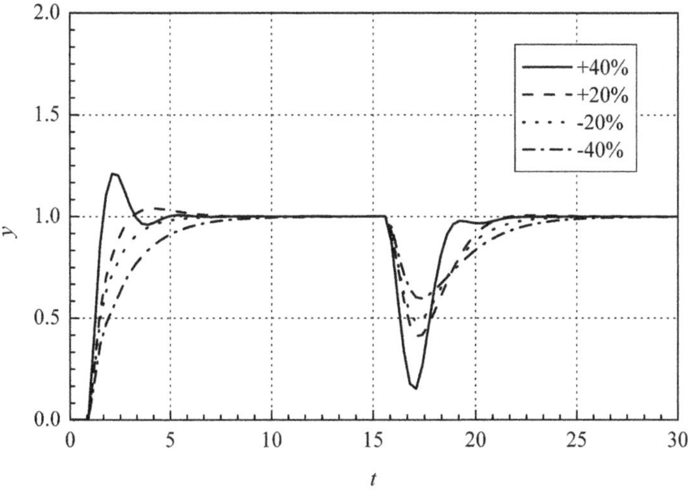

FIGURE 3.5.2
Effect of the gain uncertainty.

1. With the gain increasing, the rise time decreases and the overshoot increases. Contrarily, with the gain decreasing, the rise time increases and the overshoot decreases or disappears.
2. With the time constant increasing, the rise time increases and the response tends to be steady. If the time constant decreases, the rise time decreases and the response begins to oscillate.
3. With the time delay increasing, the overshoot increases. The overshoot decreases or disappears when the time delay decreases.

It is noticed that the gain and the time delay have a larger effect on the closed-loop response than the time constant. When the time constant decreases and the gain and the time delay increase, the change of the closed-loop response is the largest.

Theoretically, the effect of the uncertainty on the stability and performance of the closed-loop system can be analyzed with the necessary and sufficient

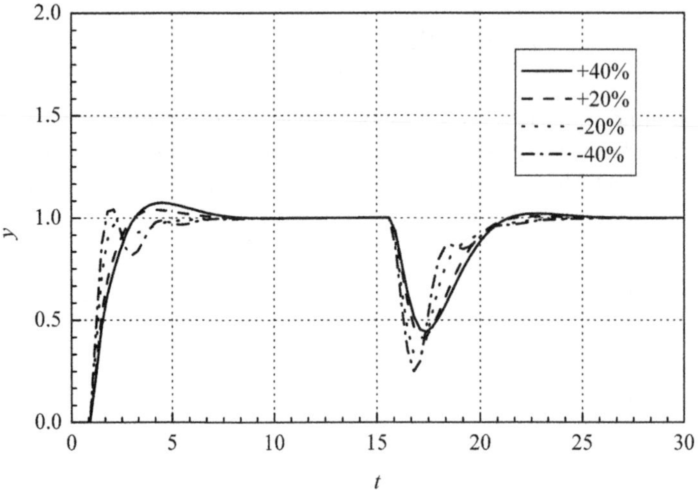

FIGURE 3.5.3
Effect of the time constant uncertainty.

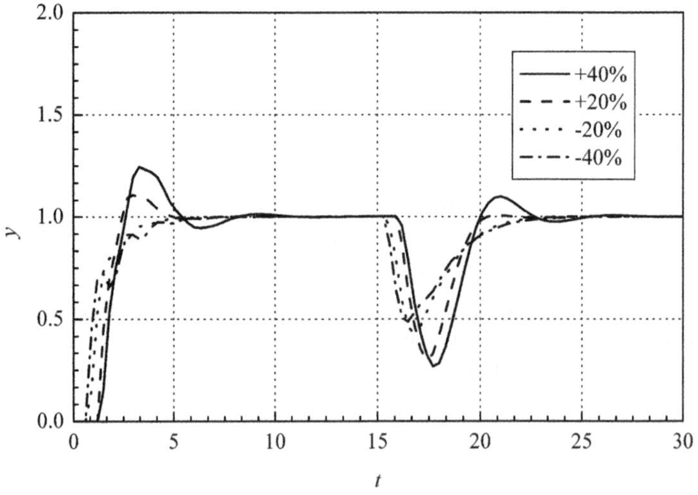

FIGURE 3.5.4
Effect of the time delay uncertainty.

condition derived in the last section. To use the result, one has to convert the parameter uncertainty of the first-order plant with time delay into the unstructured uncertainty. Rewrite the uncertain model in the form of

$$\widetilde{G}(s) = \frac{Ke^{-\theta s}}{\tau s + 1}[1 + \delta_m(s)], \tag{3.5.4}$$

where

$$\delta_m(s) = \frac{K + \delta K}{K} \frac{\tau s + 1}{(\tau + \delta\tau)s + 1} e^{-\delta\theta s} - 1. \tag{3.5.5}$$

Let $|\Delta_m(j\omega)|$ be the bound of $|\delta_m(j\omega)|$, namely $|\delta_m(j\omega)| \leq |\Delta_m(j\omega)|$. $|\Delta_m(j\omega)|$ is equal to the radius of the smallest disk containing the parameter uncertainty boundary.

When there are simultaneous uncertainties on the gain, time constant, and time delay, the following analytical expression for the uncertainty profile is obtained:

$$|\Delta_m(j\omega)| = \begin{cases} \left| \dfrac{|K| + \Delta K}{|K|} \dfrac{j\tau\omega + 1}{j(\tau - \Delta\tau)\omega + 1} e^{j\Delta\theta\omega} - 1 \right|, & \omega < \omega^* \\ \left| \dfrac{|K| + \Delta K}{|K|} \dfrac{j\tau\omega + 1}{j(\tau - \Delta\tau)\omega + 1} \right| + 1, & \omega \geq \omega^* \end{cases}, \tag{3.5.6}$$

where ω^* is determined by

$$\Delta\theta\omega^* + \arctan \frac{\Delta\tau\omega^*}{1 + \tau(\tau - \Delta\tau)\omega^{*2}} = \pi, \frac{\pi}{2} \leq \Delta\theta\omega^* \leq \pi. \tag{3.5.7}$$

Figure 3.5.5 shows the uncertainty profile where all of the three parameters are uncertain.

In particular, when only the gain is uncertain, that is, $\Delta\tau = \Delta\theta = 0$, the expression simplifies to

$$|\Delta_m(j\omega)| = \Delta K/|K|.$$

When only the time constant is uncertain, that is, $\Delta K = \Delta\theta = 0$, the expression simplifies to

$$|\Delta_m(j\omega)| = \left| \frac{j\tau\omega + 1}{j(\tau - \Delta\tau)\omega + 1} - 1 \right|.$$

When only the time delay is uncertain, $\Delta\tau = \Delta K = 0$ and $\omega^* = \pi/\Delta\theta$. In this case,

$$|\Delta_m(j\omega)| = \begin{cases} |e^{j\Delta\theta\omega} - 1|, & \omega < \pi/\Delta\theta \\ 2, & \omega \geq \pi/\Delta\theta \end{cases}.$$

Figure 3.5.6 shows the plot of the time delay uncertainty.

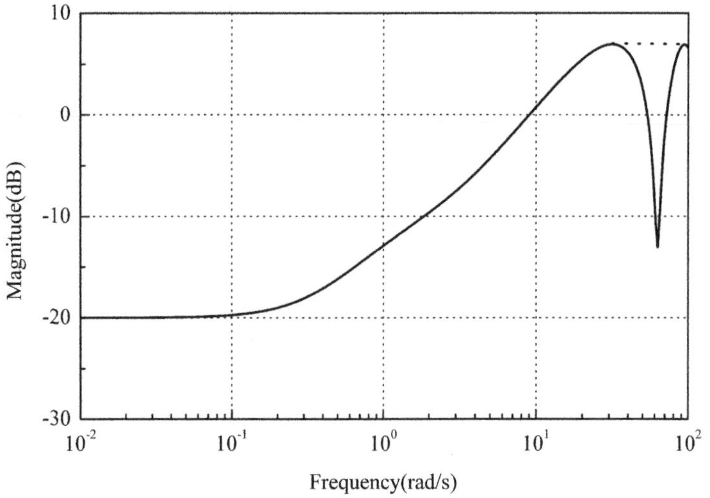

FIGURE 3.5.5
Unstructured uncertainty profile.

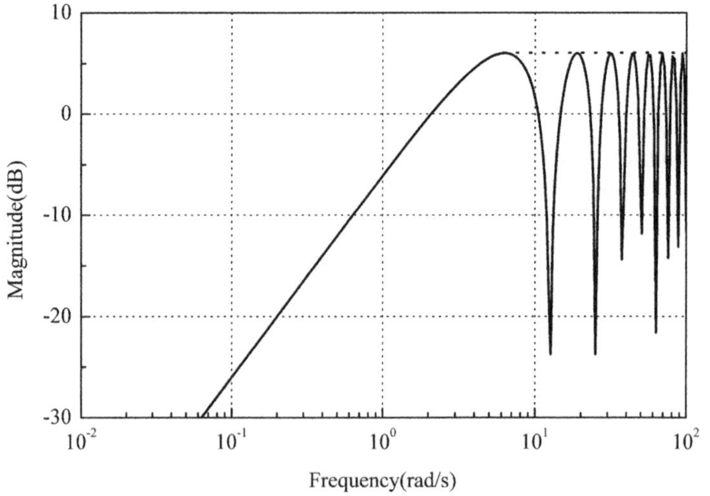

FIGURE 3.5.6
Uncertainty profile for the time delay uncertainty.

The procedure in which the unstructured uncertainty of the parameter uncertainty is derived is sketched here briefly. For all δK, $\delta\tau$, and $\delta\theta$, $|\Delta_m(j\omega)|$ is the least upper bound that makes $|\delta_m(j\omega)| \le |\Delta_m(j\omega)|$. If the maximum distance from (1,0) to $\delta_m(s) + 1$ is determined, the distance is $|\Delta_m(j\omega)|$.

For $|\delta K| \le \Delta K$, $|\delta|\tau \le \Delta\tau$, $|\delta\theta| \le \Delta\theta$, and some frequency $s = j\omega$, all possible points of $\delta_m(j\omega) + 1$ are located in a certain region. Based on the geometrical relationship, it can be proved that the point $\delta K = \Delta K, \delta\tau = -\Delta\tau$, and $\delta\theta = \Delta\theta$ is the farthest one away from (1,0) for all $\omega < \omega^*$, where ω^* is the frequency at which the angle of this point equals π. Furthermore, the result for $\omega \ge \omega^*$ can be derived by utilizing the triangular inequality.

In practice, it is not convenient to test the robust performance of systems with time delay by using the necessary and sufficient condition given in the last section. Since $\delta K = \Delta K, \delta\tau = -\Delta\tau$, and $\delta\theta = \Delta\theta$ is the worst case, a simple engineering method is to examine whether the internal stability and performance hold for the worst case.

3.6 Summary

In this chapter, some basic concepts in robust control theory are surveyed. Based on these concepts, the necessary and sufficient conditions for testing robust stability and robust performance are given.

In robust control theory, the controller parameterization is a fundamental concept. According to Youla parameterization, all stabilizing controllers for a stable plant can be expressed as

$$C(s) = \frac{Q(s)}{1 - G(s)Q(s)}.$$

The controller provides robust stability if and only if

$$\|\Delta_m(s)T(s)\|_\infty < 1,$$

and provides robust performance if and only if

$$|W(j\omega)S(j\omega)| + |\Delta_m(j\omega)T(j\omega)| < 1, \forall\omega.$$

It will be seen in the subsequent chapters that some perfect results can be derived based on controller parameterization.

At the end of this chapter, the robustness of a typical system with time delay is studied. The unstructured uncertainty profile of the parameter uncertainty is obtained. A simple rule for testing robust stability and robust performance is proposed based on the worst case plant.

Exercises

1. Show that the 2-norm of transfer functions is not sub-multiplicative.
2. Consider the plant

$$G(s) = \frac{as+1}{bs+1},$$

where $a, b > 0$. Compute the ∞-norm by utilizing its Bode magnitude plot.

3. Write a MATLAB program to compute the ∞-norm of

$$G(s) = \frac{1}{s^2+1}$$

with the searching method. Test your program on the function

$$G(s) = \frac{1}{s^2+10^{-6}s+1},$$

and analyze the result.

4. Assume that $G(s)$ is a strictly proper rational transfer function and $D(s)$ is an all-pass rational transfer function. Does the 2-norm or ∞-norm satisfy the following equation?

$$\|D(s)G(s)\| = \|G(s)\|.$$

5. Consider the control system with multiple loops in Figure E3.1, where $G_{1m}(s)$ and $G_{2m}(s)$ are the models of $G_1(s)$ and $G_2(s)$ respectively, $d_1(s)$ is the disturbance, and $y(s)$ is the system output. Assume that the models are exact. Compute the transfer function from $d_1(s)$ to $y(s)$.
6. Assume that the unity feedback loop is internally stable. Prove that
   (a) If $G(s)$ has a RHP zero at z, $T(s)$ will have a RHP zero at z.
   (b) If $G(s)$ has a RHP pole at p, $S(s)$ will have a RHP zero at p.

Notes and References

Sections 3.1–3.4 are based on Doyle et al. (1992) and Morari and Zafiriou (1989). Related content can also be found in, for example, Skogestad and Postlethwaite (2005).

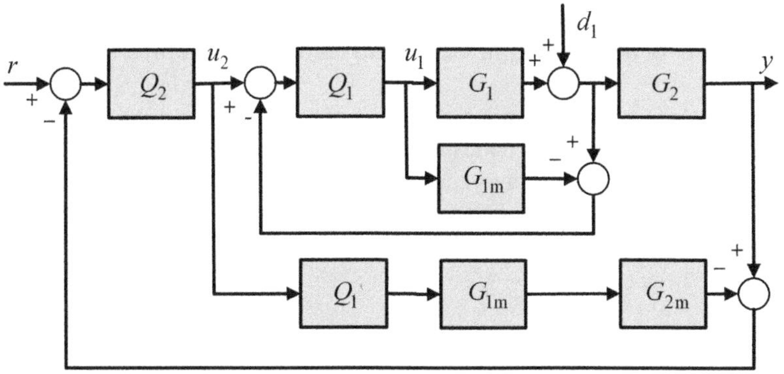

FIGURE E3.1
Control system with multiple loops.

The proofs for the system gains in Section 3.1 belong to Doyle et al. (1992, Section 2.5). Theorem 3.1.1 is from Doyle et al. (1992, Section 2.2) and Theorem 3.1.4 is from Doyle et al. (1992, Section 10.4).

There was a great deal of interest in 2-norm optimization and related topics in the late 1950s and the 1960s. See, for example, Newton et al. (1957) and Astrom (1970).

The discussion about testing internal stability in Section 3.2 is from Morari and Zafiriou (1989). The explanation about the sensitivity function is from Doyle et al. (1992).

The H_2 design problem generally refers to the one based on 2-norm optimization. This book focuses on the basic problem given in Section 3.2. The H_∞ optimal control problem was first formulated by Zames (1981). Further motivation for this problem was offered in Zames and Francis (1983).

The parameterization of all stabilizing controllers introduced in Section 3.3 was developed by Youla et al. (1976) and Kucera (1979). The IMC structure introduced in this section was studied in detail in Morari and Zafiriou (1989). This structure is particularly important in the design method of this book.

Theorem 3.4.1 was originally proposed by Kharitonov (1978). The proofs for Theorem 3.4.1 and Theorem 3.4.2 are based on Morari and Zafiriou (1989, Sections 2.5–2.6).

The unstructured uncertainty profile of the first-order plant with time delay introduced in Section 3.5 was given by Laughlin et al. (1987) (Laughlin D. L., D. E. Rivera, and M. Morari. Smith predictor design for robust performance, *Int. J. Control*, 1987, 46(2), 477–504. ©Taylor & Francis Ltd.). Figure 3.5.1 is drawn based on Morari and Zafiriou (1989, Figure 4.6.7) with different parameters and frequency points.

Exercise 1 is from Doyle et al. (1992, Chapter 2, Exercises).

Exercise 2 is based on Doyle et al. (1992, Section 2.2).

Exercise 3 is from Doyle et al. (1992, Chapter 2, Exercises).

Exercise 6 is adapted from Skogestad and Postlethwaite (2005, Section 4.7.1).

There are many books about robust control theory, among which Morari and Zafiriou (1989), Doyle et al. (1992), and Dorato (2000) are accessible for beginners. Some other references are Vidyasagar (1985), Francis (1987), Zhou et al. (1996), Goodwin et al. (2001), Skogestad and Postlethwaite (2005), and Mackenroth (2010).

4

H_∞ PID Controllers for Stable Plants

CONTENTS

Many plants involve time delays. The control of plants with time delay presents a continuing challenge to control theory. In past decades, many control strategies have been proposed to solve the problem. Among them two widely used control strategies are the PID controller and the Smith predictor.

In this chapter the design problem of H_∞ PID controllers is studied, as well as its relation with the Smith predictor. Two aspects are emphasized. One aspect is the control of plants with time delay. In traditional PID design methods, the ratio of the time delay to the time constant is restricted. For the plant with large time delay, they may result in bad performance or unstable systems. The design method introduced in this chapter does not have such a restriction. The other aspect is how to design a PID controller for quantitative performance and robustness, and how to tradeoff between the two objectives. The design method developed in this chapter is based on the optimal control theory. Most controllers designed with the optimal methods have fixed parameters. They cannot be designed or tuned for quantitative responses. This chapter will study the quantitative design problem of the optimal PID controller.

Given a plant, an important problem related to the design of PID controllers is the characterization of all PID controllers that stabilize the closed-loop system. This problem will be studied in the last section of this chapter.

4.1 Traditional Design Methods

If there is not any disturbance, feedback control is unnecessary once the steady state is reached. However, there always exist disturbances in a real system. Disturbances make the system output deviate from the desired value given by the reference. In order to keep the system output close to the reference, the plant input must be changed. To achieve this objective, the prerequisite is a control strategy.

Despite the continual advances in control theory, the most popular control strategy used in practice is still the PID controller. Let $e(t)$ denote the error and $u(t)$ denote the controller output. An ideal PID controller can be described by the following equation:

$$u(t) = K_c \left[e(t) + \frac{1}{T_I} \int e(t)dt + T_D \frac{de(t)}{dt} \right], \tag{4.1.1}$$

where K_c is the gain, T_I is the integral constant, and T_D is the derivative constant. Assume that $C(s)$ is the transfer function from $e(s)$ to $u(s)$. Using the Laplace transform, we have

$$C(s) = K_c \left(1 + \frac{1}{T_I s} + T_D s \right). \tag{4.1.2}$$

The ideal PID controller is improper. It has a pure differentiator and therefore is not physically realizable. In a practical PID controller, an approximate derivative term is used:

$$C(s) = K_c \left(1 + \frac{1}{T_I s} + \frac{T_D s}{T_F s + 1} \right). \tag{4.1.3}$$

The derivative term can also be constructed with an lead-lag element:

$$C(s) = K_c \left(1 + \frac{1}{T_I s} \right) \frac{T_D s + 1}{T_F s + 1}, \tag{4.1.4}$$

or by rolling off the response of the whole controller at high frequencies:

$$C(s) = K_c \left(1 + \frac{1}{T_I s} + T_D s \right) \frac{1}{T_F s + 1}. \tag{4.1.5}$$

Here T_F is a small positive real number. It is usually taken as $0.1T_D$.

It is observed that in all of the three forms, a low-pass transfer function is introduced to roll off the high frequency response. This is an important method for realizing an improper transfer function.

After a control system has been installed, the three parameters of the PID controller need to be adjusted until a satisfactory closed-loop performance is obtained. Even today, the widely adopted tuning methods for PID controllers are empirical methods. A limitation of these methods is that only partial information is utilized. As the tuning is a trial-and-error procedure, it is difficult for the designer to know to what extent the resulting controller approaches the optimal solution, how to tune the controller for quantitative performance and robustness, and how to reach a reasonable tradeoff among competing performance objectives.

The best-known tuning method is the Z-N method. The reaction curve method (R-C method) is developed to give a closed-loop response with the decay ratio of 1/4. Another well-known tuning method is the C-C method, which is empirically developed to give the minimum offset. Assume that the plant model is

$$G(s) = \frac{K}{\tau s + 1} e^{-\theta s}, \tag{4.1.6}$$

where K is the gain, τ is the time constant, θ is the time delay; the ultimate gain is K_u and the ultimate period is T_u. The PID controller parameters given by the three methods are listed in Table 4.1.1.

TABLE 4.1.1
Frequently used tuning methods.

Tuning methods	R-C method	C-C method	Z-N method
KK_c	$1.2(\theta/\tau)^{-1}$	$1.35(\theta/\tau)^{-1} + 0.27$	$0.6KK_u$
T_I/τ	$2(\theta/\tau)$	$\frac{2.5(\theta/\tau)[1 + (\theta/\tau)/5]}{1 + 0.6(\theta/\tau)}$	$0.5T_u/\tau$
T_D/τ	$0.5(\theta/\tau)$	$\frac{0.37(\theta/\tau)}{1 + 0.2(\theta/\tau)}$	$0.125T_u/\tau$

Although the Z-N method is the most widely used method, it has an evident disadvantage; the resulting PID controller usually gives excessive overshoot. To overcome this disadvantage, a refined Z-N method called the RZN method is developed, in which the overshoot is reduced by weighting the reference and the integral constant. The modified PID controller is

$$u(t) = K_c \left\{ [\beta r(t) - y(t)] + \frac{1}{\mu T_I} \int e(t)dt + T_D \frac{de(t)}{dt} \right\}. \tag{4.1.7}$$

The constants β and μ are determined by extensive simulation studies.

Define the normalized gain as the product of the ultimate gain and the steady-state gain of the plant:

$$K_n = KK_u, \tag{4.1.8}$$

and define the normalized time delay as the ratio of the time delay to time constant:

$$\theta_n = \frac{\theta}{T}. \tag{4.1.9}$$

When $2.25 < K_n < 15$ and $0.16 < \theta_n < 0.57$, the Z-N formulas are retained. Only the reference weight β is applied. For a 10% overshoot:

$$\beta = \frac{15 - K_n}{15 + K_n}, \tag{4.1.10}$$

and for a 20% overshoot:

$$\beta = \frac{36}{27 + 5K_n}. \tag{4.1.11}$$

If $1.5 < K_n < 2.25$ and $0.57 < \theta_n < 0.96$, both the reference weight and the new integral constant should be applied. For a 20% overshoot, the formula for β is

$$\beta = \frac{8}{17}\left(\frac{4}{9}K_n + 1\right). \tag{4.1.12}$$

μ is given as

$$\mu = \frac{4}{9}K_n. \tag{4.1.13}$$

The controller tuned by the Z-N or RZN method usually gives very bad response for the plant with large time delay. Hence, some designers believe that the PID controller cannot be used for the control of the plant with large time delay. It will be shown in the next several sections that, with proper design methods, the PID controller can actually be applied to such systems.

4.2 H_∞ PID Controller for the First-Order Plant

The popularity of the PID controller can be attributed partly to their robust performance under a wide range of operating conditions and partly to their functional simplicity, which allows engineers to operate them in a straightforward manner. To implement such a controller, three parameters must be determined for the given plant. In the traditional design procedure, first the control structure is fixed to be the PID structure and then the parameters are determined by empirical tuning rules. In this section, an alternative is developed. The optimal performance index is defined first, and then both the PID control structure and parameters are analytically derived.

Consider the unity feedback control system shown in Figure 4.2.1, where

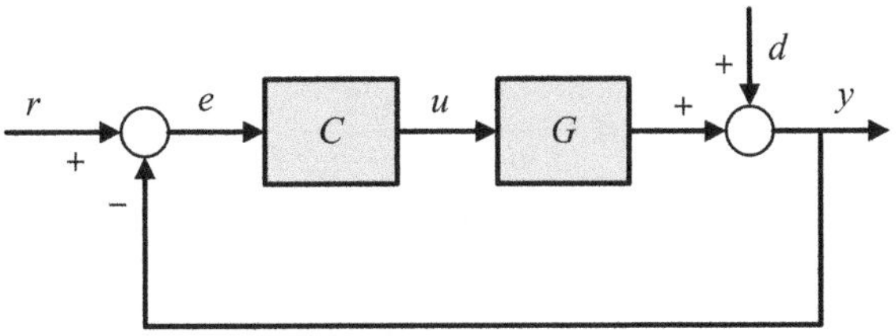

FIGURE 4.2.1
Unity feedback control loop.

$C(s)$ is the controller, $G(s)$ is the stable plant, $r(s)$ is the reference, $y(s)$ is the output, $d(s)$ is the disturbance at the plant output, $u(s)$ is the controller output, and $e(s)$ is the error. According to Youla parameterization, all stabilizing controllers can be expressed as

$$C(s) = \frac{Q(s)}{1 - G(s)Q(s)}, \tag{4.2.1}$$

where $Q(s)$ is a stable transfer function. If the model is exact, the transfer function from $d(s)$ to $y(s)$ is given by

$$S(s) = 1 - G(s)Q(s). \tag{4.2.2}$$

Take the performance index as

$$\min \|W(s)S(s)\|_\infty,$$

where $W(s)$ is a weighting function. It should be chosen so that the 2-norm of system input is bounded by unity.

It is impossible to design a controller for any inputs. Assume that the system input is a step, that is, $d(s) = 1/s$. In light of the discussion in Section 3.2, the weighting function in the H_∞ optimal control should satisfy that $\|d(s)/W(s)\|_2 \leq 1$. Then the weighting function can be simply taken as $W(s) = 1/s$.

In practice, ease of use is one of the important requirements. Since only two or three parameters can be tuned in a PID controller, it is natural to use a simple model in design. Consider the model in the form of the first-order plant with time delay:

$$G(s) = \frac{Ke^{-\theta s}}{\tau s + 1}. \tag{4.2.3}$$

Many plants can be described by this model. With the 1/1 Pade approximant,

$$e^{-\theta s} \approx \frac{1 - \theta s/2}{1 + \theta s/2},$$

the approximate plant is

$$G(s) \approx K\frac{1-\theta s/2}{(\tau s+1)(1+\theta s/2)}. \tag{4.2.4}$$

The basic idea is designing the controller for the approximate plant and then using it for the control of the original plant.

The following theorem is a fundamental fact in complex analysis.

Theorem 4.2.1 (Maximum Modulus Theorem). *Assume that $F(s)$ is a function that has no poles in Ω. If $F(s)$ is not a constant, then $|F(s)|$ does not attain its maximum value at an interior point of Ω.*

Let Ω equal the open RHP. $W(s)S(s)$ denotes the transfer function from the normalized disturbance to the system output. Evidently, $W(s)S(s)$ should be stable; that is, it has no poles in Ω. By Theorem 4.2.1 we have

$$\begin{aligned} & \|W(s)S(s)\|_\infty \\ = & \|W(s)[1-G(s)Q(s)]\|_\infty \\ = & \sup_{\mathrm{Re}s>0} |W(s)[1-G(s)Q(s)]|. \end{aligned} \tag{4.2.5}$$

$G(s)$ has a zero at $s=2/\theta$ in the open RHP. $s=2/\theta$ is an interior point of Ω. Accordingly

$$\begin{aligned} & \sup_{\mathrm{Re}s>0} |W(s)[1-G(s)Q(s)]| \\ \geq & |W(s)[1-G(s)Q(s)]|_{s=2/\theta}| = \frac{\theta}{2}. \end{aligned} \tag{4.2.6}$$

To solve the above equation, the following constraints should be considered:

1. $Q(s)$ should be stable for internal stability.
2. To make the controller physically realizable, $Q(s)$ should be proper.
3. To have a finite ∞-norm, $Q(s)$ should satisfy

$$\lim_{s\to 0} S(s) = \lim_{s\to 0}[1-G(s)Q(s)] = 0. \tag{4.2.7}$$

This constraint is also required for asymptotic tracking.

It will complicate the design problem to consider these constraints simultaneously. To get a controller that makes the closed-loop system possess desired properties, the idea is loosening the requirement of properness first and finding the optimal $Q(s)$, namely $Q_{opt}(s)$. A proper $Q(s)$ can then be obtained by rolling $Q_{opt}(s)$ off at high frequencies. This technique was used in the last section to implement a practical PID controller.

From (4.2.6) it is known that the minimum of $\|W(s)S(s)\|_\infty$ is $\theta/2$. This gives the following unique optimal solution:

$$Q_{opt}(s) = \frac{W(s)-\theta/2}{W(s)G(s)}$$

$$= \frac{(\tau s+1)(1+\theta s/2)}{K}. \quad (4.2.8)$$

$Q_{opt}(s)$ is improper. A low-pass filter must be introduced to roll $Q_{opt}(s)$ off at high frequencies. Choose the following filter:

$$J(s) = \frac{\beta_0}{(\lambda s+1)^2}, \quad (4.2.9)$$

where β_0 is a constant and λ is a positive real number. The filter should not violate the constraint of asymptotic tracking:

$$\lim_{s\to 0}[1 - G(s)Q_{opt}(s)J(s)] = 0.$$

Elementary computations give $\beta_0 = 1$. Then the suboptimal proper $Q(s)$ is

$$\begin{aligned} Q(s) &= Q_{opt}(s)J(s) \\ &= \frac{(\tau s+1)(1+\theta s/2)}{K(\lambda s+1)^2}. \end{aligned} \quad (4.2.10)$$

Here, λ is an adjustable parameter. It closely relates to the closed-loop performance. A small λ gives a fast response, while a large λ slows down the response. As $\lambda \to 0$, $\|W(s)S(s)\|_\infty$ tends to be optimal. Therefore, λ can be used as a metric of performance. It is called the performance degree in this book. In view of (4.2.1), the controller of the corresponding unity feedback loop is

$$\begin{aligned} C(s) &= \frac{Q(s)}{1-G(s)Q(s)} \\ &= \frac{1}{K}\frac{(\tau s+1)(1+\theta s/2)}{\lambda^2 s^2 + (2\lambda+\theta/2)s}. \end{aligned} \quad (4.2.11)$$

This is a PID controller. An important feature of this PID controller is that it cancels two poles of the approximate model, or equivalently, two dominant poles of the original model.

With the above formula, the parameters of PID controller can be directly calculated by using the plant parameters. Compare the H_∞ PID controller with the practical PID controller of the form

$$C(s) = K_C\left(1 + \frac{1}{T_I s} + T_D s\right)\frac{1}{T_F s+1},$$

the parameters of the H_∞ PID controller are

$$\begin{aligned} T_F &= \frac{\lambda^2}{2\lambda+\theta/2}, \quad T_I = \frac{\theta}{2}+\tau, \\ T_D &= \frac{\theta\tau}{2T_I}, \quad K_C = \frac{T_I}{K(2\lambda+\theta/2)}. \end{aligned} \quad (4.2.12)$$

If the practical PID is in the form of

$$C(s) = K_C\left(1 + \frac{1}{T_I s} + \frac{T_D s}{T_F s + 1}\right),$$

the parameters of the H_∞ PID controller are

$$\begin{aligned} T_F &= \frac{\lambda^2}{2\lambda + \theta/2}, \quad T_I = \frac{\theta}{2} + \tau - T_F, \\ T_D &= \frac{\theta\tau}{2T_I} - T_F, \quad K_C = \frac{T_I}{K(2\lambda + \theta/2)}. \end{aligned} \tag{4.2.13}$$

When the practical PID controller is

$$C(s) = K_C\left(1 + \frac{1}{T_I s}\right)\frac{T_D s + 1}{T_F s + 1},$$

the parameters of the H_∞ PID controller are

$$\begin{aligned} T_F &= \frac{\lambda^2}{2\lambda + \theta/2}, \quad T_I = \tau(\text{or } \frac{\theta}{2}), \\ T_D &= \frac{\theta}{2}(\text{or } \tau), \quad K_C = \frac{T_I}{K(2\lambda + \theta/2)}. \end{aligned} \tag{4.2.14}$$

In practice, a low-order controller is preferred to a high-order controller. Usually, there are two ways to obtain a low-order controller:

1. Design a controller for the high-order model and then reduce the order of the resulting controller.
2. Reduce the order of the model and then design a controller for the low-order model.

Although the two design procedures are different, the obtained responses are similar.

This section adopts the latter. If the plant model is of high order, it should be reduced to the model of first order and then the design procedure introduced in this section can be used.

4.3 The H_∞ PID Controller and the Smith Predictor

Since the PID controller and the Smith predictor were proposed, the two controllers have been widely studied and applied. Even today, they are still the dominant means in industrial control systems. For a very long time, the two controllers had been regarded as two irrelevant methods: the Smith predictor is an efficient scheme for plants with large time delay, while the PID controller is not. In this section, the internal relationship between the two controllers will be discussed.

Consider the unity feedback control system shown in Figure 4.2.1. Assume that $\tilde{G}(s)$ is the real plant, whose model is described by

$$G(s) = G_o(s)e^{-\theta s}, \tag{4.3.1}$$

where $G_o(s)$ is the delay-free part of $G(s)$. When the model is exact and there is no disturbance, the system output is

$$y(s) = C(s)G_o(s)e^{-\theta s}e(s). \tag{4.3.2}$$

This signal is delayed, whereas the desired feedback signal is

$$y_o(s) = C(s)G_o(s)e(s). \tag{4.3.3}$$

This is possible if we substitute $C(s)$ for $R(s)$ and add the following signal to the open-loop response $y(s)$:

$$y_s(s) = R(s)G_o(s)e(s) - R(s)G_o(s)e^{-\theta s}e(s). \tag{4.3.4}$$

The implication of adding $y_s(s)$ to the signal $y(s)$ is shown in Figure 4.3.1. This structure is the so-called Smith predictor. It is seen that the signal $y_s(s)$ is generated by introducing a simple local loop. The new feedback signal is as follows:

$$y_s(s) + y(s) = y_o(s). \tag{4.3.5}$$

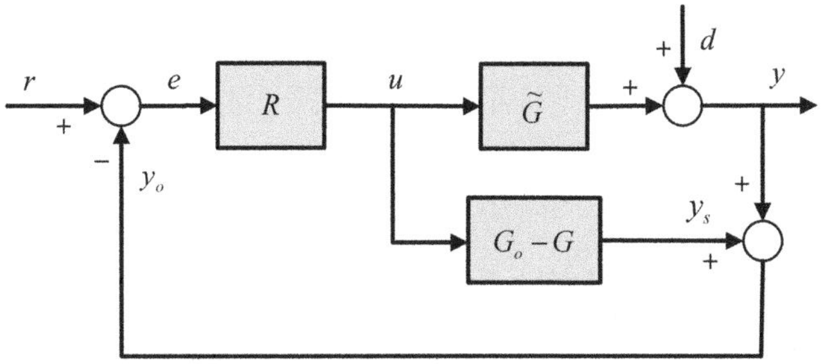

FIGURE 4.3.1
Structure of the Smith predictor.

The controller $R(s)$ in the Smith predictor differs from the controller $C(s)$ in the unity feedback loop. The Smith predictor can be related to the unity feedback loop through

$$C(s) = \frac{R(s)}{1 + [G_o(s) - G(s)]R(s)}. \tag{4.3.6}$$

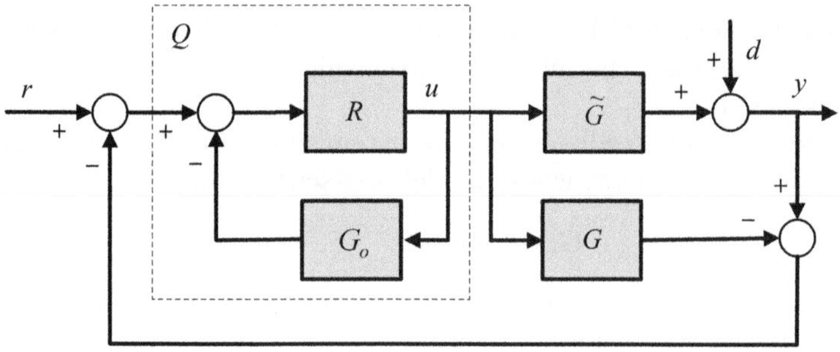

FIGURE 4.3.2
An equivalent structure of the Smith predictor.

If the plant is rational, $R(s)$ reduces to $C(s)$.

For the H_∞ PID controller given in the last section, we have

$$Q(s) = \frac{(\tau s + 1)(1 + \theta s/2)}{K(\lambda s + 1)^2}.$$

Rearrange Figure 4.3.1 to obtain Figure 4.3.2. It is easy to verify that the Smith predictor and $Q(s)$ are related through

$$Q(s) = \frac{R(s)}{1 + G_o(s)R(s)}. \tag{4.3.7}$$

The controller of the Smith predictor can be obtained by inverse operation:

$$R(s) = \frac{Q(s)}{1 - G_o(s)Q(s)} \tag{4.3.8}$$

$$= \frac{1}{K} \frac{(\tau s + 1)(1 + \theta s/2)}{\lambda^2 s^2 + (2\lambda - \theta/2)s}. \tag{4.3.9}$$

$R(s)$ is a PID controller when $\lambda > \theta/4$.

The analysis shows that the Smith predictor and PID controller can be approximately equivalent. The gap between them comes from the error caused by the Pade approximation. This implies that a PID controller can also be used to control plants with large time delay, provided it is appropriately designed.

Example 4.3.1. *Consider the paper-making machine shown in Figure 4.3.3. The paper-making machine is divided into five sections: head, table and pressing, drying, calender stack, and reel. Not shown in this figure is the stock preparation system, in which fibers are dispersed in water. The suspension is delivered to the mixing tank. In the mixing tank and the head-box, the thick stock is mixed with the recycled water. Then the head-box delivers the diluted suspension of fibers to the wire with small fine holes. The wire continuously* *moves over the table, where most of the water is removed by draining through the wire. This produces a wet mat of fibers on the wire. After being pressed and dried, the wet mat becomes a sheet of finished paper.*

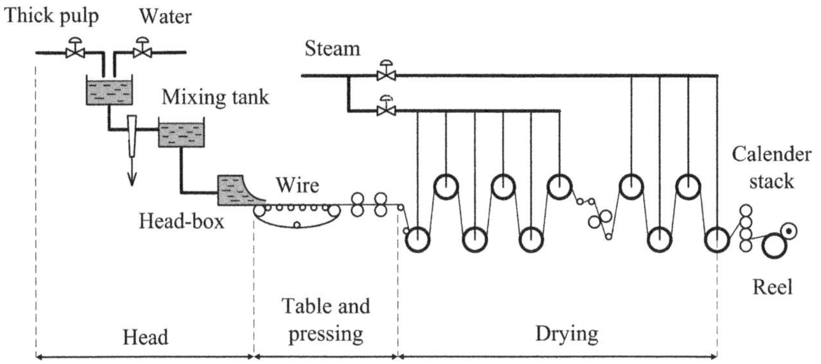

FIGURE 4.3.3
A paper-making process.
(From Zhang et al., 2001. Reprinted by permission of John Wiley & Sons)

In the system, there are many control objectives such as basis weight, moisture content, steam pressure, and consistency, among which the most important is the basis weight. By mechanistic analysis and identification, a low-order model has been developed for basis weight control:

$$\boxed{G(s) = \frac{5.15}{1.8s+1}e^{-2.8s}.}$$

That is, $K = 5.15, \tau = 1.8, \theta = 2.8$. *From (4.2.10) we have the following* H_∞ *controller:*

$$Q(s) = \frac{(1.8s+1)(1.4s+1)}{5.15(\lambda s+1)^2}.$$

A little algebra yields the following PID controller:

$$\boxed{C(s) = \frac{(1.8s+1)(1.4s+1)}{5.15[\lambda^2 s^2 + (2\lambda + 1.4)s]}.}$$

The equivalent Smith predictor is

$$R(s) = \frac{(1.8s+1)(1.4s+1)}{5.15[\lambda^2 s^2 + (2\lambda - 1.4)s]}.$$

Take $\lambda = 0.4\theta$ *(In the rest of this book, "*θ*" will be directly used in examples without repeating its meaning). A unit step reference is added at* $t = 0$ *and a step load with the magnitude of* -0.1 *is added at* $t = 50$*. The nominal response of the closed-loop system is shown in Figure 4.3.4. It is seen that the response of the system is fast and steady.*

Now assume that there exists a 50% error in estimating θ*; that is,* θ *varies within [1.4, 4.2]. Figure 4.3.5 shows the system responses. If the performance degree is* $\lambda = 0.7\theta$*, the response becomes slightly slower, but a better robustness is obtained (Figure 4.3.6).*

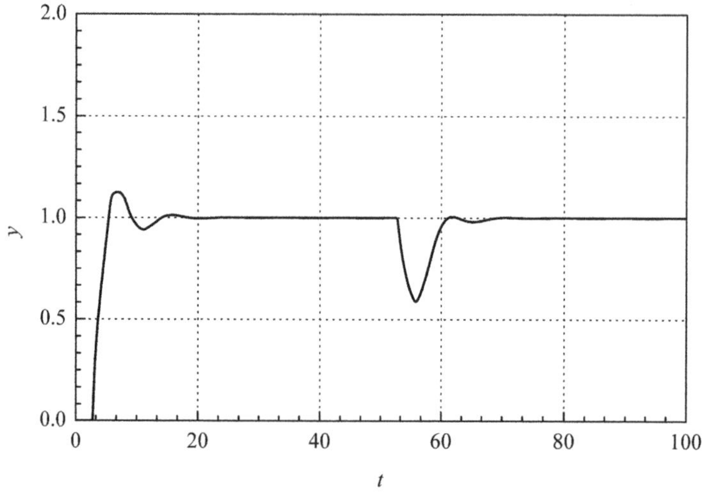

FIGURE 4.3.4
Nominal response of the closed-loop system.

4.4 Quantitative Performance and Robustness

There is an adjustable parameter, performance degree λ, in the H_∞ PID controller. The parameter directly relates to the nominal performance and robustness of the closed-loop system. It will be shown in this section how a quantitative performance or robustness can be obtained by adjusting the performance degree.

1. If the real plant were (4.2.4), the closed-loop transfer function of the system would be

$$T(s) = \frac{1 - \theta s/2}{(\lambda s + 1)^2}. \tag{4.4.1}$$

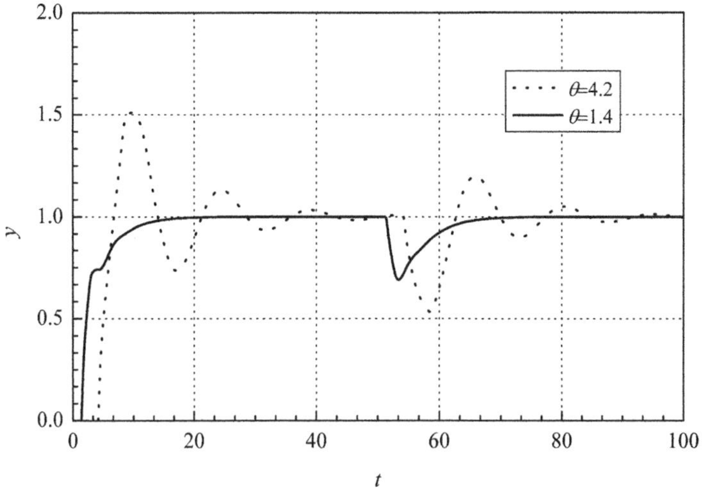

FIGURE 4.3.5
Response of the uncertain system with $\lambda = 0.4\theta$.

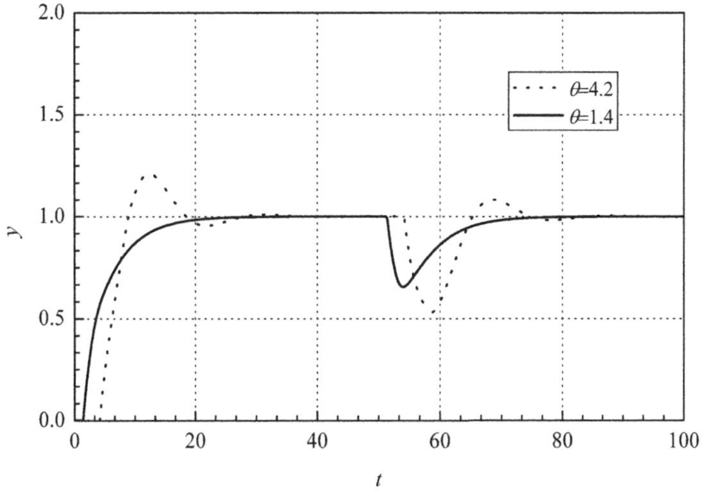

FIGURE 4.3.6
Response of the uncertain system with $\lambda = 0.7\theta$.

The disturbance transfer function of the system would be

$$S(s) = \frac{\lambda^2 s^2 + (2\lambda + \theta/2)s}{(\lambda s + 1)^2}. \tag{4.4.2}$$

The closed-loop system would have smooth and steady frequency responses (Figure 4.4.1). In this case, the performance degree could be freely selected. When $\lambda \to 0$, the system would tend to be optimal: $\|W(s)S(s)\|_\infty \to \theta/2$.

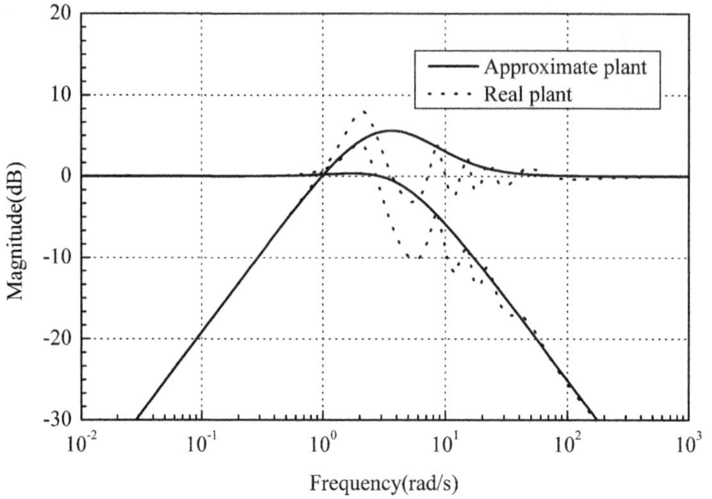

FIGURE 4.4.1
Frequency response of the closed-loop system.

2. In the last section, the real plant was in the form of (4.2.3) and the Pade approximation was used to treat the time delay in it. When the obtained controller is applied to the real plant, the response of the closed-loop system fluctuates near the break frequency, which is caused by the error from the Pade approximation (Figure 4.4.1). Regard the error as a kind of known uncertainty and let

$$|\Delta_m(j\omega)| \geq K \left| \frac{e^{-\theta j\omega}}{\tau j\omega + 1} - \frac{1 - \theta j\omega/2}{(\tau j\omega + 1)(1 + \theta j\omega/2)} \right|. \tag{4.4.3}$$

The robust stability of the closed-loop system can be tested by

$$\|\Delta_m(s)T(s)\|_\infty < 1. \tag{4.4.4}$$

It can be verified that the performance degree relates to the stability and performance of the closed-loop system in a monotonous manner:

(a) When the performance degree decreases, $|T(j\omega)|$ increases in the high frequency range and $|S(j\omega)|$ decreases in the low frequency range. Such a system has a larger bandwidth (Figure 4.4.2). According to the discussion in Chapter 3, this implies better performance and poorer robustness.
(b) When the performance degree increases, $|T(j\omega)|$ decreases in the high frequency range and $|S(j\omega)|$ increases in the low frequency range. The system has a smaller bandwidth. Its performance is sacrificed for the robustness.

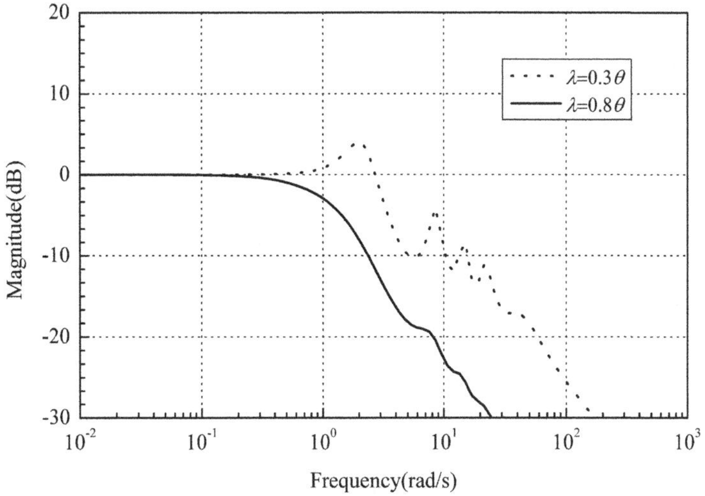

FIGURE 4.4.2
Relationship between the closed-loop frequency response and λ.

The nominal performance and robustness of a system conflict with each other. By choosing an appropriate performance degree, one can easily tradeoff between the nominal performance and the robustness. The monotonicity of the performance degree implies that the tradeoff procedure, or the controller tuning procedure, is very simple.

In Section 2.2, an example was given to illustrate that the direct use of a rational approximation for stability analysis might lead to an incorrect result. Since the approximate model is not exactly the original model, there exists a possibility that the controller stabilizes the approximate model, but cannot stabilize the original model.

The use of the performance degree overcomes the problem. In the method in Section 4.2, the controller is designed for the approximate model, and then used for the original model. That is, the approximate model is regarded as the nominal plant and the approximation error is regarded as the uncertainty. The existence of the approximation error imposes a lower bound on the performance degree for stability. As long as the performance degree is greater than the lower bound, the closed-loop system is stable. With numerical methods, it is obtained that the lower bound is about 0.07350θ.

3. A frequently encountered situation is that the real plant is uncertain. Assume that the real plant is

$$\widetilde{G}(s) = \frac{\widetilde{K}\exp(-\widetilde{\theta}s)}{\widetilde{\tau}s+1}. \tag{4.4.5}$$

Then the uncertainty profile is

$$\Delta_m(\omega) \geq \left| \frac{\widetilde{K}\exp(-\widetilde{\theta}j\omega)}{\widetilde{\tau}j\omega+1} - \frac{K(1-\theta j\omega/2)}{(\tau j\omega+1)(1+\theta j\omega/2)} \right|, \tag{4.4.6}$$

which consists of two parts: one is the approximation error and the other is the real uncertainty. Then the closed-loop system is stable if and only if

$$\|\Delta_m(s)T(s)\|_\infty < 1. \tag{4.4.7}$$

Now consider the quantitative tuning problem for nominal performance and robustness.

As stated in Section 3.5, there are two classes of design specifications. In one class the design specification involving the requirement on robustness is given for the nominal system, while in the other class the design specification is given for the uncertain system.

First, assume that the quantitative design specification involving the requirement on robustness is proposed for the nominal system. In this case, only the nominal performance is considered. In the design method in Section 4.2, the error introduced by the Pade approximation is clear. Hence, the performance degree has a definite effect on the nominal performance. It is complicated to analytically compute the effect. However, with the help of numerical methods the effect can be obtained easily.

Figures 4.4.3–4.4.5 provide an estimation about the performance. It can be seen that overshoot, rise time and resonance peak are determined only by λ/θ. The sudden change in Figure 4.4.4 is due to the different definitions for systems with overshoot and without overshoot.

With these curves, one can design the H_∞ PID controller for quantitative nominal performance. For example, if the required overshoot is 5%, one can simply take $\lambda = 0.5\theta$ according to Figure 4.4.3. In Example 4.3.1, $\lambda = 0.4\theta$ was taken. The corresponding overshoot is about 15%. Similarly, 2dB resonance peak can be reached by taking $\lambda = 0.37\theta$ in view of Figure 4.4.5. In many cases, the value of λ corresponding to the practical design requirements falls

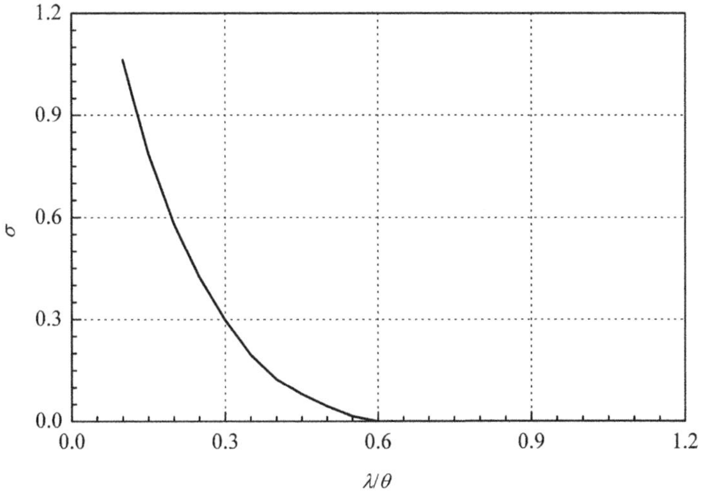

FIGURE 4.4.3
Effect of the performance degree on the overshoot.

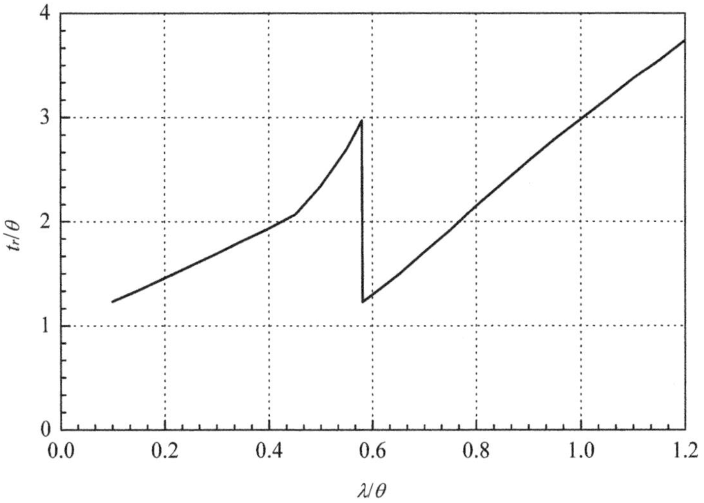

FIGURE 4.4.4
Effect of the performance degree on the rise time.

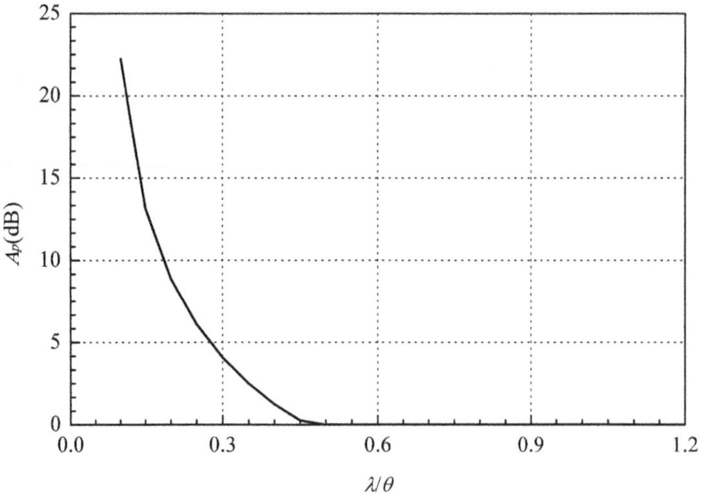

FIGURE 4.4.5
Effect of the performance degree on the resonance peak.

into the interval $0.1\theta - 1.2\theta$. Hence, the maximal x-coordinate is 1.2 in Figures 4.4.3–4.4.5.

There is a limit to the performance in any control system. For instance, in the system with an H_∞ PID controller, there is a minimal rise time when the overshoot is fixed. Higher requirements on the rise time can be achieved in two ways: improving the design methods or increasing the controller order. Theoretically, it may be possible to reach an arbitrarily fast rise time if the controller order is not restricted.

As discussed in Section 2.3, an important measure for the ability of rejecting disturbance is the perturbation peak. Since the transfer function from the disturbance at the plant input to the system output is $G(s)S(s)$, the disturbance response relates not only to λ/θ, but also to θ/τ (Figure 4.4.6). If θ/τ is fixed, then there is a simple relationship between the perturbation peak and λ/θ. Similarly, when λ/θ is fixed, the perturbation peak can also be estimated by θ/τ.

In classical control theory, the frequency response method is one of the main methods for controller design. With the Bode plot and Nyquist plot, one can analyze the stability of the closed-loop system and design the controller. Figure 4.4.7 and 4.4.8 provide the Bode plot and Nyquist plot of the system. It can be seen that the open-loop frequency response has a particular feature: with the increase of the frequency, both the magnitude and the phase decrease monotonically. The combined effect makes part of the Nyquist plot on a line between $(-1, 0)$ and the origin and at the same time parallel to the imaginary axis.

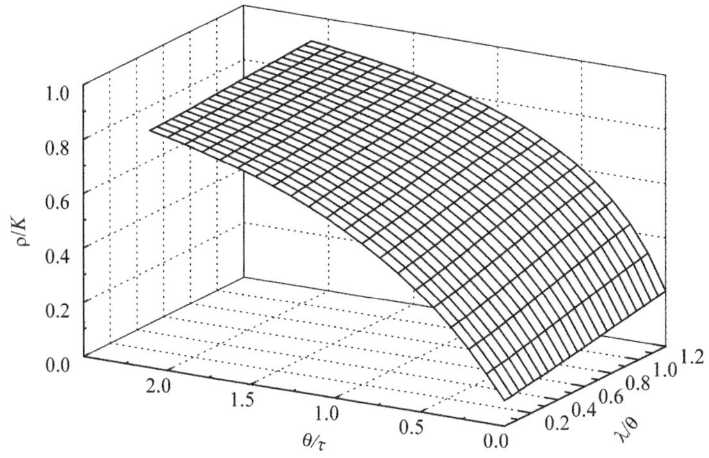

FIGURE 4.4.6
Effect of the performance degree on the perturbation peak.

Some other performance indices also relate with λ/θ monotonically. The relationship between the magnitude margin and λ/θ is shown in Figure 4.4.9 and the relationship between the phase margin and λ/θ is shown in Figure 4.4.10. It can be observed that the changes of the magnitude margin and the phase margin are monotonous. The relationship between the ISE and λ/θ is shown in Figure 4.4.11. When $\lambda/\theta < 0.3$, the smaller the λ/θ, the larger the ISE; when $\lambda/\theta \geq 0.3$, the larger the λ/θ, the larger the ISE.

Now assume that the quantitative design specification is given for the uncertain system. In this case, there exists uncertainty in addition to the approximation error. If the uncertainty profile is obtained, an exact performance degree can be calculated with the necessary and sufficient condition for robust performance. Unfortunately, the uncertainty profile is not always exactly known, owing to technical and economical reasons. Even if the uncertainty profile is available, the calculation is complicated. Therefore, it is most desirable to develop a simple tuning method to determine the performance degree. Based on the discussion in this section, a simple tuning procedure is developed here.

Without loss of generality, assume that the closed-loop system is required to have an overshoot less than 5% for all uncertain plants, that is, the worst case overshoot is 5%. The tuning procedure is as follows:

1. Design the controller for the nominal plant. For a 5% overshoot, $\lambda = 0.5\theta$.

2. Substitute the worst case plant for the nominal plant (that is, the

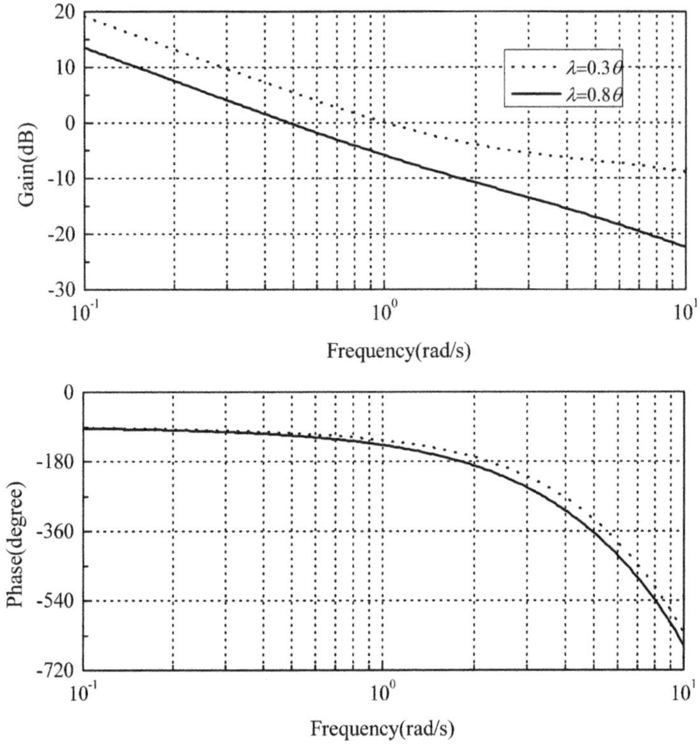

FIGURE 4.4.7
Bode plot of the H_∞ PID control system.

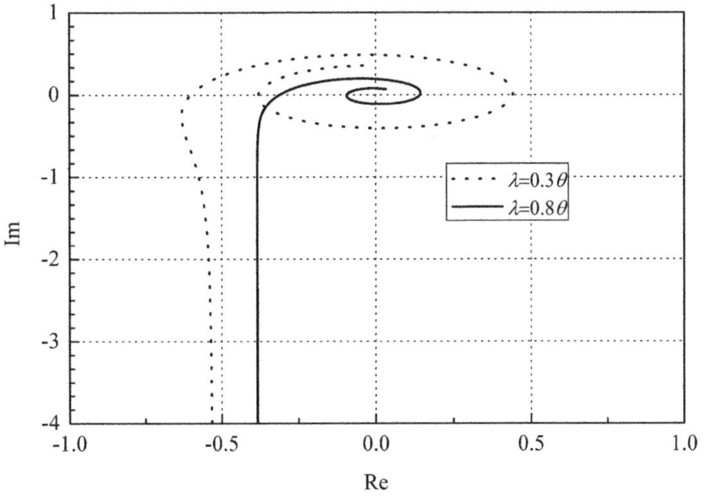

FIGURE 4.4.8
Nyquist plot of the H_∞ PID control system.

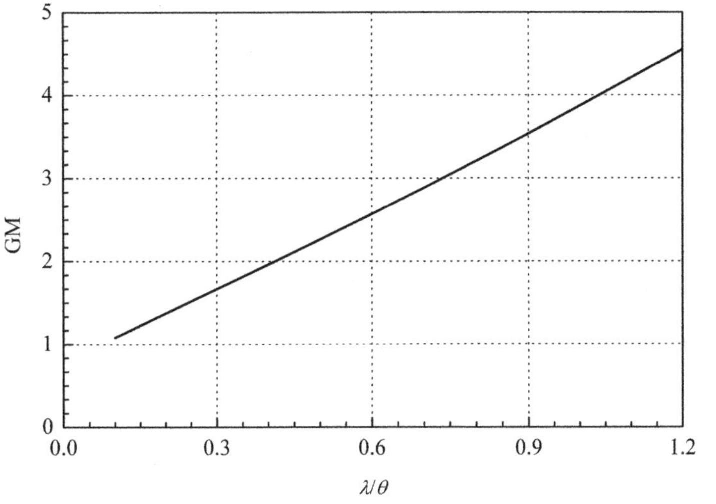

FIGURE 4.4.9
Effect of the performance degree on the gain margin.

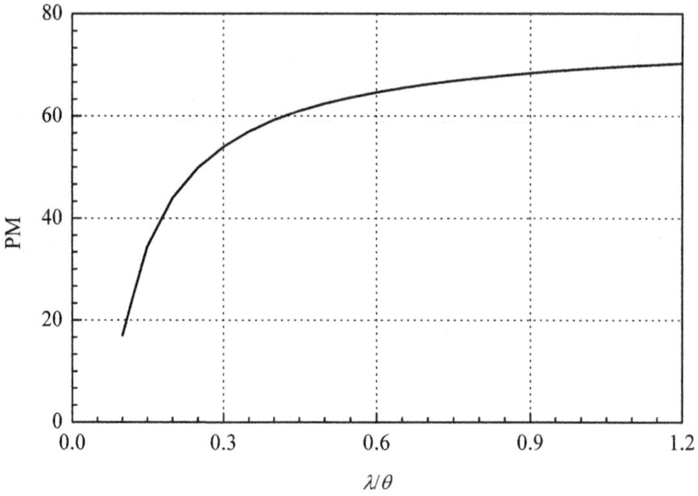

FIGURE 4.4.10
Effect of the performance degree on the phase margin.

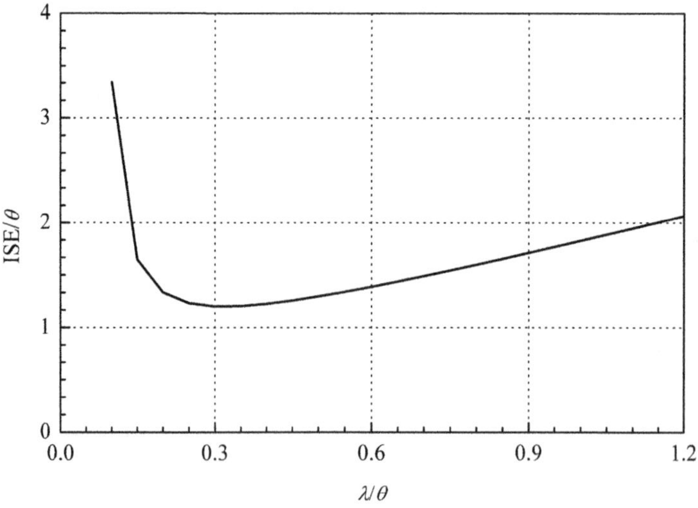

FIGURE 4.4.11
Effect of the performance degree on the ISE.

gain and time delay take their maximum value and the time constant takes its minimum value).

3. Increase the performance degree monotonically with a small step until the overshoot reaches 5%.

The first step can be omitted. In this case, the initial value of the performance degree is set to be 0. A typical step is 0.01θ or smaller. If the time delay is very small, for instance, $\theta \leq 0.1\tau$, the time constant can be used to determine the step. For example, the step can be taken as 0.01τ or smaller.

To summarize, both the nominal performance and the robust performance can be quantitatively tuned through such a procedure: Increase the performance degree monotonically until the required response is obtained.

4.5 H_∞ PID Controller for the Second-Order Plant

The design in the preceding sections was carried out on the basis of the 1/1 Pade approximant. In this section, the first-order Taylor series expansion (equivalently the 1/0 Pade approximant) is used to design the H_∞ PID controller. Despite having lower accuracy than the 1/1 Pade approximant, the first-order Taylor series expansion allows us to design a PID controller for the second-order plant with time delay.

Assume that the plant model is

$$G(s) = \frac{Ke^{-\theta s}}{(\tau_1 s + 1)(\tau_2 s + 1)}, \tag{4.5.1}$$

where τ_1 and τ_2 are two time constants. The poles of the plant are $-1/\tau_1$ and $-1/\tau_2$. If both $1/\tau_1$ and $1/\tau_2$ are positive real numbers, the dynamics of the plant is similar to that of the first-order plant. One can reduce the model to the first-order one and then design the controller. When $1/\tau_1$ and $1/\tau_2$ are conjugate imaginary roots, the dynamics of the plant cannot be well approximated by the first-order plant. In this case, it is not recommended to reduce the order of the model. The controller should be designed for the second-order model.

Using the first-order Taylor series expansion, we have

$$e^{-\theta s} \approx 1 - \theta s.$$

The approximate model is

$$G(s) \approx \frac{K(1 - \theta s)}{(\tau_1 s + 1)(\tau_2 s + 1)}. \tag{4.5.2}$$

Take the performance index as

$$\min \|W(s)S(s)\|_\infty.$$

Assume that the system input is a unit step. Then $W(s) = 1/s$. By Theorem 4.2.1,

$$\begin{aligned} \|W(s)S(s)\|_\infty &= \|W(s)[1 - G(s)Q(s)]\|_\infty \\ &\geq |W(1/\theta)| \end{aligned} \tag{4.5.3}$$

for all $Q(s)$s. Substituting (4.5.2) into (4.5.3) and minimizing the left-hand side of the equation yields:

$$\left\| \frac{1}{s}\left[1 - \frac{K(1-\theta s)}{(\tau_1 s+1)(\tau_2 s+1)}Q(s)\right]\right\|_\infty = \theta. \tag{4.5.4}$$

It is now clear that the unique optimal solution is

$$Q_{opt}(s) = \frac{(\tau_1 s+1)(\tau_2 s+1)}{K}. \tag{4.5.5}$$

The degree of the numerator polynomial of $Q_{opt}(s)$ is higher by two than that of the denominator polynomial. Since the asymptotic tracking property requires that

$$\lim_{s\to 0}[1 - G(s)Q(s)] = 0, \tag{4.5.6}$$

the following filter is introduced to roll $Q_{opt}(s)$ off at high frequencies:

$$J(s) = \frac{1}{(\lambda s+1)^2}.$$

A proper $Q(s)$ is then obtained:

$$\begin{aligned} Q(s) &= Q_{opt}(s)J(s) \\ &= \frac{(\tau_1 s+1)(\tau_2 s+1)}{K(\lambda s+1)^2}. \end{aligned} \tag{4.5.7}$$

The controller of the unity feedback loop is:

$$\begin{aligned} C(s) &= \frac{Q(s)}{1 - G(s)Q(s)} \\ &= \frac{1}{K}\frac{(\tau_1 s+1)(\tau_2 s+1)}{\lambda^2 s^2 + (2\lambda+\theta)s}. \end{aligned} \tag{4.5.8}$$

This is a PID controller. If it is realized in the form of

$$C(s) = K_C\left(1 + \frac{1}{T_I s} + T_D s\right)\frac{1}{T_F s+1},$$

the controller parameters are as follows:

$$\begin{aligned} T_F &= \frac{\lambda^2}{2\lambda+\theta}, \quad T_I = \tau_1 + \tau_2, \\ T_D &= \frac{\tau_1\tau_2}{\tau_1+\tau_2}, \quad K_C = \frac{\tau_1+\tau_2}{K(2\lambda+\theta)}. \end{aligned} \tag{4.5.9}$$

Usually λ falls into the interval $0.2\theta - 1.2\theta$.

The H_∞ PID controller of the second-order plant possesses similar features to that of the first-order plant. Since there are two time constants in the plant model, the relationship between the performance degree and the system response depends not only on λ/θ but also on time constants. The nominal performance and robust performance of the system with a second-order plant can also be quantitatively tuned through the procedure given in Section 4.4: Increase the performance degree monotonically until the required response is obtained.

Example 4.5.1. *The task of heat exchangers is to transfer heat from one flow of medium to another — without any physical contact. Heat transfer takes place through the thermally conductive material used to separate the two media, one cold and the other hot. Figure 4.5.1 describes an industrial heat exchanger in which steam is used to heat the liquid product. The requirement on the control system is to retain the product temperature at 55 degrees centigrade. To cater for downstream process requirements, the flow rate of product regularly alters within the range* $1.5 - 3.0$ *L/min. Fix the flow rate of product at* 2.1 *L/min. The transfer function from the steam flow rate to the product temperature is obtained by carrying out step tests:*

$$\boxed{G(s) = \frac{0.54e^{-15s}}{(15s+1)^2}.}$$

The time delay depends on the flow rate of the product. When the flow rate alters over the confined range, the time delay varies between 10 *and* 20 *seconds.*

The design requirement is that the overshoot should not exceed 10% for the worst case. With (4.5.8) the controller of the second-order plant is obtained as follows:

$$\boxed{C(s) = \frac{1}{0.54}\frac{(15s+1)^2}{\lambda^2 s^2 + (2\lambda + 15)s}.}$$

The parameter is taken as $\lambda = 0.9\theta$. *For the sake of comparison, a plant of reduced order is computed:*

$$G(s) = \frac{0.54e^{-21s}}{25s+1},$$

and $\lambda = 0.78\theta$ *is taken for the controller of the first-order plant given by (4.2.11):*

$$\boxed{C(s) = \frac{1}{0.54}\frac{(25s+1)(11.5s+1)}{\lambda^2 s^2 + (2\lambda + 11.5)s}.}$$

A unit step reference is added at $t = 0$ *and a unit step load is added at*

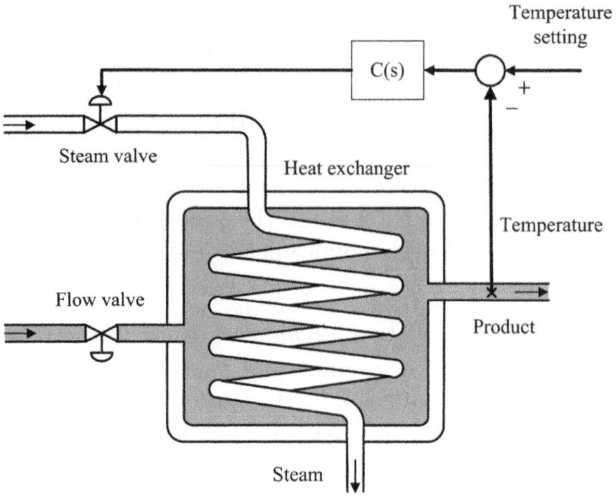

FIGURE 4.5.1
An industrial heat exchanger.

$t = 200$. *The nominal responses of the closed-loop system are shown in Figure 4.5.2. Since the two poles of the plant are real, the responses given by the two controllers are similar.*

Assume that the flow rate of the product decreases to the lowest so that the time delay becomes 20 seconds. Responses of this worst case are shown in Figure 4.5.3. The overshoot of the closed-loop system increases to 10%.

4.6 All Stabilizing PID Controllers for Stable Plants

In this section, the PID controller is discussed from another angle. One might encounter such a case in practice: even when the parameters of a PID controller are chosen in random, the closed-loop system still works well. Unfortunately, not every time can one find appropriate parameters, since the range of the PID parameters, in which the feedback system is stable, is not clear. This seems to be a simple problem. However, simple problems do not always have simple solutions. Because of the time delay involved in the characteristic equation, it is fairly difficult to analyze this problem.

The goal of this section is to determine the set of controller parameters that guarantees the stability of the closed-loop system. Evidently, the set is independent of design methods.

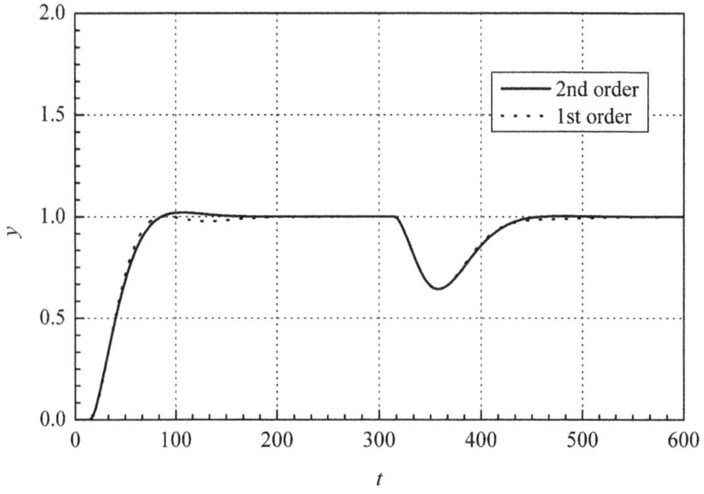

FIGURE 4.5.2
Responses of the nominal plant.

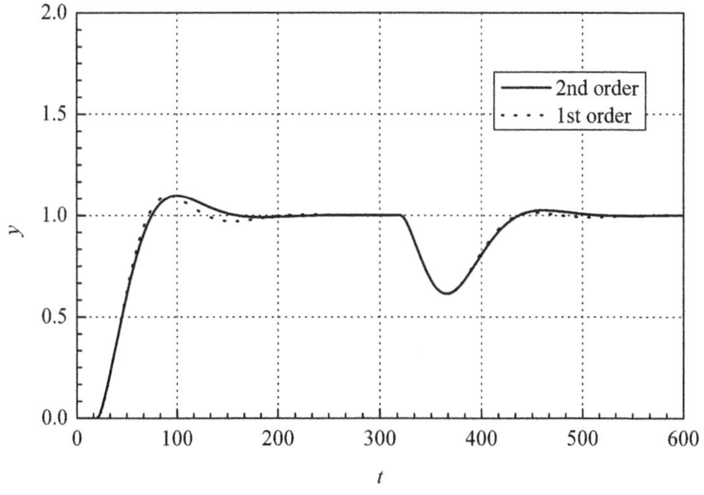

FIGURE 4.5.3
Responses of the worst case.

The attention here is put on the first-order plant with time delay:

$$G(s) = \frac{Ke^{-\theta s}}{\tau s + 1}. \tag{4.6.1}$$

To simplify presentation, the standard PID controller is considered:

$$C(s) = K_C + \frac{K_I}{s} + K_D s, \tag{4.6.2}$$

where $K_I = K_C/T_I$ and $K_D = K_C T_D$.

Theorem 4.6.1. *The plant (4.6.1) can be stabilized by the PID controller (4.6.2) if and only if the controller parameters satisfy*

$$-\frac{1}{K} < K_C < K_T,$$

where

$$K_T = \frac{1}{K}\left[\frac{\tau}{\theta}\alpha_1 \sin(\alpha_1) - \cos(\alpha_1)\right],$$

and α_1 is the solution to the equation

$$\tan(\alpha) = -\frac{\tau}{\tau + \theta}\alpha$$

in the interval $(0, \pi)$. The complete stabilizing region is given as follows:

1. *For $K_C \in (-1/K, 1/K]$, the stabilizing region of the integral constant and the derivative constant is the trapezoid in Figure 4.6.1.*
2. *For $K_C \in (1/K, K_T)$, the stabilizing region of the integral constant and the derivative constant is the quadrilateral in Figure 4.6.2.*

Here

$$\begin{aligned}
z &= \theta\omega, \\
m(z) &= \frac{\theta^2}{z^2}, \\
b(z) &= -\frac{\theta}{Kz}\left[\sin(z) + \frac{\tau}{\theta} z\cos(z)\right], \\
w(z) &= \frac{z}{K\theta}\left\{\sin(z) + \frac{\tau}{\theta} z[\cos(z) + 1]\right\},
\end{aligned}$$

and $z_j (j = 1, 2, ...)$ are the positive real roots of

$$KK_C + \cos(z) - \frac{\tau}{\theta} z \sin(z) = 0.$$

These roots are arranged in an increasing order of magnitude.

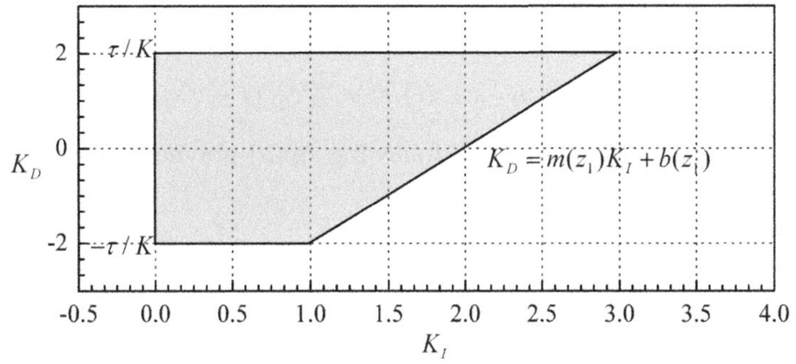

FIGURE 4.6.1
Stabilizing region for $K_C \in (-1/K, 1/K]$.
(From Silva et al., 2002. Reprinted by permission of the IEEE)

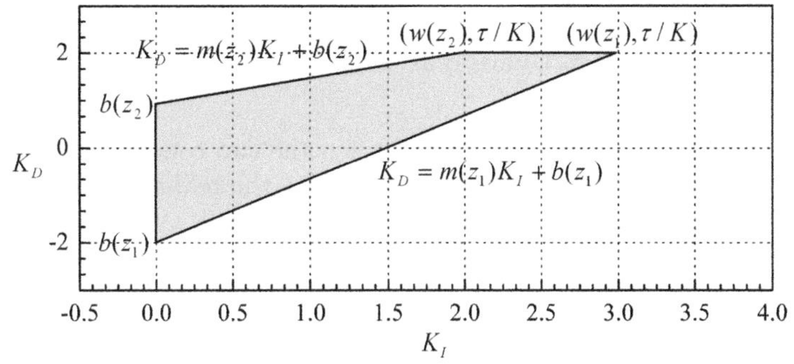

FIGURE 4.6.2
Stabilizing region for $K_C \in (1/K, K_T)$.
(From Silva et al., 2002. Reprinted by permission of the IEEE)

Proof. The conclusion in this theorem is simple while the proof is a bit complicated. Hence, the proof is only sketched. Those application-oriented readers can skip the proof.

The characteristic polynomial of the system is in the form of a quasi-polynomial:

$$\delta(s) = K(K_I + K_C s + K_D s^2)e^{-\theta s} + (1+\tau s)s. \tag{4.6.3}$$

Since $e^{\theta s}$ has no finite zeros, the following quasi-polynomial is considered instead:

$$\delta^*(s) = K(K_I + K_C s + K_D s^2) + (1+\tau s)se^{\theta s}. \tag{4.6.4}$$

$\delta^*(s)$ and $\delta(s)$ are equivalent for stability analysis. Rewrite $\delta^*(s)$ as

$$\delta^*(j\omega) = \delta_r(\omega) + j\delta_i(\omega), \tag{4.6.5}$$

where $\delta_r(\omega)$ and $\delta_i(\omega)$ represent the real part and the imaginary part of $\delta^*(j\omega)$, respectively.

$$\begin{aligned} \delta_r(\omega) &= KK_I - KK_D\omega^2 - \omega\sin(\theta\omega) - \tau\omega^2\cos(\theta\omega), \\ \delta_i(\omega) &= \omega[KK_C + \cos(\theta\omega) - \tau\omega\sin(\theta\omega)]. \end{aligned} \tag{4.6.6}$$

It can be seen that the controller gain K_C only affects the imaginary part of $\delta^*(j\omega)$ whereas the integral constant K_I and the derivative constant K_D only affect the real part of $\delta^*(j\omega)$. It can be proved that $\delta^*(s)$ is stable if and only if

1. $E(\omega_0) := \delta_i'(\omega_0)\delta_r(\omega_0) - \delta_i(\omega_0)\delta_r'(\omega_0) > 0$ for some ω_0 in $(-\infty, +\infty)$.
2. $\delta_r(\omega)$ and $\delta_i(\omega)$ have only simple real roots and these roots interlace.

In what follows, it will be examined when the two conditions hold.

First, check the first condition. Since $z = \theta\omega$, the real part and the imaginary part of $\delta^*(j\omega)$ can be, respectively, expressed as

$$\begin{aligned} \delta_r(z) &= KK_I - \frac{KK_D}{\theta^2}z^2 - \frac{1}{\theta}z\sin(z) - \frac{\tau}{\theta^2}z^2\cos(z), \\ \delta_i(z) &= \frac{z}{\theta}[KK_C + \cos(z) - \frac{\tau}{\theta}z\sin(z)]. \end{aligned}$$

Take $\omega_0 = z_0 = 0$. Then $\delta_r(z_0) = KK_I$ and $\delta_i(z_0) = 0$. On the other hand,

$$E(z_0) = \frac{KK_C + 1}{\theta}KK_I. \tag{4.6.7}$$

If we pick

$$K_I > 0, K_C > -\frac{1}{K},$$

or

$$K_I < 0, K_C < -\frac{1}{K},$$

then $E(z_0) > 0$.

Next, check the second condition. Plotting the terms involved in the equation $\delta_i(z) = 0$ and graphically examining the nature of the solution, it can be concluded that the roots are all real if and only if $K_C \in (-1/K, K_T)$.

Furthermore, compute the roots of the imaginary part by letting $\delta_i(z) = 0$. Evidently, one root is $z_0 = 0$. Other roots $z_j (j = 1, 2, ...)$ are given by the equation

$$KK_C + \cos(z) - \frac{\tau}{\theta} z \sin(z) = 0.$$

Arrange these roots in an increasing order of magnitude. By evaluating $\delta_r(z)$ at $z_j (j = 0, 1, ...)$, it can be proved that the K_I and K_D for the roots of $\delta_r(z)$ and $\delta_i(z)$ to interlace are determined by the following infinite set of inequalities:

$$\begin{aligned} &K_I > 0, \\ &(-1)^j K_D < (-1)^j m(z_j) K_I + (-1)^j b(z_j), \quad j = 1, 2, ... \end{aligned} \tag{4.6.8}$$

Now it will be shown that all these regions do have a nonempty intersection. Notice that the slopes $m(z_j)$ of the boundary lines of these regions decrease with z_j. The limit is

$$\lim_{j \to \infty} m(z_j) = 0. \tag{4.6.9}$$

With this in mind, the following observations are obtained:

1. When $K_C \in (-1/K, 1/K]$, the intersection is given by the trapezoid sketched in Figure 4.6.1. This region is obtained by utilizing the following properties:

(a) $b(z_j) < b(z_{j+2}) < -\tau/K$ for odd values of j.
(b) $b(z_j) > \tau/K$ and $b(z_j) \to \tau/K$ as $j \to \infty$ for even values of j.
(c) $0 < v(z_j) < v(z_{j+2})$ for odd values of j, where

$$v(z) = \frac{z}{K\theta}\left\{\sin(z) + \frac{\tau}{\theta} z[\cos(z) - 1]\right\}.$$

2. When $K_C \in (1/K, 1/K_T)$, the intersection is given by the quadrilateral sketched in Figure 4.6.2. This region is obtained by using the following properties:

(a) $b(z_j) > b(z_{j+2}) > -\tau/K$ for odd values of j.
(b) $b(z_j) < b(z_{j+2}) < \tau/K$ for even values of j.

(c) $w(z_j) > w(z_{j+2}) > 0$ for even values of j.

(d) $b(z_1) < b(z_2)$, $w(z_1) > w(z_2)$.

So far, the interlacing property, as well as that the roots of $\delta_i(z) = 0$ are all real for $K_C \in (-1/K, K_T)$, has been proven. The two conditions can be used to prove that $\delta_r(z) = 0$ only has real roots.

Therefore, for $(-1/K, K_T)$ there is a solution to the PID stabilization problem of the first-order plant with time delay. For those values of K_C beyond this range, the PID stabilization problem does not have a solution. □

4.7 Summary

In this chapter, an analytical method for PID controller design is proposed based on the H_∞ optimal control theory. If the plant is in the form of

$$G(s) = \frac{Ke^{-\theta s}}{\tau s + 1},$$

the controller is

$$C(s) = \frac{1}{K} \frac{(\tau s + 1)(1 + \theta s/2)}{\lambda^2 s^2 + (2\lambda + \theta/2)s}.$$

For the second-order plant with time delay,

$$G(s) = \frac{Ke^{-\theta s}}{(\tau_1 s + 1)(\tau_2 s + 1)},$$

the controller is

$$C(s) = \frac{1}{K} \frac{(\tau_1 s + 1)(\tau_2 s + 1)}{\lambda^2 s^2 + (2\lambda + \theta)s}.$$

The controller order is the same as that of the approximate plant. When using these design formulas, the designer is not required to choose weighting functions.

Although many design methods have been developed, little work has been done on the quantitative design of PID controller. As we know, practical design requirements on control systems are usually specified in terms of time domain responses or frequency domain responses. By using the performance degree, the proposed PID controller can provide quantitative closed-loop responses. The performance degree can be determined by computing or tuning. The tuning procedure is very simple: Increase the performance degree monotonically until the required response is obtained.

The design method in this chapter provides a smooth transition between the classical design requirement and the optimal design result. On one hand, the controller is analytically derived from the optimal performance index. On the other hand, the controller parameters directly relate to the classical design requirement.

The relationship between the H_∞ PID controller and the Smith predictor is also investigated in this chapter. It is shown that the two controllers are approximately equivalent. This clarifies why the H_∞ PID controller can be used to control systems with large time delay.

The characterization of all stabilizing PID controllers is studied in the last section of this chapter. This problem has important implications on both theoretical analysis and application. The analyzing procedure is rather complicated. The result, however, is quite simple.

Exercises

1. Which rational approximation for a time delay can be utilized to design an H_∞ PID controller besides the 1/1 Pade approximant and the first-order Taylor series expansion? Give one example.
2. Let $\|G(s)\|_\infty = \|1/(s+1)\|_\infty$. Is it true or false: $G(s) = 1/(s+1)$?
3. Assume that the plant is in the form of
$$G(s) = \frac{s-3}{s+1},$$
and the closed-loop transfer function of the unity feedback loop is
$$T(s) = \frac{5}{s+5}.$$
Compute the controller and analyze the internal stability of the closed-loop system.
4. Show that
$$|W(j\omega)S(j\omega)| + |\Delta_m(j\omega)T(j\omega)| < 1, \forall\omega,$$
if
$$|W(j\omega)S(j\omega)|^2 + |\Delta_m(j\omega)T(j\omega)|^2 < 1/2, \forall\omega.$$
5. A measure of the closed-loop performance is
$$1/\max|\mathrm{Re}\,(G(j\omega)C(j\omega))\,|.$$
Sketch the relationship between this measure and λ/θ for the H_∞ PID controller in Section 4.2.

6. The aim of the strip casting process is to produce hot strip directly from the molten steel. Since the hot rolling process is eliminated, substantial reduction in investment and operating cost would be possible. The schematic diagram of a strip caster is given in Figure E4.1. Molten steel, fed from the tundish, flows through a nozzle into the sump comprising two casting rolls and side dams, solidifies in a short time, and is rolled out to a thin strip between the two counter-rotating rolls.

   The control loop of the molten steel height is one of the most important control loops in the strip casting process. A very high precision is required. The height of molten steel is captured by a CCD video camera. Based on the difference in brightness of the two materials, the interface between the molten steel and the roll surface (that is, the height of molten steel) can be distinguished. The transfer function from the flow rate to the height of molten steel is

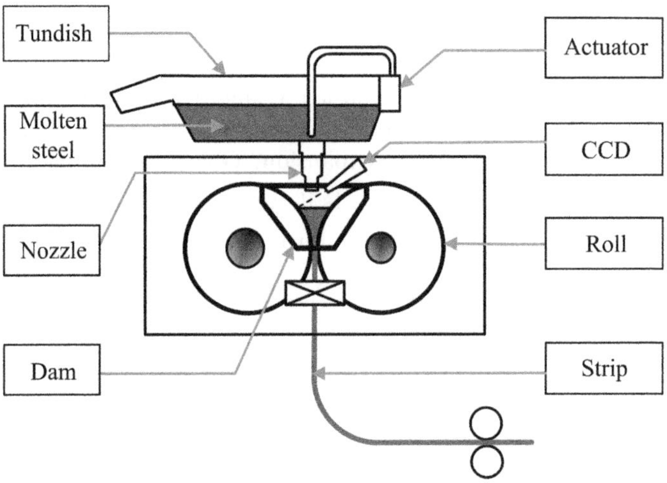

FIGURE E4.1
Control of the molten steel level.

$$G(s) = \frac{0.42}{0.78s+1}e^{-0.15s}.$$

   There is a 10% error in estimating the gain, and a 25% error in estimating the time constant and time delay. The worst-case overshoot is required to be 10%. Design a PID controller for the control loop.

Notes and References

The Z-N method in Section 4.1 was proposed by Ziegler and Nichols (1942) and the C-C method can be found in Cohen and Coon (1953). The refined Z-N method belongs to Hang et al. (1991) and Astrom et al. (1992). For some recent advances in PID controller design, please refer to Astrom and Hagglund (2005), O'Dwyer (2006), and Visioli (2006).

The comparison study of the three practical forms of PID controller can be found in Luyben (2001).

The material in Sections 4.2–4.4 comes from Zhang (1996) and Zhang et al. (2002).

The introduction to the constructing of the Smith predictor can be found in Stephanopoulos (1984, Section 19.2). Morari and Zafiriou (1989, Section 6.2) pointed out that the Smith predictor can be designed by the IMC design method (that is, by $Q(s)$).

The plant in Example 4.3.1 and Figure 4.3.3 are from Zhang (1996) and Zhang et al. (2001).

Section 4.5 closely follows the paper by Zhang and Sun (1997).

The plant in Example 4.5.1 is from Golten and Verwer (1991, Section 9.7).

The discussion in Section 4.6 is adapted from Bhattacharyya et al. (2009, Section 3.6) and Silva et al. (2002) (Silva G. J., A. Datta, and S. P. Bhattachcharyya. New results on the synthesis of PID controllers, *IEEE Trans. Auto. Control*, 2002, 47(2), 241–252. ©IEEE). The results for the general plant can be found in Ou et al. (2009).

Exercise 4 gives a special robust performance problem called the mixed sensitivity problem. See, for example, Doyle et al. (1992, Chapter 12).

The measure in Exercise 5 was discussed by Wang and Shao (2000).

The plant in Example 6 is from Zhu (2005).

5

H_2 PID Controllers for Stable Plants

CONTENTS

In Chapter 4, an analytical method was presented for PID controller design based on the H_∞ optimal control theory. The question studied in this chapter is whether an analytical PID controller can be designed with other methods, and if so, whether the resulting PID controller has similar features to that of the H_∞ PID controller.

An analog of the H_∞ optimal control theory is the H_2 optimal control theory. This chapter is devoted to the design of H_2 PID controllers for the plant with time delay. Analytical design methods are developed based on the H_2 optimal control theory. It will be shown that H_2 PID controllers can also be tuned for quantitative performance and robustness. To illustrate the difference between the H_2 PID controller and the H_∞ PID controller, an inverse response process is employed to compare them.

When controller structure is fixed, there is a limit to the achievable performance. One interesting problem about the PID controller is how to design a PID controller to reach the limit. The problem is studied in this chapter by utilizing a rational transfer function to approximate the obtained suboptimal controller. An analytical design method is developed to derive the PID controller with the best achievable performance.

5.1 H_2 PID Controller for the First-Order Plant

Consider the unity feedback loop shown in Figure 4.2.1. Assume that the plant is

$$G(s) = \frac{Ke^{-\theta s}}{\tau s + 1}. \tag{5.1.1}$$

Using Youla parameterization, we have

$$C(s) = \frac{Q(s)}{1 - G(s)Q(s)}, \tag{5.1.2}$$

where $Q(s)$ is a stable transfer function. It is difficult to treat $e^{-\theta s}$ analytically. Approximate it with the 1/1 Pade approximant:

$$G(s) \approx K\frac{1 - \theta s/2}{(\tau s + 1)(1 + \theta s/2)}. \tag{5.1.3}$$

The design procedure for the H_2 PID controller is similar to that for the H_∞ PID controller. The controller is first designed for the approximate plant and then is used to control the original plant. The approximation error is regarded as a kind of uncertainty.

The H_2 optimal index is

$$\min \|W(s)S(s)\|_2,$$

where $W(s)$ is the weighting function. Assume that the system input is a unit step. In view of the discussion in Section 3.2, the weighting function in the H_2 optimal control should be chosen so that the normalized input is the impulse, that is, $d(s)/W(s) = 1$. Then, $W(s) = 1/s$. $W(s)$ has a pole on the imaginary axis. To guarantee a finite 2-norm and the asymptotic tracking property, a constraint has to be imposed on the design:

$$\lim_{s\to 0} S(s) = \lim_{s\to 0}[1 - G(s)Q(s)] = 0. \tag{5.1.4}$$

In other words, $S(s)$ must have a zero at the origin to cancel the pole of $W(s)$. This gives

$$Q(0) = \frac{1}{G(0)} = \frac{1}{K}. \tag{5.1.5}$$

It should be emphasized that this constraint is also required for asymptotic tracking. The set of all $Q(s)$s satisfying the constraint can be written as

$$Q(s) = \frac{1}{K} + sQ_1(s), \tag{5.1.6}$$

where $Q_1(s)$ is stable. The function to be minimized is

$$\begin{aligned}
&\|W(s)S(s)\|_2^2 \\
= &\left\| W(s)\left\{1 - G(s)\left[\frac{1}{K} + sQ_1(s)\right]\right\}\right\|_2^2 \\
= &\left\| \frac{1}{s}\left\{1 - \frac{K(1-\theta s/2)}{(\tau s+1)(1+\theta s/2)}\left[\frac{1}{K} + sQ_1(s)\right]\right\}\right\|_2^2 \\
= &\left\| \frac{\theta\tau s/2 + (\theta+\tau)}{(\tau s+1)(\theta s/2+1)} - \frac{K(1-\theta s/2)}{(\tau s+1)(1+\theta s/2)}Q_1(s)\right\|_2^2 \\
= &\left\| \frac{1-\theta s/2}{1+\theta s/2}\left[\frac{\theta\tau s/2 + (\theta+\tau)}{(\tau s+1)(1-\theta s/2)} - \frac{K}{\tau s+1}Q_1(s)\right]\right\|_2^2 .
\end{aligned}$$

$(1 - \theta s/2)/(1 + \theta s/2)$ in this equation is an all-pass transfer function. By the definition of 2-norm, it is easy to verify that the 2-norm of a transfer function keeps its value after an all-pass transfer function is introduced to it. Therefore,

$$\|W(s)S(s)\|_2^2 = \left\| \frac{\theta\tau s/2 + (\theta+\tau)}{(\tau s+1)(1-\theta s/2)} - \frac{K}{\tau s+1}Q_1(s)\right\|_2^2 . \tag{5.1.7}$$

As we know, by partial fraction expansion, a strictly proper transfer function without poles on the imaginary axis can always be uniquely expressed as a stable part (which has no poles in Re $s > 0$) and an unstable part (which has no poles in Re $s < 0$):

$$\frac{\theta\tau s/2 + (\theta+\tau)}{(\tau s+1)(1-\theta s/2)} = \frac{\theta}{1-\theta s/2} + \frac{\tau}{\tau s+1}.$$

By Theorem 3.1.4, we have

$$\|W(s)S(s)\|_2^2 = \left\| \frac{\theta}{1-\theta s/2}\right\|_2^2 + \left\| \frac{\tau}{\tau s+1} - \frac{K}{\tau s+1}Q_1(s)\right\|_2^2 . \tag{5.1.8}$$

Temporarily relax the requirement on the properness of $Q(s)$. To obtain the minimum, the only choice is

$$Q_{1opt}(s) = \frac{\tau}{K}. \tag{5.1.9}$$

Substitute this into (5.1.6). The optimal $Q(s)$ is

$$Q_{opt}(s) = \frac{\tau s+1}{K}. \tag{5.1.10}$$

$Q(s)$ should be proper. Use the following filter to roll off the improper solution:

$$J(s) = \frac{1}{\lambda s+1},$$

where λ is the performance degree. It is a positive real number. The suboptimal $Q(s)$ is

$$Q(s) = Q_{opt}(s)J(s) = \frac{\tau s+1}{K(\lambda s+1)}. \tag{5.1.11}$$

Since $Q(0) = 1/K$, $Q(s)$ satisfies the constraint of asymptotic tracking. The unity feedback loop controller is

$$C(s) = \frac{Q(s)}{1-G(s)Q(s)} = \frac{1}{K}\frac{(\tau s+1)(1+\theta s/2)}{\theta\lambda s^2/2+(\lambda+\theta)s}. \tag{5.1.12}$$

Comparing the controller with

$$C = K_C\left(1+\frac{1}{T_I s}+T_D s\right)\frac{1}{T_F s+1}$$

gives that

$$\begin{aligned} T_F &= \frac{\theta\lambda}{2(\lambda+\theta)}, \quad T_I = \tau+\frac{\theta}{2}, \\ T_D &= \frac{\theta\tau}{2T_I}, \quad K_C = \frac{T_I}{K(\lambda+\theta)}. \end{aligned} \tag{5.1.13}$$

If the following form is chosen:

$$C(s) = K_C\left(1+\frac{1}{T_I s}+\frac{T_D s}{T_F s+1}\right),$$

the parameters of the PID controller are

$$\begin{aligned} T_F &= \frac{\theta\lambda}{2(\lambda+\theta)}, \quad T_I = \tau+\frac{\theta}{2}-T_F, \\ T_D &= \frac{\theta\tau}{2T_I}-T_F, \quad K_C = \frac{T_I}{K(\lambda+\theta)}. \end{aligned} \tag{5.1.14}$$

When the PID controller is in the form of

$$C(s) = K_C\left(1+\frac{1}{T_I s}\right)\frac{T_D s+1}{T_F s+1},$$

its parameters are

$$\begin{aligned} T_F &= \frac{\theta\lambda}{2(\lambda+\theta)}, \quad T_I = \tau(\text{or } \frac{\theta}{2}), \\ T_D &= \frac{\theta}{2}(\text{or } \tau), \quad K_C = \frac{T_I}{K(\lambda+\theta)}. \end{aligned} \tag{5.1.15}$$

It can be verified that the H_2 PID controller can also be equivalent to the Smith predictor.

With the optimal controller, the optimal performance for (5.1.3) can be obtained as follows:

$$\min \|W(s)S(s)\|_2 = \left\|\frac{\theta}{1-\theta s/2}\right\|_2 = \sqrt{\theta}. \tag{5.1.16}$$

5.2 Quantitative Tuning of the H_2 PID Controller

In the method given in the last section, the Youla parameterization is used to design a PID controller. The internal stability of the nominal system is automatically guaranteed and the suboptimal controller is obtained. This section analyzes the quantitative relationship between the closed-loop response and the performance degree.

Consider the nominal stability first. Normally, the existence of time delay pushes the system close to instability. The larger the time delay is, the more difficult it is to stabilize the closed-loop system. Nevertheless, the design methods in the last chapter and this chapter can always guarantee the stability of the closed-loop system. Here, the ratio of the performance degree to the time delay is the key. Because of the error introduced by the Pade approximation, there is a lower bound for the ratio. By following the discussion in Section 4.4, it is concluded that, as long as the ratio is greater than the lower bound, the nominal closed-loop system is internally stable.

In addition to the stability problem, one has to consider the performance problem. The existence of time delay adversely affects the performance of the closed-loop system. The performance is getting worse and worse with the increase of time delay. For the plant with a large time delay, the methods in the last chapter and this chapter provide better performance than the traditional methods introduced in Section 4.1. However, for the plant with a small time delay or without any time delay, the PID controllers designed by the traditional methods can also provide acceptable performance, even though these methods are empirical ones.

The performance degree of the H_2 PID controller has a similar function to that of the H_∞ PID controller. When there is no modeling error, the performance degree can be used to tune the response shape of the nominal closed-loop system quantitatively. The relationships between the performance degree and the overshoot, rise time, resonance peak, and perturbation peak are shown in Figure 5.2.1–Figure 5.2.4, respectively. For example, a 12% overshoot can be obtained by taking $\lambda = 0.3\theta$ according to Figure 5.2.1; if the performance specification is the resonance peak of 2dB, one can take $\lambda = 0.22\theta$ based on Figure 5.2.3. Normally, the value of λ corresponding to the practical design requirements falls into the interval $0.1\theta - 1.2\theta$.

Simple computations give that

$$L(s) = \frac{(1+\theta s/2)e^{-\theta s}}{\theta\lambda s^2/2 + (\lambda+\theta)s}. \tag{5.2.1}$$

The Bode plot and Nyquist plot of $L(s)$ are shown in Figure 5.2.5 and Figure 5.2.6, respectively. The relationships between the performance degree and the gain margin, phase margin, and ISE are shown in Figure 5.2.7–Figure 5.2.9. It can be seen that the curves describing the relationships between the perfor-

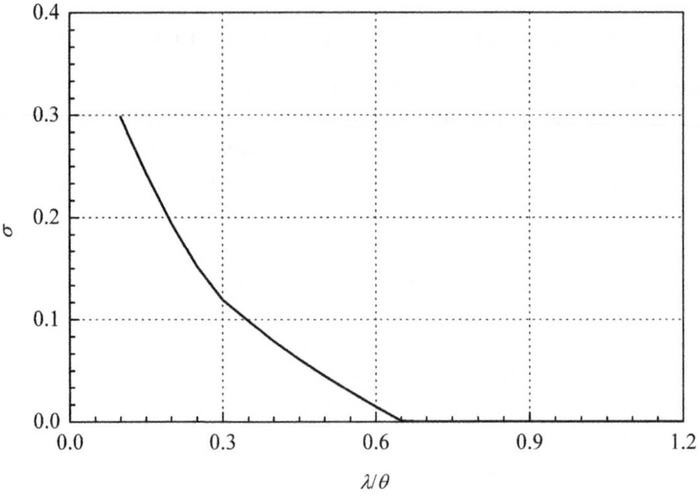

FIGURE 5.2.1
Relationship between the performance degree and the overshoot.

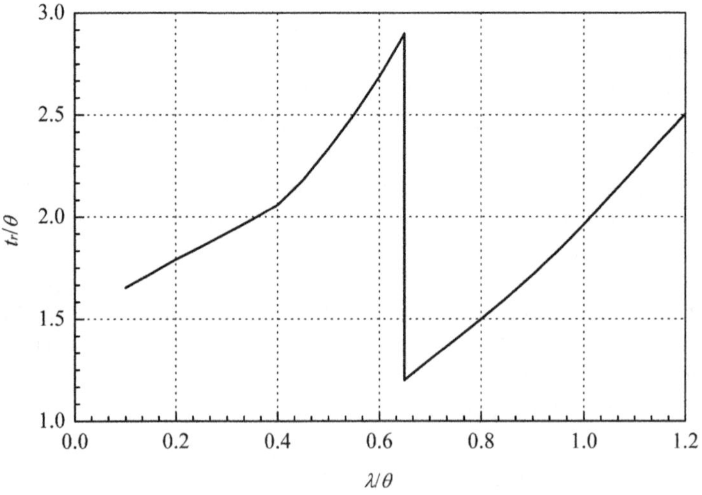

FIGURE 5.2.2
Relationship between the performance degree and the rise time.

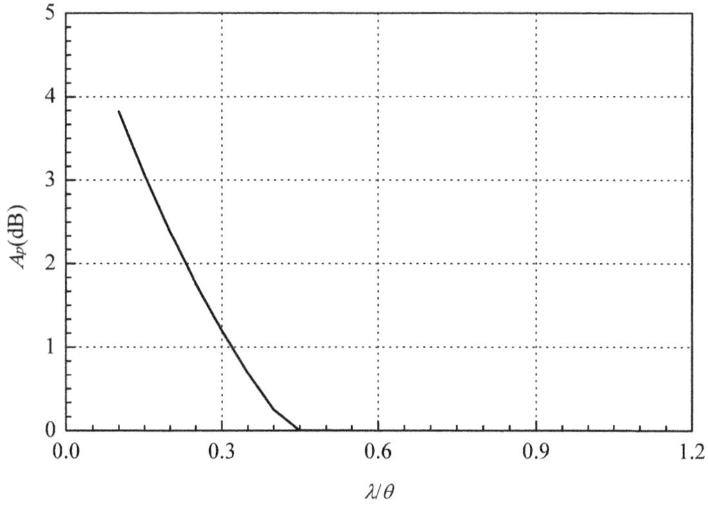

FIGURE 5.2.3
Relationship between the performance degree and the resonance peak.

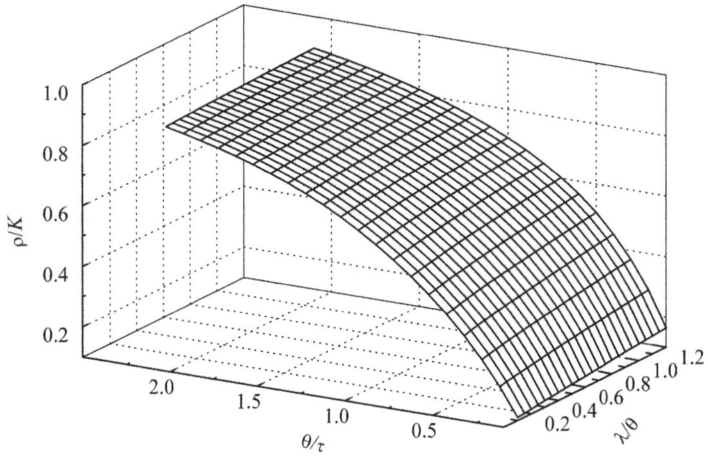

FIGURE 5.2.4
Relationship between the performance degree and the perturbation peak.

mance degree and the gain margin, phase margin, and ISE are almost straight lines.

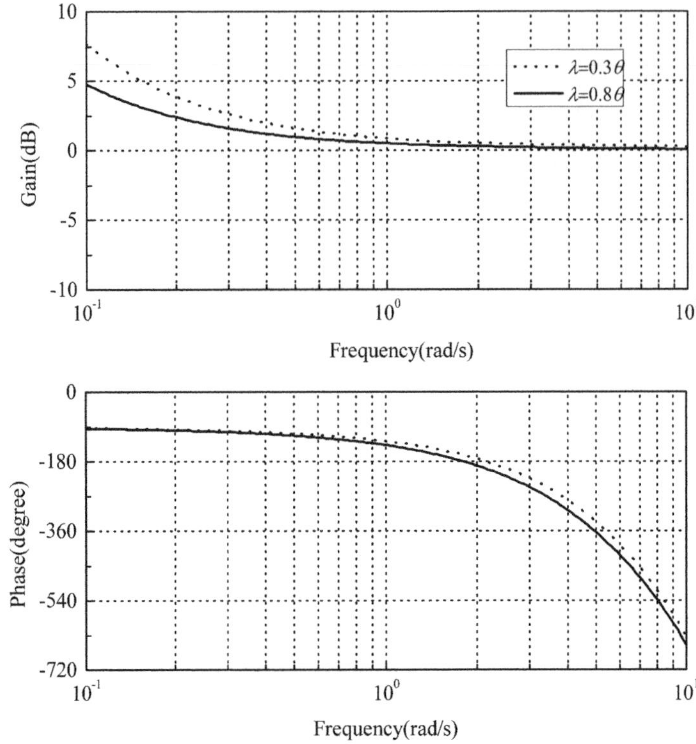

FIGURE 5.2.5
Bode plot of the H_2 PID control system.

Now consider the uncertain system. The robust performance problem is to design a controller such that the feedback system is internally stable and the performance objective is satisfied for all uncertain plants. With the performance degree of the H_2 PID controller, one can easily tradeoff between the nominal performance and the robustness. The determination of the performance degree is similar to that for an H_∞ PID controller: Increase the performance degree monotonically until the required response is obtained.

The robust performance problem is not always solvable, because the desired performance objective may be too stringent for the given nominal plant and the associated uncertainty. The design methods in this book provide easy check solutions to this problem. By adjusting the performance, it is easy for designers to estimate whether or not the required performance is achievable for some uncertain plant.

In traditional PID controllers, T_F is fixed. It is usually chosen as $0.1T_D$.

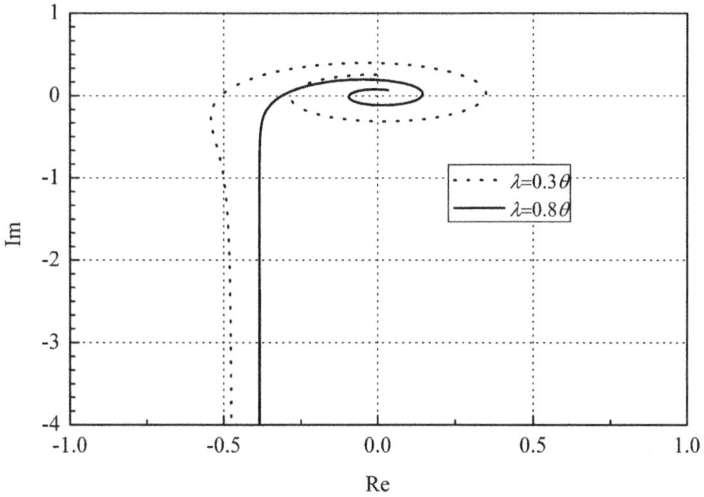

FIGURE 5.2.6
Nyquist plot of the H_2 PID control system.

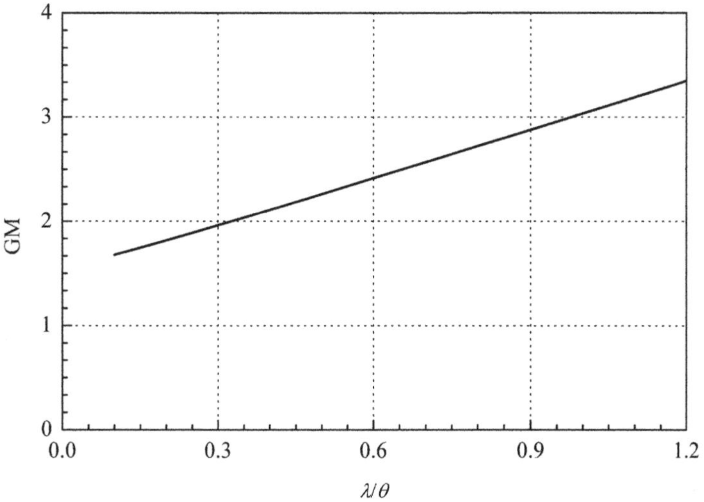

FIGURE 5.2.7
Relationship between the performance degree and the gain margin.

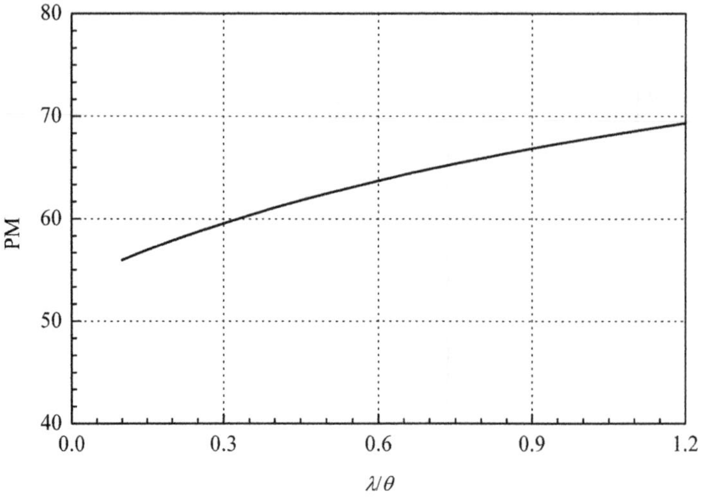

FIGURE 5.2.8
Relationship between the performance degree and the phase margin.

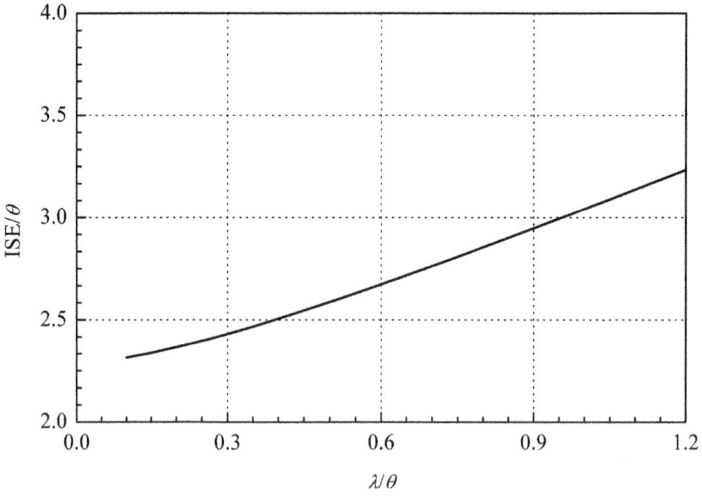

FIGURE 5.2.9
Relationship between the performance degree and the ISE.

However, the T_F is not a fixed value in the H_∞ PID controller and the H_2 PID controller. If a traditional PID controller has been installed in a system and one desires to use the tuning method here, then the T_F in the analytical formulas can be omitted; only the other three parameters are used for tuning. The responses are similar.

Example 5.2.1. *Consider a strip thickness control system. A typical tandem hot strip mill is depicted in Figure 5.2.10. The metal slab is first heated to certain temperature in the reheating furnace. Its thickness is then reduced in the roughing mill stand and finally refined in the finishing mill stand. At the exit, the strip is cooled and coiled by the down coiler. One main quantity to be controlled in the process is the thickness of the strip. The thickness is controlled through the roll force of finishing mill.*

It is known that the distance from the thickness meter to the finishing mill stand is 4.9m, the speed of the strip is 0.7m/s, and the time constant of the actuator is 3s. Then the transfer function of the plant can be written as

$$\boxed{G(s) = \frac{0.2e^{-7s}}{3s+1}.}$$

From (5.1.12) the H_2 PID controller is

$$\boxed{C(s) = \frac{1}{0.2}\frac{(3s+1)(3.5s+1)}{3.5\lambda s^2 + (\lambda+7)s}.}$$

The performance degree is taken to be $\lambda = 0.3\theta$, which corresponds to about 12% overshoot according to Figure 5.2.1. A unit step reference is added at $t = 0$ and a unit step load is added at $t = 100$. The nominal response of the closed-loop system is shown in Figure 5.2.11. The controller provides fast and steady response for this plant with a large time delay.

Now take $T_F = 0.1T_D$ in the H_2 PID controller. It is seen in Figure 5.2.11 that the response given by the approximate H_2 PID controller is similar to that given by the original H_2 PID controller.

5.3 H_2 PID Controller for the Second-Order Plant

This section considers the second-order plant with time delay. Assume that the plant is given by

$$G(s) = \frac{Ke^{-\theta s}}{(\tau_1 s+1)(\tau_2 s+1)}. \tag{5.3.1}$$

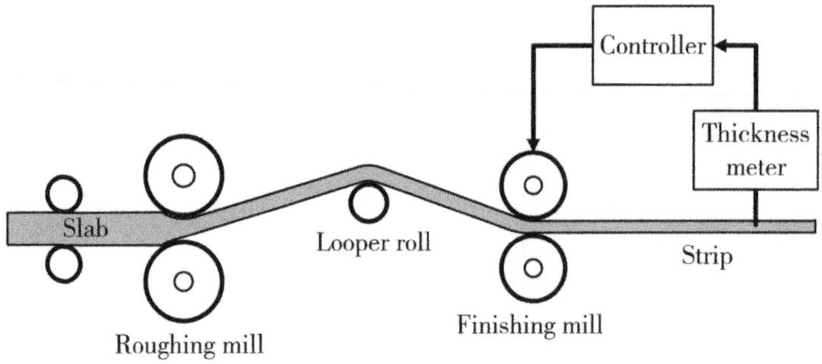

FIGURE 5.2.10
Control system for the strip thickness.

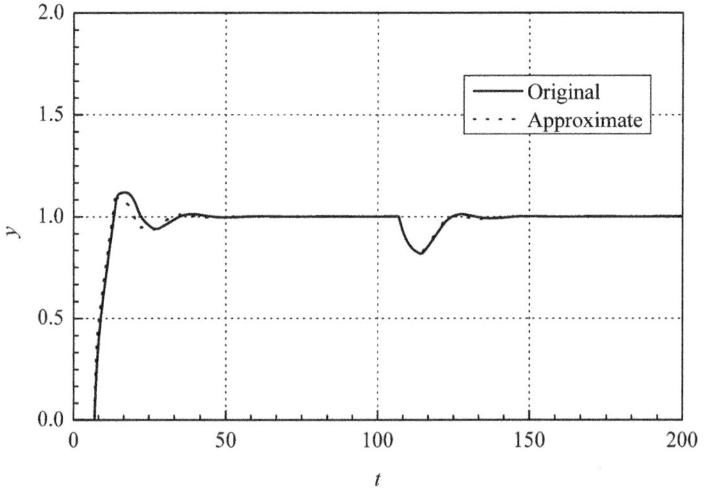

FIGURE 5.2.11
System response for the H_2 PID controller.

With the first-order Taylor series expansion, the following approximate plant is obtained:

$$G(s) \approx \frac{K(1-\theta s)}{(\tau_1 s+1)(\tau_2 s+1)}. \tag{5.3.2}$$

Define the optimal performance index as

$$\min \|W(s)S(s)\|_2 .$$

If the system input is a unit step, $W(s) = 1/s$ is taken. To guarantee a finite 2-norm and the asymptotic tracking property, the following constraint should be satisfied:

$$\lim_{s\to 0}[1 - G(s)Q(s)] = 0. \tag{5.3.3}$$

It follows that

$$Q(0) = \frac{1}{G(0)} = \frac{1}{K}. \tag{5.3.4}$$

All $Q(s)$s that satisfy the constraint are in the form of

$$Q(s) = \frac{1}{K} + sQ_1(s), \tag{5.3.5}$$

where $Q_1(s)$ is a stable transfer function. Then

$$\begin{aligned}
&\|W(s)S(s)\|_2^2 \\
&= \left\| W(s)\left\{1 - G(s)\left[\frac{1}{K} + sQ_1(s)\right]\right\}\right\|_2^2 \\
&= \left\| \frac{\tau_1\tau_2 s + \tau_1 + \tau_2 + \theta}{(\tau_1 s+1)(\tau_2 s+1)} - \frac{K(1-\theta s)Q_1(s)}{(\tau_1 s+1)(\tau_2 s+1)}\right\|_2^2 \\
&= \left\| \frac{(\tau_1\tau_2 s + \tau_1 + \tau_2 + \theta)(1+\theta s)}{(\tau_1 s+1)(\tau_2 s+1)(1-\theta s)} - \frac{K(1+\theta s)Q_1(s)}{(\tau_1 s+1)(\tau_2 s+1)}\right\|_2^2 \\
&= \left\| \frac{2\theta}{1-\theta s} + \frac{(\tau_1\tau_2 s + \tau_1 + \tau_2 - \theta)}{(\tau_1 s+1)(\tau_2 s+1)} - \frac{K(1+\theta s)Q_1(s)}{(\tau_1 s+1)(\tau_2 s+1)}\right\|_2^2 .
\end{aligned}$$

Expanding the right-hand side by Theorem 3.1.2 gives that

$$\begin{aligned}
&\|W(s)S(s)\|_2^2 \\
&= \left\| \frac{2\theta}{1-\theta s}\right\|_2^2 + \\
&\quad \left\| \frac{(\tau_1\tau_2 s + \tau_1 + \tau_2 - \theta)}{(\tau_1 s+1)(\tau_2 s+1)} - \frac{K(1+\theta s)Q_1(s)}{(\tau_1 s+1)(\tau_2 s+1)}\right\|_2^2 .
\end{aligned} \tag{5.3.6}$$

Minimize $\|W(s)S(s)\|_2$. The unique optimal solution is

$$Q_{1opt}(s) = \frac{\tau_1\tau_2 s + \tau_1 + \tau_2 - \theta}{K(1+\theta s)}. \tag{5.3.7}$$

Consequently,

$$Q_{opt}(s) = \frac{(\tau_1 s + 1)(\tau_2 s + 1)}{K(1+\theta s)}. \tag{5.3.8}$$

Introduce the following filter to roll $Q_{opt}(s)$ off at high frequencies:

$$J(s) = \frac{1}{\lambda s + 1}.$$

We have

$$Q(s) = Q_{opt}(s)J(s) = \frac{(\tau_1 s + 1)(\tau_2 s + 1)}{K(1+\theta s)(\lambda s + 1)}. \tag{5.3.9}$$

$Q(s)$ satisfies the constraint of asymptotic tracking. The unity feedback loop controller is

$$C(s) = \frac{Q(s)}{1 - G(s)Q(s)} = \frac{1}{K}\frac{(\tau_1 s + 1)(\tau_2 s + 1)}{\lambda\theta s^2 + (\lambda + 2\theta)s}. \tag{5.3.10}$$

Compare it with

$$C(s) = K_C\left(1 + \frac{1}{T_I s} + T_D s\right)\frac{1}{T_F s + 1},$$

the parameters of the PID controller are

$$\begin{aligned} T_F &= \frac{\lambda\theta}{2\lambda + \theta}, \quad T_I = \tau_1 + \tau_2, \\ T_D &= \frac{\tau_1\tau_2}{\tau_1 + \tau_2}, \quad K_C = \frac{\tau_1 + \tau_2}{K(\lambda + 2\theta)}. \end{aligned} \tag{5.3.11}$$

Normally, the value of λ falls into the interval 0.2θ–1.2θ.

Example 5.3.1. *Consider the plant given in Example 4.5.1:*

$$\boxed{G(s) = \frac{0.54e^{-15s}}{(15s+1)^2}.}$$

Take $\lambda = 0.9\theta$ for the H_∞ PID controller given by (4.5.8):

$$\boxed{C(s) = \frac{1}{0.54}\frac{(15s+1)^2}{\lambda^2 s^2 + (2\lambda + 15)s}.}$$

The H_2 PID controller given by (5.3.10) is

$$\boxed{C(s) = \frac{1}{0.54} \frac{(15s+1)^2}{15\lambda s^2 + (\lambda + 30)s}.}$$

The parameter of this H_2 PID controller is chosen in such a way that the closed-loop system has the same overshoot as that with the above H_∞ PID controller. In this case, $\lambda = 0.78\theta$. A unit step reference is added at $t = 0$ and a unit step load is added at $t = 300$. The nominal responses of the closed-loop system are shown in Figure 5.3.1. The two controllers provide similar responses.

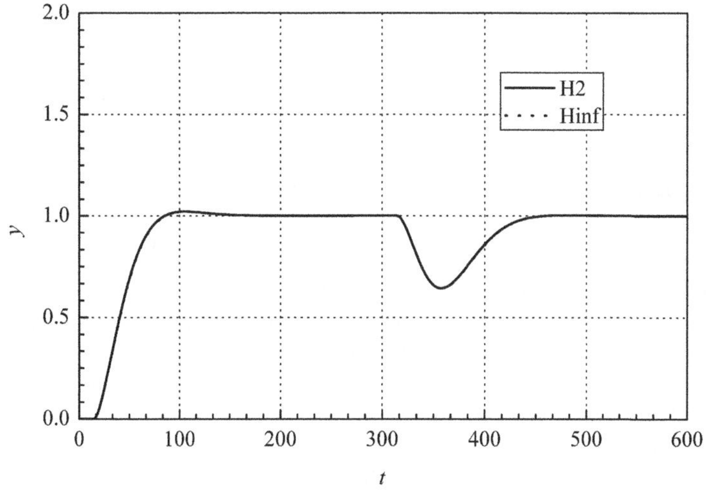

FIGURE 5.3.1
Responses of the H_∞ PID controller and H_2 PID controller.

Note that the disturbance is always added at the plant input in simulations when the ability of rejecting disturbances is considered. Why is the disturbance at the plant output not considered? This is because the transfer function from the reference $r(s)$ to the output $y(s)$ is $T(s)$, the transfer function from the output disturbance $d(s)$ to the output $y(s)$ is $S(s)$, and $S(s) + T(s) = 1$. In other words, the closed-loop response and the output disturbance response are complementary (Figure 5.3.2).

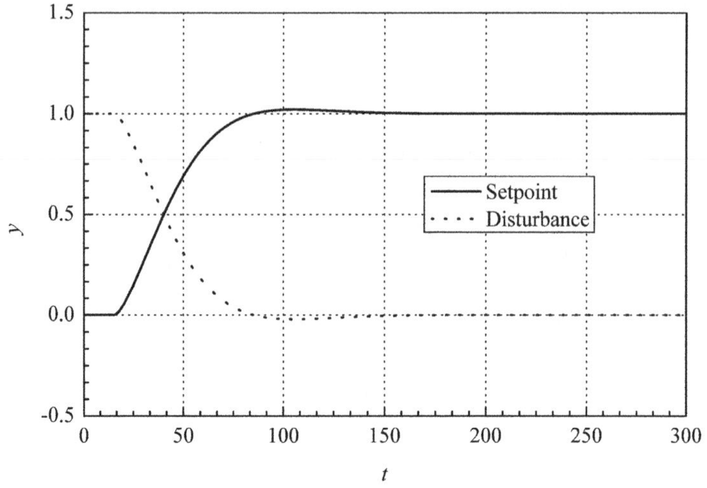

FIGURE 5.3.2
The closed-loop response and the output disturbance response.

5.4 Control of Inverse Response Processes

So far, two similar but different analytical methods have been developed for PID controller design based on the H_∞ optimal control theory and its H_2 counterpart. Is there any relationship between the two controllers? The aim of this section is to compare the features of the H_∞ controller and the H_2 controller by utilizing a simple plant.

A special NMP plant called the inverse response process is used here. The terminology NMP can be well explained with stable plants. With the same magnitude, there exist plants exhibiting less phase than the NMP plant. One example is the following plant:

$$G(s) = \frac{1 - s}{1 + s}.$$

The magnitude of the plant is $|G(j\omega)| = 1$ and the phase is $\angle G(j\omega) = \arctan 2\omega / \left(\omega^2 - 1\right)$. Obviously, there exist other plants with the same magnitude and less phase. For example, the magnitude of $G(s) = 1$ is 1 and the phase is 0. NMP plants are difficult to control.

In an inverse response process, the initial response of the process to a step input is in the opposite direction of its final response. The phenomenon arises from the competing dynamic effects. For example, an inverse response may occur in a distillation column, when the steam pressure to the reboiler suddenly rises. Usually, the initial effect is the increasing in the amount of frothing on the trays above the reboiler, causing a rapid spillover of liquid from these trays into the reboiler. This effect results in an initial increase in the reboiler liquid level. However, the increase in steam pressure will ultimately decrease the reboiler liquid level by boiling off more liquid.

The feature of the inverse response process is that its transfer function has one zero or an odd number of zeros in the open RHP. The simplest inverse response process can be obtained by combining two first-order plants with opposing effects, as shown in Figure 5.4.1. The transfer function of the whole plant is

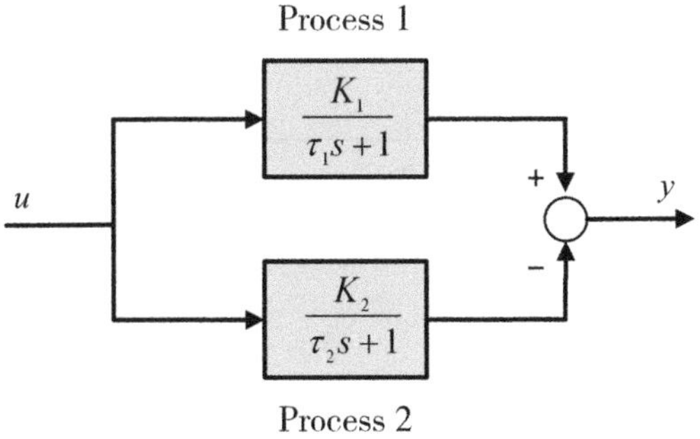

FIGURE 5.4.1
Two opposing first-order processes.

$$G(s) = \frac{K_1}{\tau_1 s + 1} - \frac{K_2}{\tau_2 s + 1}, \tag{5.4.1}$$

or

$$G(s) = \frac{(K_1\tau_2 - K_2\tau_1)s + (K_1 - K_2)}{(\tau_1 s + 1)(\tau_2 s + 1)}. \tag{5.4.2}$$

The inverse response will be obtained when $\tau_1/\tau_2 > K_1/K_2 > 1$; that is, Process 1 initially reacts faster than Process 2, but Process 2 ultimately reaches a higher steady state value than Process 1 (Figure 5.4.2). The transfer function of the plant has a zero in the open RHP:

$$z_r = \frac{K_2 - K_1}{K_1\tau_2 - K_2\tau_1} > 0.$$

Let the performance index be H_∞ optimal, which implies that the worst ISE resulting from a set of energy-bounded inputs is minimized:

$$\min \sup_{r(t)} \|e(t)\|_2,$$

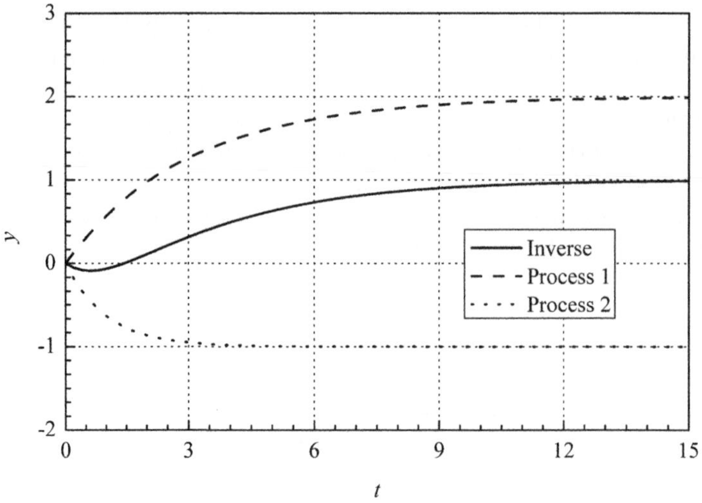

FIGURE 5.4.2
Overall response of the opposing processes for $\tau_1/\tau_2 > K_1/K_2 > 1$.

or equivalently, the ∞-norm of the weighted sensitivity function is minimized:

$$\min \|W(s)S(s)\|_{\infty}.$$

By Theorem 4.2.1, we have

$$\|W(s)S(s)\|_{\infty} \geq 1/z_r.$$

Following the design procedure in Section 4.2, one readily gets

$$C(s) = \frac{1}{(K_1 - K_2)} \frac{(\tau_1 s + 1)(\tau_2 s + 1)}{\lambda^2 s^2 + (2\lambda + z_r^{-1})s}. \tag{5.4.3}$$

Thus, the H_{∞} optimal solution can be realized by only using a PID controller. The closed-loop transfer function is

$$T(s) = \frac{-z_r^{-1}s + 1}{(\lambda s + 1)^2}. \tag{5.4.4}$$

Notice that no poles of the plant appear in $T(s)$. All of them are canceled by the H_{∞} controller.

Differing from the H_{∞} optimal control, the H_2 optimal control minimizes the ISE resulting from the impulse input:

$$\min \|e(t)\|_2,$$

or equivalently, the 2-norm of the weighted sensitivity function is minimized:

$$\min \|W(s)S(s)\|_2 .$$

With the design procedure in Section 5.1, it is easy to obtain the following H_2 controller:

$$C(s) = \frac{1}{(K_1 - K_2)} \frac{(\tau_1 s + 1)(\tau_2 s + 1)}{\lambda z_r^{-1} s^2 + (2z_r^{-1} + \lambda)s}, \tag{5.4.5}$$

which is also a PID controller. The closed-loop transfer function is

$$T(s) = \frac{-z_r^{-1} s + 1}{(z_r^{-1} s + 1)(\lambda s + 1)}. \tag{5.4.6}$$

Factorize the plant into the MP part and the all-pass part:

$$G(s) = (K_1 - K_2) \frac{(z_r^{-1} s + 1)}{(\tau_1 s + 1)(\tau_2 s + 1)} \frac{-z_r^{-1} s + 1}{z_r^{-1} s + 1}.$$

It is seen that the H_2 controller only cancels the poles in the MP part of plant, while those poles in the all-pass part are retained in $T(s)$.

The filter that makes the controller proper is not unique, since the only constraint imposed on it is that it should be a low-pass transfer function satisfying the requirement of asymptotic tracking. If the following filter is chosen for the H_2 controller:

$$J(s) = \frac{z_r^{-1} s + 1}{(\lambda s + 1)^2}, \tag{5.4.7}$$

that is, a zero corresponding to the pole of the all-pass part is introduced factitiously, then the H_2 controller will be identical to the H_∞ controller. Certainly, the H_2 controller can also be made equivalent to some other controllers by selecting appropriate filters. However, such a filter is seldom used, since it introduces additional dynamics.

As there exists no time delay in the plant, the response of the closed-loop system can be computed easily. For example, when the reference is the unit step, the time domain response of the H_∞ controller is

$$y(t) = 1 - \left(1 + \frac{t}{\lambda} + \frac{t z_r^{-1}}{\lambda^2}\right) e^{-t/\lambda}. \tag{5.4.8}$$

The response has no overshoot. Let $dy(t)/dt = 0$. One can get the time when the peak of the inverse response happens:

$$t = \frac{\lambda}{1 + \lambda z_r}.$$

Substituting this into $y(t)$ gives the peak of the inverse response:

$$1 - \frac{1+\lambda z_r}{\lambda z_r} e^{-1/(1+\lambda z_r)}. \tag{5.4.9}$$

Let $y(t) = 0.9$. We have

$$e^{t/\lambda} = 10\left(1 + \frac{t}{\lambda} + \frac{t z_r^{-1}}{\lambda^2}\right). \tag{5.4.10}$$

The rise time is the solution to the above equation.

On the surface, the H_2 optimal control aims at a known specific input and thus can only be used within confined scope, while the H_∞ optimal control aims at all energy-bounded inputs and has a much wider application scope. However, this is not the case.

In the design procedure of the H_2 optimal control, the input is normalized as the impulse. The goal is to express the design procedure in a unified form. Due to the introducing of weighting function, the H_2 optimal control is applicable to the input that differs from the impulse.

The idea behind the analysis is that there is no evident difference between the applicable scopes of the two controllers. Similar insight comes from the system gain, too. Recall the discussion in Chapter 3:

$$\|y(t)\|_\infty \le \|T(s)\|_2 \|r(t)\|_2 .$$

Thus, another objective of the H_2 optimal control is to minimize the maximum amplitude of output for energy-bounded inputs.

Example 5.4.1. *Magnetic Levitation (maglev) trains may replace the airplanes on routes of several hundred kilometers (Figure 5.4.3), because it can offer the environmental and safety advantages of a train and the speed of an airplane. In a maglev system, vehicles are suspended on a guideway and driven by magnetic forces instead of relying on wheels or aerodynamic forces. There is an electronic pas de deux between the vehicle's weight and the repelling force of the electromagnets. The gap between each arm and the guideway is measured 100,000 times per second. This distance is fed to a control system, in which the current in the support magnets is continually adjusted so as to reach an equilibrium point at which the weight of the vehicle is supported by the magnet repellence. The result is that the vehicle suspends and the gap between each arm and the underside of the guideway is kept at* $10 \pm 2mm$.

The dynamic model for controlling the gap is

$$\boxed{G(s) = \frac{s-4}{(s+1.5)(s+2.5)}}.$$

From (5.4.3) we know that the H_∞ *controller is*

$$\boxed{C(s) = -\frac{(s+1.5)(s+2.5)}{4\lambda^2 s^2 + (8\lambda+1)s}}.$$

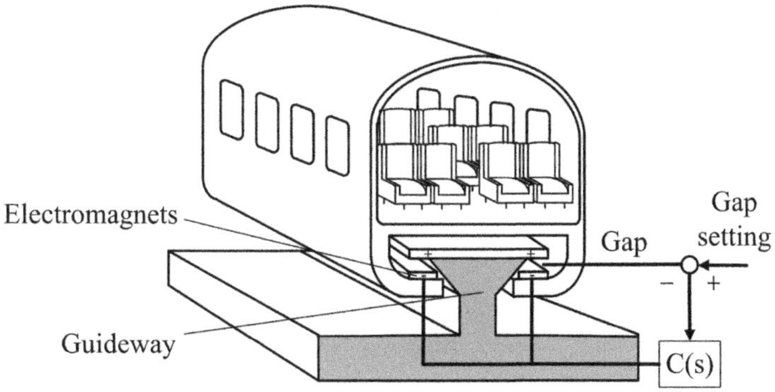

FIGURE 5.4.3
Control of a maglev train.

Take the performance degree as $\lambda = 1$. According to (5.4.5) the H_2 controller is

$$\boxed{C(s) = -\frac{(s+1.5)(s+2.5)}{\lambda s^2 + (4\lambda + 2)s}.}$$

Its performance degree is tuned so that the two controllers have the same rise time. In this case, the performance degree of the H_2 controller is $\lambda = 1.6$. A unit step reference is added at $t = 0$ and a unit step load is added at $t = 30$. The responses of the closed-loop system are shown in Figure 5.4.4. The two controllers both provide fast and steady response, while the H_2 controller has a smaller perturbation peak.

5.5 PID Controller Based on the Maclaurin Series Expansion

The preceding design was carried out in the following way: a rational plant was obtained by employing the rational approximation to expand time delay, and then the controller was designed for the rational plant. In this section, an alternative design will be developed, in which the desired controller is first designed for the plant with time delay, and then a PID controller is derived by approximating the obtained controller. Compared with the preceding design, this method has two important features:

1. It provides better performance.

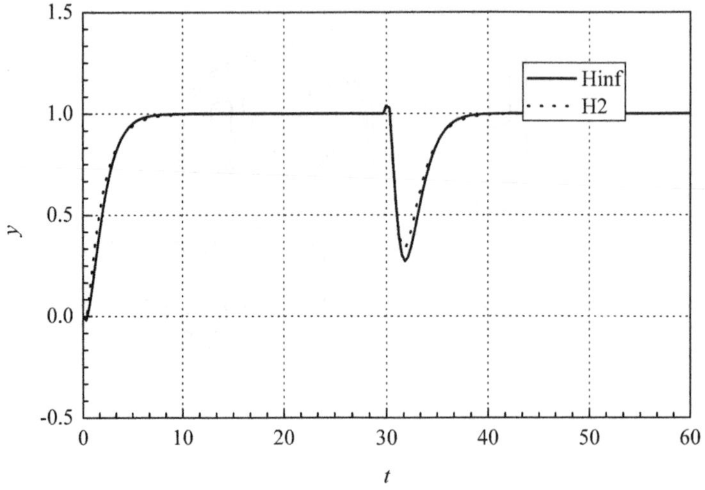

FIGURE 5.4.4
Responses of the gap control system.

2. It can be used for high-order plants directly.

Consider a stable MP plant described by

$$G(s) = \frac{KN_-(s)}{M_-(s)} e^{-\theta s}, \tag{5.5.1}$$

where $N_-(s)$ and $M_-(s)$ are the polynomials with roots in the open LHP, $N_-(0) = M_-(0) = 1$, and $\deg\{N_-(s)\} \leq \deg\{M_-(s)\}$. The desired closed-loop transfer function is chosen as follows:

$$T(s) = \frac{e^{-\theta s}}{(\lambda s + 1)^{n_j}}, \tag{5.5.2}$$

where $n_j = \deg\{M_-(s)\} - \deg\{N_-(s)\}$ for strictly proper plants and $n_j = 1$ for bi-proper plants. In the next chapter, it will be proved that this desired closed-loop transfer function is suboptimal. Since $Q(s) = T(s)/G(s)$ is stable, the closed-loop system is internally stable. The desired closed-loop transfer function results in the following desired controller:

$$C(s) = \frac{1}{G(s)} \frac{T(s)}{1 - T(s)}. \tag{5.5.3}$$

Here, only stable MP plants are considered. The design procedure for NMP plants is similar, but some knowledge in the next chapter is needed in choosing the desired closed-loop transfer function.

Even though the controller is physically realizable, it is not in the form of a PID controller. The problem now reduces to the one of finding a PID controller to approximate the desired controller. A method that works is to use the Maclaurin series expansion. Since

$$\lim_{s\to 0}\left[(\lambda s+1)^{n_j}-e^{-\theta s}\right]=0,$$

$C(s)$ has a pole at the origin. Expanding $C(s)$ in a Maclaurin series gives

$$C(s)=\frac{f(s)}{s} \tag{5.5.4}$$

with

$$f(s)=f(0)+f'(0)s+\frac{f''(0)}{2!}s^2+\dots. \tag{5.5.5}$$

It can be seen that the resulting controller has a proportional term, an integral term, and a derivative term in addition to an infinite number of high-order derivative terms. If all terms are realized, the desired controller can be perfectly achieved. In practice, however, it is impossible to realize the desired controller because of the infinite number of derivative terms. Here, only the first three terms are taken to approximate the desired controller. The three terms form a PID controller:

$$C(s)=K_C\left(1+\frac{1}{T_Is}+T_Ds\right),$$

where

$$K_C=f'(0),\quad T_I=f'(0)/f(0),\quad T_D=\frac{f''(0)}{2f'(0)}. \tag{5.5.6}$$

Certainly, one can also take the first two terms to form a PI controller.

Now let us see how to compute the controller parameters in (5.5.6) for a given plant. For convenience of presentation, let

$$\begin{aligned} N(s)&=\frac{M_-(s)}{KN_-(s)},\\ M(s)&=\frac{(\lambda s+1)^{n_j}-e^{-\theta s}}{s}. \end{aligned} \tag{5.5.7}$$

The values of $f(s)$ and its first-order and second-order derivatives at the origin are

$$\begin{aligned} f(0)&=\frac{N(0)}{M(0)},\\ f'(0)&=\frac{N'(0)M(0)-M'(0)N(0)}{M(0)^2},\\ f''(0)&=\frac{\begin{array}{c}N''(0)M(0)^2-M''(0)N(0)M(0)-\\ 2M'(0)N'(0)M(0)+2M'(0)^2N(0)\end{array}}{M(0)^3}. \end{aligned} \tag{5.5.8}$$

If the plant is of first order; that is,

$$G(s) = \frac{Ke^{-\theta s}}{\tau s + 1}, \tag{5.5.9}$$

we have

$$N(0) = \frac{1}{K},\ N'(0) = \frac{\tau}{K},\ N''(0) = 0,$$
$$M(0) = \lambda + \theta,\ M'(0) = -\frac{\theta^2}{2},\ M''(0) = \frac{\theta^3}{3}.$$

The function $f(s)$ and its first and second derivatives at the origin are given by

$$f(0) = \frac{1}{K(\lambda+\theta)},$$
$$f'(0) = \frac{\theta^2 + 2\lambda\tau + 2\theta\tau}{2K(\lambda+\theta)^2},$$
$$f''(0) = \frac{\theta^2(-2\lambda\theta + \theta^2 + 6\lambda\tau + 6\theta\tau)}{6K(\lambda+\theta)^3}.$$

Consequently, the PID controller parameters are

$$\begin{aligned} T_I &= \tau + \frac{\theta^2}{2(\lambda+\theta)}, \\ K_C &= \frac{T_I}{K(\lambda+\theta)}, \\ T_D &= \frac{\theta^2(3T_I - \theta)}{6T_I(\lambda+\theta)}. \end{aligned} \tag{5.5.10}$$

When a second-order model is used:

$$G(s) = \frac{Ke^{-\theta s}}{(\tau_1 s + 1)(\tau_2 s + 1)}, \tag{5.5.11}$$

utilizing the Maclaurin series expansion, we have

$$N(0) = \frac{1}{K},\ N'(0) = \frac{\tau_1 + \tau_2}{K},\ N''(0) = \frac{2\tau_1\tau_2}{K},$$
$$M(0) = 2\lambda + \theta,\ M'(0) = \lambda^2 - \frac{\theta^2}{2},\ M''(0) = \frac{\theta^3}{3}.$$

The function $f(s)$ and its first and second derivatives are given by

$$f(0) = \frac{1}{K(2\lambda+\theta)},$$
$$f'(0) = \frac{-2\lambda^2+\theta^2+2(2\lambda+\theta)(\tau_1+\tau_2)}{2K(2\lambda+\theta)^2},$$
$$f''(0) = \frac{\begin{array}{c}2\tau_1\tau_2(2\lambda+\theta)^2-\theta^3(2\lambda+\theta)/3-\\(\tau_1+\tau_2)(2\lambda^2-\theta^2)(2\lambda+\theta)+2(\lambda^2-\theta^2/2)^2\end{array}}{K(2\lambda+\theta)^3}.$$

The PID controller parameters are

$$\begin{aligned} T_I &= \tau_1+\tau_2-\frac{2\lambda^2-\theta^2}{2(2\lambda+\theta)}, \\ K_C &= \frac{T_I}{K(2\lambda+\theta)}, \\ T_D &= T_I-\tau_1-\tau_2+\frac{12\tau_1\tau_2\lambda+6\tau_1\tau_2\theta-\theta^3}{T_I(12\lambda+6\theta)}. \end{aligned} \tag{5.5.12}$$

The integral constant and derivative constant computed based on the above formulas might be negative for some plants. When this occurs, the designer can take the first two terms to form a PI controller, or take the first four terms to form a controller.

In the Maclaurin PID controller, the effect of the performance degree on the closed-loop response is similar to that in the H_∞ PID controller and the H_2 PID controller. Since more complicated formulas are used to compute PID controller parameters, it is not surprising that the Maclaurin PID controller can provide better performance than the H_∞ PID controller and the H_2 PID controller.

5.6 PID Controller with the Best Achievable Performance

It is always desirable to enhance the performance of a control system by improving the design method. However, there is a limit to the achievable performance after the controller structure is fixed; it is an important issue to explore the performance limit of a controller. This section discusses the problem for the PID controller. More precisely, the following problems are studied:

1. What a performance limit does the PID controller have?
2. Is it possible to analytically design the PID controller with the best achievable performance?

3. How can this PID controller be tuned for quantitative performance and robustness?

Depending on the way the rational approximation is used, two different design procedures have been developed in the foregoing sections to analytically design PID controllers. In the first method, the PID controller is designed by reducing the order of the plant, while in the second method the PID controller is designed by employing the Maclaurin series expansion to reduce the order of the desired controller. Although the second method approximates the desired controller better than the first one, it still does not reach the performance limit. An improvement on the performance is possible.

Since the Pade approximation can provide higher precision than the Maclaurin series expansion, the PID controller will be designed in this section by applying the Pade approximation for controller reduction. The property of the Pade approximation guarantees that the resulting controller, compared with other PID controllers designed with analytical methods, provides the best performance.

The solving of this problem follows the design procedure of the Maclaurin PID controller. First, choose the desired closed-loop transfer function (5.5.2) for the plant (5.5.1). Next, expand the desired controller (5.5.3):

$$\begin{aligned} C(s) &= \frac{f(s)}{s} \\ &= \frac{1}{s}\left[f(0) + f'(0)s + \frac{f''(0)}{2!}s^2 + \frac{f^{(3)}(0)}{3!}s^3 + \ldots\right]. \end{aligned} \tag{5.6.1}$$

As we know, the ideal PID controller has a pure derivative term in it and thus is not physically realizable. Realizable PID controllers are usually in three forms, which have been listed in Section 4.1. All of the three can be expressed in a unified form:

$$C(s) = \frac{a_2 s^2 + a_1 s + a_0}{s(b_1 s + 1)}, \tag{5.6.2}$$

where a_0, a_1, a_2 and b_1 are positive real numbers. Let the Pade approximation of $f(s)$ be

$$f(s) = \frac{a_2 s^2 + a_1 s + a_0}{b_1 s + 1}. \tag{5.6.3}$$

In light of the discussion in Section 2.2, we have

$$\begin{aligned} &\begin{bmatrix} a_0 \\ a_1 \\ a_2 \end{bmatrix} = \begin{bmatrix} f(0) & 0 \\ f'(0) & f(0) \\ f''(0)/2! & f'(0) \end{bmatrix} \begin{bmatrix} 1 \\ b_1 \end{bmatrix}, \\ &b_1 f''(0)/2! = -f^{(3)}(0)/3!. \end{aligned} \tag{5.6.4}$$

It follows that

$$\begin{aligned} a_0 &= f(0), \\ a_1 &= b_1 f(0) + f'(0), \\ a_2 &= b_1 f'(0) + f''(0)/2!, \\ b_1 &= -\frac{f^{(3)}(0)}{3f''(0)}. \end{aligned} \tag{5.6.5}$$

Assume that the PID controller is in the form of

$$C = K_C\left(1 + \frac{1}{T_I s} + T_D s\right)\frac{1}{T_F s + 1}. \tag{5.6.6}$$

The controller parameters are

$$K_C = a_1, \quad T_I = \frac{a_1}{a_0}, \quad T_D = \frac{a_2}{a_1}, \quad T_F = b_1. \tag{5.6.7}$$

In model reduction, the major disadvantage of the Pade approximation is that it may be unstable even if the original transfer function is stable. This problem can be easily overcome by choosing an appropriate controller form:

1. Suppose that there is a strict requirement on controller form: the controller must be a PID controller whose order is less than three. One could choose the first two or three terms of the Maclaurin series expansion as the PID controller.
2. If the designer does not have a strict requirement on controller form, one could choose a PID controller with a second-order lag or a higher order controller.

Consider the following first-order plant with time delay:

$$G(s) = \frac{Ke^{-\theta s}}{\tau s + 1}. \tag{5.6.8}$$

The function $f(s)$ and its derivatives at the origin are given by

$$\begin{aligned} f(0) &= \frac{1}{K(\lambda + \theta)}, \\ f'(0) &= \frac{\theta^2 + 2\lambda\tau + 2\theta\tau}{2K(\lambda + \theta)^2}, \\ f''(0) &= \frac{\theta^2(-2\lambda\theta + \theta^2 + 6\lambda\tau + 6\theta\tau)}{6K(\lambda + \theta)^3}, \\ f^{(3)}(0) &= \frac{\theta^3(-2\tau\lambda\theta + 2\tau\theta^2 - 4\lambda^2\tau - 2\theta^2\lambda + \theta\lambda^2)}{4K(\lambda + \theta)^4}. \end{aligned} \tag{5.6.9}$$

The parameters in (5.6.7) are

$$\begin{aligned} a_0 &= \frac{1}{K(\lambda+\theta)}, \\ a_1 &= \frac{\theta^3+6\tau\theta^2-\theta^2\lambda+12\tau^2\theta+12\tau^2\lambda}{2K(-2\lambda\theta+\theta^2+6\tau\lambda+6\tau\theta)(\lambda+\theta)}, \\ a_2 &= \frac{\theta(\theta^3+6\tau\theta^2+24\tau^2\theta-6\tau\theta\lambda+24\tau^2\lambda)}{12K(-2\lambda\theta+\theta^2+6\tau\lambda+6\tau\theta)(\lambda+\theta)}, \\ b_1 &= -\frac{\theta(-2\tau\lambda\theta+2\tau\theta^2-4\lambda^2\tau-2\theta^2\lambda+\theta\lambda^2)}{2(-2\lambda\theta+\theta^2+6\tau\lambda+6\tau\theta)(\lambda+\theta)}. \end{aligned} \tag{5.6.10}$$

While the above PID controller provides the best achievable performance as compared with the H_∞ PID controller and the H_2 PID controller, the corresponding formulas are in the most complicated form.

5.7 Choice of the Filter

In the design procedure in the last chapter and this chapter, the controller $Q_{opt}(s)$ is always augmented by a low-pass filter $J(s)$: $Q(s) = Q_{opt}(s)J(s)$. The filter has several functions:

1. The optimal controller $Q_{opt}(s)$ is usually improper. One main function of the filter is to make $Q_{opt}(s)$ proper. Certainly, the controller becomes suboptimal after the filter is introduced.
2. Since $S(s) = 1 - G(s)Q(s)$ and $T(s) = G(s)Q(s)$, the filter parameter can be utilized to tune the nominal performance and robustness, and to quantitatively tradeoff between the two objectives.
3. There is a direct relationship between the filter parameter and the control variable, since $u(s) = Q(s)r(s)$. If the control structure is not permitted to modify, one can confine the magnitude of control variable by adjusting the filter parameter.

How to choose the filter? The filter should at least satisfy the following requirements:

1. The closed-loop system is internally stable.
2. The controller $Q(s) = Q_{opt}(s)J(s)$ is proper.
3. Asymptotic tracking is achieved.

The first condition is easy to meet. When the plant is stable, the closed-loop system is internally stable as long as $Q(s)$ is stable.

The second condition can also be easily satisfied. As it is known, an improper rational transfer function implies that the degree of its numerator is greater than that of its denominator. To make it proper, one can simply introduce a filter whose numerator degree is less than its denominator degree. As the filter is stable, it is of low-pass.

Now consider the third condition. As a basic requirement on the closed-loop performance of a control system, the tracking error should vanish asymptotically. Recall that for asymptotic tracking a Type m system should satisfy

$$\lim_{s\to 0} \frac{1 - G(s)Q_{opt}(s)J(s)}{s^k} = 0, \quad k = 0, 1, ..., m-1. \tag{5.7.1}$$

or

$$\lim_{s\to 0} \frac{d^k}{ds^k}[1 - G(s)Q_{opt}(s)J(s)] = 0, \quad k = 0, 1, ..., m-1. \tag{5.7.2}$$

However, these conditions are still not enough for determining the structure and parameter of a filter. For example, one can choose a filter with either a single parameter or multiple parameters and multiple zeros. The introduction of zeros will complicate the response of the closed-loop system, which can be utilized to satisfy some special design objectives, such as tracking a complex input. Nevertheless, zeros are seldom introduced to a filter unless it is necessary, since this will change the performance in a way difficult to grasp and may limit the performance as well.

The usual structure of a filter consists of one or more first-order lags in series. To simplify design task, it is desirable that the filter should have as few parameters as possible, for example, there is only one parameter in the filter. Typically the single parameter filter is in the form of

$$J(s) = \frac{\beta_{m-1}s^{m-1} + ... + \beta_1 s + \beta_0}{(\lambda s + 1)^{n_j}}, \tag{5.7.3}$$

where λ is the performance degree, n_j should be chosen large enough to make $Q(s) = Q_{opt}(s)J(s)$ proper, for a stable plant m equals the number of poles that the input has at the origin, and β_i $(i = 0, 1, ..., m-1)$ are chosen to satisfy the requirement of asymptotic tracking. If the plant is stable, $m = 1$ for the step input. When

$$\lim_{s\to 0}[1 - G(s)Q_{opt}(s)J(s)] = 0,$$

$\beta_0 = 1$.

The order of the single parameter filter can be freely chosen. However, the higher the order, the more complicated the controller. Due to this reason, the order of the filter should be chosen as low as possible. It should be chosen such that $Q(s)$ is bi-proper for a strictly proper plant, or the degree of its denominator is higher by one than that of its numerator for a bi-proper plant. Thus far, the structure of the filter has been determined.

The single parameter filter has been used in the last chapter and this chapter. It is seen that, the single parameter filter provides a simple and good way to describe the distance between the optimal system and the suboptimal system.

There is a performance degree in the filter, which is adjustable. As discussed, the performance degree directly relates to the closed-loop response. For an improved model, a better performance can be obtained by decreasing the performance degree. When the uncertainty increases, one has to increase the performance degree to obtain better robustness. In this way, a reasonable tradeoff between the two competing objectives can be reached easily.

In the H_2 optimal control of a stable rational MP plant, $G(s)Q_{opt}(s) = 1$. From (5.7.1) the Type 1 filter is as follows:

$$J(s) = \frac{1}{(\lambda s + 1)^{n_j}}. \tag{5.7.4}$$

Such a system can track inputs of step type without offset. If the input is a ramp, then a Type 2 filter must be used:

$$J(s) = \frac{n_j \lambda s + 1}{(\lambda s + 1)^{n_j}}. \tag{5.7.5}$$

An input with higher order than a ramp is seldom used. Typical responses of the two filters are shown in Figure 5.7.1.

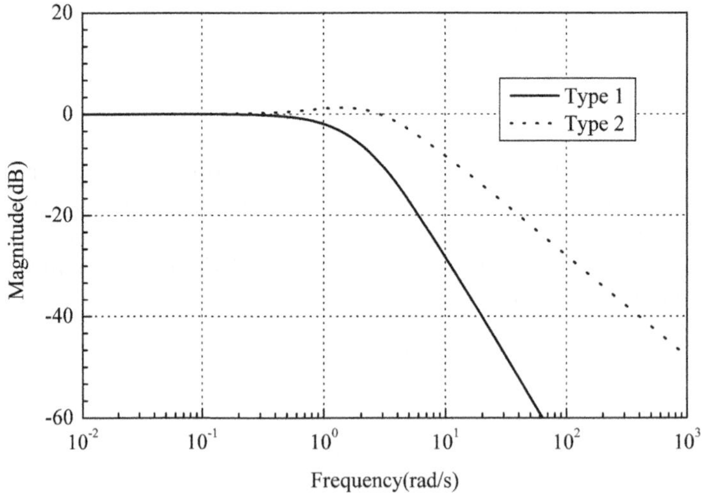

FIGURE 5.7.1
Typical responses of the filters.

Now consider the constraint on the control variable. H_2 controllers try to minimize the ISE index:

$$\min \int_0^{\infty} e^2(t)dt. \tag{5.7.6}$$

There are two internally related problems in such a design. The first is that the optimal controller is improper, which is not physically realizable. The second is that the magnitude peak of the control variable is usually large. To solve the two problems, one has to use a suboptimal controller.

Different methods have been proposed to design a suboptimal H_2 controller, which implies different kinds of degradations in the optimal performance. For example, a suboptimal controller can be obtained by introducing weights to the performance index:

$$\min \int_0^{\infty} [Qe^2(t) + Ru^2(t)]dt, \tag{5.7.7}$$

where Q and R are constant weights. The optimal controller designed by (5.7.7) is suboptimal for the original ISE index (5.7.6). The shortcoming of the design is that it is not known how to determine the weights; moreover, the computation complexity is relatively high.

In some literature, to avoid choosing the weights, Q and R are simply taken as 1:

$$\min \int_0^{\infty} [e^2(t) + u^2(t)]dt. \tag{5.7.8}$$

Evidently, this is not a good choice, since the system performance depends on the weights.

In the design method of this book, the problem is solved by choosing an appropriate filter $J(s)$. This works because $u(s) = Q_{opt}(s)J(s)r(s)$. The comparison of the two design ideas is shown in Figure 5.7.2.

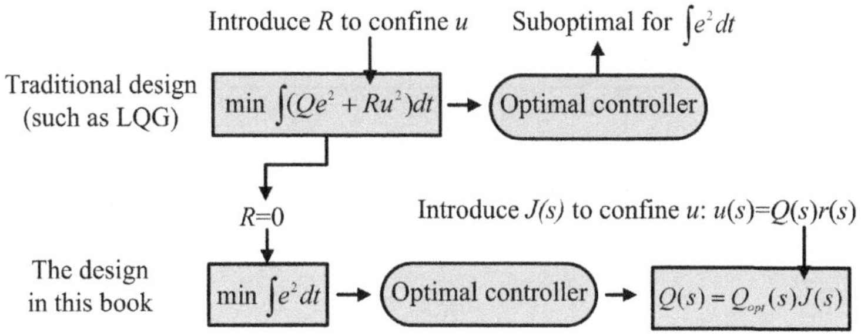

FIGURE 5.7.2
Two different optimizing procedures.

5.8 Summary

Most PID controllers in practical systems are still tuned by empirical methods. Mathematical elegance has not been applied to these methods due to their complexity. Aimed at this problem, several simple analytical methods were proposed for PID controller design based on the H_∞ optimal control theory in the last chapter and the H_2 optimal control theory in this chapter. Assume that the plant is described by

$$G(s) = \frac{Ke^{-\theta s}}{\tau s + 1}.$$

The H_2 PID controller is

$$C(s) = \frac{1}{K}\frac{(\tau s + 1)(1 + \theta s/2)}{\lambda\theta s^2/2 + (\lambda + \theta)s},$$

the Maclaurin PID controller is

$$C(s) = \frac{\dfrac{6\tau\theta^2(\lambda+\theta) + \theta^4 - 2\lambda\theta^3}{12(\lambda+\theta)^2}s^2 + \dfrac{2\tau(\lambda+\theta)+\theta^2}{2(\lambda+\theta)}s + 1}{K(\lambda+\theta)s},$$

and the PID controller with the best achievable performance is

$$K_C = a_1, \quad T_I = \frac{a_1}{a_0}, \quad T_D = \frac{a_2}{a_1}, \quad T_F = b_1,$$

where

$$\begin{aligned}
a_0 &= \frac{1}{K(\lambda+\theta)},\\
a_1 &= \frac{\theta^3 + 6\tau\theta^2 - \theta^2\lambda + 12\tau^2\theta + 12\tau^2\lambda}{2K(-2\lambda\theta + \theta^2 + 6\tau\lambda + 6\tau\theta)(\lambda+\theta)},\\
a_2 &= \frac{\theta(\theta^3 + 6\tau\theta^2 + 24\tau^2\theta - 6\tau\theta\lambda + 24\tau^2\lambda)}{12K(-2\lambda\theta + \theta^2 + 6\tau\lambda + 6\tau\theta)(\lambda+\theta)},\\
b_1 &= -\frac{\theta(-2\tau\lambda\theta + 2\tau\theta^2 - 4\lambda^2\tau - 2\theta^2\lambda + \theta\lambda^2)}{2(-2\lambda\theta + \theta^2 + 6\tau\lambda + 6\tau\theta)(\lambda+\theta)} > 0.
\end{aligned}$$

If the plant is

$$G(s) = \frac{Ke^{-\theta s}}{(\tau_1 s + 1)(\tau_2 s + 1)},$$

the H_2 PID controller can be expressed as

$$C(s) = \frac{1}{K}\frac{(\tau_1 s + 1)(\tau_2 s + 1)}{\lambda\theta s^2 + (\lambda + 2\theta)s},$$

and the Maclaurin PID controller is

$$C(s) = \frac{\left[\tau_1 + \tau_2 - \frac{2\lambda^2 - \theta^2}{2(2\lambda + \theta)}\right]\left[-\frac{2\lambda^2 - \theta^2}{2(2\lambda + \theta)} + \frac{\tau_1\tau_2 - \theta^3/(12\lambda + 6\theta)}{\tau_1 + \tau_2 - (2\lambda^2 - \theta^2)/2/(2\lambda + \theta)}\right] s^2 + \left[\tau_1 + \tau_2 - \frac{2\lambda^2 - \theta^2}{2(2\lambda + \theta)}\right] s + 1}{K(2\lambda + \theta)s}.$$

Despite its wide use, the PID controller is frequently badly tuned in practice. There might be two reasons. The external reason is that the real plant is too complex. The internal one is that the tuning procedure is not easy to fulfill. The design methods in the last chapter and this chapter provide an easy-to-use alternative for designers. Due to the use of analytical design formulas, the design task is significantly simplified. By defining the performance degree, the quantitative performance indices can be reached easily with simple tuning.

Exercises

1. A solar-powered lunar rover is shown in Figure E5.1, which is able to move across an obstacle of 18cm high. There is a TV camera and a manipulator installed on the vehicle. The manipulator can be controlled remotely from the Earth. The average distance between the Earth and the Moon is 384,000 km. The time delay of a signal in single journal transmission is 1.28s. The dynamics of the manipulator is given by

$$G(s) = \frac{1}{(s+1)^2}.$$

Design a unity feedback controller such that the overshoot is less than 5% and the closed-loop response is as fast as possible.

2. The control of the fuel-to-air ratio in an automobile carburetor became of prime importance in the 1980s as automakers strived to reduce exhaust pollutant. Operation of an engine at or near a particular air-to-fuel ratio requires management of both air and fuel flow into the manifold system. Choose the fuel command as the input and the engine speed as the output. The dynamics of the engine can be described by

$$G(s) = \frac{2}{4.35s + 1}.$$

FIGURE E5.1
Lunar rover with a manipulator.

Assume that the change of the air flow rate can be equivalent to a disturbance at the plant output, which is generated by using a unit step to impluse a linear system $1/(s+1)$; that is, the disturbance at the plant output is $1/s/(s+1)$. Design a controller.

3. The simplest inverse response process can also be obtained by combining two second-order plants:

$$G(s) = \frac{-z_r^{-1}s+1}{(\tau s+1)^2}.$$

Derive the H_∞ controller and compare it with (5.4.3).

4. It is assumed that the plant model is

$$G(s) = \frac{-s+z_1}{s+\bar{z}_1}.$$

Prove that for a unit step input

$$\min \|e\|_2^2 = 2\text{Re}(z_1)/|z_1|^2.$$

5. In the design of a Maclaurin PID controller, assume that the PID controller is

$$C(s) = K_C\left(1+\frac{1}{T_I s}+T_D s\right)\frac{1}{T_F s+1}.$$

Write the controller in the form of

$$C(s) = \frac{f(s)}{s} = \frac{f(s)(T_F s+1)}{s(T_F s+1)}.$$

Use the Maclaurin series to expand $f(s)(T_F s + 1)$. Choose the parameter T_F such that the third-order term in the expansion becomes zero. Then a PID controller can be obtained. Give the analytical expression of the PID controller and compare it with the one in Section 5.6.

6. When is the b_1 in (5.6.10) positive?

Notes and References

The results in Section 5.1 and 5.3 are equivalent to those in Rivera et al. (1986) and Morari and Zafiriou (1989). They showed that the IMC method can be used to derive PID controllers for a wide variety of models. The derivation here can be found in Zhang (1996), Zhang and Sun (1996a), and Zhang and Sun (1997).

Section 5.2 is drawn from Zhang (1998).

The plant in Example 5.3.1 is from Wang and Ren (1986, p. 156). Similar plants can also be found in Goodwin et al. (2001).

Section 5.4 follows closely the discussion in Zhang et al. (2000) and Zhang (1998). They studied the control problem of inverse response process in detail.

Waller and Nygardas (1975) discussed the PID control problem for inverse response processes. Iinoya and Altpeter (1962) developed an inverse response compensator. The two methods were introduced in Stephanopoulos (1984). The feature of inverse response process was explored by de la Barra S. (1994) and Howell (1996).

The plants in Example 5.4.1 and Exercise 3 are based on Dorf and Bishop (2001, p. 725). Figure 5.4.3 is drawn based on Dorf and Bishop (2001, Figure P12.3). Introduction to the maglev control problem can also be found in Bittar and Sales (1994).

The Maclaurin PID controller in Section 5.5 was proposed by Lee et al. (1998) (Lee Y., M. Lee and S. Park and C. Brisilow. PID controller tuning for desired closed loop responses for SISO systems, *AIChE J.*, 1998, 44(1), 106–115. ©AIChE).

The basis of Section 5.6 is Zhang et al. (2005). The stability problem of the Pade approximation was discussed by Shamash (1975) for model reduction.

The first half of Section 5.7 is adapted from Morari and Zafiriou (1989). The filter form and the requirements on the filter were well studied in this book.

The background about the control of the fuel-to-air ratio in Exercise 2 can be found in Dorf and Bishop (2001, p. 627) and Powell et al. (1998).

Exercise 4 is adapted from Morari and Zafiriou (1989, Section 4.1.3).

The problem in Exercise 5 is based on Lee et al. (1998).

6

Control of Stable Plants

CONTENTS

Thus far, several optimal methods for PID controller design have been developed. It is seen that the resulting PID controllers provide good performance for the control of plants with large time delay. An interesting question is whether the performance can be further improved. Actually, improved performance is possible through modifying the control structure, for example, using time delay compensation techniques.

The Smith predictor, developed several decades ago, is a well-known technique for time delay compensation. For a long time, the technique failed to offer its performance advantage over the PID control technique due to the lack of effective design and tuning methods. As we know, most established methods were developed for rational plants and the unity feedback loop. They cannot be directly used for plants with time delay and the Smith predictor.

Although the PID controller design method in the preceding chapters can be used to derive a Smith predictor, the result is not exact because of the low-order rational approximation used. In this chapter, a systematic design procedure will be developed based on the H_∞ control theory and the H_2 control theory. With a rigorous treatment on time delay, the H_∞ Smith predictor and H_2 Smith predictor are analytically derived. It is shown that the design of the Smith predictors is the same as that of the controller in the unity feedback loop. The Smith predictors can be exactly implemented in the unity feedback loop, provided that the unity feedback loop controller is allowed to be irrational. When the plant has no time delay, this unity feedback loop controller reduces to a rational one.

It is also shown that there exists a close relationship between the H_∞/H_2 Smith predictors and many other well-known control strategies, such as the

IMC, Dahlin algorithm, deadbeat control, inferential control, and model predictive control.

6.1 The Quasi-H_∞ Smith Predictor

Usually, there is more than one method to solve a problem. In Chapter 4 and Chapter 5, it was seen how a controller was designed by minimizing the weighted sensitivity function. The controller can also be analytically designed by specifying the desired closed-loop response, which will be discussed in this section. Actually, a simplified version of this method has already been used in Sections 5.5 and 5.6.

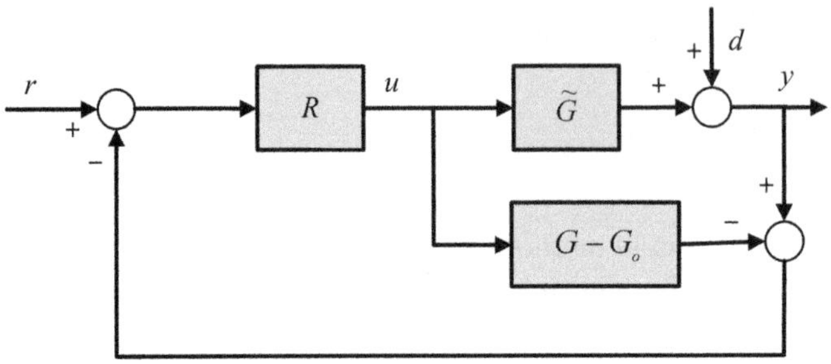

FIGURE 6.1.1
Diagram of the Smith predictor.

As the starting point of design, consider the diagram of the Smith predictor in Figure 6.1.1, where $R(s)$ is the controller, $\tilde{G}(s)$ is the plant, $G(s)$ is the model, and $G_o(s)$ is the delay-free part of $G(s)$. If the closed-loop transfer function $T(s)$ is known, the controller of the Smith predictor is

$$R(s) = \frac{T(s)}{G(s) - T(s)G_o(s)}. \tag{6.1.1}$$

How to choose the desired closed-loop transfer function? This is the key to the design. To introduce the idea clearly, the simple cases of the H_∞ control are analyzed first. Then the general result is inductively derived.

Consider the following stable rational MP plant:

$$G(s) = \frac{KN_-(s)}{M_-(s)}, \tag{6.1.2}$$

where K is the gain, $N_-(s)$ and $M_-(s)$ are the polynomials with roots in the open LHP, $N_-(0) = M_-(0) = 1$, and $\deg\{N_-\} \leq \deg\{M_-\}$. Assume that the performance index is min $\|W(s)S(s)\|_\infty$ and the weighting function is $W(s) = 1/s$. Following the discussion in Section 4.2, we have

$$\begin{aligned}\|W(s)S(s)\|_\infty &= \|W(s)[1 - G(s)Q(s)]\|_\infty \\ &\geq 0.\end{aligned}$$

The following controller is optimal:

$$Q_{opt}(s) = \frac{M_-(s)}{KN_-(s)}. \tag{6.1.3}$$

Introduce the filter

$$J(s) = \frac{1}{(\lambda s + 1)^{n_j}},$$

where λ is the performance degree. According to the discussion in Section 5.7, n_j is chosen as follows:

$$n_j = \begin{cases} \deg\{M_-\} - \deg\{N_-\} & \deg\{M_-\} > \deg\{N_-\} \\ 1 & \deg\{M_-\} = \deg\{N_-\} \end{cases}.$$

The suboptimal proper controller is

$$Q(s) = \frac{M_-(s)}{KN_-(s)(\lambda s + 1)^{n_j}}. \tag{6.1.4}$$

The closed-loop transfer function is

$$T(s) = \frac{1}{(\lambda s + 1)^{n_j}}. \tag{6.1.5}$$

Consider a more complex case. Assume that the plant has a zero in the RHP:

$$G(s) = \frac{KN_-(s)(-z_r^{-1}s + 1)}{M_-(s)}, \tag{6.1.6}$$

where $z_r > 0$, $N_-(0) = M_-(0) = 1$, and $\deg\{N_-\} + 1 \leq \deg\{M_-\}$. Solve the weighted sensitivity problem again:

$$\begin{aligned}\|W(s)S(s)\|_\infty &= \|W(s)[1 - G(s)Q(s)]\|_\infty \\ &\geq |W(z_r)|.\end{aligned}$$

The optimal controller is obtained as follows:

$$Q_{opt}(s) = \frac{M_-(s)}{KN_-(s)}. \tag{6.1.7}$$

Introduce the following filter:

$$J(s) = \frac{1}{(\lambda s + 1)^{n_j}},$$

where

$$n_j = \deg\{M_-\} - \deg\{N_-\}.$$

The suboptimal proper controller is

$$Q(s) = \frac{M_-(s)}{KN_-(s)(\lambda s + 1)^{n_j}}. \tag{6.1.8}$$

The closed-loop transfer function can be written as

$$T(s) = \frac{-z_r^{-1}s + 1}{(\lambda s + 1)^{n_j}}. \tag{6.1.9}$$

Now, consider the general stable rational plant described by

$$G(s) = \frac{KN_+(s)N_-(s)}{M_-(s)}, \tag{6.1.10}$$

where $N_-(s)$ and $M_-(s)$ are the polynomials with roots in the open LHP, $N_+(s)$ is a polynomial with roots in the closed RHP, $N_+(0) = N_-(0) = M_-(0) = 1$, and $\deg\{N_+\} + \deg\{N_-\} \leq \deg\{M_-\}$. Motivated by the results in the foregoing simple cases, the following function is chosen as the desired closed-loop transfer function:

$$T(s) = N_+(s)J(s), \tag{6.1.11}$$

where $J(s)$ is a filter:

$$J(s) = \frac{1}{(\lambda s + 1)^{n_j}}. \tag{6.1.12}$$

In light of the discussion in Section 5.7, n_j is chosen as follows:

$$n_j = \begin{cases} \deg\{M_-\} - \deg\{N_-\} & \deg\{M_-\} > \deg\{N_-\} \\ 1 & \deg\{M_-\} = \deg\{N_-\} \end{cases}.$$

The feature of the closed-loop transfer function is that it has the same RHP zeros as the plant.

Once the desired $T(s)$ is determined, the controller of the Smith predictor can be analytically derived through

$$R(s) = \frac{T(s)}{G(s) - T(s)G_o(s)}$$

$$= \frac{1}{K}\frac{M_-(s)}{N_-(s)[(\lambda s+1)^{n_j} - N_+(s)]}. \tag{6.1.13}$$

For rational plants, $G_o(s) = G(s)$. The unity feedback loop controller $C(s)$ is identical to $R(s)$. The controller has the same order as that of the plant. The corresponding $Q(s)$ is

$$Q(s) = \frac{T(s)}{G(s)} = \frac{M_-(s)}{N_-(s)}. \tag{6.1.14}$$

When there is a time delay in the plant, the basic idea of designing the Smith predictor is moving the time delay out from the feedback loop, so that the controller can be designed for the rational part of plant. Along this line, the above design procedure can be extended to the plant with time delay. Assume that the plant with time delay is

$$G(s) = \frac{KN_+(s)N_-(s)}{M_-(s)}e^{-\theta s}, \tag{6.1.15}$$

where θ is the time delay. The desired closed-loop transfer function is chosen as

$$T(s) = N_+(s)J(s)e^{-\theta s}, \tag{6.1.16}$$

where $J(s)$ is identical to (6.1.12). The $R(s)$ and $Q(s)$ corresponding to this desired closed-loop transfer function is the same as those in (6.1.13) and (6.1.14) respectively, but $C(s)$ contains a time delay:

$$\begin{aligned} C(s) &= \frac{Q(s)}{1 - G(s)Q(s)} \\ &= \frac{1}{K}\frac{M_-(s)}{N_-(s)[(\lambda s+1)^{n_j} - N_+(s)e^{-\theta s}]}. \end{aligned} \tag{6.1.17}$$

This unity feedback loop controller is irrational.

Stability is a basic requirement of control system design. A question associated with the design is whether the closed-loop system is internally stable.

Theorem 6.1.1. *Given the plant (6.1.15) and the closed-loop transfer function (6.1.16), the closed-loop system is internally stable.*

Proof. This conclusion follows directly from the Youla parameterization for stable plants. □

The design method here is in fact an improved pole placement method. Since the method is developed based on special H_∞ solutions, it is named the quasi-H_∞ control. A frequently encountered case is that the plant is MP or has only one zero in the RHP. Then an exact H_∞ controller can be obtained by the method. If the plant has more than one zero in the RHP or the plant contains time delay, the results of the H_∞ control and quasi-H_∞ control are different. The quasi-H_∞ control is a compromise: the solution may not be an exact H_∞ controller, but the design is significantly simplified.

The analytical design formula for the quasi-H_∞ controller has been given. When the nominal plant is known, the quasi-H_∞ controller can be obtained by substituting the plant parameters into the formula directly. With regard to the plant (6.1.15), one can also design the quasi-H_∞ controller through the following steps:

1. If the plant does not contain any time delay, turn to 3.
2. If the plant contains a time delay, take the rational part of the plant as the nominal plant.
3. If the nominal plant has no zeros in the RHP, take its inverse as $Q_{opt}(s)$ and turn to 5.
4. If the nominal plant has zeros in the RHP, remove the factor that contains these zeros and take the inverse of the remainder as $Q_{opt}(s)$.
5. Introduce a filter to $Q_{opt}(s)$, compute the controller by $R(s) = Q(s)/[1 - G_o(s)Q(s)]$ and $C(s) = Q(s)/[1 - G(s)Q(s)]$.

If it is necessary, the desired closed-loop transfer function in quasi-H_∞ control can be chosen as complex as desired. For example, in some applications, the design may impose constraints on $T(s)$. One can use the quasi-H_∞ design method by choosing a more complex $T(s)$ as the desired closed-loop transfer function.

6.2 The H_2 Optimal Controller and the Smith Predictor

The subject of this section is to design the unity feedback loop controller and the Smith predictor that minimize the 2-norm of the weighted sensitivity function. Consider the general plant used in the last section:

$$G(s) = \frac{KN_+(s)N_-(s)}{M_-(s)}e^{-\theta s}. \tag{6.2.1}$$

It is assumed that the performance index is $\min \|W(s)S(s)\|_2$, the input is a unit step, and the weighting function is $W(s) = 1/s$. For asymptotic tracking, the following constraint must be satisfied:

$$\lim_{s\to 0}[1 - G(s)Q(s)] = 0. \tag{6.2.2}$$

The $Q(s)$ that satisfies the condition can be expressed as

$$Q(s) = \frac{1}{K} + sQ_1(s), \tag{6.2.3}$$

where $Q_1(s)$ is stable. Therefore,

$$\begin{aligned}
&\|W(s)S(s)\|_2^2 \\
= &\left\| W(s)\left\{1 - G(s)\left[\frac{1}{K} + sQ_1(s)\right]\right\}\right\|_2^2 \\
= &\left\| \frac{1}{s}\left[1 - \frac{N_+(s)N_-(s)}{M_-(s)}e^{-\theta s} - \frac{KN_+(s)N_-(s)s}{M_-(s)}e^{-\theta s}Q_1(s)\right]\right\|_2^2 \\
= &\left\| \frac{M_-(s) - N_+(s)N_-(s)e^{-\theta s}}{sM_-(s)} - \frac{KN_+(s)N_-(s)}{M_-(s)}e^{-\theta s}Q_1(s)\right\|_2^2 \\
= &\left\| \frac{N_+(s)}{N_+(-s)}e^{-\theta s}\left[\begin{array}{l}\frac{M_-(s)N_+(-s)e^{\theta s} - N_+(s)N_-(s)N_+(-s)}{sM_-(s)N_+(s)} - \\ \frac{KN_+(-s)N_-(s)}{M_-(s)}Q_1(s)\end{array}\right]\right\|_2^2 \\
= &\left\| \begin{array}{l}\frac{M_-(s)N_+(-s)e^{\theta s} - N_+(s)N_-(s)N_+(-s)}{sM_-(s)N_+(s)} - \\ \frac{KN_+(-s)N_-(s)}{M_-(s)}Q_1(s)\end{array}\right\|_2^2 \\
= &\left\| \begin{array}{l}\frac{N_+(-s)e^{\theta s} - N_+(s)}{sN_+(s)} + \\ \frac{M_-(s) - N_-(s)N_+(-s)}{sM_-(s)} - \frac{KN_-(s)N_+(-s)}{M_-(s)}Q_1(s)\end{array}\right\|_2^2 .
\end{aligned}$$

Since $M_-(0) = N_+(0) = N_-(0) = 1$, s must be a factor of

$$N_+(-s)e^{\theta s} - N_+(s)$$

and

$$M_-(s) - N_-(s)N_+(-s).$$

Then we have

$$\begin{aligned}
&\|W(s)S(s)\|_2^2 \\
= &\left\| \frac{N_+(-s)e^{\theta s} - N_+(s)}{sN_+(s)}\right\|_2^2 + \\
&\left\| \frac{M_-(s) - N_-(s)N_+(-s)}{sM_-(s)} - \frac{KN_-(s)N_+(-s)}{M_-(s)}Q_1(s)\right\|_2^2 .
\end{aligned}$$

Minimizing the right-hand side of the equality gives the optimal performance:

$$\min \|W(s)S(s)\|_2^2 = \left\| \frac{N_+(-s)e^{\theta s} - N_+(s)}{sN_+(s)}\right\|_2^2 . \tag{6.2.4}$$

There are three important implications about the result:

1. The optimal performance is obtained with only the input-output information.
2. This performance is the limit of H_2 control for the given index and input, no matter what design method is used.
3. Better performance can only be obtained by modifying the plant itself.

The unique optimal $Q_{1opt}(s)$ is

$$Q_{1opt}(s) = \frac{M_-(s) - N_-(s)N_+(-s)}{KsN_-(s)N_+(-s)}.$$

Substitute this into (6.2.3). The optimal controller is

$$Q_{opt}(s) = \frac{M_-(s)}{KN_-(s)N_+(-s)}. \tag{6.2.5}$$

It is observed that the optimal controller is in the form of the inverse of the plant rational part. If the plant is strictly proper, the controller is improper. It is necessary to introduce the following filter to roll off the optimal controller:

$$J(s) = \frac{1}{(\lambda s + 1)^{n_j}},$$

where λ is the performance degree,

$$n_j = \begin{cases} \deg\{M_-\} - \deg\{N_+\} - \deg\{N_-\} & \deg\{M_-\} > \deg\{N_+\} + \deg\{N_-\} \\ 1 & \deg\{M_-\} = \deg\{N_+\} + \deg\{N_-\} \end{cases}.$$

The suboptimal controller is

$$Q(s) = Q_{opt}(s)J(s) = \frac{M_-(s)}{KN_-(s)N_+(-s)(\lambda s + 1)^{n_j}}. \tag{6.2.6}$$

The Smith predictor is

$$R(s) = \frac{1}{K}\frac{M_-(s)}{N_-(s)[(\lambda s + 1)^{n_j}N_+(-s) - N_+(s)]}. \tag{6.2.7}$$

Notice that the order of this controller is identical to that of the rational part of the plant. The unity feedback loop controller is

$$C(s) = \frac{1}{K}\frac{M_-(s)}{N_-(s)[(\lambda s + 1)^{n_j}N_+(-s)e^{-\theta s} - N_+(s)]}. \tag{6.2.8}$$

It can be verified that (6.2.7) and (6.1.13) are identical when $N_+(s) = 1$.

In addition to designing the H_2 controller by means of the analytical formula, one can also design it through the following steps:

1. If the plant does not contain any time delay, turn to 3.
2. If the plant contains a time delay, take the rational part of the plant as the nominal plant.
3. If the nominal plant has no zeros in the RHP, take its inverse as $Q_{opt}(s)$ and turn to 5.
4. If the nominal plant has zeros in the RHP, construct an all-pass transfer function with the factor that contains these zeros and then remove the all-pass transfer function. Take the inverse of the remainder as $Q_{opt}(s)$.
5. Introduce a filter to $Q_{opt}(s)$, compute $R(s)$ and $C(s)$.

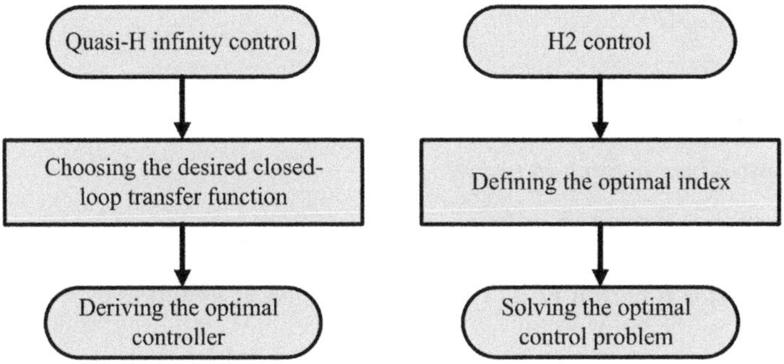

FIGURE 6.2.1
Different philosophies of the quasi-H_∞ control and the H_2 control.

Comparing the discussion in the last section with that in this section, it can be found that the quasi-H_∞ control and the H_2 control have thoroughly different philosophies, which are shown in Figure 6.2.1. The procedures of designing the quasi-H_∞ controller and the H_2 controller are illustrated in the following example.

Example 6.2.1. *Consider the control system of maglev gap described in the last chapter. The dynamic model of the gap is*

$$G(s) = \frac{s-4}{(s+2)^2}.$$

Normalize the plant so that the constant terms of all factors are 1:

$$G(s) = -\frac{-s/4+1}{(s/2+1)^2}.$$

First, the quasi-H_∞ controller is designed. There is no time delay in the *plant, but there is a RHP zero. Remove the factor containing the zero and take the inverse of the remainder as $Q_{opt}(s)$:*

$$Q_{opt}(s) = \frac{(s/2+1)^2}{-1}.$$

A proper $Q(s)$ can be obtained by introducing a filter. When $Q(s)$ is known, it is trivial to compute $R(s)$ and $C(s)$.

Next, design the H_2 controller. First, an all-pass transfer function has to be constructed by utilizing the factor that contains the RHP zero:

$$G(s) = -\frac{s/4+1}{(s/2+1)^2}\frac{-s/4+1}{s/4+1}.$$

Then, remove the all-pass transfer function and take the inverse of the remainder as $Q_{opt}(s)$:

$$Q_{opt}(s) = -\frac{(s/2+1)^2}{s/4+1}.$$

Finally, introduce a filter to $Q_{opt}(s)$.

The construction of an all-pass transfer function is very simple for SISO plants. Assume that an open RHP zero of plant is $z_r = \mathrm{a} + \mathrm{b}i, \mathrm{a} > 0$. The all-pass transfer function $G_A(s)$ can be constructed as follows:

$$G_A(s) = \frac{-s+z_r}{s+\bar{z}_r}. \tag{6.2.9}$$

6.3 Equivalents of the Optimal Controller

Although the derivation of the optimal controller is somewhat complicated, the result is rather simple. It will be shown that there is a simple and reasonable explanation for the controller. Actually, the result was already adopted by some engineers in early practice except that they did not know this was the optimal result.

Rearrange the diagram of the Smith predictor. An equivalent is obtained, which is in fact the IMC structure (Figure 6.3.1). Assume that the rational part of the stable plant $G(s)$, namely $G_o(s)$, is MP. The design objective of this control system is to find a controller $R(s)$ or $C(s)$ such that the closed-loop system is internally stable and the output $y(s)$ can track the reference $r(s)$ as closely as possible. Assume that the model is exact (that is, $\widetilde{G}(s) = G(s)$) and there is no disturbance. Then the feedback signal is zero. A natural idea is to take

$$Q(s) = G(s)^{-1} \tag{6.3.1}$$

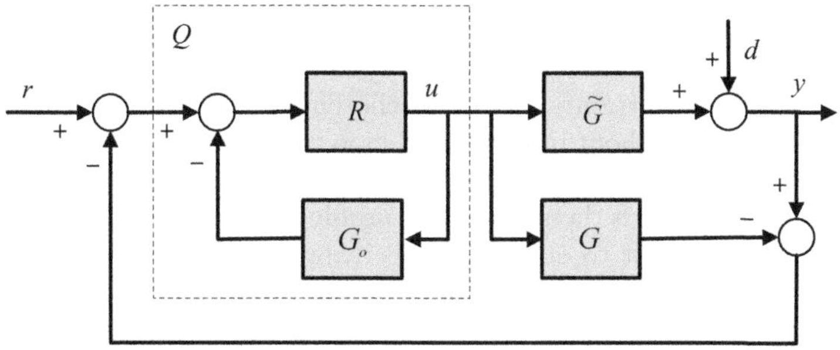

FIGURE 6.3.1
Rearrangement of the Smith predictor.

as the controller. Then the closed-loop transfer function is

$$T(s) = G(s)Q(s) = 1. \tag{6.3.2}$$

This implies that the output can track the reference instantaneously without any error. This situation, referred to as the perfect control in Section 3.2, is impossible in a real system, since the inverse of a time delay is non-causal. A non-causal transfer function is not physically realizable. An alternative is to take

$$Q(s) = G_o(s)^{-1} \tag{6.3.3}$$

as the controller. The closed-loop transfer function becomes

$$T(s) = G(s)Q(s) = e^{-\theta s}, \tag{6.3.4}$$

which implies that the output can track the reference perfectly after the time delay θ. Such a result is reasonable. To explain this, let us consider a shower control system (Figure 2.1.1), an example encountered in everyday life. Assume that the temperature of outlet water is controlled by adjusting the flow rate of inlet hot water. When the valve of hot water is increased by a small percentage (so that the pressure change can be omitted), the increased temperature can only be detected at the outlet after a period of time. No matter what control method is used, it is impossible to eliminate the time delay.

$G_o(s)^{-1}$ is improper. Since $G_o(s)^{-1}$ is rational, it can be arbitrarily approached by a proper transfer function of finite order. The simplest way to approximate $G_o(s)^{-1}$ is to let

$$Q(s) = \frac{G_o(s)^{-1}}{(\lambda s + 1)^{n_j}}, \tag{6.3.5}$$

where λ is the performance degree and n_j is the positive integer that makes

$Q(s)$ bi-proper. Obviously, this is exactly the result obtained in the optimal design.

If there is no uncertainty, λ can be chosen as any positive real number. There is not any overshoot in the closed-loop response. The rise time can be arbitrarily fast as λ tends to be zero. When there is uncertainty, theoretically λ can be calculated with the uncertainty profile and Theorem 3.4.2. However, in practice it is difficult to obtain an uncertainty profile with high precision, and the profile may vary with the use of equipments. In this case, the tuning method introduced in the last two chapters can be used: Increase the performance degree monotonically until the required response is obtained.

Many different design methods have been developed in the past decades. Some of them have been applied to real systems and provide satisfactory performance. On the other hand, the optimal solution is unique in mathematics. Then, one may ask such a question: are these methods really independent?

If the plant has a stable rational part of MP, the quasi-H_∞ controller is identical to the H_2 controller. Furthermore, it can be seen that the two controllers are equivalent to several well-known controllers on certain premises.

Dahlin Algorithm and Deadbeat Control

The Dahlin algorithm is a distinctive algorithm for the control of the plant with time delay. The attractiveness of this technique consists in the fact that it is easy to use and can provide good performance. This algorithm has been included in many textbooks. It was presented for the first-order plant with time delay:

$$G(s) = \frac{Ke^{-\theta s}}{\tau s + 1}.$$

The basic idea is to specify the desired closed-loop transfer function $T(s)$ as a first-order transfer function with its time delay equal to that of the plant $G(s)$; that is,

$$T(s) = \frac{e^{-\theta s}}{\lambda s + 1}, \tag{6.3.6}$$

from which a unity feedback loop controller $C(s)$ can be derived. Since it is difficult to treat a time delay in the Laplace domain, the design procedure is performed in the discrete domain. The time delay is a finite-dimension function in the discrete domain. Evidently, the Dahlin algorithm is a discrete domain version of the quasi-H_∞ controller and H_2 controller for the first-order plant with time delay. This makes it clear that some of the methods developed in early times are effective indeed, although their optimalities were not proved.

When $\lambda \to 0$, the Dahlin controller reduces to the deadbeat controller (also referred to as the minimal prototype controller). Therefore, the deadbeat controller is a special case of the quasi-H_∞ controller or the H_2 controller as well.

Inferential Control and IMC

In some situations, the plant output to be controlled cannot be measured online because of the lack of reliable and economical measuring devices. This is the control problem to which the inferential control is the solution. The control scheme is later extended to the plant whose output can be measured.

Consider the diagram of the inferential control in Figure 6.3.2. A plant is given to the right-hand side of the dotted line, with one unmeasured output $\widetilde{y}(s)$ and one measured auxiliary output $y(s)$. The manipulated variable $u(s)$ and the disturbance $d(s)$ affect both outputs. The disturbance is considered to be unmeasured. $Q(s)$ is the controller and $G(s)$ is the model of a stable MP plant. Since

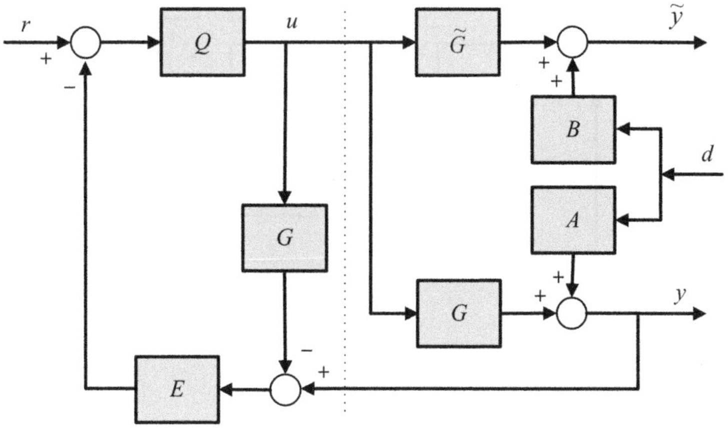

FIGURE 6.3.2
The inferential control system.

$$y(s) = G(s)u(s) + A(s)d(s), \tag{6.3.7}$$

the disturbance can be written as

$$d(s) = \frac{y(s)}{A(s)} - \frac{G(s)}{A(s)}u(s).$$

Define an estimator

$$E(s) := \frac{B(s)}{A(s)}. \tag{6.3.8}$$

The estimated value of the unmeasured output is

$$\begin{aligned} \widetilde{y}(s) &= \widetilde{G}(s)u(s) + B(s)d(s) \\ &= \widetilde{G}(s)u(s) + E(s)[y(s) - G(s)u(s)]. \end{aligned} \tag{6.3.9}$$

The function of $E(s)$ is predicting the effect of the unmeasured disturbance on the plant output.

Now assume that the output $\widetilde{y}(s)$ can be measured. If the model is exact, that is, $\widetilde{G}(s) = G(s)$, $\widetilde{y}(s) = y(s)$, $A(s) = B(s)$, then $E(s) = 1$. The inferential control structure reduces to the IMC structure (Figure 6.3.3). The signal entering the estimator is $d(s)A(s)$. To reject its effect on the plant output, the controller should make a control effort in the opposite direction; that is,

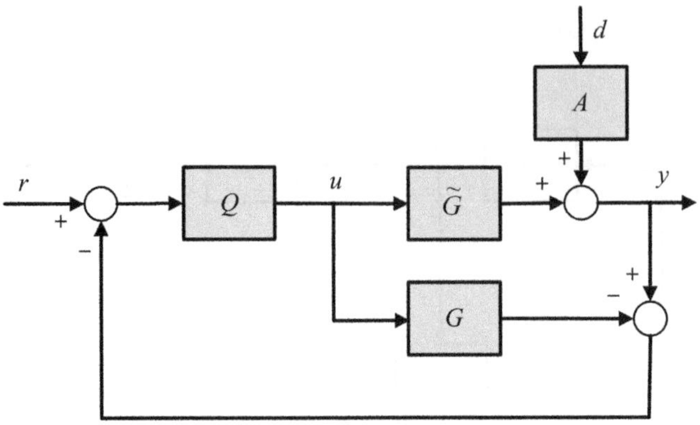

FIGURE 6.3.3
Reduced inferential control system.

$$u(s) = -d(s)A(s)Q(s).$$

It can be seen from (6.3.7) that the cancellation is perfect when $Q(s) = 1/G(s)$.

For the general plant

$$G(s) = \frac{KN_+(s)N_-(s)}{M_-(s)},$$

the controller is

$$Q(s) = \frac{M_-(s)}{KN_-(s)N_+(-s)}.$$

This controller contains the element that is not physically realizable. The problem can be solved by introducing a filter to the controller. Then, the result is identical to the H_2 controller.

Consequently, the quasi-H_∞ control, H_2 control, inferential control with measured output, and IMC are equivalent for the plant whose rational part is stable MP; the H_2 control, inferential control scheme with measured output, and IMC are equivalent for the plant whose rational part is stable.

Model Predictive Control

The model predictive control is a general designation of a variety of control algorithms developed for computer control systems, rather than a single control algorithm. The most widely-used predictive control techniques are those based on the optimization of quadratic objective functions. The basic design methods in this category include the Dynamic Matrix Control (DMC) and the Model Algorithmic Control (MAC). Both of the techniques have been applied in industrial systems.

Let us look at the design idea of model predictive control. Assume that the plant is rational stable and MP, and T_s denotes the sampling time. The values of a unit step response at the sampling instants $t = T_s, 2T_s, ..., NT_s$ are given by $a_1, a_2, ..., a_N$. Different methods were developed to predict the plant output on the basis of this model. The control objective is that the predicted output $y_p(k)$ on the considered horizon L follows the desired output trajectory $y_r(k)$. The desired output trajectory with respect to the reference $r(k)$ is normally given by

$$y_r(k+1) = \alpha y(k) + (1-\alpha)r(k). \tag{6.3.10}$$

Here $\alpha = e^{-T_s/\lambda}$, λ is the time constant of the desired output trajectory, and $y(s)$ is the real output of the plant. The objective function of the control system is as follows:

$$\min \sum_{i=1}^{L} [y_p(k+i) - y_r(k+i)]^2. \tag{6.3.11}$$

P control variables, $u(k)$s, can be calculated by minimizing the objective function. If there exists a time delay in the plant, then the control sequence $u(k)$ used for the plant with time delay is the same as that calculated for the delay-free plant.

Compare the predictive model with the step response model in the form of a transfer function, the desired output trajectory with the desired closed-loop transfer function, and the objective function with the optimal performance index. It is seen that the design idea of model predictive control is very similar to those of the quasi-H_∞ Smith predictor and the H_2 Smith predictor. Certainly, the model predictive control involves many algorithms. Each predictive algorithm possesses its specific form. Not every predictive algorithm is exactly equivalent to the quasi-H_∞ Smith predictor and H_2 Smith predictor.

Assume that the plant is of first order. Then $N = 1$ is enough to express the dynamic characteristic of plant. Consider the one step MAC, that is, $P = L = 1$. Let the model output be $y_m(k)$. The predicted output of the plant, $y_p(k)$, is

$$y_p(k+1) = y_m(k+1) + [y(k) - y_m(k)]. \tag{6.3.12}$$

When the system is optimal, we have $y_p(k+1) = y_r(k+1)$ and $y(k) = y_r(k)$.

Combining (6.3.10) with (6.3.12) yields

$$\alpha y_r(k) + (1-\alpha)r(k) = y_r(k) + y_m(k+1) - y_m(k).$$

Taking the Z-transform, one gets

$$(1-\alpha)r(z) - (1-\alpha)y_r(z) = (z-1)y_m(z).$$

This equation, together with the Z-transforms of the model output and the desired output trajectory

$$y_m(z) = u(z)G(z), y_r(z) = \frac{1-\alpha}{z-\alpha}r(z),$$

shows that

$$\frac{u(z)}{r(z)} = \frac{1-\alpha}{(z-\alpha)G(z)}. \tag{6.3.13}$$

Computing the Laplace domain version of this controller, one can find it is identical to the quasi-H_∞ controller and H_2 controller.

Similarly, it can be proved that the DMC with $P = L$ is identical to the quasi-H_∞ controller and H_2 controller.

6.4 The PID Controller and High-Order Controllers

It was shown in Section 4.3 that an approximate Smith predictor could be derived by utilizing the obtained PID controller. This section discusses how a PID controller can be derived by utilizing the quasi-H_∞ Smith predictor or H_2 Smith predictor, which will show how several different design methods are internally related. For simplicity of presentation, it might as well let the rational part of plant be stable MP and have no zeros. In this case, the quasi-H_∞ control and the H_2 control result in the same controller.

First of all, consider the feature of a Smith predictor in the framework of the unity feedback loop. Rearranging the diagram of the Smith predictor, one can obtain an equivalent unity feedback loop with the following controller:

$$C(s) = \frac{R(s)}{1 + [G_o(s) - G(s)]R(s)}. \tag{6.4.1}$$

Substitute the nominal plant

$$G(s) = \frac{Ke^{-\theta s}}{M_-(s)} \tag{6.4.2}$$

and the Smith predictor

$$R(s) = \frac{M_-(s)}{K(\lambda s + 1)^{n_j} - K} \tag{6.4.3}$$

into (6.4.1). The obtained controller is

$$C(s) = \frac{1}{K} \frac{M_-(s)}{(\lambda s + 1)^{n_j} - e^{-\theta s}}. \tag{6.4.4}$$

When $\lambda \to 0$, the $C(s)$ tends to be optimal. The optimal controller is unique.

Assume that there is a first-order plant with time delay, that is, $M_-(s) = \tau s + 1$ and $n_j = 1$. The controller $C(s)$ is

$$C(s) = \frac{1}{K} \frac{\tau s + 1}{\lambda s + 1 - e^{-\theta s}}. \tag{6.4.5}$$

The open-loop transfer function of the system is

$$\begin{aligned} L(s) &= G(s)C(s) \\ &= \frac{e^{-\theta s}}{\lambda s + 1 - e^{-\theta s}}. \end{aligned} \tag{6.4.6}$$

The Nyquist plot and the Bode plot are given in Figure 6.4.1 and Figure 6.4.2, respectively.

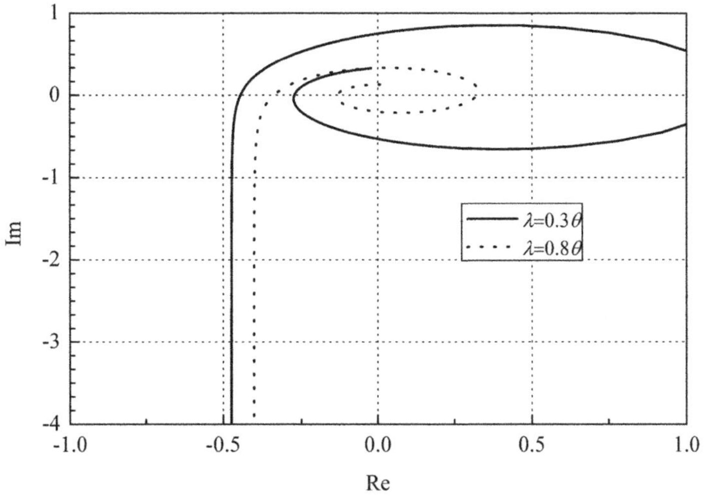

FIGURE 6.4.1
Nyquist plot of the system with the first-order plant.

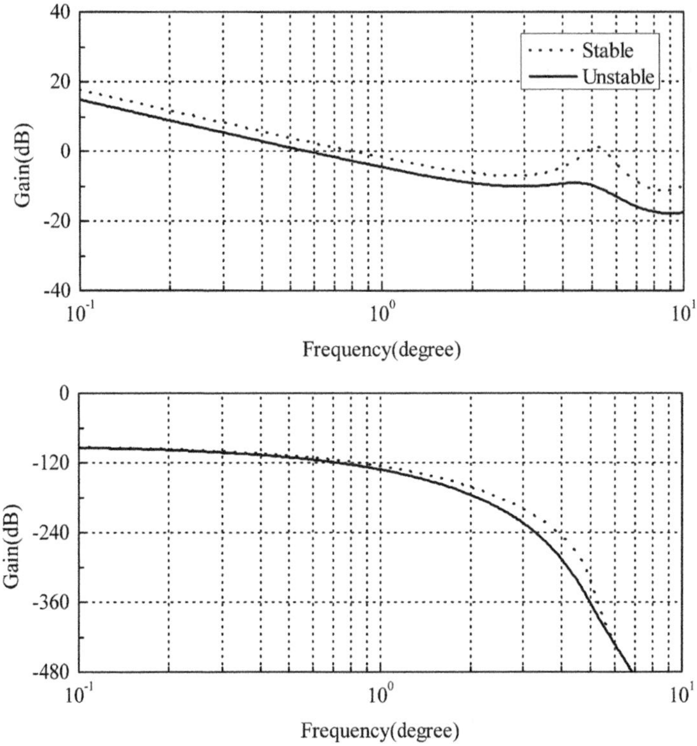

FIGURE 6.4.2
Bode plot of the system with the first-order plant.

With this controller, the closed-loop response can be computed analytically. When the reference is the unit step, the reference response is

$$y(t) = \begin{cases} 0 & 0 < t < \theta \\ 1 - e^{-(t-\theta)/\lambda} & t \geq \theta \end{cases} . \tag{6.4.7}$$

The load response to the unit step disturbance can be written as

$$y(t) = \begin{cases} 0 & 0 < t < \theta \\ K\left[1 - e^{-(t-\theta)/\tau}\right] & \theta \leq t < 2\theta \\ K\left[\begin{array}{l} \dfrac{\lambda}{\lambda-\tau}e^{-(t-2\theta)/\lambda} - \\ \qquad \dfrac{\tau}{\lambda-\tau}e^{-(t-2\theta)/\tau} - e^{-(t-\theta)/\tau} \end{array}\right] & t \geq 2\theta \end{cases} . \tag{6.4.8}$$

Since the feedback acts after $t \geq 2\theta$, the error appearing during $t < 2\theta$ can never be overcome. This error will not be less than $K(1 - e^{-\theta/\tau})$. The larger the θ/τ, the larger the error.

Evidently, the PID controller cannot be used as an exact substitute for this optimal controller. There are two reasons:

1. When the time delay equals zero, if the plant order is greater than 3, $C(s)$ is of high order. The PID controller is not able to realize the dynamics of $C(s)$ exactly.
2. When the time delay is not zero, $C(s)$ involves a time delay and thus is of infinite dimension. Since the PID controller is of finite dimension, it cannot realize the dynamics of $C(s)$ exactly.

One way to derive a PID controller from the high-order controller is to expand the term $e^{-\theta s}$ with the 1/1 Pade approximation. The resulting controller is the H_2 PID controller developed in Section 5.1. If the overall controller is approximated by a low-order rational function, the Maclaurin PID controller or the PID controller with the best achievable performance can be obtained, which has been introduced in Sections 5.5 and 5.6, respectively.

Besides these analytical design methods, one may utilize the pole-placement method or other numerical algorithms to design PID controllers. This is not recommended in this book due to the following reasons:

1. Numerical methods are tedious.
2. The controller has fixed parameters. It cannot be tuned for quantitative responses.
3. Almost no performance improvement can be obtained.

As it is seen, the difference between the PID controller and the optimal Smith predictor lies in the approximation error. The larger the time delay, the larger the error.

The following example is used to compare the responses of different controllers.

Example 6.4.1. *The primary control loop of a nuclear power plant is shown in Figure 6.4.3. The goal is to control the temperature of water by adjusting the reaction speed, which is determined by the depth of the control rods in reactor. Since the water has to be transported from the reactor to the measurement point, there is a time delay in the plant. The transfer function of this plant is obtained by carrying out experiments:*

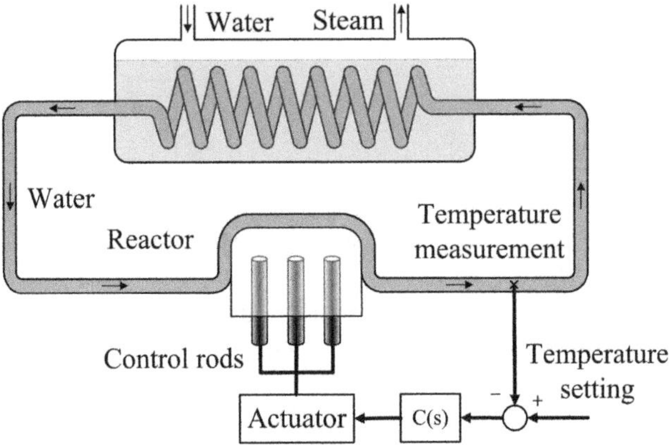

FIGURE 6.4.3
Control system of a nuclear reactor.

$$\boxed{G(s) = \frac{e^{-0.4s}}{0.2s+1}.}$$

In this example, $K = 1, \theta = 0.4, N_+(s) = 1, N_-(s) = 1, M_-(s) = 0.2s+1,$ *and* $n_j = 1$. *The Smith predictor given by (6.1.13) or (6.2.7) is*

$$\boxed{R(s) = \frac{0.2s+1}{\lambda s}.}$$

The H_2 *PID controller given by (5.1.12) is*

$$\boxed{C(s) = \frac{(0.2s+1)^2}{s(0.2\lambda s+\lambda+0.4)}.}$$

The Maclaurin PID controller is given by (5.5.10):

$$\boxed{C(s) = \frac{T_I}{\lambda+0.4}\left[1+\frac{1}{T_I s}+\frac{0.08(3T_I-0.4)}{3T_I(\lambda+0.4)}\right],}$$

where

$$T_I = 0.2 + \frac{0.08}{\lambda + 0.4}.$$

The PID controller with the best achievable performance is given by (5.6.7) and (5.6.10):

$$\boxed{C(s) = a_1\left[1 + \frac{a_0}{a_1 s} + \frac{a_2}{a_1}s\right]\frac{1}{b_1 s + 1},}$$

where

$$\begin{aligned} a_0 &= \frac{1}{2(\lambda + 0.4)}, \\ a_1 &= \frac{0.224 + 0.16\lambda}{0.64 + 0.4\lambda}, \\ a_2 &= 0.0333\frac{0.64 - 0.48\lambda}{0.64 + 0.4\lambda}, \\ b_1 &= -0.2\frac{0.064 - 0.48\lambda - 0.4\lambda^2}{0.64 + 0.4\lambda}. \end{aligned}$$

Take $\lambda = 0.41\theta$ for these controllers. A unit step reference is added at $t = 0$ and a unit step load is added at $t = 100$. The nominal responses of the closed-loop system are shown in Figure 6.4.4.

In practice, the reference has a limited bandwidth. When the bandwidth of reference is restricted to be 5 rad/s, the responses of these methods are closer to each other than those without restriction (Figure 6.4.5).

It is assumed that there are 10% uncertainties on the three parameters of the plant, respectively. The worst case is that the gain and the time delay become the maximum and the time constant becomes the minimum. In this case, the closed-loop responses are given in Figure 6.4.6. The responses of these controllers are still steady when there exist such uncertainties.

6.5 Choice of the Weighting Function

There are two problems that closely relate to the design of an optimal controller: The first is the choice of filter; the second is the choice of weighting function, which is the subject of this section.

The basis of the design methods in this book is the system gain introduced in Section 3.1. Let us first examine what constraints are imposed on the weighting function by the system gain. To use the system gains in Table 3.1.1, the weighting function should be chosen in accordance with the system

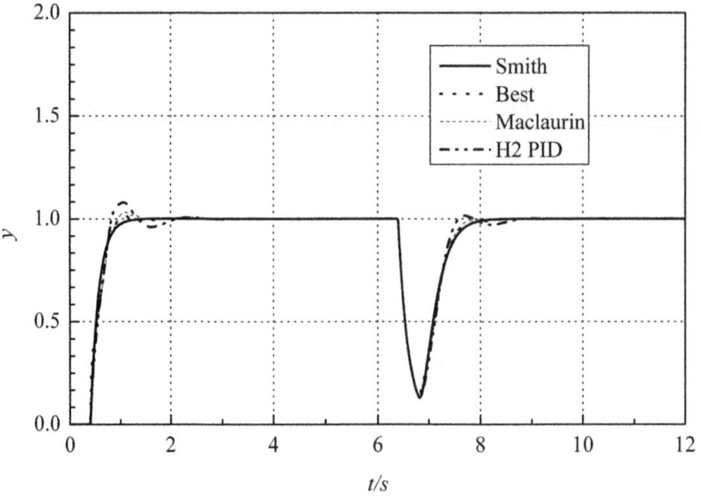

FIGURE 6.4.4
Responses for full frequency range.

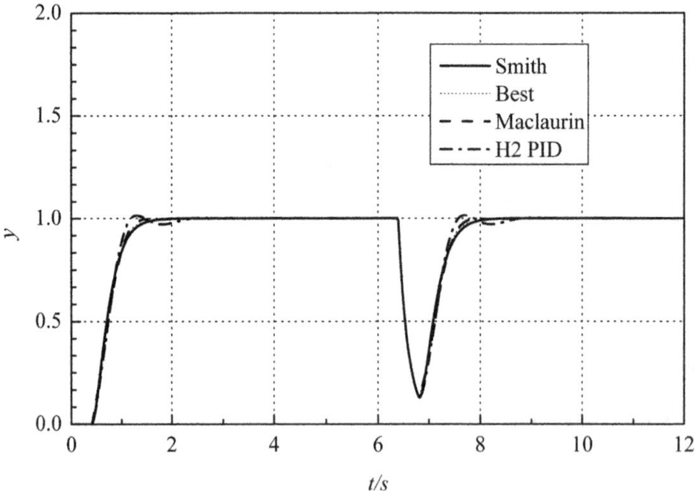

FIGURE 6.4.5
Responses for limited frequency range.

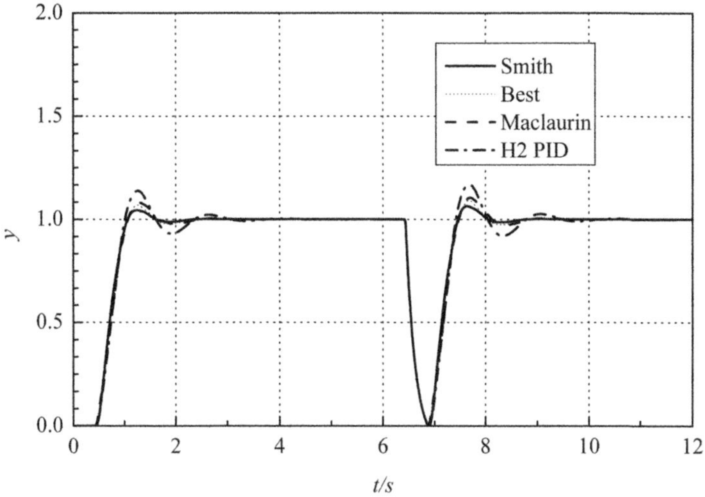

FIGURE 6.4.6
Worst-case responses with 10% uncertainties.

input (reference or disturbance). In the H_2 optimal control, the controller is designed for the impulse. The weighting function should be chosen so that

$$\frac{r(s)}{W(s)} = 1, \tag{6.5.1}$$

or equivalently, the weighting function should equal the system input. The H_∞ optimal control requires that

$$\left\| \frac{r(s)}{W(s)} \right\|_2 \leq 1. \tag{6.5.2}$$

To cover all energy-bounded inputs, a reasonable choice for the weighting function is to make it equal the input.

Now consider the choice problem of weighting function in several frequently encountered cases. In most methods, the controller is designed for ideal step inputs or step-like signals (that is, an ideal step signal with a lag). The transfer function of the input signal can be directly chosen as the weighting function for both the H_2 optimal control and the H_∞ optimal control.

In statistic control, the input is a random signal. Normally, the statistics feature or the spectrum of the input is known. Then an equivalent transfer function with regard to the input can be obtained. The weighting function $W(s)$ can be taken as this transfer function.

In some applications, designers have acquired through experience the desired shape of the Bode magnitude plot of $S(s)$. In particular, a good performance is known to be reached if the plot of $S(s)$ lies under some curve. In this case, the weighting function $W(s)$ can be chosen as the transfer function corresponding to the curve. A known $S(s)$ has two implications:

1. Since $S(s) + T(s) = 1$, the desired shape of $T(s)$ is determined.
2. The bandwidth of the system is restricted.

Nevertheless, the above method may not be a good one for the design problem. The reason is that the system input is complex in some applications. Consequently, the weighting function $W(s)$ is in a complex form. For example, the weighting function might be

$$W(s) = \frac{1}{s(\tau_1 s + 1)(\tau_2 s + 1)}.$$

The design of an optimal controller for such a weighting function is tedious. To avoid this problem, the choosing of the weighting function can be simplified further. One can fix the weighting function as the standard input. For example, when the input is a step-like signal, take $W(s) = 1/s$. This works because the ideal step is a signal abstracted from a large variety of real step-like signals. These real signals can be viewed as the ideal step with high frequency content being rolled off. The controller designed for the ideal step has the potential to work well for step-like signals.

The simplified design procedure is as follows:

1. Design the controller for the ideal step.
2. Choose an appropriate filter according to design requirements.

In the next chapter, it will be seen that such a design method is also applicable to other input signals, such as a ramp.

Since $W(s)$ is fixed, the obtained closed-loop response may not reflect the required feature. This problem can be simply solved by tuning the filter (Figure 6.5.1).

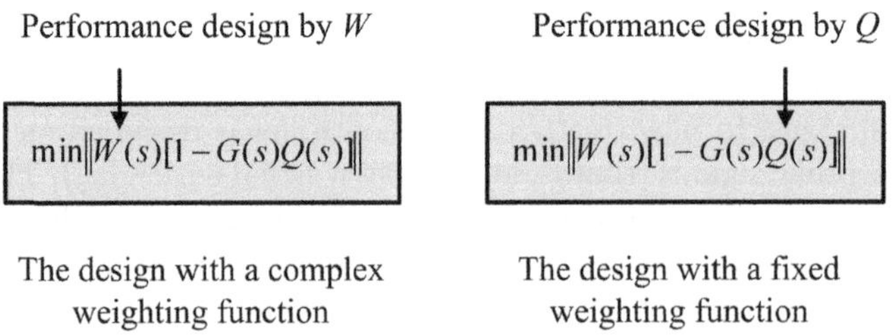

FIGURE 6.5.1
Different design procedures.

The question of interest is whether the role of the weighting function can be thoroughly substituted by a filter. To analyze this problem, suppose that the complex weighting function and the associated controller are $W_{p1}(s)$ and $Q_1(s)$ respectively, and the fixed weighting function and the associated controller are $W_{p2}(s)$ and $Q_2(s)$ respectively. In most cases, both of the weighting functions are MP and do not have poles in the open RHP. For example, the weighting functions may be

$$W_{p1}(s) = \frac{1}{s(\tau_1 s + 1)(\tau_2 s + 1)} \quad \text{and} \quad W_{p2}(s) = \frac{1}{s}.$$

Let

$$W_{p1}(s)[1 - G(s)Q_1(s)] = W_{p2}(s)[1 - G(s)Q_2(s)].$$

Then

$$Q_2(s) = G^{-1}(s)[1 - W_{p2}^{-1}(s)W_{p1}(s)] + W_{p2}^{-1}(s)W_{p1}(s)Q_1(s). \qquad (6.5.3)$$

In view of the feature of $W_{p1}(s)$ and $W_{p2}(s)$, $W_{p2}^{-1}(s)W_{p1}(s)$ is stable. If $G(s)$ is MP, then for a given $W_{p1}(s)$ one can always find a stable $Q_2(s)$, which can reach the same performance as $W_{p1}(s)$.

With regard to an NMP $G(s)$, if $W_{p1}(s)$ is chosen such that $G^{-1}(s)[1 - W_{p2}^{-1}(s)W_{p1}(s)]$ is stable, one can always find an equivalent $Q_2(s)$ for $W_{p1}(s)$. In other cases, the effect of the weighting function cannot be exactly substituted by a filter. They affect the closed-loop response in their own ways. Since $T(s) = G(s)Q_{opt}(s)J(s)$, the filter provides an alternative for the designer to achieve the required closed-loop response.

The design method of using a filter has several advantages:

1. In many advanced design methods, the determination of weighting functions is a difficult problem. The "no-weight" design procedure given in this book does not require the designer to choose weighting functions. The design task is significantly simplified.

2. In some methods, the weighting function is determined by empirical methods, which implies that different designers will obtain different controllers, even when the same design method is used. With the design procedure in this book, different designers will obtain the same optimal controller.

3. The filter is closely related to the closed-loop response. Compared with the method using weighting functions for performance design, the new method provides a simple direct means for adjusting the closed-loop response.

6.6 Simplified Tuning for Quantitative Robustness

Many methods for robust controller design need the prior information about the uncertainty profile. The controller is designed based on both the nominal plant and the uncertainty profile. If the uncertainty profile varies, the controller has to be redesigned. Such methods are inconvenient in practice.

The design methods introduced in this book do not depend on the prior information about the uncertainty profile. When the uncertainty profile is known, quantitative performance and robustness can be obtained by computation or tuning. When the uncertainty profile is unknown, quantitative performance and robustness can roughly be achieved by tuning. If the uncertainty profile varies, it is not necessary to redesign the controller. The robust performance can be obtained by tuning the performance degree. The tuning procedure is simple: Increase the performance degree monotonically until the required response is obtained.

For online tuning, the above procedure is still a bit inconvenient. In many applications, the requirement on dynamic performance is not very strict, while the convenience is very important. Sometimes, the dynamic performance is even sacrificed for the convenience. The goal of this section is to simplify the tuning procedure and to provide an engineering tuning method that can be applied in control software or hardware.

Assume that the plant model is

$$G(s) = \frac{Ke^{-\theta s}}{\tau s + 1}, \tag{6.6.1}$$

and the uncertainties of the three parameters have the same amplitude. The parameter uncertainty profile, e_{pu}, is expressed in the percentage of the nominal value. The unstructured uncertainty of the system can be expressed as

$$\Delta_m(j\omega) = \begin{cases} \left| \dfrac{|K| + e_{pu}K}{|K|} \dfrac{j\tau\omega + 1}{j(\tau - e_{pu}\tau)\omega + 1} e^{je_{pu}\theta\omega} - 1 \right| & \omega < \omega^* \\ \left| \dfrac{|K| + e_{pu}K}{|K|} \dfrac{j\tau\omega + 1}{j(\tau - e_{pu}\tau)\omega + 1} \right| + 1 & \omega \geq \omega^* \end{cases}, \tag{6.6.2}$$

where ω^* is determined by

$$e_{pu}\theta\omega^* + \arctan \frac{e_{pu}\tau\omega^*}{1 + \tau(\tau - e_{pu}\tau)\omega^{*2}} = \pi, \frac{\pi}{2} \leq e_{pu}\theta\omega^* \leq \pi. \tag{6.6.3}$$

It should be pointed out that the above assumption does not require the three parameters to change simultaneously, or that the three parameters must have the same level of uncertainty. The real case could be that the uncertainty of one parameter is larger than that of another one or the other two; nevertheless, the unstructured uncertainty profile is still within the same scope.

Now let us see how to simplify the tuning procedure. Split the overall uncertainty scope e_{pu} into three ranges: small, middle, and large. For engineering applications, one can roughly regard $e_{pu} = 10\%$ as "small," $e_{pu} = 20\%$ as "middle," and $e_{pu} = 35\%$ as "large." Certainly, it can also be split into more ranges for finer tuning. The three ranges are marked on the performance degree knob, which is set up on the panel of a regulator (Figure 6.6.1). An alternative way to realize the performance degree knob is a button or a keyboard.

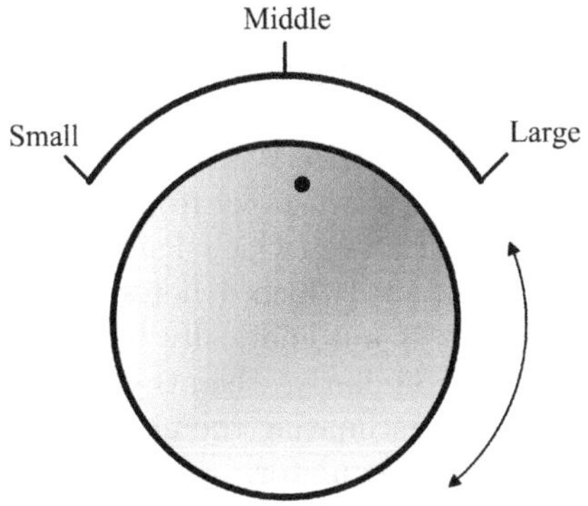

FIGURE 6.6.1
The three-range knob.

When using the regulator, the operator first estimates the uncertainty roughly, and then sets the performance degree knob at an appropriate position. For example, when an industrial process runs, if the uncertainty is small, the operator should set the performance degree knob at the position of "small." When the performance requirement changes or the uncertainty profile varies, the performance and the robustness can be adjusted by resetting the performance degree knob.

Two cases may be encountered in practical operation. One is that the closed-loop system is unstable or the response oscillates intensely. This implies that the actual uncertainty is larger than estimated. The performance degree knob should be reset at a larger value. The other is that the response of the closed-loop system is always slow. This implies that the actual uncertainty is smaller than estimated. The performance degree knob should be reset at a smaller value.

After the parameter uncertainty scope, e_{pu}, is given, the question that arises is how to determine the relationship between e_{pu} and λ: $\lambda = f(e_{pu})$. This is the key to the split-range tuning method. For engineering applications, λ can be chosen according to Table 6.6.1. The table is from the rule of thumb.

As indicated in Section 6.3, many widely applied methods are equivalent

TABLE 6.6.1
Values of λ/θ.

	Small	Middle	Large
H_∞ PID	0.35	0.55	0.80
H_2 PID	0.20	0.55	0.95

to each other. Hence, the split-range tuning method can be applied not only to the design method in this book, but also to many other methods, for example, the Dahlin controller and the model predictive control.

The operation of the controller with the split-range tuning is somewhat similar to that of an automatic camera. When using an automatic camera, the operator has to roughly estimate the distance from the objective to the camera. There are usually three choices: flower (which denotes "near"), portrait (which denotes "middle"), and mountain (which denotes "far"). The focus range is then determined. Other work is automatically finished by the camera after the shutter button is pressed. Similarly, when the split-range method is used, the only work to do is roughly estimating the uncertainty profile. Other work is automatically finished by the regulator.

6.7 Summary

This chapter concentrates on the optimizing technique of designing a controller for general stable plants with time delay. With a rigorous treatment on time delay, the quasi-H_∞ controller and H_2 controller are analytically obtained. They have the same order as that of the rational part of plant.

Assume that the plant is

$$G(s) = \frac{Ke^{-\theta s}}{\tau s + 1}.$$

The controller $R(s)$ is

$$R(s) = \frac{1}{K}\frac{\tau s + 1}{\lambda s}.$$

If the plant is of second order,

$$G(s) = \frac{Ke^{-\theta s}}{(\tau_1 s + 1)(\tau_2 s + 1)},$$

then the controller is

$$R(s) = \frac{1}{K}\frac{(\tau_1 s + 1)(\tau_2 s + 1)}{(\lambda^2 s + 2\lambda)s}.$$

There is no overshoot in the resulting nominal response and the rise time can be arbitrarily fast.

For a long time, the H_∞ optimal control, H_2 optimal control, Smith predictor, Dahlin algorithm, deadbeat control, IMC, inferential control, and model predictive control have been studied as independent control strategies. This chapter studies the relationship among them. It is shown that these methods are equivalent to each other on certain premises. This conclusion provides insight into designing control systems in a unified framework.

The study of this chapter shows that the PID controller can be derived from the Smith predictor with different approximations. The PID controller with better performance is related to a complex design formula.

This chapter also discusses the choice of the weighting function, and an engineering tuning method is presented. The discussion makes the design methods introduced in this chapter more practical.

Exercises

1. Assume that the design specification is as follows: the settling time is 8s and the overshoot is less than 10%. Compute the rational transfer function of the corresponding second-order system.

2. Given a rational plant $G(s)$, the input disturbance is a zero-mean stationary random signal with power spectral density $W(-s)W(s)$. The objective of minimum variance control is to minimize the variance of the plant output $y(t)$. Prove that the variance of $y(t)$ is $\|W(s)G(s)S(s)\|_2^2$.

3. When analog controllers are used, all of the signals in a system are continuous with time. The introduction of digital computers changed the picture, because the digital computer can only handle information on a discrete-time basis. Derive the quasi-H_∞ controller in the $Z-$domain for the first-order plant with time delay.

4. Consider a blending process (Figure E6.1). The stream of hot water and the stream of cold water enter a column, respectively. It is assumed that the two streams are mixed thoroughly in the column and the temperature of the blend is controlled by adjusting the flow rate of hot water. The outlet flow rate is kept constant by automatically adjusting the flow rate of cold water. The flow control loop is arranged to act at a much faster speed than the temperature control loop, so that the interaction between the two loops is negligible. For changes small enough to allow the assumption of linear operation, the transfer function of the temperature plant can be written as

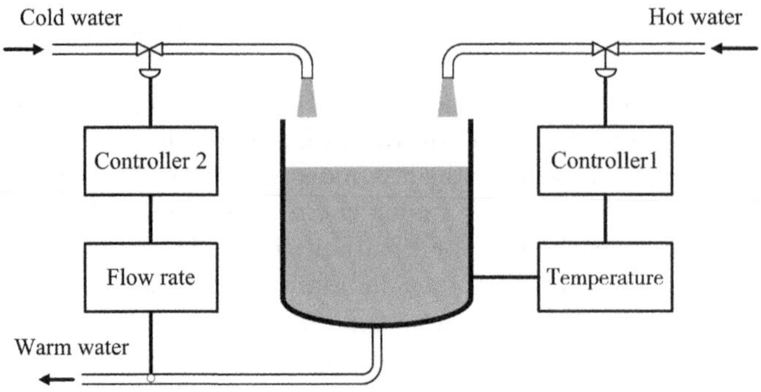

FIGURE E6.1
A blending process.

$$G(s) = \frac{0.4e^{-24s}}{(13.8s+1)(6.1s+1)(3.9s+1)}.$$

Design the quasi-H_∞ Smith predictor and H_2 Smith predictor.

5. Consider the following plant:

$$G(s) = \frac{s}{s+1}.$$

Is it possible to design an H_∞ controller or an H_2 controller?

6*. Assume that the LQ method is used to design the unity feedback loop controller $C(s)$. Give the constructing procedure of $C(s)$ with the output feedback.

Notes and References

Section 6.1 is adapted from Zhang et al. (2002). The idea of designing a controller by specifying the desired closed-loop transfer function was used in many other methods, such as the Dahlin controller (Dahlin, 1968) and the model predictive control (Camacho and Bordons, 1999).

The design in Section 6.2 is in the spirit of the work in Morari and Zafiriou (1989). The derivation can be found in Zhang (1996) and Zhang et al. (2006).

Section 6.3 is adapted from Zhang et al. (1998b) and Zhang and Xu (2003). The relationships among some controllers were also studied by Garcia and Morari (1982) and Jin (1993). The Dahlin algorithm was developed by Dahlin (1968). For the deadbeat control, see, for example, Stephanopoulos (1984).

The inferential control was proposed by Brosilow and Tong (1978). Two important original references about model predictive control are Culter and Ramaker (1979) and Richalet et al. (1978).

The plant in Example 6.4.1 is from Dorf and Bishop (2001, p. 612). Figure 6.4.3 is drawn based on Dorf and Bishop (2001, Figure P10.7).

Regarding the control of systems with time delay, readers can refer to Foias et al. (1996), Brosilow and Joseph (2002), Gu et al. (2003), Sun et al. (2010), Wu et al. (2010), and Atay (2010).

In Lin and Fang (1997), the non-overshoot and monotone non-decreasing responses were studied for the three-order system. It is easy to obtain such a response with the methods introduced in this book.

The plant in Exercise 4 is from Singh and Mcewan (1975).

7

Control of Integrating Plants

CONTENTS

The performance of a control system depends on both the controller and the plant. For some plants it is easy to achieve good performance, while for some others it is not. This chapter considers a class of plants referred to as integrating plants, which is difficult to control. The basic feature of an integrating plant is that its model contains integrators. The integrating plant is a specific type of unstable plant.

For stable plants, many control strategies have been developed, like the PID controller and the Smith predictor. However, there are far fewer methods for the control of integrating plants. An interesting problem is whether the PID controller and the Smith predictor are still effective for integrating plants. Early studies have shown that a PID controller can be used to control integrating plants, but often results in an excessive overshoot. For the Smith predictor, the closed-loop system is internally unstable, because the plant is unstable. This chapter discusses the design problems of H_∞ controllers and H_2 controllers for integrating plants.

7.1 Feature of Integrating Systems

Before discussing how to design control systems for integrating plants, it is helpful to examine the feature of integrating systems. As introduced in Section 2.1, integrating plants in this book do not have any open RHP poles. Those with poles in the open RHP are included in unstable plants. This assumption is made solely for simplicity of presentation.

The minimal requirement on a control system is internal stability. Consider the feedback control loop in Figure 7.1.1, where $G(s)$ is an integrating plant and $C(s)$ is the controller. The closed-loop system is internally stable if and only if all elements in the transfer matrix $\boldsymbol{H}(s)$are stable:

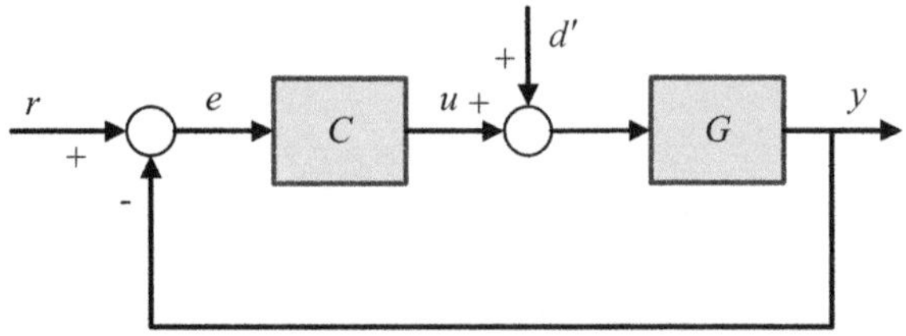

FIGURE 7.1.1
Control system for integrating plants.

$$\begin{bmatrix} y(s) \\ u(s) \end{bmatrix} = \boldsymbol{H}(s) \begin{bmatrix} r(s) \\ d'(s) \end{bmatrix}, \tag{7.1.1}$$

where

$$\boldsymbol{H}(s) = \begin{bmatrix} \dfrac{G(s)C(s)}{1+G(s)C(s)} & \dfrac{G(s)}{1+G(s)C(s)} \\ \dfrac{C(s)}{1+G(s)C(s)} & \dfrac{-G(s)C(s)}{1+G(s)C(s)} \end{bmatrix}.$$

Since the Youla parameterization for stable plants cannot be used for integrating plants, the following transfer function is defined:

$$Q(s) = \frac{C(s)}{1+G(s)C(s)}. \tag{7.1.2}$$

The transfer function $Q(s)$ is in fact the IMC controller. Then the transfer matrix $\boldsymbol{H}(s)$ becomes

$$\boldsymbol{H}(s) = \begin{bmatrix} G(s)Q(s) & [1-G(s)Q(s)]G(s) \\ Q(s) & -G(s)Q(s) \end{bmatrix}. \tag{7.1.3}$$

Evidently, the conclusion about the internal stability of such a system is different from that of the system with a stable plant. Since $G(s)$ is not stable, the stability of $Q(s)$ cannot guarantee the stability of the closed-loop system.

Theorem 7.1.1. *Assume that $G(s)$ is an integrating plant. The unity feedback loop shown in Figure 7.1.1 is internally stable if and only if*

1. $Q(s)$ is stable.

2. $[1 - G(s)Q(s)]G(s)$ is stable.

Proof. Necessity is obvious. Consider sufficiency. Assume that the two conditions hold. It remains to show that $G(s)Q(s)$ is stable. If $G(s)Q(s)$ is unstable, $1 - G(s)Q(s)$ is unstable, which implies that $[1 - G(s)Q(s)]G(s)$ must be unstable. This contradicts the assumption. □

The conclusion of Theorem 7.1.1 is drawn for the unity feedback loop. It may not be applicable to other structures. Consider the IMC structure shown in Figure 7.1.2. When the model is exact, the system is open-loop for $G(s)$ and $Q(s)$. Since $G(s)$ is unstable and $G(s)Q(s)$ is stable, there must exist closed RHP zero-pole cancellations between $G(s)$ and $Q(s)$. In this case, even the condition in Theorem 7.1.1 is satisfied, the closed-loop system is not internally stable. Consequently, the IMC structure can not be used for the control of integrating plants.

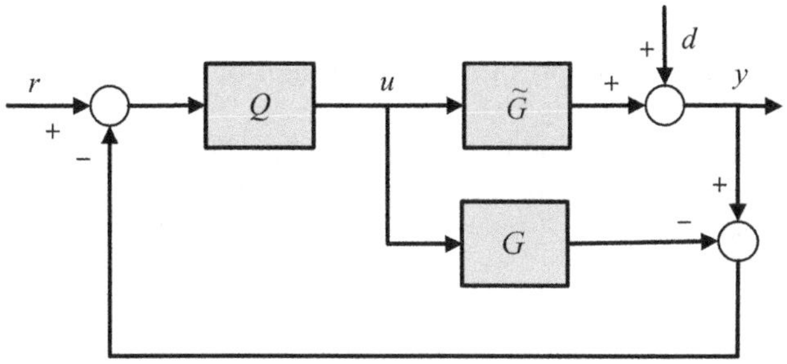

FIGURE 7.1.2
IMC control system for integrating plants.

A further requirement on an internally stable system is the steady-state performance. Besides the reference, there are various disturbances in a system. Two kinds of disturbances are frequently considered. The first is the disturbance at the plant output, which has a similar effect to the reference. The consideration on this disturbance has been involved in that on the reference. The second is the disturbance at the plant input. Special attention must be paid to this kind of disturbance in the system with an integrating plant. In what follows, the effect of such a disturbance is illustrated by utilizing a simple integrating plant.

Consider the first-order integrating plant, which consists of an integrator and a time delay:

$$G(s) = \frac{K}{s}e^{-\theta s}, \tag{7.1.4}$$

where K is the gain, θ is the time delay. Assume that the disturbance at the plant input is $d'(s) = 1/s$. The effect of $d'(s)$ on the plant input can be equivalent to that of a disturbance $d(s)$ on the plant output:

$$d(s) = d'(s)G(s) = \frac{K}{s^2}e^{-\theta s}. \tag{7.1.5}$$

This implies that the system is in fact of Type 2. Only when the controller is designed for ramps, can the steady-state error caused by the disturbance $d'(s)$ vanish asymptotically.

In general, if the plant has m poles at the origin, the system should be of Type $m+1$ for asymptotic tracking; equivalently, the controller has to satisfy

$$\lim_{s\to 0} \frac{1 - G(s)Q(s)}{s^k} = 0, k = 0, 1, ..., m, \tag{7.1.6}$$

or

$$\lim_{s\to 0} \frac{d^k}{ds^k}[1 - G(s)Q(s)] = 0, k = 0, 1, ..., m. \tag{7.1.7}$$

This conclusion is very important in the design of the system with an integrating plant.

Derivatives of function are frequently calculated in the design of the system with an integrating plant. To avoid complicated computation, two algebra results are given here.

Theorem 7.1.2. *The kth ($k = 0, 1, ..., q$) coefficient of a q-order polynomial $N(s)$ is $d^kN(0)/ds^k/k!$.*

Proof. Follows directly from its Taylor series expansion. □

Theorem 7.1.3. *Given the transfer function $N(s)/M(s)$, $N(s)$ and $M(s)$ are polynomials, and $q = \deg\{N(s)\} \le p = \deg\{M(s)\}$. Let $m \le q$ be any nonnegative integer. Then*

$$\lim_{s\to 0} \frac{d^k}{ds^k}\left[1 - \frac{N(s)}{M(s)}\right] = 0, k = 0, 1, ..., m$$

holds if and only if the coefficients of the first $k(k = 0, 1, ..., m)$ terms of $N(s)$ are the same as those of $M(s)$, respectively.

Proof. Sufficiency is obvious. To prove necessity, assume that

$$N(s) = \beta_q s^q + ... + \beta_k s^k + ... + \beta_1 s + \beta_0,$$
$$M(s) = \alpha_p s^p + ... + \alpha_k s^k + ... + \alpha_1 s + \alpha_0,$$

where $\beta_i (i = 0, 1, ..., q)$ and $\alpha_i (i = 0, 1, ..., p)$ are positive real numbers. Let

$$F(s) = 1 - \frac{N(s)}{M(s)}.$$

Then

$$M(s)F(s) = M(s) - N(s).$$

The inductive method is used to prove the result: First, the $k = 0$ and $k = 1$ cases are shown to be true; then the k case is shown to be true if the $(k-1)$ case is true.

When $k = 0$,

$$\lim_{s\to 0} F(s) = \frac{\alpha_0 - \beta_0}{\alpha_0}.$$

Let the right-hand side be 0. We have

$$\alpha_0 = \beta_0.$$

When $k = 1$,

$$\begin{aligned}\frac{d}{ds}[M(s)F(s)] &= F(s)\frac{d}{ds}M(s) + M(s)\frac{d}{ds}F(s),\\ \frac{d}{ds}[M(s) - N(s)] &= \frac{d}{ds}M(s) - \frac{d}{ds}N(s).\end{aligned}$$

Since

$$\lim_{s\to 0} F(s) = 0,$$

the derivative of $F(s)$ is

$$\begin{aligned}\lim_{s\to 0}\frac{d}{ds}F(s) &= \frac{\frac{d}{ds}M(s) - \frac{d}{ds}N(s)}{M(s)}\\ &= \frac{\alpha_1 - \beta_1}{\alpha_0}.\end{aligned}$$

Let $\lim_{s\to 0}\frac{d}{ds}F(s) = 0$. This yields

$$\alpha_1 = \beta_1.$$

Now assume that the conclusion holds for the first $(k-1)$ times differentiating. To prove the theorem, it suffices to prove that the conclusion holds for the kth differentiating. Consider the following fact:

$$\begin{aligned}\frac{d^k}{ds^k}[M(s)F(s)] &= \frac{d^k}{ds^k}M(s)F(s) + C_k^1\frac{d^{k-1}}{ds^{k-1}}M(s)\frac{d}{ds}F(s) + \ldots\\ &\quad + C_k^{k-1}\frac{d}{ds}M(s)\frac{d^{k-1}}{ds^{k-1}}F(s) + M(s)\frac{d^k}{ds^k}F(s)\\ \frac{d^k}{ds^k}[M(s) - N(s)] &= \frac{d^k}{ds^k}M(s) - \frac{d^k}{ds^k}N(s),\end{aligned}$$

where

$$C_k^i = \frac{k!}{i!(k-i)!}.$$

With the assumption, we have

$$\begin{aligned}\lim_{s\to 0}\frac{d^k}{ds^k}F(s) &= \frac{\frac{d^k}{ds^k}M(s) - \frac{d^k}{ds^k}N(s)}{M(s)} \\ &= \frac{\alpha_k - \beta_k}{\alpha_0}.\end{aligned}$$

The left-hand side is 0. Therefore,

$$\alpha_k = \beta_k.$$

This completes the proof. □

Corollary 7.1.4. *Given the transfer function $N(s)/M(s)$.*

$$\lim_{s\to 0}\left[1 - \frac{N(s)}{M(s)}\right] = 0 \text{ and } \lim_{s\to 0}\frac{d}{ds}\left[1 - \frac{N(s)}{M(s)}\right] = 0$$

hold if and only if the coefficients of the first 2 terms of $N(s)$ are the same as those of $M(s)$, respectively.

If the quasi-polynomial containing a time delay is encountered when these results are used, the time delay should be substituted by its Taylor series expansion.

Example 7.1.1. *There are two polynomials: $N(s) = (1 - \theta s/2)(\beta_1 s + 1)$ and $M(s) = (1 + \theta s/2)(\lambda s + 1)^2$. Compute the constant β_1 that makes*

$$\lim_{s\to 0}\left[1 - \frac{N(s)}{M(s)}\right] = 0 \text{ and } \lim_{s\to 0}\frac{d}{ds}\left[1 - \frac{N(s)}{M(s)}\right] = 0$$

hold.

According to Corollary 7.1.4, the zeroth-order and the first-order coefficients of $N(s)$ and $M(s)$ should equal, respectively. Both the zeroth-order coefficients of $N(s)$ and $M(s)$ are 1. The first-order coefficient of $N(s)$ is $\beta_1 - \theta/2$ and the first-order coefficient of $M(s)$ is $2\lambda + \theta/2$. This yields

$$\beta_1 = 2\lambda + \theta.$$

Example 7.1.2. *It is known that $N(s) = (\beta_1 s + 1)e^{-\theta s}$ and $M(s) = (\lambda s + 1)^{n_j}$. Compute the constant β_1 that makes*

$$\lim_{s\to 0}\left[1 - \frac{N(s)}{M(s)}\right] = 0 \text{ and } \lim_{s\to 0}\frac{d}{ds}\left[1 - \frac{N(s)}{M(s)}\right] = 0$$

hold.

Again, the zeroth-order and the first-order coefficients of $N(s)$ and $M(s)$ should equal, respectively. Both the zeroth-order coefficients of $N(s)$ and $M(s)$ are 1. The first-order coefficient of $N(s)$ is $\beta_1 - \theta$ and the first-order coefficient of $M(s)$ is $n_j\lambda$. Let them equal. One readily obtains

$$\beta_1 = n_j\lambda + \theta.$$

7.2 H$_\infty$ PID Controller for Integrating Plants

In principle, the design idea of the PID controllers for stable plants is also applicable to integrating plants. Assume that the coprime factorization of $G(s)$ is $G(s) = V(s)/U(s)$, where $U(s)$ and $V(s)$ are stable proper real rational. According to the discussion in Section 3.3, all stabilizing controllers for integrating plants can be expressed as

$$C(s) = \frac{X(s) + U(s)Q(s)}{Y(s) - V(s)Q(s)},$$

where $Q(s)$ is stable, and $X(s)$ and $Y(s)$ are stable proper real rational functions that satisfy the following equation:

$$V(s)X(s) + U(s)Y(s) = 1.$$

The design procedure is as follows. First, expand the time delay with the Pade approximation. Then, take the performance index as min $\|W(s)S(s)\|_\infty$. Finally, design the controller by

1. Calculating the coprime factorization of the plant: $G(s) = V(s)/U(s)$ (then $S(s) = U(s)[Y(s) - V(s)Q(s)]$).
2. Deriving $Q_{opt}(s)$ by minimizing $\|W(s)S(s)\|_\infty$.
3. Introducing a filter to roll $Q_{opt}(s)$ off at high frequencies.
4. Computing the controller $C(s)$ with $Q(s)$.

Such a design procedure will not be adopted in this section, since it is tedious to obtain a coprime factorization. There are only numerical algorithms available. A simple design procedure is developed here.

First, following the Youla parameterization of stable plants, a transfer function $Q(s)$ is defined:

$$Q(s) = \frac{C(s)}{1 + G(s)C(s)}. \tag{7.2.1}$$

This was already done in the last section. To guarantee the internal stability, $Q(s)$ has to satisfy that

1. $Q(s)$ is stable.
2. $G(s)[1 - G(s)Q(s)]$ is stable.

Next, design the optimal controller $Q_{opt}(s)$ for step inputs. Finally, introduce an appropriate filter for the internal stability and asymptotic tracking requirements, and compute $C(s)$.

Consider the first-order integrating process described by

$$G(s) = \frac{K}{s}e^{-\theta s}. \tag{7.2.2}$$

An approximate plant is obtained by employing the 1/1 Pade approximant:

$$G(s) \approx \frac{K(1 - \theta s/2)}{s(1 + \theta s/2)}. \tag{7.2.3}$$

The internal stability requires that

$$\lim_{s \to 0}[1 - G(s)Q(s)] = 0. \tag{7.2.4}$$

To satisfy the condition, s must be a factor of $Q(s)$, and the constant term of the remainder must be $1/K$; that is,

$$Q(s) = \frac{s[1 + sQ_1(s)]}{K}, \tag{7.2.5}$$

where $Q_1(s)$ is a stable transfer function.

Assume that the input is a unit step. Take $W(s) = 1/s$. The approximate plant has a RHP zero at $2/\theta$. By Theorem 4.2.1 we have

$$\begin{aligned} \|W(s)S(s)\|_\infty &= \|W(s)[1 - G(s)Q(s)]\|_\infty \\ &\geq \left|W\left(\frac{2}{\theta}\right)\right|. \end{aligned}$$

Minimizing the left-hand side of the equality yields

$$\min \left\| W(s)\left\{1 - G(s)\frac{s[1 + sQ_1(s)]}{K}\right\}\right\|_\infty = \frac{\theta}{2}. \tag{7.2.6}$$

Then the optimal controller is

$$Q_{1opt}(s) = \frac{\theta}{2}.$$

Substituting this into (7.2.5) yields

$$Q_{opt}(s) = \frac{s}{K}\left(1 + \frac{\theta}{2}s\right). \tag{7.2.7}$$

Similar to the design for stable plants, a filter $J(s)$ is introduced to $Q_{opt}(s)$

so that a proper $Q(s)$ can be obtained: $Q(s) = Q_{opt}(s)J(s)$. For asymptotic tracking, the system with the first-order integrating plant has to be of Type 2, which imposes the following constraint on $Q(s)$:

$$\lim_{s\to 0} \frac{d^k}{ds^k}[1 - G(s)Q(s)] = 0, \quad k = 0, 1, \tag{7.2.8}$$

or equivalently,

$$\lim_{s\to 0}[1 - G(s)Q(s)] = 0, \tag{7.2.9}$$

$$\lim_{s\to 0} \frac{d}{ds}[1 - G(s)Q(s)] = 0. \tag{7.2.10}$$

To achieve this, the filter has a more complex form than that in the system with a stable plant. It must have a zero:

$$J(s) = \frac{\beta s + 1}{(\lambda s + 1)^{n_j}},$$

where λ is the performance degree, n_j should be large enough to make $Q(s)$ bi-proper, and β is a positive real number and is chosen to satisfy (7.2.8).

It is easy to verify that a first-order or second-order filter can not satisfy the asymptotic tracking requirement. It might as well choose a third-order filter. Utilizing Corollary 7.1.4, the filter satisfying the requirement can be obtained as follows:

$$J(s) = \frac{(3\lambda + \theta/2)s + 1}{(\lambda s + 1)^3}. \tag{7.2.11}$$

Accordingly, the suboptimal controller is

$$Q(s) = \frac{s(1 + \theta s/2)\left[(3\lambda + \theta/2)s + 1\right]}{K(\lambda s + 1)^3}. \tag{7.2.12}$$

A little algebra gives

$$\begin{aligned} C(s) &= \frac{Q(s)}{1 - G(s)Q(s)} \\ &= \frac{1}{K}\frac{s(1 + \theta s/2)\left[(3\lambda + \theta/2)s + 1\right]}{(\lambda s + 1)^3 - (1 - \theta s/2)\left[(3\lambda + \theta/2)s + 1\right]} \\ &= \frac{1}{K}\frac{\left(3\lambda\theta/2 + \theta^2/4\right)s^2 + (3\lambda + \theta)s + 1}{\lambda^3 s^2 + \left(3\lambda^2 + 3\lambda\theta/2 + \theta^2/4\right)s}. \end{aligned} \tag{7.2.13}$$

This is a PID controller. Compare it with the following PID controller:

$$C(s) = K_C\left(1 + \frac{1}{T_I s} + T_D s\right)\frac{1}{T_F s + 1}.$$

The controller parameters are

$$\begin{aligned} T_F &= \frac{\lambda^3}{3\lambda^2 + 3\lambda\theta/2 + \theta^2/4}, \\ T_I &= 3\lambda + \theta, \\ T_D &= \frac{3\lambda\theta/2 + \theta^2/4}{T_I}, \\ K_C &= \frac{1}{K}\frac{T_I}{3\lambda^2 + 3\lambda\theta/2 + \theta^2/4}. \end{aligned} \tag{7.2.14}$$

One main feature of the PID controller designed for stable plants is that the controller can be tuned for quantitative performance and robustness. This feature remains in the PID controller for integrating plants. Relationships between the performance degree and the overshoot, rise time, perturbation peak, and resonance peak are shown in Figure 7.2.1–Figure 7.2.4, respectively. The stability margin is not given here, but it can also be computed.

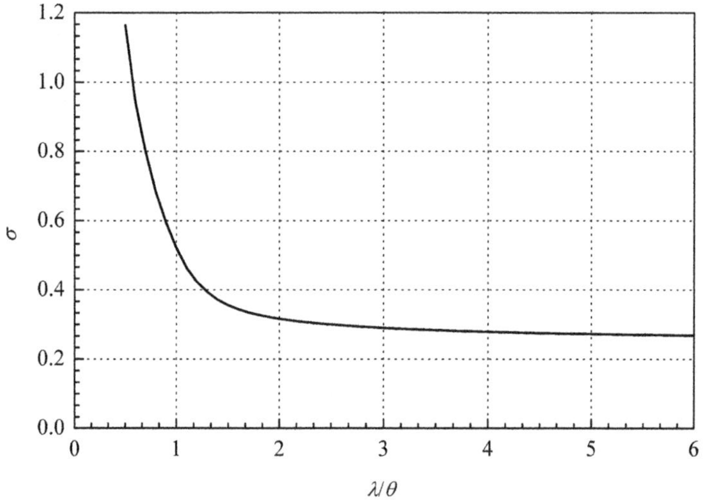

FIGURE 7.2.1
Overshoot of the H_∞ control system with an integrating plant.

Now consider the H_∞ PID controller design for a second-order integrating plant. Assume the second-order integrating plant is expressed as

$$G(s) = \frac{K}{s(\tau s + 1)}e^{-\theta s}, \tag{7.2.15}$$

where τ is the time constant. By utilizing the first-order Taylor series expan-

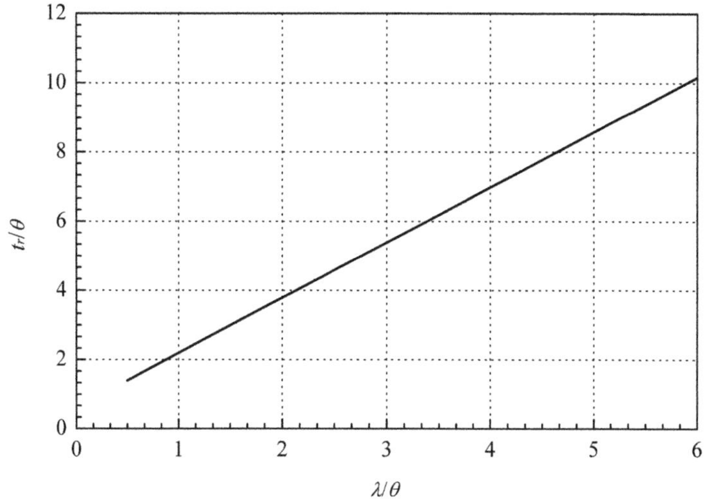

FIGURE 7.2.2
Rise time of the H_∞ control system with an integrating plant.

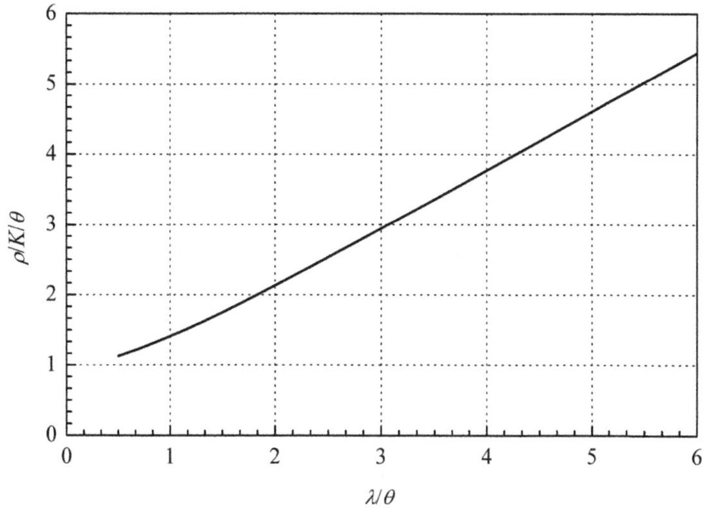

FIGURE 7.2.3
Perturbation peak of the H_∞ control system with an integrating plant.

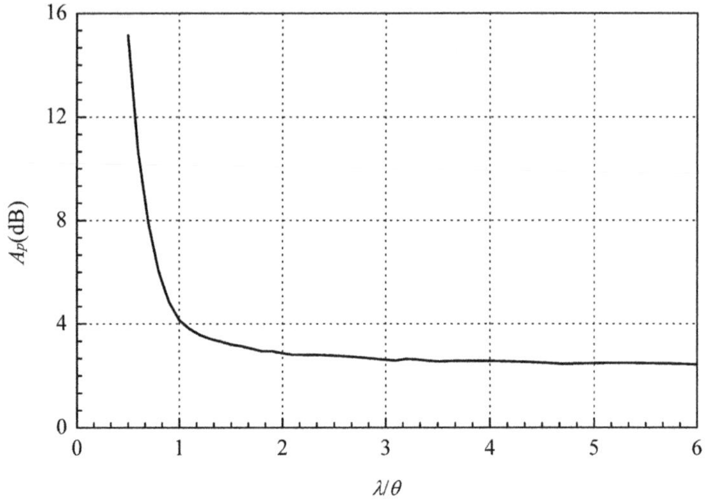

FIGURE 7.2.4
Resonance peak of the H_∞ control system with an integrating plant.

sion, the plant can be rewritten as follows:

$$G(s) = \frac{K(1-\theta s)}{s(\tau s+1)}. \tag{7.2.16}$$

Apply the H_∞ design procedure. The solution is

$$Q_{opt}(s) = \frac{s(\tau s+1)}{K}. \tag{7.2.17}$$

To get a proper $Q(s)$, the following filter is introduced:

$$J(s) = \frac{(3\lambda+\theta)s+1}{(\lambda s+1)^3}. \tag{7.2.18}$$

Then the unity feedback loop controller is

$$C(s) = \frac{1}{K}\frac{(\tau s+1)[(3\lambda+\theta)s+1]}{s(\lambda^3 s+3\lambda^2+3\lambda\theta+\theta^2)}. \tag{7.2.19}$$

If the PID controller is in the form of

$$C(s) = K_C\left(1+\frac{1}{T_I s}+T_D s\right)\frac{1}{T_F s+1},$$

the following controller parameters can readily be obtained:

$$\begin{aligned} T_F &= \frac{\lambda^3}{3\lambda^2 + 3\lambda\theta + \theta^2}, \\ T_I &= 3\lambda + \theta + \tau, \\ T_D &= \frac{(3\lambda + \theta)\tau}{T_I}, \\ K_C &= \frac{1}{K}\frac{T_I}{3\lambda^2 + 3\lambda\theta + \theta^2}. \end{aligned} \tag{7.2.20}$$

7.3 H_2 PID Controller for Integrating Plants

This section discusses the design of the H_2 PID Controller. The design procedure is similar to that in the last section: First, the optimal controller is designed for a step input. Then, a filter is introduced for asymptotic tracking.

Consider the approximate first-order integrating plant obtained by utilizing the 1/1 Pade approximant:

$$G(s) \approx \frac{K(1 - \theta s/2)}{s(1 + \theta s/2)}. \tag{7.3.1}$$

The performance index is chosen as $\min \|W(s)S(s)\|_2$. The internal stability requires that

$$\lim_{s \to 0}[1 - G(s)Q(s)] = 0.$$

The $Q(s)$ that satisfies this requirement can be expressed as

$$Q(s) = \frac{s[1 + sQ_1(s)]}{K}, \tag{7.3.2}$$

where $Q_1(s)$ is stable. This leads to

$$\begin{aligned} &\|W(s)S(s)\|_2^2 \\ = &\left\| \frac{1}{s}\left\{1 - \frac{1 - \theta s/2}{1 + \theta s/2}[1 + sQ_1(s)]\right\} \right\|_2^2 \\ = &\left\| \frac{\theta}{1 + \theta s/2} - \frac{1 - \theta s/2}{1 + \theta s/2}Q_1(s) \right\|_2^2 \\ = &\left\| \frac{\theta}{1 - \theta s/2} - Q_1(s) \right\|_2^2 \\ = &\left\| \frac{\theta}{1 - \theta s/2} \right\|_2^2 + \|Q_1(s)\|_2^2. \end{aligned}$$

Evidently, $Q_{1opt}(s) = 0$ gives the optimal solution, which implies

$$Q_{opt}(s) = s/K. \tag{7.3.3}$$

To satisfy the constraints of asymptotic tracking:

$$\lim_{s\to 0}[1 - G(s)Q(s)] = 0 \text{ and } \lim_{s\to 0}\frac{d}{ds}[1 - G(s)Q(s)] = 0,$$

introduce the following filter:

$$J(s) = \frac{(2\lambda + \theta)s + 1}{(\lambda s + 1)^2}.$$

The suboptimal solution is $Q(s) = Q_{opt}(s)J(s)$. The unity feedback loop controller is

$$\begin{aligned} C(s) &= \frac{Q(s)}{1 - G(s)Q(s)} \\ &= \frac{1}{K}\frac{s[(2\lambda + \theta)s + 1](1 + \theta s/2)}{(\lambda s + 1)^2(1 + \theta s/2) - (1 - \theta s/2)[(2\lambda + \theta)s + 1]} \\ &= \frac{1}{K}\frac{(1 + \theta s/2)[(2\lambda + \theta)s + 1]}{\theta\lambda^2 s^2/2 + (\lambda^2 + 2\lambda\theta + \theta^2/2)s}. \end{aligned} \tag{7.3.4}$$

Compare the obtained controller with the PID controller

$$C(s) = K_C\left(1 + \frac{1}{T_I s} + T_D s\right)\frac{1}{T_F s + 1}.$$

The controller parameters are

$$\begin{aligned} T_F &= \frac{\lambda^2\theta}{2\lambda^2 + 4\lambda\theta + \theta^2}, \\ T_I &= 2\lambda + \frac{3\theta}{2}, \\ T_D &= \frac{(2\lambda + \theta)\theta}{2T_I}, \\ K_C &= \frac{1}{K}\frac{T_I}{\lambda^2 + 2\lambda\theta + \theta^2/2}. \end{aligned} \tag{7.3.5}$$

The relationships between the performance degree and the closed-loop responses are shown in Figure 7.3.1–Figure 7.3.4, respectively.

Suppose the plant is of second-order. By applying the first-order Taylor series expansion, the approximate plant is obtained as follows:

$$G(s) = \frac{K(1 - \theta s)}{s(\tau s + 1)}. \tag{7.3.6}$$

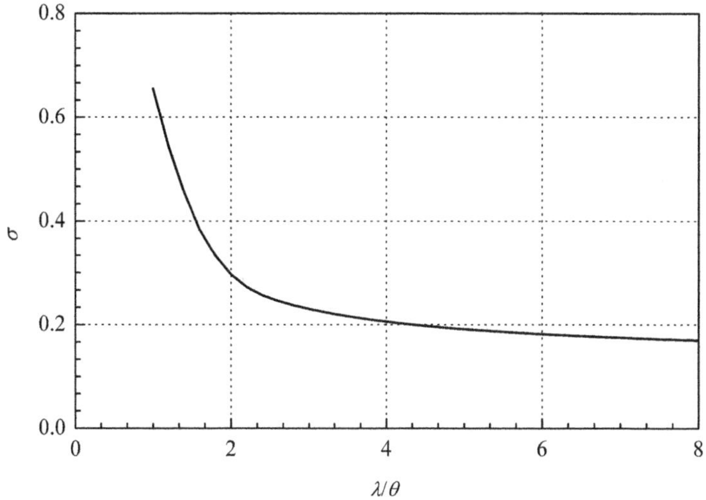

FIGURE 7.3.1
Overshoot of the H_2 control system with an integrating plant.

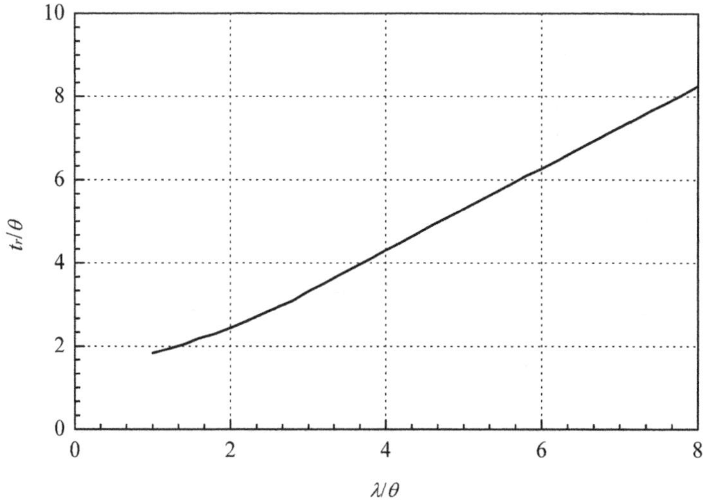

FIGURE 7.3.2
Rise time of the H_2 control system with an integrating plant.

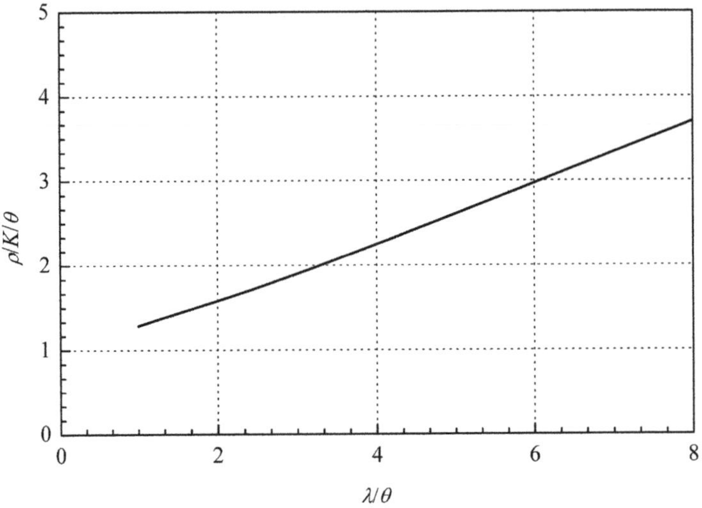

FIGURE 7.3.3
Perturbation peak of the H_2 control system with an integrating plant.

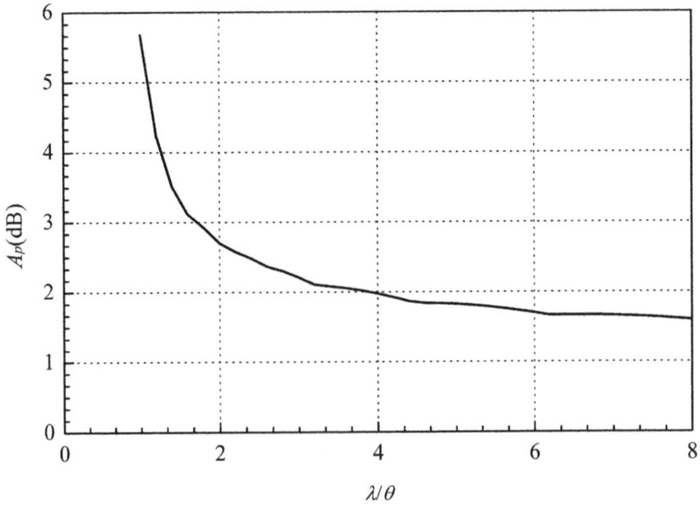

FIGURE 7.3.4
Resonance peak of the H_2 control system with an integrating plant.

With a similar design procedure, the optimal controller is

$$Q_{opt}(s) = \frac{s(\tau s + 1)}{K(1 + \theta s)}. \tag{7.3.7}$$

The filter is taken as

$$J(s) = \frac{(2\lambda + 2\theta)s + 1}{(\lambda s + 1)^2}. \tag{7.3.8}$$

It follows that

$$C(s) = \frac{1}{K}\frac{(\tau s + 1)[(2\lambda + 2\theta)s + 1]}{\lambda^2\theta s^2 + (\lambda^2 + 4\lambda\theta + 2\theta^2)s}. \tag{7.3.9}$$

For the PID controller in the form of

$$C(s) = K_C\left(1 + \frac{1}{T_I s} + T_D s\right)\frac{1}{T_F s + 1},$$

the parameters are as follows:

$$\begin{aligned} T_F &= \frac{\lambda^2\theta}{\lambda^2 + 4\lambda\theta + 2\theta^2}, \\ T_I &= 2\lambda + 2\theta + \tau, \\ T_D &= \frac{(2\lambda + 2\theta)\tau}{T_I}, \\ K_C &= \frac{1}{K}\frac{T_I}{\lambda^2 + 4\lambda\theta + 2\theta^2}. \end{aligned} \tag{7.3.10}$$

Example 7.3.1. *Consider the middle column temperature control of a high-purity distillation column. The column of interest is 110 ft high. The material to be separated is the mixture of three isomers and a small amount of other heavy components. To increase the production rate, the distillation column is designed so that there is very little excess separation ability. In other words, the design has a very small safety margin. This makes tight control very important in maintaining product quality.*

The control strategy is illustrated in Figure7.3.5. The control objective is to keep the distillate composition nearly pure while maintaining it at a very low level in the bottom product. The heat to the column is fixed, because the heat source is a vapor boiler that runs best at a fixed rate. The feed is set to fix the overall production rate for this part of the process. Under these conditions, the composition control can be accomplished by manipulating the middle column temperature.

In the control loop, the output of the process is the temperature and the input is the distillate flow rate. With open-loop step tests, the process model is obtained as follows:

$$\boxed{G(s) = \frac{0.01}{s}e^{-5.5s}.}$$

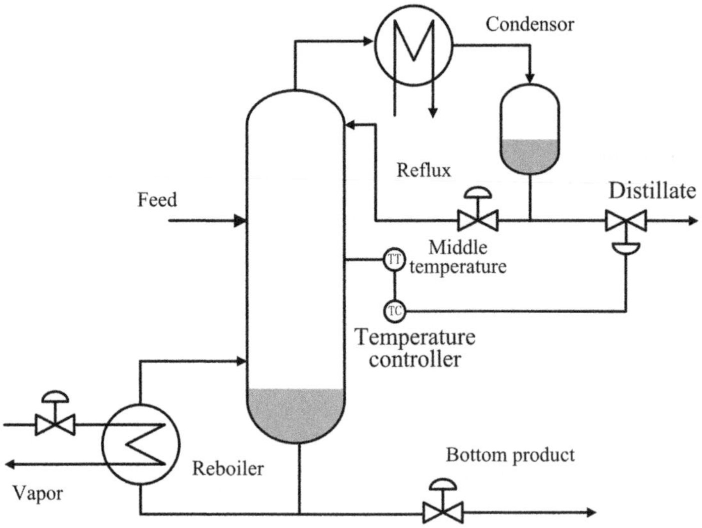

FIGURE 7.3.5
Control system of a high-purity distillation column.

Take $\lambda = 25$ *for the* H_2 *PID controller given by (7.3.4):*

$$\boxed{C(s) = \frac{100(138.75s^2 + 58s + 1)}{s(2.75\lambda^2 s + \lambda^2 + 11\lambda + 15.125)},}$$

and $\lambda = 16$ *for the* H_∞ *PID controller given by (7.2.13):*

$$\boxed{C(s) = \frac{100[(8.25\lambda + 7.5625)s^2 + (3\lambda + 5.5)s + 1]}{s(\lambda^3 s + 3\lambda^2 + 8.25\lambda + 7.5625)}.}$$

A unit step reference is added at $t = 0$ *and a unit step load is added at* $t = 200$. *The nominal responses of the closed-loop system are shown in Figure 7.3.6. It is observed that both controllers give large overshoots and long settling times. This is an evident feature of the system with an integrating plant.*

7.4 Controller Design for General Integrating Plants

In the system with an integrating plant, the IMC structure must be abandoned for the implementation of the control system, since there always exist zero-pole cancellations at the origin, which will cause the internal instability problem.

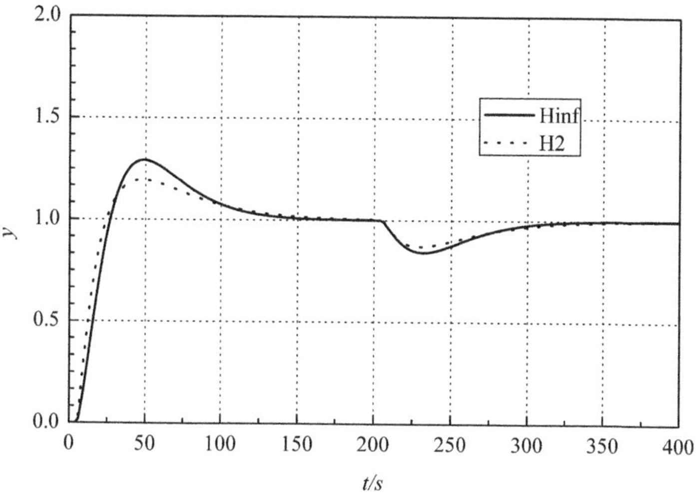

FIGURE 7.3.6
Nominal responses of the H_2 PID and H_∞ PID controllers.

Nevertheless, as it was shown in the preceding sections, even for the system with an integrating plant, the IMC controller $Q(s)$ could be utilized to design the unity feedback loop controller. In this section, $Q(s)$ is employed to discuss the controller design problem for general integrating plants.

A general integrating plant may contain one or more poles at the origin. The analysis procedure is complicated. To simplify the presentation, only the integrating plant with a simple pole at the origin is considered in this section. A pole is simple if its multiplicity is one. The analysis for the integrating plant with multiple poles at the origin will be introduced in the next section.

Without loss of generality, assume that the integrating plant is expressed by

$$G(s) = \frac{KN_+(s)N_-(s)}{sM_-(s)}e^{-\theta s}, \tag{7.4.1}$$

where K is the gain, θ is the time delay, $N_-(s)$ and $M_-(s)$ are the polynomials with roots in the open LHP, $N_+(s)$ is a polynomial with roots in the closed RHP, $N_+(0) = N_-(0) = M_-(0) = 1$, and $\deg\{N_+(s)\} + \deg\{N_-(s)\} \leq \deg\{M_-(s)\} + 1$.

First, consider the quasi-H_∞ control. By following the quasi-H_∞ controller design procedure for stable plants, the desired closed-loop transfer function is chosen as

$$T(s) = N_+(s)J(s)e^{-\theta s}, \tag{7.4.2}$$

where $J(s)$ is a filter:

$$J(s) = \frac{(\beta s + 1)}{(\lambda s + 1)^{n_j}},$$

λ is the performance degree,

$$n_j = \begin{cases} 2 + \deg\{M_-\} - \deg\{N_-\} & \deg\{M_-\} + 1 > \deg\{N_-\} \\ 2 & \deg\{M_-\} + 1 = \deg\{N_-\} \end{cases},$$

and β is determined by the following constraints:

$$\lim_{s\to 0} \frac{d^k}{ds^k}[1 - T(s)] = 0, k = 0, 1, \tag{7.4.3}$$

or equivalently,

$$\begin{aligned} \lim_{s\to 0}[1 - N_+(s)J(s)e^{-\theta s}] &= 0, \\ \lim_{s\to 0} \frac{d}{ds}[1 - N_+(s)J(s)e^{-\theta s}] &= 0. \end{aligned}$$

Notice that one zero is introduced to $T(s)$.

Since both $G(s)$ and $T(s)$ are known, the controller can be analytically derived:

$$\begin{aligned} Q(s) &= \frac{T(s)}{G(s)} \\ &= \frac{1}{K} \frac{sM_-(s)(\beta s + 1)}{N_-(s)(\lambda s + 1)^{n_j}}. \end{aligned} \tag{7.4.4}$$

The unity feedback loop controller is

$$\begin{aligned} C(s) &= \frac{T(s)}{1 - T(s)} \frac{1}{G(s)} \\ &= \frac{1}{K} \frac{sM_-(s)(\beta s + 1)}{N_-(s)[(\lambda s + 1)^{n_j} - (\beta s + 1)N_+(s)e^{-\theta s}]}. \end{aligned} \tag{7.4.5}$$

The reason why the IMC structure cannot be used to control integrating plants is that there are zero-pole cancellations at the origin between $Q(s)$ and $G(s)$. One may want to solve the problem by combining $Q(s)$ and $G(s)$ into $C(s)$. Unfortunately, the problem cannot be overcome so easily. It can be verified that

$$\lim_{s\to 0}[(\lambda s + 1)^{n_j} - (\beta s + 1)N_+(s)e^{-\theta s}] = 0, \tag{7.4.6}$$

which implies that the denominator of $C(s)$ has a root at the origin. Since there is a time delay in the denominator, the root cannot be removed directly. As a result, the obtained $C(s)$ cannot guarantee the internal stability of the closed-loop system. Only after the RHP root is removed by employing a rational approximation, can $C(s)$ guarantee the internal stability of the closed-loop system.

This brings on a subtle problem; that is, can Theorem 7.1.1 be used to test the internal stability of the system with an integrating plant? Actually, the statement of Theorem 7.1.1 is rigorous. The second condition of Theorem 7.1.1 is in fact equivalent to

1. $\lim_{s\to 0}[1 - G(s)Q(s)] = 0$.
2. all RHP zero-pole cancellations in $[1 - G(s)Q(s)]G(s)$ are removed.

The design for quasi-H_∞ controllers can also be carried out as follows:

1. If the plant does not have a time delay, turn to 3.
2. If the plant contains a time delay, take the rational part of the plant as the nominal plant.
3. If the nominal plant has no zeros in the RHP, take its inverse as $Q_{opt}(s)$ and turn to 5.
4. If the nominal plant has zeros in the RHP, remove the factor that contains these zeros and take the inverse of the remainder as $Q_{opt}(s)$.
5. Introduce a filter to $Q_{opt}(s)$, compute the controller $C(s)$ and remove the RHP zero-pole cancellations in $C(s)$.

Now consider the H_2 control. The integrating plant is

$$G(s) = \frac{KN_+(s)N_-(s)}{sM_-(s)}e^{(-\theta s)}. \tag{7.4.7}$$

The performance index is $\min \|W(s)S(s)\|_2$. The internal stability imposes a constraint on $Q(s)$:

$$\lim_{s\to 0}[1 - G(s)Q(s)] = 0.$$

The $Q(s)$ that satisfies this constraint has the following expression:

$$Q(s) = \frac{s[1 + sQ_1(s)]}{K}, \tag{7.4.8}$$

where $Q_1(s)$ is stable. Then

$$\begin{aligned}
&\|W(s)S(s)\|_2^2 \\
= &\left\| \frac{1}{s}\left\{1 - \frac{N_+(s)N_-(s)}{M_-(s)}e^{-\theta s}[1 + sQ_1(s)]\right\}\right\|_2^2 \\
= &\left\| \frac{1}{s}\left\{\frac{N_+(-s)}{N_+(s)}e^{\theta s} - \frac{N_+(-s)N_-(s)}{M_-(s)}[1 + sQ_1(s)]\right\}\right\|_2^2
\end{aligned}$$

$$= \left\| \frac{e^{\theta s}N_+(-s) - N_+(s)}{sN_+(s)} \right\|_2^2 + \left\| \frac{1}{s} - \frac{N_+(-s)N_-(s)}{sM_-(s)}[1 + sQ_1(s)] \right\|_2^2.$$

Solving the optimal problem yields

$$Q_{1opt}(s) = \frac{M_-(s) - N_+(-s)N_-(s)}{sN_+(-s)N_-(s)}. \quad (7.4.9)$$

Hence

$$Q_{opt}(s) = \frac{sM_-(s)}{KN_+(-s)N_-(s)}. \quad (7.4.10)$$

Introduce the following filter:

$$J(s) = \frac{\beta s + 1}{(\lambda s + 1)^{n_j}}, \quad (7.4.11)$$

where λ is the performance degree,

$$n_j = \begin{cases} 2 + \deg\{M_-\} - \deg\{N_+\} - \deg\{N_-\} & \deg\{M_-\} + 1 > \deg\{N_+\} + \deg\{N_-\} \\ 2 & \deg\{M_-\} + 1 = \deg\{N_+\} + \deg\{N_-\} \end{cases}.$$

Consequently,

$$Q(s) = Q_{opt}(s)J(s) = \frac{sM_-(s)(\beta s + 1)}{KN_+(-s)N_-(s)(s + 1)^{n_j}}. \quad (7.4.12)$$

β is determined by the constraints of asymptotic tracking:

$$\lim_{s \to 0} \frac{d^k}{ds^k}[1 - G(s)Q(s)] = 0, k = 0, 1,$$

or equivalently,

$$\begin{aligned} \lim_{s \to 0}[1 - \frac{N_+(s)}{N_+(-s)}J(s)e^{-\theta s}] &= 0, \\ \lim_{s \to 0} \frac{d}{ds}[1 - \frac{N_+(s)}{N_+(-s)}J(s)e^{-\theta s}] &= 0. \end{aligned}$$

The design procedure for H_2 controllers can be described in a way similar to that for quasi-H_∞ controllers, except that Step 4 is modified as follows.

> When the nominal plant has zeros in the RHP, construct an all-pass transfer function with the factor that contains these zeros and then remove the all-pass transfer function, take the inverse of the remainder as $Q_{opt}(s)$.

In the above design procedures, a Type 2 controller was derived by modifying a Type 1 controller. A question of interest is what difference exists between the two controllers. As a comparison, a Type 2 system is directly designed here; that is, take $W(s) = 1/s^2$.

The first step is to determine the form of the $Q(s)$ for a Type 2 system. $Q(s)$ should satisfy the following two conditions:

$$\lim_{s\to 0}[1 - G(s)Q(s)] = 0, \tag{7.4.13}$$

$$\lim_{s\to 0}\frac{d}{ds}[1 - G(s)Q(s)] = 0. \tag{7.4.14}$$

The $Q(s)$ that satisfies the first condition can be expressed as

$$Q(s) = \frac{s[1 + sQ_1(s)]}{K}, \tag{7.4.15}$$

where $Q_1(s)$ is stable. Substituting this into the left-hand side of the second condition, we have

$$\begin{aligned}&\lim_{s\to 0}\frac{d}{ds}\left\{1 - \frac{N_+(s)N_-(s)}{M_-(s)}e^{-\theta s}[1 + sQ_1(s)]\right\}\\ = \quad & -\lim_{s\to 0}\frac{d}{ds}\left[\frac{N_+(s)N_-(s)}{M_-(s)}e^{-\theta s} + s\frac{N_+(s)N_-(s)}{M_-(s)}e^{-\theta s}Q_1(s)\right].\end{aligned}$$

The second condition gives

$$Q_1(0) = \theta + \frac{d}{ds}M_-(0) - \frac{d}{ds}N_+(0) - \frac{d}{ds}N_-(0). \tag{7.4.16}$$

Then, the $Q(s)$ that satisfies the two conditions can be written as

$$Q(s) = \frac{s[1 + sQ_1(0) + s^2Q_2(s)]}{K}, \tag{7.4.17}$$

where $Q_2(s)$ is stable. Therefore,

$$\begin{aligned}&\|W(s)S(s)\|_2^2\\ = \quad & \left\|\frac{1}{s^2}\left\{1 - \frac{N_+(s)N_-(s)}{M_-(s)}e^{-\theta s}[1 + sQ_1(0) + s^2Q_2(s)]\right\}\right\|_2^2\\ = \quad & \left\|\frac{1}{s^2}\left\{\frac{N_+(-s)}{N_+(s)}e^{\theta s} - \frac{N_+(-s)N_-(s)}{M_-(s)}[1 + sQ_1(0) + s^2Q_2(s)]\right\}\right\|_2^2\\ = \quad & \left\|\frac{e^{\theta s}N_+(-s) - N_+(s)[1 + \theta s - 2\frac{d}{ds}N_+(0)s]}{s^2N_+(s)}\right\|_2^2 +\end{aligned}$$

$$\left\| \frac{1+\theta s-2\frac{d}{ds}N_+(0)s}{s^2} - \frac{N_+(-s)N_-(s)}{s^2M_-(s)}[1+sQ_1(0)+s^2Q_2(s)] \right\|_2^2.$$

It is evident that when

$$Q_{2opt}(s)=\frac{M_-(s)[1+\theta s-2\frac{d}{ds}N_+(0)s]M_-(s)}{s^2N_+(-s)N_-(s)}-\frac{1+Q_1(0)s}{s^2}, \quad (7.4.18)$$

the right-hand side is minimum. The optimal controller is

$$Q_{opt}(s)=\frac{sM_-(s)[1+\theta s-2\frac{d}{ds}N_+(0)s]}{KN_+(-s)N_-(s)}. \quad (7.4.19)$$

Introduce a Type 2 filter. The suboptimal controller is

$$Q(s)=\frac{sM_-(s)[1+\theta s-2\frac{d}{ds}N_+(0)s]\{[n_j\lambda+2\frac{d}{ds}N_+(0)]s+1\}}{KN_+(-s)N_-(s)(\lambda s+1)^{n_j}}, \quad (7.4.20)$$

where

$$n_j=\begin{cases} 3+\deg\{M_-\} \\ \quad -\deg\{N_+\}-\deg\{N_-\} & \deg\{M_-\}+1>\deg\{N_+\}+\deg\{N_-\} \\ 3 & \deg\{M_-\}+1=\deg\{N_+\}+\deg\{N_-\} \end{cases}.$$

It is easy to verify that the optimal performance of the Type 2 controller designed for $W(s)=1/s$ is

$$\min\|W(s)S(s)\|_2=\left\|\frac{N_+(-s)-N_+(s)e^{-\theta s}}{sN_+(-s)}\right\|_2, \quad (7.4.21)$$

while the optimal performance of the Type 2 controller designed for $W(s)=1/s^2$ is

$$\min\|W(s)S(s)\|_2 = \left\| \frac{N_+(-s)-N_+(s)[1+\theta s-2\frac{d}{ds}N_+(0)s][2\frac{d}{ds}N_+(0)s+1]e^{-\theta s}}{s^2N_+(-s)} \right\|_2. \quad (7.4.22)$$

When there exists uncertainty, the robustness of the system with an integrating plant can be analyzed with the necessary and sufficient condition in Theorem 3.4.2:

$$\|\Delta_m(s)T(s)\|_\infty<1. \quad (7.4.23)$$

With a similar tuning procedure to that in the system with a stable plant, the quantitative performance and robustness can also be achieved: Increase the performance degree monotonically until the required response is obtained. In comparison to that in the system with a stable plant, the performance degree in the system with an integrating plant is usually large.

7.5 Maclaurin PID Controller for Integrating Plants

To design a PID controller by utilizing the Maclaurin series expansion, a closed-loop transfer function must be determined first. Assume that the integrating plant has m poles at the origin:

$$G(s) = \frac{KN_+(s)N_-(s)}{s^m M_-(s)} e^{-\theta s}. \tag{7.5.1}$$

The closed-loop transfer function at least should be a stable transfer function. A reasonable form of $T(s)$ is

$$T(s) = N_+(s)J(s)e^{-\theta s}, \tag{7.5.2}$$

where $J(s)$ is a filter:

$$J(s) = \frac{N_x(s)}{(\lambda s + 1)^{n_j}},$$

λ is the performance degree, and

$$n_j = \begin{cases} 2m + \deg\{M_-\} - \deg\{N_-\} & \deg\{M_-\} + m > \deg\{N_-\} \\ m + 1 & \deg\{M_-\} + m = \deg\{N_-\} \end{cases}.$$

$N_x(s)$ is a polynomial with its roots in the open LHP, $N_x(0) = 1$, and $\deg\{N_x(s)\} = m$. $N_x(s)$ is determined by the asymptotic tracking constraints:

$$\lim_{s \to 0} \frac{d^k}{ds^k}[1 - T(s)] = 0, k = 0, 1, ..., m. \tag{7.5.3}$$

This $T(s)$ corresponds to the quasi-H_∞ control. If the closed-loop transfer function is in the form of

$$T(s) = \frac{N_+(s)}{N_+(-s)} J(s)e^{-\theta s}, \tag{7.5.4}$$

where

$$J(s) = \frac{N_x(s)}{(\lambda s + 1)^{n_j}},$$

λ is the performance degree,

$$n_j = \begin{cases} 2m + \deg\{M_-\} \\ \quad - \deg\{N_+\} - \deg\{N_-\} & \deg\{M_-\} + m > \deg\{N_+\} + \deg\{N_-\} \\ m+1 & \deg\{M_-\} + m = \deg\{N_+\} + \deg\{N_-\} \end{cases},$$

and the determination of $N_x(s)$ is similar to that for the quasi-H_∞ control, then, the H_2 control is achieved.

Both the quasi-H_∞ controller and the H_2 controller can be computed by

$$C(s) \quad = \quad \frac{1}{G(s)} \frac{T(s)}{1 - T(s)}. \tag{7.5.5}$$

Since $T(0) = 1$, $C(s)$ has a pole at the origin. Write $C(s)$ in the form of

$$C(s) = \frac{f(s)}{s}.$$

The Maclaurin series expansion of $C(s)$ is

$$C(s) = \frac{1}{s}\left[f(0) + f'(0)s + \frac{f''(0)}{2!}s^2 + ...\right]. \tag{7.5.6}$$

Omit those high-order terms. Only the first three terms are taken to approximate the ideal controller. The three terms form a PID controller:

$$C(s) = K_C\left(1 + \frac{1}{T_I s} + T_D s\right),$$

whose parameters are

$$K_C = f'(0), \quad T_I = \frac{f'(0)}{f(0)}, \quad T_D = \frac{f''(0)}{2f'(0)}. \tag{7.5.7}$$

Furthermore, define

$$f(s) = \frac{N(s)}{M(s)}, \tag{7.5.8}$$

where $M(s)$ and $N(s)$ are polynomials. The values of $f(s)$ and its derivatives at the origin are

$$\begin{aligned} f(0) &= \frac{N(0)}{M(0)}, \\ f'(0) &= \frac{N'(0)M(0) - M'(0)N(0)}{M(0)^2}, \\ f''(0) &= \frac{\begin{array}{c} N''(0)M(0)^2 - M''(0)N(0)M(0) - \\ 2M'(0)N'(0)M(0) + 2M'(0)^2N(0) \end{array}}{M(0)^3}. \end{aligned} \tag{7.5.9}$$

Two cases are considered: the plant is of first order and the plant is of second order. First, assume that the plant is

$$G(s) = \frac{K}{s}e^{-\theta s}. \tag{7.5.10}$$

As $N_+(s) = 1$, (7.5.2) is identical to (7.5.4). The closed-loop transfer function with the asymptotic tracking property is

$$T(s) = \frac{(2\lambda + \theta)s + 1}{(\lambda s + 1)^2}e^{-\theta s}. \tag{7.5.11}$$

Then

$$\begin{aligned} N(s) &= \frac{(2\lambda + \theta)s + 1}{K}, \\ M(s) &= \frac{(\lambda s + 1)^2 - [(2\lambda + \theta)s + 1]e^{-\theta s}}{s^2}, \end{aligned}$$

which yields

$$\begin{aligned} N(0) &= \frac{1}{K}, \\ N'(0) &= \frac{2\lambda + \theta}{K}, \\ N''(0) &= 0, \\ M(0) &= \frac{2\lambda^2 + 4\lambda\theta + \theta^2}{2}, \\ M'(0) &= \frac{-3\lambda\theta^2 - \theta^3}{3}, \\ M''(0) &= \frac{3\theta^4 + 8\theta^3\lambda}{12}. \end{aligned}$$

The values of $f(s)$ and its first and second derivatives at the origin are

$$\begin{aligned} f(0) &= \frac{2}{K(2\lambda^2 + 4\lambda\theta + \theta^2)}, \\ f'(0) &= \frac{2(12\lambda^3 + 30\lambda^2\theta + 24\lambda\theta^2 + 5\theta^3)}{3K(2\lambda^2 + 4\lambda\theta + \theta^2)^2}, \\ f''(0) &= \frac{\theta^2(288\lambda^4 + 768\lambda^3\theta + 702\lambda^2\theta^2 + 252\lambda\theta^3 + 31\theta^4)}{9K(2\lambda^2 + 4\lambda\theta + \theta^2)^3}. \end{aligned} \tag{7.5.12}$$

Consequently, the controller parameters are

$$\begin{aligned} T_I &= 2\lambda + \theta + \frac{2\theta^3 + 6\lambda\theta^2}{3(2\lambda^2 + 4\lambda\theta + \theta^2)}, \\ K_C &= \frac{2T_I}{K(2\lambda^2 + 4\lambda\theta + \theta^2)}, \\ T_D &= \frac{\theta^2(288\lambda^4 + 768\lambda^3\theta + 702\lambda^2\theta^2 + 252\lambda\theta^3 + 31\theta^4)}{36T_I(2\lambda^2 + 4\lambda\theta + \theta^2)^2}. \end{aligned} \tag{7.5.13}$$

The formula seems a bit complicated. However, the computation is not difficult, since all parameters of the plant are known.

Next, consider the second-order plant:

$$G(s) = \frac{K}{s(\tau s + 1)} e^{-\theta s}. \tag{7.5.14}$$

The closed-loop transfer function with the asymptotic tracking property can be written as

$$T(s) = \frac{(3\lambda + \theta)s + 1}{(\lambda s + 1)^3} e^{-\theta s}. \tag{7.5.15}$$

Then

$$\begin{aligned} N(s) &= \frac{(\tau s + 1)[(3\lambda + \theta)s + 1]}{K}, \\ M(s) &= \frac{(\lambda s + 1)^3 - [(3\lambda + \theta)s + 1]e^{-\theta s}}{s^2}. \end{aligned}$$

This yields

$$\begin{aligned} N(0) &= \frac{1}{K}, \\ N'(0) &= \frac{3\lambda + \tau + \theta}{K}, \\ N''(0) &= \frac{2\tau(3\lambda + \theta)}{K}, \\ M(0) &= \frac{6\lambda^2 + 6\lambda\theta + \theta^2}{2}, \\ M'(0) &= \frac{6\lambda^3 - 9\lambda\theta^2 - 2\theta^3}{6}, \\ M''(0) &= \frac{\theta^4 + 4\lambda\theta^3}{4}. \end{aligned}$$

The values of $f(s)$ and its first and second derivatives at the origin are

$$\begin{aligned} f(0) &= \frac{2}{K(6\lambda^2 + 6\lambda\theta + \theta^2)}, \\ f'(0) &= \frac{2(18\tau\lambda^2 + 3\theta^2\tau + 18\tau\lambda\theta + 72\lambda^2\theta + 5\theta^3 + 36\lambda\theta^2 + 48\lambda^3)}{3K(6\lambda^2 + \theta^2 + 6\theta\lambda)^2}, \\ f''(0) &= \frac{\begin{matrix} 8352\tau\lambda^3\theta^2 + 31\theta^6 - 1152\lambda^6 + 3456\tau\lambda^5 + 60\tau\theta^5 + \\ 1602\lambda^2\theta^4 + 378\lambda\theta^5 + 2640\lambda^3\theta^3 + 864\lambda^4\theta^2 - \\ 1728\lambda^5\theta + 8640\tau\lambda^4\theta + 792\tau\lambda\theta^4 + 3816\tau\lambda^2\theta^3 \end{matrix}}{9K(6\lambda^2 + 6\lambda\theta + \theta^2)^3}. \end{aligned}$$

Therefore, the controller parameters are

$$\begin{aligned}
T_I &= \tau + \frac{5\theta^3 + 36\lambda\theta^2 + 48\lambda^3 + 72\theta\lambda^2}{3(6\lambda^2 + 6\lambda\theta + \theta^2)}, \\
K_C &= \frac{2T_I}{K(6\lambda^2 + 6\lambda\theta + \theta^2)}, \\
T_D &= \frac{\begin{array}{r} 240\tau\lambda^4\theta + 22\tau\lambda\theta^4 + 232\tau\lambda^3\theta^2 + 106\tau\lambda^2\theta^3 + \\ 5\tau\theta^5/3 + 96\tau\lambda^5 + 89\theta^4\lambda^2/2 + 31\theta^6/36 - 48\lambda^5\theta + \\ 24\lambda^4\theta^2 + 21\lambda\theta^5/2 + 220\lambda^3\theta^3/3 - 32\lambda^6 \end{array}}{T_I(6\lambda^2 + 6\lambda\theta + \theta^2)^2}.
\end{aligned} \tag{7.5.16}$$

7.6 The Best Achievable Performance of a PID Controller

There exists a compromise between good performance and computation complexity in choosing the design method. On one hand, a good result can be expected if the Pade approximation is applied to the reduction of the whole controller. The result is, however, relatively complicated. On the other hand, it will give a simple but not so good result to expand time delay by applying the first-order rational approximation first and then designing the controller for the approximate plant. In this section, the former method is employed to design the PID controller with the best achievable performance.

A general integrating plant can be described as

$$G(s) = \frac{KN_+(s)N_-(s)}{s^m M_-(s)} e^{-\theta s}. \tag{7.6.1}$$

Suppose that the closed-loop transfer function $T(s)$ is given as that in (7.5.2) or (7.5.4). The controller can be computed by

$$C(s) = \frac{1}{G(s)} \frac{T(s)}{1 - T(s)}. \tag{7.6.2}$$

$C(s)$ has a pole at the origin. Express it as

$$C(s) = \frac{f(s)}{s}. \tag{7.6.3}$$

Using the Maclaurin series expansion, we have

$$f(s) = f(0) + f'(0)s + \frac{f''(0)}{2!}s^2 + \frac{f^{(3)}(0)}{3!}s^3 + ... \tag{7.6.4}$$

A practical PID controller can be written as

$$C(s) = \frac{a_2 s^2 + a_1 s + a_0}{s(b_1 s + 1)}. \tag{7.6.5}$$

Let the Pade approximation of $f(s)$ be

$$\frac{a_2 s^2 + a_1 s + a_0}{b_1 s + 1}. \tag{7.6.6}$$

Then

$$\begin{bmatrix} a_0 \\ a_1 \\ a_2 \end{bmatrix} = \begin{bmatrix} f(0) & 0 \\ f'(0) & f(0) \\ f''(0)/2! & f'(0) \end{bmatrix} \begin{bmatrix} 1 \\ b_1 \end{bmatrix}, \tag{7.6.7}$$

$$b_1 f''(0)/2! = -f^{(3)}(0)/3!. \tag{7.6.8}$$

This yields

$$\begin{aligned} a_0 &= f(0), \\ a_1 &= b_1 f(0) + f'(0), \\ a_2 &= b_1 f'(0) + f''(0)/2!, \\ b_1 &= -\frac{f^{(3)}(0)}{3f''(0)}. \end{aligned} \tag{7.6.9}$$

If the PID controller is in the form of

$$C = K_C \left(1 + \frac{1}{T_I s} + T_D s\right) \frac{1}{T_F s + 1},$$

a little computation gives

$$K_C = a_1, \quad T_I = \frac{a_1}{a_0}, \quad T_D = \frac{a_2}{a_1}, \quad T_F = b_1. \tag{7.6.10}$$

As discussed in Section 5.6, all of these parameters should be chosen as positive numbers.

Consider the plant

$$G(s) = \frac{Ke^{-\theta s}}{s}. \tag{7.6.11}$$

The value of $f(s)$ and its derivatives at the origin are

$$\begin{aligned}
f(0) &= \frac{2}{K(2\lambda^2+\theta^2+4\lambda\theta)},\\
f'(0) &= \frac{2(12\lambda^3+24\lambda\theta^2+30\lambda^2\theta+5\theta^3)}{3K(2\lambda^2+\theta^2+4\lambda\theta)^2},\\
f''(0) &= \frac{\theta^2(768\theta\lambda^3+252\lambda\theta^3+702\theta^2\lambda^2+31\theta^4+288\lambda^4)}{9K(2\lambda^2+\theta^2+4\lambda\theta)^3},\\
f^{(3)}(0) &= \frac{\begin{array}{c}\theta^3(121\theta^6-2880\lambda^6+1044\theta^2\lambda^4+\\ 5040\theta\lambda^5+1248\lambda\theta^5+4620\lambda^2\theta^4+6696\lambda^3\theta^3)\end{array}}{45K(2\lambda^2+\theta^2+4\lambda\theta)^4}.
\end{aligned} \tag{7.6.12}$$

The controller parameters are obtained as follows:

$$\begin{aligned}
a_0 &= \frac{2}{K(2\lambda^2+\theta^2+4\lambda\theta)},\\
a_1 &= \frac{\begin{array}{c}4(109\theta^5+1026\lambda\theta^4+3648\lambda^2\theta^3+\\ 6090\lambda^3\theta^2+4800\theta\lambda^4+1440\lambda^5)\end{array}}{\begin{array}{c}5K(252\theta^3\lambda+702\theta^2\lambda^2+768\theta\lambda^3+\\ 31\theta^4+288\lambda^4)(2\lambda^2+\theta^2+4\lambda\theta)\end{array}},\\
a_2 &= \frac{\begin{array}{c}(265\theta^5+2496\lambda\theta^4+9000\lambda^2\theta^3+\\ 15408\theta^2\lambda^3+12480\theta\lambda^4+3840\lambda^5)\end{array}}{\begin{array}{c}10K(768\theta\lambda^3+252\theta^3\lambda+702\theta^2\lambda^2+\\ 31\theta^4+288\lambda^4)(2\lambda^2+\theta^2+4\lambda\theta)\end{array}},\\
b_1 &= -\frac{\begin{array}{c}\theta(121\theta^6+1248\theta^5\lambda-2880\lambda^6+6696\theta^3\lambda^3-\\ 5040\theta\lambda^5+1044\theta^2\lambda^4+4620\theta^4\lambda^2)\end{array}}{\begin{array}{c}15(768\theta\lambda^3+252\theta^3\lambda+702\theta^2\lambda^2+31\theta^4+\\ 288\lambda^4)(2\lambda^2+\theta^2+4\lambda\theta)\end{array}}.
\end{aligned} \tag{7.6.13}$$

7.7 Summary

In traditional textbooks, the control problem of integrating plants is seldom discussed. This chapter discusses the problem. A conclusion drawn in this chapter is that, if the integrating plant in the unity feedback loop has m poles at the origin, the control system should be designed as a Type $m+1$ system. This constraint must be considered in controller design; otherwise, the resulting controller may not reject disturbances asymptotically.

Several controllers are analytically designed in this chapter based on the optimal control theory. The obtained controllers, similar to those for stable plants, can also be tuned for quantitative performance and robustness.

Assume that the plant is

$$G(s) = \frac{K}{s}e^{-\theta s}.$$

The H_∞ PID controller is

$$\begin{aligned}
T_F &= \frac{\lambda^3}{3\lambda^2 + 3\lambda\theta/2 + \theta^2/4},\\
T_I &= 3\lambda + \theta,\\
T_D &= \frac{3\lambda\theta/2 + \theta^2/4}{T_I},\\
K_C &= \frac{1}{K}\frac{T_I}{3\lambda^2 + 3\lambda\theta/2 + \theta^2/4}.
\end{aligned}$$

The H_2 PID controller is

$$\begin{aligned}
T_F &= \frac{\lambda^2\theta}{2\lambda^2 + 4\lambda\theta + \theta^2},\\
T_I &= 2\lambda + \frac{3\theta}{2},\\
T_D &= \frac{(2\lambda + \theta)\theta}{2T_I},\\
K_C &= \frac{1}{K}\frac{T_I}{\lambda^2 + 2\lambda\theta + \theta^2/2}.
\end{aligned}$$

The Maclaurin PID controller is

$$\begin{aligned}
T_I &= 2\lambda + \theta + \frac{2\theta^3 + 6\lambda\theta^2}{3(2\lambda^2 + 4\lambda\theta + \theta^2)},\\
K_C &= \frac{2T_I}{K(2\lambda^2 + 4\lambda\theta + \theta^2)},\\
T_D &= \frac{\theta^2(288\lambda^4 + 768\lambda^3\theta + 702\lambda^2\theta^2 + 252\lambda\theta^3 + 31\theta^4)}{36T_I(2\lambda^2 + 4\lambda\theta + \theta^2)^2}.
\end{aligned}$$

The PID controller with the best achievable performance is

$$\begin{aligned}
K_C &= a_1,\\
T_I &= \frac{a_1}{a_0},\\
T_D &= \frac{a_2}{a_1},\\
T_F &= b_1.
\end{aligned}$$

Sometimes, the plant is described by

$$G(s) = \frac{K}{s(\tau s + 1)} e^{-\theta s}.$$

The H_∞ PID controller is

$$\begin{aligned}
T_F &= \frac{\lambda^3}{3\lambda^2 + 3\lambda\theta + \theta^2}, \\
T_I &= 3\lambda + \theta + \tau, \\
T_D &= \frac{(3\lambda + \theta)\tau}{T_I}, \\
K_C &= \frac{1}{K} \frac{T_I}{3\lambda^2 + 3\lambda\theta + \theta^2}.
\end{aligned}$$

The H_2 PID controller is

$$\begin{aligned}
T_F &= \frac{\lambda^2\theta}{\lambda^2 + 4\lambda\theta + 2\theta^2}, \\
T_I &= 2\lambda + 2\theta + \tau, \\
T_D &= \frac{(2\lambda + 2\theta)\tau}{T_I}, \\
K_C &= \frac{1}{K} \frac{T_I}{\lambda^2 + 4\lambda\theta + 2\theta^2}.
\end{aligned}$$

The Maclaurin controller is

$$\begin{aligned}
T_I &= \tau + \frac{5\theta^3 + 36\lambda\theta^2 + 48\lambda^3 + 72\theta\lambda^2}{3(6\lambda^2 + 6\lambda\theta + \theta^2)}, \\
K_C &= \frac{2T_I}{K(6\lambda^2 + 6\lambda\theta + \theta^2)}, \\
T_D &= \frac{\begin{array}{r} 240\tau\lambda^4\theta + 22\tau\lambda\theta^4 + 232\tau\lambda^3\theta^2 + 106\tau\lambda^2\theta^3 + \\ 5\tau\theta^5/3 + 96\tau\lambda^5 + 89\theta^4\lambda^2/2 + 31\theta^6/36 - 48\lambda^5\theta + \\ 24\lambda^4\theta^2 + 21\lambda\theta^5/2 + 220\lambda^3\theta^3/3 - 32\lambda^6 \end{array}}{T_I(6\lambda^2 + 6\lambda\theta + \theta^2)^2}.
\end{aligned}$$

It is seen that the formulas for integrating plants are more complicated than those for stable plants. This is because integrating plants are more difficult to control.

One main feature of the control system with an integrating plant is that the closed-loop response has a large overshoot and long settling time. This problem can be solved well in a 2 DOF control system. The 2 DOF control problem will be studied in Chapter 9.

Exercises

1. Assume that $G(s)$ is MP and rational, but not necessarily stable. Prove that all stabilizing controllers can be parameterized as

$$C(s) = \frac{1}{Q'(s)} - \frac{1}{G(s)},$$

where $Q'(s)$ is a stable nonzero transfer function.

2. The very large spacecraft often extends solar panels to generate electrical energy (Figure E7.1). As a result, the spacecraft is very flexible. When a thruster fires to push such a spacecraft to change position, the entire vehicle moves (rigid motion) about some center of gravity, but individual parts of the vehicle bend and oscillate (flexible motion) in the same way that a taut guitar string oscillates when plucked. Most aircrafts and satellites have both rigid and flexible behaviors. The model of a spacecraft with flexible behaviors is given by

$$G(s) = \frac{1}{s^2}(\text{Rigid}) - \frac{1}{s^2+s+1}(\text{Flexible}).$$

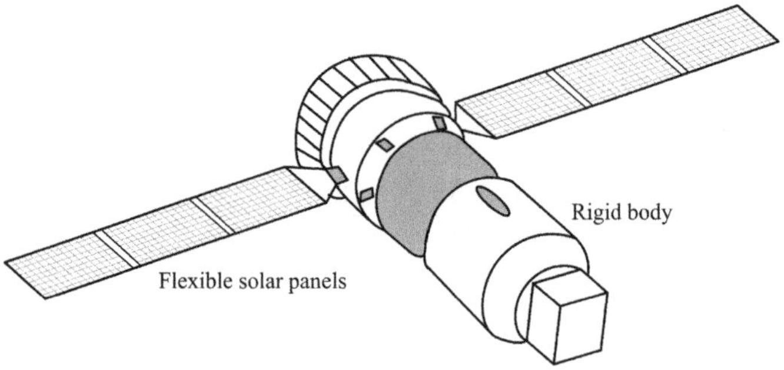

FIGURE E7.1
A flexible spacecraft.

Design a quasi-H_∞ controller for the spacecraft.

3. The following model can be used to describe the behavior of some distillation columns:

$$G(s) = \frac{(1 - ke^{-\theta s})}{s}.$$

The system input is a step. Design a $Q(s)$ with the H_2 method.

4. Disk drives are widely used in computers of all sizes. The goal of the disk drive reader device is to position the reader head in order to read the data stored in a track on the disk (Figure E7.2). The disk drive reader uses a voice coil motor to rotate the reader arm. The error signal is provided by reading a prerecorded index track. The model of a disk drive is given by

$$G(s) = \frac{5}{s(s+20)}.$$

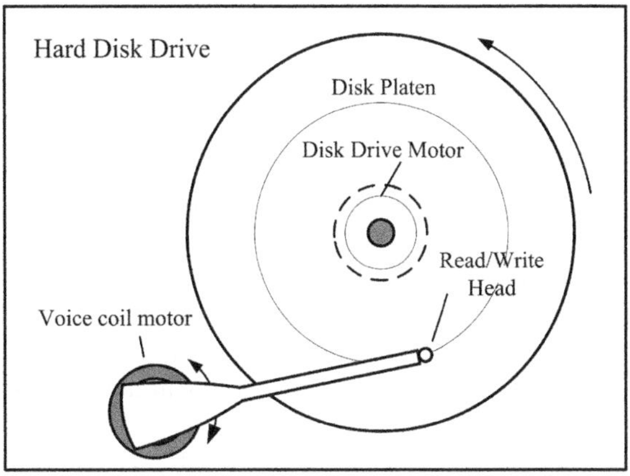

FIGURE E7.2
A disk drive.

The time unit here is ms. Design a PID controller to reach the requirements that the overshoot is less than 5% and the settling time is less than 50 ms.

5. Let $N(s)$ be a polynomial and

$$N(s_i) = \gamma_i, \quad i = 1, 2, ..., r.$$

Then $N(s)$ can be obtained by using the Lagrange's interpolation formula:

$$N(s) = \gamma_1 \prod_{j \neq 1} \frac{s - s_j}{s_1 - s_j} + \gamma_2 \prod_{j \neq 2} \frac{s - s_j}{s_2 - s_j} + ... + \gamma_r \prod_{j \neq r} \frac{s - s_j}{s_r - s_j}.$$

The formula can be used to solve many problems, for example, the model match problem. Find a polynomial $N(s)$ such that $N(1) = 0.5$ and $N(2) = 4$.

6*. The rotational dynamics of a s spacecraft rigid body can be modeled as

$$G(s) = \frac{1}{Js^2},$$

where $J = 5700$ denotes the polar moment of inertia. The design goal is to find a stabilizing controller $C(s)$ that has a control loop bandwidth 10 rad/s. Design the controller by state space methods and compare the design procedure with that in this chapter.

Notes and References

The material in Section 7.1 is based on Zhang (1998).

The design method for the H_∞ PID controller in Section 7.2 is drawn from Zhang (1998) and Zhang et al. (1999).

The H_2 PID controller in Section 7.3 was given in Zhang (1998).

The design problem for systems with integrating plants was also studied by Morari and Zafiriou (1989), Normey-Rico and Camacho (2002), Kwak et al. (2001), Chien et al. (2002), etc.

The plant in Example 7.3.1 is from Chien and Fruehauf (1990). Figure 7.3.5 is drawn based on Chien and Fruehauf (1990, Figure 10).

The presentation in Section 7.4 is based on Zhang et al. (2006).

The parameterization in Exercise 1 is the main result of Glaria and Goodwin (1994). For further discussion one can refer to Zhang et al. (2002).

The plant in Exercise 2 was given by Stefani et al. (2002, p. 480).

The plant in Exercise 4 is adapted from Dorf and Bishop (2001, p. 719).

Exercise 5 is from Dorato et al. (1992, p. 14).

Exercise 6 is adapted from Mathworks (2001, p. 76). This plant is selected because it is from a typical example used to illustrate the H_∞ control. The discussion about the double-integrator plant can also be found in Liu et al. (2004).

8

Control of Unstable Plants

CONTENTS

This chapter concentrates on the control problem of unstable plants. In SISO design, all plants are categorized into three types: stable plants, integrating plants, and unstable plants. In this way, controllers can be designed aiming at the reduced scope of plants, so that the design is more effective and simple controllers are easier to obtain.

Most practical plants are stable. Unstable dynamics can only be found in a few plants. Such plants are difficult to control. There are two main challenges in the design of control systems with an unstable plant. First, the existence of RHP poles makes the stabilization of the closed-loop system difficult to reach. Next, the combined effect of RHP poles and the time delay greatly limits the achievable performance. The study of this topic is of great theoretical significance, since stable plants and integrating plants can be viewed as the special case of unstable plants.

In the control system with an unstable plant, there exists a limit to the ratio of the time constant to the time delay. If no further explanation is given, it is assumed that the condition is satisfied.

8.1 Controller Parameterization for General Plants

The Youla parameterization is an important basis of many optimization-based design methods, since it automatically guarantees the internal stability of the obtained closed-loop system and thus simplifies the searching procedure for optimal controllers. Nevertheless, while this tool possesses considerable advantages, it suffers from several limitations in the design methods of this book:

1. The parameterization cannot be directly used for the plant with time delay.
2. To obtain the parameterization, one has to compute the coprime factorization of plant. As no analytical methods are available, the computation is involved.
3. The $Q(s)$ in the general parameterization no longer corresponds to the IMC controller.

In this section, a parameterization for the plant with time delay will be developed based on algebra theory. As a matter of fact, the special cases of the new parameterization have already been used in the preceding chapters.

Consider the unity feedback loop, in which the transfer function of the plant is given by

$$G(s) = \frac{KN_+(s)N_-(s)}{M_+(s)M_-(s)} e^{-\theta s}, \tag{8.1.1}$$

where K is the gain, θ is the time delay, $N_-(s)$ and $M_-(s)$ are the polynomials with roots in the open LHP, $N_+(s)$ and $M_+(s)$ are the polynomials with roots in the closed RHP, $N_+(0) = N_-(0) = M_-(0) = M_+(0) = 1$, and $\deg\{N_+\} + \deg\{N_-\} \leq \deg\{M_-\} + \deg\{M_+\}$. Assume that $G(s)$ has r_p unstable poles and the unstable pole $p_j(\mathrm{Re}(p_j) \geq 0; j = 1, 2, ..., r_p)$ is of l_j multiplicity; that is,

$$M_+(s) = \prod_{j=1}^{r_p} (s - p_j)^{l_j}. \tag{8.1.2}$$

$N_+(s)$ and $M_+(s)$ do not have common roots; that is, there is no RHP zero-pole cancellation in $G(s)$.

Define

$$Q(s) = \frac{C(s)}{1 + G(s)C(s)}, \tag{8.1.3}$$

which corresponds to the IMC controller. The closed-loop system is internally stable, if and only if all elements in the transfer matrix $\boldsymbol{H}(s)$ are stable:

$$\boldsymbol{H}(s) = \begin{bmatrix} G(s)Q(s) & G(s)[1 - G(s)Q(s)] \\ Q(s) & -G(s)Q(s) \end{bmatrix}. \tag{8.1.4}$$

Theorem 8.1.1. *The unity feedback system with a general plant $G(s)$ is internally stable if and only if*

1. $Q(s)$ is stable.

2. $[1 - G(s)Q(s)]G(s)$ is stable.

Or equivalently,

1. $Q(s)$ is stable.

2. $1 - G(s)Q(s)$ has zeros wherever $G(s)$ has unstable poles.

3. All RHP zero-pole cancellations in $[1 - G(s)Q(s)]G(s)$ are removed.

Proof. Similar to that of Theorem 7.1.1. □

Example 8.1.1. *This example is used to illustrate that the third condition in Theorem 8.1.1 is necessary.*

Consider the plant with the transfer function

$$G(s) = \frac{1}{s-1}.$$

$G(s)$ has one simple RHP pole at $s = 1$. Construct a controller

$$C(s) = \frac{s-1}{e^{0.1s}(e^{-0.1}s - 0.1s + 0.1) - 1}.$$

The $Q(s)$ corresponding to this $C(s)$ is

$$Q(s) = \frac{s-1}{e^{0.1s}(e^{-0.1}s - 0.1s + 0.1)}.$$

$Q(s)$ is stable. The first condition is satisfied. Furthermore,

$$1 - G(s)Q(s) = \frac{e^{0.1s}(e^{-0.1}s - 0.1s + 0.1) - 1}{e^{0.1s}(e^{-0.1}s - 0.1s + 0.1)}.$$

It has zeros where $G(s)$ has unstable poles. The second condition is also satisfied.

However, the closed-loop system is internally unstable, because there exists a RHP zero-pole cancellation in $[1-G(s)Q(s)]G(s)$, which cannot be removed.

The case associated with the third condition occurs only in the system where the plant or the controller contains a time delay. If both the plant and the controller are rational, it is not necessary to consider the third condition.

In control system design, $G(s)Q(s)$ is always stable. Since $[1 - G(s)Q(s)]G(s) = C^{-1}(s)Q(s)G(s)$, the third condition can be satisfied by removing the RHP zero-pole cancellation in $C(s)$ with a rational approximation.

Theorem 8.1.2. *All controllers that make the unity feedback control system internally stable can be parameterized as*

$$C(s) = \frac{Q(s)}{1 - G(s)Q(s)},$$

where

$$Q(s) = \frac{Q_1(s)M_+(s)}{K}.$$

$Q_1(s)$ *is any stable transfer function that makes* $Q(s)$ *proper and satisfies*

$$\lim_{s \to p_j} \frac{d^k}{ds^k}\left[1 - \frac{Q_1(s)N_+(s)N_-(s)e^{-\theta s}}{M_-(s)}\right] = 0, k = 0, 1, ..., l_j - 1,$$

and all RHP zero-pole cancellations in $[1 - G(s)Q(s)]G(s)$ *are removed.*

Proof. To guarantee the internal stability of the closed-loop system, firstly, $Q(s)$ should be stable. This implies that $Q(s)$ should be proper and $Q_1(s)$ should be stable.

Secondly, $[1 - G(s)Q(s)]G(s)$ should be stable. This condition has three implications: $Q(s)$ cancels all RHP poles of $G(s)$, $1 - G(s)Q(s)$ cancels all RHP poles of $G(s)$, and all RHP zero-pole cancellations in $[1 - G(s)Q(s)]G(s)$ are removed. All stable transfer functions that have zeros wherever $G(s)$ has RHP poles can be expressed as

$$Q(s) = \frac{Q_1(s)M_+(s)}{K},$$

where $Q_1(s)$ is a stable transfer function that makes $Q(s)$ proper. It follows that

$$1 - G(s)Q(s) = 1 - \frac{Q_1(s)N_+(s)N_-(s)e^{-\theta s}}{M_-(s)}.$$

That $1 - G(s)Q(s)$ has zeros wherever $G(s)$ has RHP poles is equivalent to

$$\lim_{s \to p_j} \frac{d^k}{ds^k}\left[1 - \frac{Q_1(s)N_+(s)N_-(s)e^{-\theta s}}{M_-(s)}\right] = 0, k = 0, 1, ..., l_j - 1.$$

□

Corollary 8.1.3. *Assume that* $G(s)$ *is a stable plant. That is,* $M_+(s) = 1$. *All controllers that make the unity feedback control system internally stable can be parameterized as*

$$C(s) = \frac{Q(s)}{1 - G(s)Q(s)},$$

where $Q(s)$ *is any stable transfer function.*

Example 8.1.2. *Consider a plant with the transfer function*

$$G(s) = \frac{s-2}{(s-1)(s+2)}.$$

The plant has only one simple unstable pole at $s = 1$. By Theorem 8.1.2, we have

$$Q(s) = (s-1)Q_1(s),$$

where $Q_1(s)$ is a stable transfer function satisfying

$$\lim_{s\to 1}\left[1 - Q_1(s)\frac{s-2}{s+2}\right] = 0.$$

This is equivalent to

$$Q_1(s) = -3 + (s-1)Q_2(s),$$

where $Q_2(s)$ is any stable transfer function that makes $Q(s)$ proper. All controllers that make the unity feedback system internally stable can be parameterized as

$$C(s) = \frac{(s-1)(s+2)[-3+(s-1)Q_2(s)]}{(s+2)-(s-2)[-3+(s-1)Q_2(s)]}.$$

Example 8.1.3. *Consider the stabilizing problem of the plant*

$$G(s) = \frac{1}{(s-1)(s-2)}.$$

The plant has one unstable pole at $s = 1$ and $s = 2$, respectively. Then

$$Q(s) = (s-1)(s-2)Q_1(s),$$

where $Q_1(s)$ is a stable transfer function satisfying

$$\begin{aligned}\lim_{s\to 1}[1 - Q_1(s)] &= 0,\\ \lim_{s\to 2}[1 - Q_1(s)] &= 0.\end{aligned}$$

This is equivalent to

$$Q_1(s) = 1 + (s-1)(s-2)Q_2(s),$$

where $Q_2(s)$ is any stable transfer function that makes $Q(s)$ proper. All controllers that make the unity feedback system internally stable can be parameterized as

$$\begin{aligned}C(s) &= \frac{1+(s-1)(s-2)Q_2(s)}{-Q_2(s)}\\ &= (s-1)(s-2) - \frac{-1}{Q_2(s)}.\end{aligned}$$

When the system performance is considered, it is always desirable that the system should have asymptotic tracking property. The parameterization can be further developed to cover the requirement of asymptotic tracking.

Theorem 8.1.4. *All controllers that make the unity feedback control system internally stable and have the asymptotic tracking property for a step input can be parameterized as*

$$C(s) = \frac{Q(s)}{1 - G(s)Q(s)},$$

where

$$Q(s) = \frac{[1 + sQ_2(s)]M_+(s)}{K}.$$

$Q_2(s)$ is any stable transfer function that makes $Q(s)$ proper and satisfies

$$\lim_{s \to p_j} \frac{d^k}{ds^k}\left\{1 - \frac{[1 + sQ_2(s)]N_+(s)N_-(s)e^{-\theta s}}{M_-(s)}\right\} = 0, k = 0, 1, ..., l_j - 1,$$

and all RHP zero-pole cancellations in $[1 - G(s)Q(s)]G(s)$ are removed.

Proof. The proof of this theorem is similar to that of Theorem 8.1.2.

$Q(s)$ should be stable and have zeros wherever $G(s)$ has RHP poles. Such a transfer function can be expressed as

$$Q(s) = \frac{Q_1(s)M_+(s)}{K},$$

where $Q_1(s)$ is stable. If

$$\lim_{s \to 0}[1 - G(s)Q(s)] = 0,$$

the closed-loop system possesses the asymptotic tracking property, which implies that

$$Q_1(s) = 1 + sQ_2(s),$$

where $Q_2(s)$ is a stable transfer function that makes $Q(s)$ proper. This leads to

$$1 - G(s)Q(s) = 1 - \frac{[1 + sQ_2(s)]N_+(s)N_-(s)e^{-\theta s}}{M_-(s)}.$$

$Q_2(s)$ should satisfy

$$\lim_{s \to p_j} \frac{d^k}{ds^k}\left\{1 - \frac{[1 + sQ_2(s)]N_+(s)N_-(s)e^{-\theta s}}{M_-(s)}\right\} = 0, k = 0, 1, ..., l_j - 1.$$

The condition cannot guarantee the internal stability of the unity feedback control system, unless all RHP zero-pole cancellations in $[1 - G(s)Q(s)]G(s)$ are removed. □

It is seen that no coprime factorization is used in the new controller parameterization. Nevertheless, the properness of $Q(s)$ and the related constraints must be tested. In the design framework of this book, the parameterization is only used to derive the analytical design formula. Its computing is not necessary.

8.2 H_∞ PID Controller for Unstable Plants

Assume that the transfer function of the plant is

$$G(s) = \frac{K}{\tau s - 1} e^{-\theta s}, \tag{8.2.1}$$

where K is the gain, τ is the time constant, θ is the time delay. As we know, if the time delay in a plant is rigorously dealt with, it is impossible to design a PID controller analytically. A rational approximation has to be used to approximate the time delay or the whole controller.

In the control of unstable plants, two rational approximations are often used to expand a time delay: one is the first-order Taylor series expansion:

$$e^{-\theta s} \approx 1 - \theta s;$$

the other is the first-order lag expansion:

$$e^{-\theta s} \approx \frac{1}{1 + \theta s}.$$

Here the design is carried out only for the former. For the latter, the design procedure is similar.

With the help of the first-order Taylor series expansion, the approximate plant is obtained as follows:

$$G(s) \approx \frac{K(1 - \theta s)}{\tau s - 1}. \tag{8.2.2}$$

The plant has a RHP pole at $s = 1/\tau$. Obviously, θ and τ should not be equal; otherwise, there would be a RHP zero-pole cancellation in the model. If the closed-loop system is internally stable, then

$$Q(s) = \frac{(\tau s - 1)Q_1(s)}{K}.$$

Take the performance index as $\min \|W(s)S(s)\|_\infty$ with $W(s) = 1/s$. Then

$$\begin{aligned} W(s)S(s) &= W(s)[1 - G(s)Q(s)] \\ &= W(s)[1 - Q_1(s)(1 - \theta s)]. \end{aligned} \tag{8.2.3}$$

According to Theorem 4.2.1, we have

$$\begin{aligned} \|W(s)[1-Q_1(s)(1-\theta s)]\|_\infty &\geq |W(1/\theta)| \\ &= \theta. \end{aligned} \tag{8.2.4}$$

The left-hand side reaches its minimum value when

$$Q_{1opt}(s) = 1.$$

Then the optimal controller is

$$Q_{opt}(s) = \frac{\tau s - 1}{K}. \tag{8.2.5}$$

To guarantee that the controller is physically realizable, a filter must be introduced: $Q(s) = Q_{opt}(s)J(s)$. The closed-loop system with the filter should be internally stable:

$$\lim_{s \to 1/\tau} [1 - G(s)Q(s)] = 0, \tag{8.2.6}$$

and should possess the asymptotic tracking property:

$$\lim_{s \to 0} [1 - G(s)Q(s)] = 0. \tag{8.2.7}$$

It is evident that a first-order filter cannot satisfy these requirements. Similar to the control of integrating plants, take

$$J(s) = \frac{\beta s + 1}{(\lambda s + 1)^2}, \tag{8.2.8}$$

where λ is the performance degree and β is a positive real number. As $J(0) = 1$, (8.2.7) is satisfied. It is noticed that a zero is introduced in $J(s)$. The zero is used to satisfy (8.2.6). Elementary computation gives

$$\beta = \frac{\lambda^2 + 2\lambda\tau + \theta\tau}{\tau - \theta}. \tag{8.2.9}$$

One readily obtains the suboptimal controller

$$Q(s) = \frac{(\tau s - 1)(\beta s + 1)}{K(\lambda s + 1)^2}. \tag{8.2.10}$$

It follows that

$$\begin{aligned} C(s) &= \frac{Q(s)}{1 - G(s)Q(s)} \\ &= \frac{\alpha}{K}\left(1 + \frac{1}{\beta s}\right), \end{aligned} \tag{8.2.11}$$

where

$$\alpha = \frac{\lambda^2 + 2\lambda\tau + \theta\tau}{(\lambda+\theta)^2}. \tag{8.2.12}$$

$C(s)$ is a PI controller. When it is used to control the original plant, the closed-loop transfer function is

$$T(s) = \frac{\left(\dfrac{\lambda^2 + 2\lambda\tau + \theta\tau}{\tau - \theta}s + 1\right)e^{-\theta s}}{(\tau s - 1)\dfrac{(\lambda+\theta)^2}{\tau - \theta}s + \left(\dfrac{\lambda^2 + 2\lambda\tau + \theta\tau}{\tau - \theta}s + 1\right)e^{-\theta s}}. \tag{8.2.13}$$

The H_∞ PID controller remains the feature of the controller for stable plants. The performance and robustness can be quantitatively tuned with the performance degree. Nevertheless, in the control system with an unstable plant, the closed-loop response is affected by the time constant in addition to the time delay (Figure 8.2.1–Figure 8.2.4).

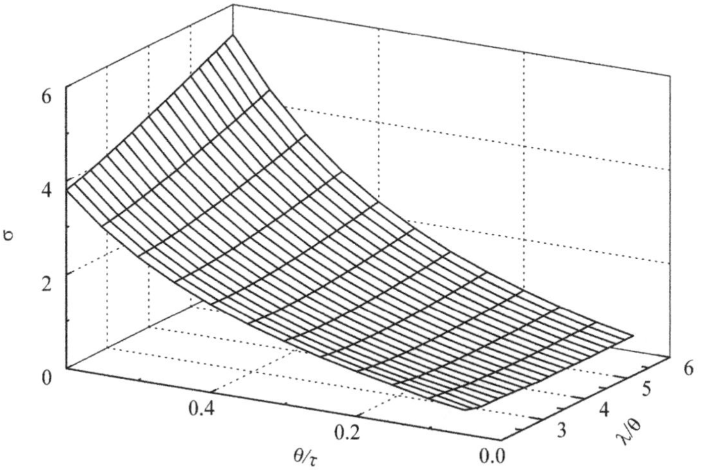

FIGURE 8.2.1
Overshoot of the H_∞ control system with an unstable plant.

Similar to the design in Section 4.2, to get a controller that makes the closed-loop system have the desired properties, the requirements on $Q(s)$ for internal stability and asymptotic tracking are temporarily relaxed when designing the optimal controller $Q_{opt}(s)$. These requirements can be satisfied by introducing a filter $J(s)$ to $Q_{opt}(s)$. It is possible to consider these requirements in designing $Q_{opt}(s)$. This, however, imposes additional constraints on the design. As a result, the obtained $Q_{opt}(s)$ is complicated. The mathematically precise readers can interpret the design in such a way: (8.2.10) is directly given as the solution to (8.2.4) without the deriving procedure in between.

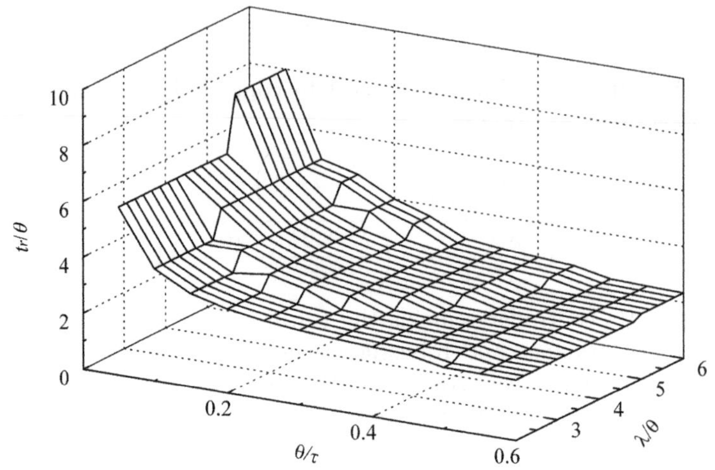

FIGURE 8.2.2
Rise time of the H_∞ control system with an unstable plant.

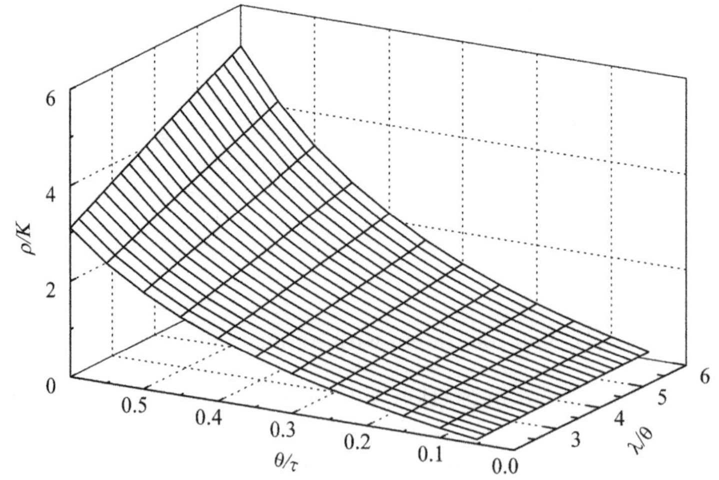

FIGURE 8.2.3
Perturbation peak of the H_∞ control system with an unstable plant.

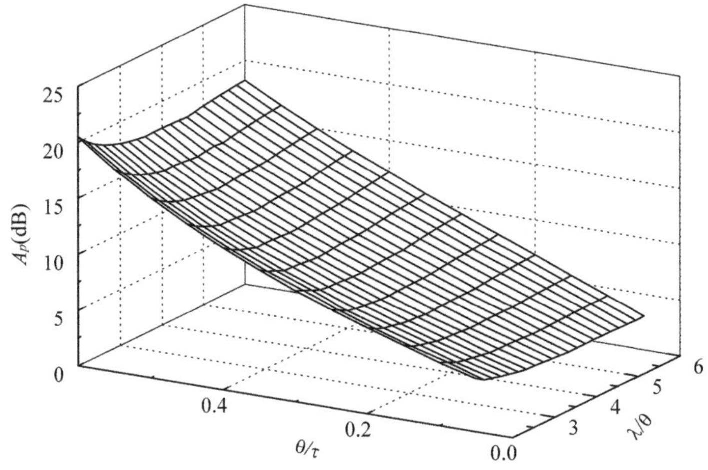

FIGURE 8.2.4
Resonance peak of the H_∞ control system with an unstable plant.

One may use the second-order model with time delay:

$$G(s) = \frac{K}{(\tau_1 s - 1)(\tau_2 s + 1)} e^{-\theta s}, \tag{8.2.14}$$

where τ_1 and τ_2 are two time constants. With the same design procedure, the following optimal controller can be obtained:

$$Q_{opt}(s) = \frac{(\tau_1 s - 1)(\tau_2 s + 1)}{K}. \tag{8.2.15}$$

A filter that can guarantee the internal stability and asymptotic tracking property is

$$J(s) = \frac{\beta s + 1}{(\lambda s + 1)^3}, \tag{8.2.16}$$

where

$$\beta = \frac{\lambda^3 + 3\lambda^2 \tau_1 + 3\lambda \tau_1^2 + \theta \tau_1^2}{\tau_1(\tau_1 - \theta)}. \tag{8.2.17}$$

A little algebra yields

$$C(s) = \frac{1}{K} \frac{(\tau_2 s + 1)(\beta s + 1)}{s(\lambda^3 s/\tau_1 + \alpha)}, \tag{8.2.18}$$

where

$$\alpha = \frac{\lambda^3 + 3\lambda^2 \tau_1 + 3\lambda \tau_1 \theta + \theta^2 \tau_1}{\tau_1(\tau_1 - \theta)}. \tag{8.2.19}$$

If the PID controller is in the form of

$$C(s) = K_C\left(1 + \frac{1}{T_I s} + T_D s\right)\frac{1}{T_F s + 1},$$

the controller parameters are

$$\begin{aligned} T_F &= \frac{\lambda^3}{\tau_1 \alpha}, \quad T_I = \tau_2 + \beta, \\ T_D &= \frac{\tau_2 \beta}{\tau_2 + \beta}, \quad K_C = \frac{\tau_2 + \beta}{K\alpha}. \end{aligned} \tag{8.2.20}$$

8.3 H_2 PID Controller for Unstable Plants

This section studies the H_2 design problem, that is, design a PID controller for the plant $G(s)$ to minimize the 2-norm of the weighted sensitivity function. The design procedure is to parameterize all stabilizing controllers first, and then derive the optimal PID controller.

The first-order Taylor series expansion is used to obtain the following plant:

$$G(s) \approx \frac{K(1 - \theta s)}{\tau s - 1}. \tag{8.3.1}$$

The controller that makes the closed-loop system internally stable and have the asymptotic tracking property can be expressed as

$$Q(s) = \frac{(\tau s - 1)[1 + sQ_1(s)]}{K}. \tag{8.3.2}$$

$G(s)$ has a pole in the RHP. $Q_1(s)$ should satisfy

$$\lim_{s \to 1/\tau} [1 - G(s)Q(s)] = 0 \tag{8.3.3}$$

for internal stability.

The performance index is taken as $\min \|W(s)S(s)\|_2$ and the weighting function is taken as $W(s) = 1/s$. Therefore,

$$\begin{aligned} \|W(s)S(s)\|_2^2 &= \|W(s)[1 - G(s)Q(s)]\|_2^2 \\ &= \left\|\frac{1}{s} - \frac{1 - \theta s}{s}[1 + sQ_1(s)]\right\|_2^2 \\ &= \|\theta - (1 - \theta s)Q_1(s)\|_2^2 \\ &= \left\|\frac{\theta(1 + \theta s)}{1 - \theta s} - (1 + \theta s)Q_1(s)\right\|_2^2 \end{aligned}$$

$$= \left\| \frac{2\theta}{1-\theta s} \right\|_2^2 + \|\theta + (1+\theta s)Q_1(s)\|_2^2 .$$

To minimize the right-hand side of the equality, one should take

$$Q_{1opt}(s) = \frac{-\theta}{1+\theta s}.$$

Consequently, the optimal controller is

$$Q_{opt}(s) = \frac{\tau s - 1}{K(1+\theta s)}. \tag{8.3.4}$$

In the system with an unstable plant, a filter more complex than that in the system with a stable plant has to be introduced to satisfy the constraints on internal stability and asymptotic tracking. Let $Q(s) = Q_{opt}(s)J(s)$ and

$$J(s) = \frac{\beta s + 1}{\lambda s + 1}, \tag{8.3.5}$$

where λ is the performance degree and β is a positive real number. According to (8.3.3), we get

$$\beta = \frac{\lambda\tau + \lambda\theta + 2\tau\theta}{\tau - \theta}.$$

Hence,

$$\begin{aligned} C(s) &= \frac{Q(s)}{1 - G(s)Q(s)} \\ &= \frac{\beta}{K\alpha}\left(1 + \frac{1}{\beta s}\right), \end{aligned} \tag{8.3.6}$$

where

$$\alpha = \frac{2\theta(\lambda + \theta)}{\tau - \theta}.$$

$C(s)$ is a PI controller. The relationships between the closed-loop response and the performance degree are shown in Figure 8.3.1–Figure 8.3.4.

Assume that the plant is of second order:

$$G(s) = \frac{K}{(\tau_1 s - 1)(\tau_2 s + 1)} e^{-\theta s}. \tag{8.3.7}$$

With the same design procedure, the optimal controller is obtained as follows:

$$Q_{opt}(s) = \frac{(\tau_1 s - 1)(\tau_2 s + 1)}{K(1+\theta s)}. \tag{8.3.8}$$

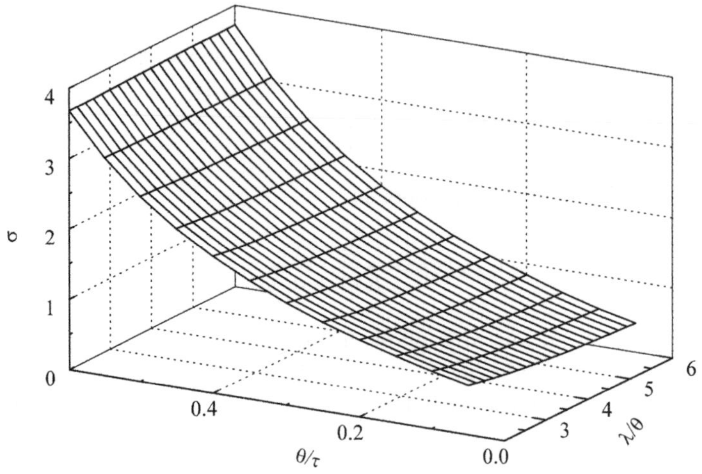

FIGURE 8.3.1
Overshoot of the H_2 control system with an unstable plant.

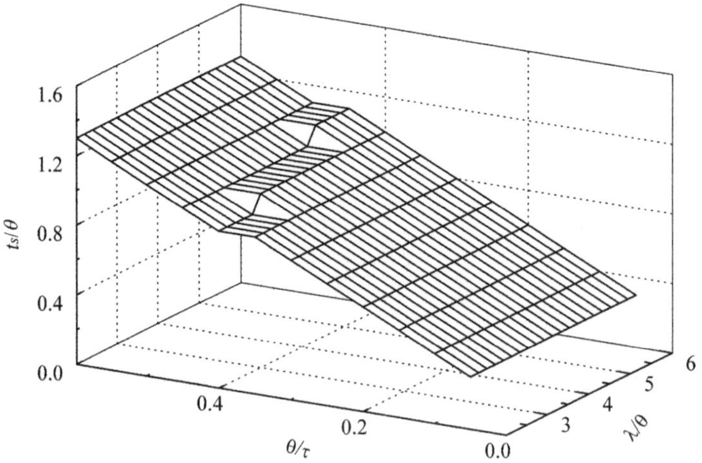

FIGURE 8.3.2
Rise time of the H_2 control system with an unstable plant.

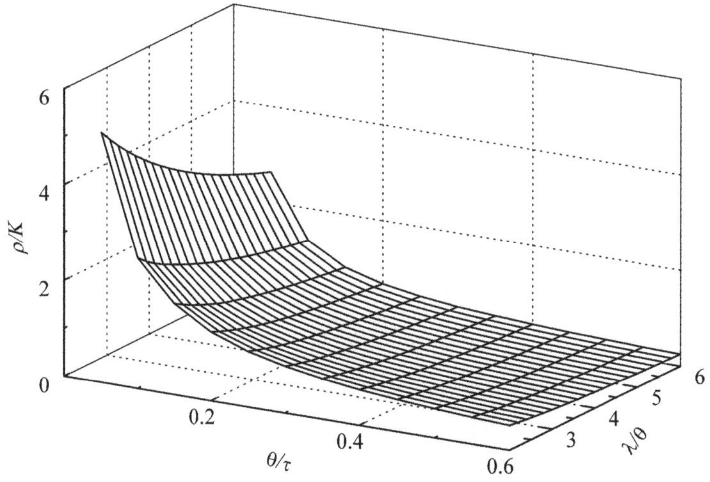

FIGURE 8.3.3
Perturbation peak of the H_2 control system with an unstable plant.

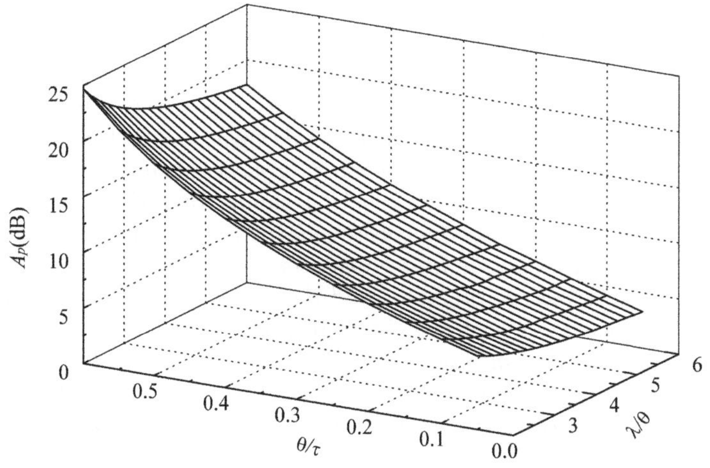

FIGURE 8.3.4
Resonance peak of the H_2 control system with an unstable plant.

The following filter can guarantee the internal stability and asymptotic tracking property:

$$J(s) = \frac{\beta s + 1}{(\lambda s + 1)^2}, \tag{8.3.9}$$

where

$$\beta = \frac{(\lambda^2 + 2\lambda\tau_1)(\tau_1 + \theta) + 2\tau_1^2\theta}{\tau_1(\tau_1 - \theta)}.$$

A little computation gives

$$C(s) = \frac{1}{K}\frac{(\tau_2 s + 1)(\beta s + 1)}{s(\lambda^2\theta s/\tau_1 + \alpha)}, \tag{8.3.10}$$

where

$$\alpha = \frac{\lambda^2(\tau_1 + \theta) + 4\lambda\tau_1\theta + 2\theta^2\tau_1}{\tau_1(\tau_1 - \theta)}.$$

Comparing it with the PID controller

$$C(s) = K_C\left(1 + \frac{1}{T_I s} + T_D s\right)\frac{1}{T_F s + 1},$$

the following parameters are obtained:

$$\begin{aligned} T_F &= \frac{\lambda^2\theta}{\tau_1\alpha}, \quad & T_I &= \tau_2 + \beta, \\ T_D &= \frac{\tau_2\beta}{\tau_2 + \beta}, \quad & K_C &= \frac{\tau_2 + \beta}{K\alpha}. \end{aligned} \tag{8.3.11}$$

Example 8.3.1. *A perfectly mixed reactor is depicted in Figure 8.3.5, in which an exothermic, irreversible reaction takes place. Heat from the reaction is removed by heat transfer to the coolant in a jacket surrounding the reactor. The reactor works at an unstable working point. After the reaction begins, the temperature in the reactor increases with the temperature of the feed. The released heat is more than the heat brought out by the coolant. Therefore, the temperature in the reactor increases and the reaction speeds up. This makes the reaction release more heat and, in return, increases the temperature in the reactor. In this case, a controller is needed to guarantee the stability. Why is the reactor controlled at an unstable working point? There are two reasons: low temperature decreases the production rate, while high temperature is not safe and the quality of product is low.*

Choose the temperature in the reactor as the output and the flow rate of coolant as the manipulated variable. The dynamics of the reactor is described by

$$\boxed{G(s) = \frac{1}{s - 1}e^{-0.5s}.}$$

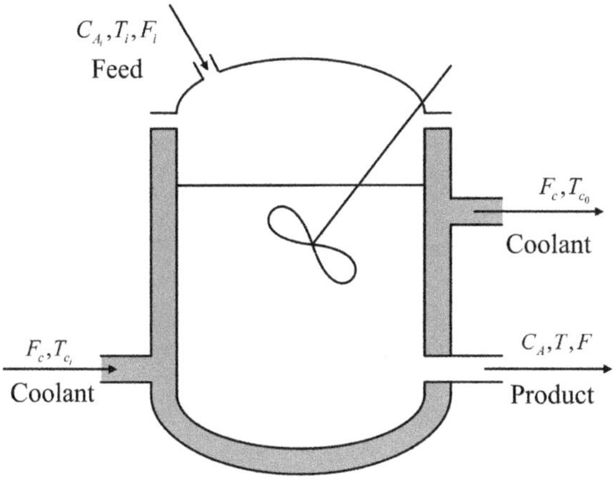

FIGURE 8.3.5
A jacket-cooled reactor.

Take $\lambda = 1.5$ for the H_∞ PI controller given by (8.2.11):

$$\boxed{C(s) = \alpha\left(1 + \frac{1}{\beta s}\right),}$$

where

$$\begin{aligned} \alpha &= \frac{2\lambda^2 + 4\lambda + 1}{2(\lambda + 0.5)^2}, \\ \beta &= 2\lambda^2 + 4\lambda + 1, \end{aligned}$$

and take $\lambda = 3$ for the H_2 PI controller given by (8.3.6):

$$\boxed{C(s) = \frac{\beta}{\alpha}\left(1 + \frac{1}{\beta s}\right),}$$

where

$$\begin{aligned} \alpha &= 2\lambda + 1, \\ \beta &= 3\lambda + 2. \end{aligned}$$

A unit step reference is added at $t = 0$ and a unit step load is added at $t = 40$. The nominal responses of the closed-loop system are shown in Figure 8.3.6. It is seen that the closed-loop responses have large overshoots. This is the common feature of control systems with unstable plants.

The first-order lag expansion can also be utilized to obtain the approximate

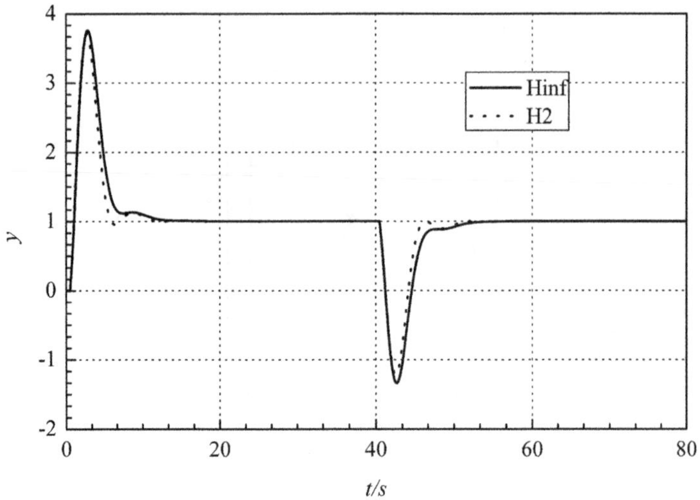

FIGURE 8.3.6
Nominal responses of the H_2 system and H_∞ system with an unstable plant.

model for PID controller design. Consider the first-order plant. Expand the time delay by employing the first-order lag. The approximate plant is

$$G(s) \approx \frac{K}{(\tau s - 1)(1 + \theta s)}. \tag{8.3.12}$$

This is an MP plant. The design results of the H_∞ control and H_2 control are the same. The optimal controller is

$$Q_{opt}(s) = \frac{(\tau s - 1)(1 + \theta s)}{K}. \tag{8.3.13}$$

It might as well take the filter

$$J(s) = \frac{\beta s + 1}{(\lambda s + 1)^3}, \tag{8.3.14}$$

where

$$\beta = \frac{\lambda^3}{\tau^2} + \frac{3\lambda^2}{\tau} + 3\lambda.$$

Then

$$C(s) = \frac{1}{Ks} \frac{(1 + \theta s)(\beta s + 1)}{\lambda^3 s/\tau^2 + \alpha}, \tag{8.3.15}$$

where

$$\alpha = \frac{\lambda^3}{\tau^2} + \frac{3\lambda^2}{\tau}.$$

Assume that the PID controller is in the form of

$$C(s) = K_C\left(1 + \frac{1}{T_I s} + T_D s\right)\frac{1}{T_F s + 1}.$$

The controller parameters are

$$\begin{aligned} T_F &= \frac{\lambda^3}{\tau\alpha} \quad , \quad T_I = \theta + \beta, \\ T_D &= \frac{\theta\beta}{\theta+\beta} \quad , \quad K_C = \frac{\theta+\beta}{K\alpha}. \end{aligned} \tag{8.3.16}$$

8.4 Performance Limitation and Robustness

Consider a control system, in which the reference signal has its energy spectrum concentrated within a known frequency range. Good performance implies that the maximum magnitude of $|S(j\omega)|$ in this frequency range is as small as possible. On the other hand, the maximum magnitude of $|S(j\omega)|$ over all frequencies, $\|S(j\omega)\|_\infty$, is not permitted to be too large. Unfortunately, the two aspects conflict. The situation is like a waterbed, and thus is called the waterbed effect. As $|S(j\omega)|$ is pushed down in one frequency range, it pops up somewhere else. NMP plants exhibit the waterbed effect.

If a plant has a zero and a pole which are close to each other in the RHP, the waterbed effect will be amplified. $|S(j\omega)|$s are then very large, both in a frequency range and over all frequencies. For example, in the following plant

$$G(s) = \frac{K(1-\theta s)}{\tau s - 1}, \tag{8.4.1}$$

if $\tau \to \theta$, then the zero and the pole of $G(s)$ are very close in the RHP. $G(s)$ tends to be internally unstable. It can be imagined that such a plant is very difficult to control.

In what follows, the performance of the system with an unstable plant is analyzed. As introduced in Section 8.1, a general unstable plant can be described by

$$G(s) = \frac{KN_+(s)N_-(s)}{M_+(s)M_-(s)}e^{-\theta s}. \tag{8.4.2}$$

Consider the quasi-H_∞ control first. The quasi-H_∞ control for stable plants provides us an insight into the choice of the desired closed-loop transfer function. The following desired closed-loop transfer function can be chosen:

$$T(s) = N_+(s)J(s)e^{-\theta s}, \tag{8.4.3}$$

where $J(s)$ is the filter:

$$J(s) = \frac{N_x(s)}{(\lambda s + 1)^{n_j}},$$

λ is the performance degree,

$$n_j = \begin{cases} \deg\{M_+\} + \deg\{M_-\} + \deg\{N_x\} - \deg\{N_-\} & \deg\{M_+\} + \deg\{M_-\} > \deg\{N_-\} \\ \deg\{N_x\} + 1 & \deg\{M_+\} + \deg\{M_-\} = \deg\{N_-\} \end{cases}.$$

$N_x(s)$ is a polynomial with roots in the open LHP, of which the order equals the number of the closed RHP poles of the plant, $N_x(0) = 1$. Since $J(0) = 1$, the closed-loop system satisfies the requirement of asymptotic tracking. $N_x(s)$ is determined by the constraint of internal stability:

$$\lim_{s \to p_j} \frac{d^k}{ds^k} [1 - T(s)] = 0, k = 0, 1, ..., l_j - 1, \tag{8.4.4}$$

or equivalently,

$$\lim_{s \to p_j} \frac{d^k}{ds^k} \left[1 - N_+(s)J(s)e^{-\theta s}\right] = 0, k = 0, 1, ..., l_j - 1.$$

When $G(s)$ and $T(s)$ are known, $Q(s)$ can be derived analytically:

$$\begin{aligned} Q(s) &= \frac{T(s)}{G(s)} \\ &= \frac{1}{K} \frac{M_+(s)M_-(s)N_x(s)}{N_-(s)(\lambda s + 1)^{n_j}}. \end{aligned} \tag{8.4.5}$$

Then the unity feedback loop controller is

$$\begin{aligned} C(s) &= \frac{T(s)}{1 - T(s)} \frac{1}{G(s)} \\ &= \frac{1}{K} \frac{M_+(s)M_-(s)N_x(s)}{N_-(s)[(\lambda s + 1)^{n_j} - N_+(s)N_x(s)e^{-\theta s}]}. \end{aligned} \tag{8.4.6}$$

From (8.4.4) it is known that $C(s)$ contains a zero and a pole at the same point in the RHP. For internal stability they must be removed by utilizing a rational approximation.

The controller can also be designed with the H_2 method. For step inputs, the $Q(s)$ that makes the closed-loop system internally stable and possesses the asymptotic tracking property can be described by

$$Q(s) = \frac{[1 + sQ_2(s)]M_+(s)}{K},$$

where $Q_2(s)$ is any stable transfer function that makes $Q(s)$ proper and satisfies

$$\lim_{s\to p_j}\frac{d^k}{ds^k}\left\{1-\frac{[1+sQ_2(s)]N_+(s)N_-(s)e^{-\theta s}}{M_-(s)}\right\}=0, k=0,1,...,l_j-1.$$

The performance index is min $\|W(s)S(s)\|_2$. Then

$$\begin{aligned}
&\|W(s)S(s)\|_2^2\\
=&\left\|W(s)\left\{1-\frac{G(s)M_+(s)}{K}[1+sQ_2(s)]\right\}\right\|_2^2\\
=&\left\|\frac{1}{s}\left[1-\frac{N_+(s)N_-(s)}{M_-(s)}e^{-\theta s}-\frac{N_+(s)N_-(s)s}{M_-(s))}e^{-\theta s}Q_2(s)\right]\right\|_2^2\\
=&\left\|\frac{M_-(s)-N_+(s)N_-(s)e^{-\theta s}}{sM_-(s)}-\frac{N_+(s)N_-(s)}{M_-(s)}e^{-\theta s}Q_2(s)\right\|_2^2\\
=&\left\|\frac{N_+(s)}{N_+(-s)}e^{-\theta s}\begin{bmatrix}\dfrac{M_-(s)N_+(-s)e^{\theta s}-N_+(s)N_-(s)N_+(-s)}{sM_-(s)N_+(s)}-\\ \dfrac{N_+(-s)N_-(s)}{M_-(s)}Q_2(s)\end{bmatrix}\right\|_2^2\\
=&\left\|\begin{matrix}\dfrac{M_-(s)N_+(-s)e^{\theta s}-N_+(s)N_-(s)N_+(-s)}{sM_-(s)N_+(s)}-\\ \dfrac{N_+(-s)N_-(s)}{M_-(s)}Q_2(s)\end{matrix}\right\|_2^2\\
=&\left\|\begin{matrix}\dfrac{N_+(-s)e^{\theta s}-N_+(s)}{sN_+(s)}+\\ \dfrac{M_-(s)-N_-(s)N_+(-s)}{sM_-(s)}-\dfrac{N_-(s)N_+(-s)}{M_-(s)}Q_2(s)\end{matrix}\right\|_2^2.
\end{aligned}$$

Since $N_+(0)=M_-(0)=N_+(0)N_-(0)=1$, s must be a factor of

$$N_+(-s)e^{\theta s}-N_+(s)$$

and

$$M_-(s)-N_-(s)N_+(-s).$$

Expanding the right-hand side of the above equality, we have

$$\begin{aligned}
&\|W(s)S(s)\|_2^2\\
=&\left\|\frac{N_+(-s)e^{\theta s}-N_+(s)}{sN_+(s)}\right\|_2^2+\\
&\left\|\frac{M_-(s)-N_-(s)N_+(-s)}{sM_-(s)}-\frac{N_-(s)N_+(-s)}{M_-(s)}Q_2(s)\right\|_2^2.
\end{aligned}$$

Minimize the right-hand side. The optimal performance is

$$\min \|W(s)S(s)\|_2^2 = \left\| \frac{N_+(-s)e^{\theta s} - N_+(s)}{sN_+(s)} \right\|_2^2. \tag{8.4.7}$$

Temporarily relax the requirement on $Q(s)$. The optimal $Q(s)$ can be derived with the help of $Q_{2opt}(s)$:

$$Q_{opt}(s) = \frac{M_+(s)M_-(s)}{KN_-(s)N_+(-s)}. \tag{8.4.8}$$

Now consider the properness problem of the controller. In order to make $Q_{opt}(s)$ proper, the filter $J(s)$ is introduced: $Q(s) = Q_{opt}(s)J(s)$. Here

$$J(s) = \frac{N_x(s)}{(\lambda s + 1)^{n_j}}, \tag{8.4.9}$$

where

$$n_j = \begin{cases} \deg\{M_+\} + \deg\{M_-\} + \deg\{N_x\} - \deg\{N_+\} - \deg\{N_-\} & \deg\{M_+\} + \deg\{M_-\} > \deg\{N_+\} + \deg\{N_-\} \\ \deg\{N_x\} + 1 & \deg\{M_+\} + \deg\{M_-\} = \deg\{N_+\} + \deg\{N_-\} \end{cases}.$$

$N_x(s)$ is a polynomial with roots in the open LHP, of which the order equals the number of the closed RHP poles of the plant. $N_x(0) = 1$. $N_x(s)$ can be derived from the constraint of internal stability:

$$\lim_{s \to p_j} \frac{d^k}{ds^k} [1 - G(s)Q_{opt}(s)J(s)] = 0, k = 0, 1, ..., l_j - 1, \tag{8.4.10}$$

or equivalently,

$$\lim_{s \to p_j} \frac{d^k}{ds^k} \left[1 - \frac{N_+(s)}{N_+(-s)} J(s)e^{-\theta s}\right] = 0, k = 0, 1, ..., l_j - 1.$$

Then

$$\begin{aligned} C(s) &= \frac{Q(s)}{1 - G(s)Q(s)} \\ &= \frac{1}{K} \frac{M_+(s)M_-(s)N_x(s)}{N_-(s)[N_+(-s)(\lambda s + 1)^{n_j} - N_+(s)N_x(s)e^{-\theta s}]}. \end{aligned} \tag{8.4.11}$$

An alternative design procedure for the quasi-H_∞ controller is as follows:

1. If the plant does not have a time delay, turn to 3.
2. If the plant contains a time delay, take the rational part of the plant as the nominal plant.

3. If the nominal plant has no zeros in the RHP, take its inverse as $Q_{opt}(s)$ and turn to 5.
4. If the nominal plant has zeros in the RHP, remove the factor that contains these zeros and take the inverse of the remainder as $Q_{opt}(s)$.
5. Introduce a filter to $Q_{opt}(s)$, compute the controller $C(s)$ and remove the RHP zero-pole cancellations in $C(s)$.

The design procedure for the H_2 controller is similar, except that Step 4 is modified as follows:

> When the nominal plant has zeros in the RHP, construct an all-pass transfer function with the factor that contains these zeros and then remove the all-pass transfer function. Take the inverse of the remainder as $Q_{opt}(s)$.

Example 8.4.1. *This example is used to illustrate the above design procedures for the quasi-H_∞ control and H_2 control.*

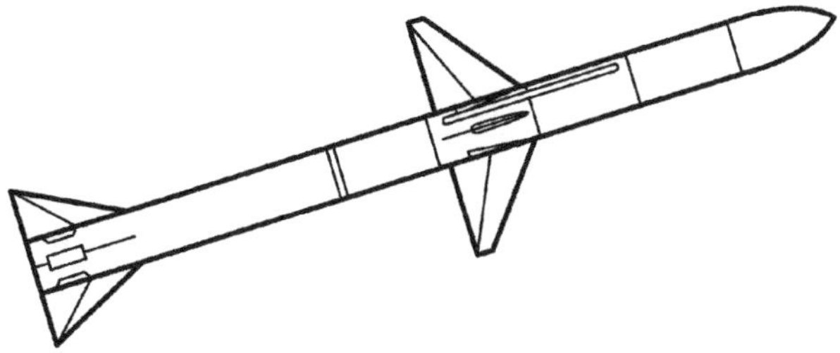

FIGURE 8.4.1
A bank-to-turn missile.

A bank-to-turn missile is controlled for yaw acceleration (Figure 8.4.1). The input is the acceleration command and the output is the acceleration. The unit of the acceleration is g. The missile dynamics is described by

$$G(s) = \frac{-0.5(s^2 - 2500)}{(s-3)(s^2 + 50s + 1000)}.$$

Normalize the plant as follows:

$$G(s) = \frac{-5(-s/50+1)(s/50+1)}{12(-s/3+1)(s^2/1000 + s/20 + 1)}.$$

The plant does not contain any time delay, but there is a RHP zero in it. If the quasi-H_∞ controller is designed, the factor that contains the zero is removed and the inverse of the remainder is taken as $Q_{opt}(s)$:

$$Q_{opt}(s) = \frac{12(-s/3+1)(s^2/1000 + s/20 + 1)}{-5(s/50+1)}.$$

For the design of the H_2 controller, an all-pass transfer function is constructed with the factor containing the zero:

$$G(s) = \frac{-5(s/50+1)^2}{12(-s/3+1)(s^2/1000+s/20+1)} \frac{-s/50+1}{s/50+1}.$$

The next step is to remove the all-pass transfer function and take the inverse of the remainder as $Q_{opt}(s)$:

$$Q_{opt}(s) = \frac{12(-s/3+1)(s^2/1000+s/20+1)}{-5(s/50+1)^2}.$$

When there exists uncertainty, the analysis of the system with an unstable plant is similar to that of the system with a stable plant. The robust stability can be tested by

$$\|\Delta_m(s)T(s)\|_\infty < 1. \tag{8.4.12}$$

Now consider the parameter uncertainty. Assume that the real plant is described by

$$\widetilde{G}(s) = \frac{\widetilde{K}\exp(-\widetilde{\theta}s)}{\widetilde{\tau}s - 1}, \tag{8.4.13}$$

where $\widetilde{K}$ is the gain, $\widetilde{\tau}$ is the time constant, and $\widetilde{\theta}$ is the time delay. The three parameters are uncertain:

$$\begin{aligned} \widetilde{K} &\in [\widetilde{K}_{min}, \widetilde{K}_{max}], \\ \widetilde{\tau} &\in [\widetilde{\tau}_{min}, \widetilde{\tau}_{max}], \\ \widetilde{\theta} &\in [\widetilde{\theta}_{min}, \widetilde{\theta}_{max}]. \end{aligned}$$

The nominal plant is constructed as follows:

$$G(s) = \frac{Ke^{-\theta s}}{\tau s - 1} \tag{8.4.14}$$

with

$$\begin{aligned} K &= \frac{\widetilde{K}_{min} + \widetilde{K}_{max}}{2}, \\ \tau &= \frac{\widetilde{\tau}_{min} + \widetilde{\tau}_{max}}{2}, \\ \theta &= \frac{\widetilde{\theta}_{min} + \widetilde{\theta}_{max}}{2}. \end{aligned}$$

The parameter uncertainty can then be expressed as

$$|\delta K| \leq \Delta K = |\widetilde{K}_{max} - K| < |K|,$$

$$\begin{aligned}|\delta\tau| \leq \Delta\tau &= |\widetilde{\tau}_{max} - \tau| < |\tau|, \\ |\delta\theta| \leq \Delta\theta &= |\widetilde{\theta}_{max} - \theta| < |\theta|.\end{aligned}$$

To use (8.4.12), one has to convert the parameter uncertainty into the unstructured uncertainty. Let the unstructured uncertain model family be

$$\widetilde{G}(s) = \frac{Ke^{-\theta s}}{\tau s - 1}\left[1 + \delta_m(s)\right], \tag{8.4.15}$$

where $|\delta_m(j\omega)| \leq |\Delta_m(j\omega)|$. When there are simultaneous uncertainties on the gain, time constant, and time delay, the following analytical expression for the unstructured uncertainty profile can be derived:

$$\Delta_m(j\omega) = \begin{cases} \left|\dfrac{|K| + \Delta K}{|K|}\dfrac{j\tau\omega - 1}{j(\tau - \Delta\tau)\omega + 1}e^{j\Delta\theta\omega} - 1\right|, & \omega < \omega^* \\ \left|\dfrac{|K| + \Delta K}{|K|}\dfrac{j\tau\omega - 1}{j(\tau - \Delta\tau)\omega + 1}\right| + 1, & \omega \geq \omega^* \end{cases}, \tag{8.4.16}$$

where ω^* is determined by

$$\begin{aligned}&-\Delta\theta\omega^* + \arctan\frac{-\Delta\tau\omega^*}{1 - \tau(-\tau + \Delta\tau)\omega^{*2}} = -\pi, \qquad (8.4.17) \\ &\frac{\pi}{2} \leq \Delta\theta\omega^* \leq \pi.\end{aligned}$$

In particular, when only the gain is uncertain (that is, $\Delta\tau = \Delta\theta = 0$), the expression simplifies to

$$\Delta_m(j\omega) = \Delta K/|K|.$$

When only the time constant is uncertain (that is, $\Delta K = \Delta\theta = 0$), the expression simplifies to

$$\Delta_m(j\omega) = \left|\frac{j\tau\omega - 1}{j(\tau - \Delta\tau)\omega - 1} - 1\right|.$$

When only the time delay is uncertain (that is, $\Delta\tau = \Delta K = 0$), $\omega^* = \pi/\Delta\theta$. In this case,

$$\Delta_m(j\omega) = \begin{cases} \left|e^{j\Delta\theta\omega} - 1\right| & \omega < \pi/\Delta\theta \\ 2 & \omega \geq \pi/\Delta\theta \end{cases}.$$

With the following tuning procedure, quantitative performance and robustness can be obtained: Increase the performance degree monotonically until the required response is obtained. Compared with the performance degrees for the stable plant and the integrating plant, the one for the unstable plant is usually large.

8.5 Maclaurin PID Controller for Unstable Plants

If the RHP zero-pole cancellation in the obtained controller cannot be removed directly, a rational approximation has to be used. There are many ways to reach this goal. In this section, the attention is paid to approximating a controller with the Maclaurin series expansion.

Consider the plant with the transfer function

$$G(s) = \frac{KN_+(s)N_-(s)}{M_+(s)M_-(s)}e^{-\theta s}. \tag{8.5.1}$$

From the discussion in the last section, it is known that the controller designed with the quasi-H_∞ method is

$$C(s) = \frac{1}{K}\frac{M_+(s)M_-(s)N_x(s)}{N_-(s)[(\lambda s+1)^{n_j} - N_+(s)N_x(s)e^{-\theta s}]}. \tag{8.5.2}$$

The controller designed with the H_2 method is

$$C(s) = \frac{1}{K}\frac{M_+(s)M_-(s)N_x(s)}{N_-(s)[N_+(-s)(\lambda s+1)^{n_j} - N_+(s)N_x(s)e^{-\theta s}]}. \tag{8.5.3}$$

It is easy to verify that $C(s)$ has a pole at the origin. Write $C(s)$ in the form of

$$C(s) = \frac{f(s)}{s}.$$

The Maclaurin series expansion of $C(s)$ is

$$C(s) = \frac{1}{s}\left[f(0) + f'(0)s + \frac{f''(0)}{2!}s^2 + ...\right]. \tag{8.5.4}$$

Take the first three terms to approximate the ideal controller. The three terms construct a PID controller:

$$C(s) = K_C\left(1 + \frac{1}{T_I s} + T_D s\right),$$

of which the parameters are

$$K_C = f'(0), \quad T_I = \frac{f'(0)}{f(0)}, \quad T_D = \frac{f''(0)}{2f'(0)}. \tag{8.5.5}$$

In this PID controller, the RHP zero-pole cancellations have been removed.

To simplify the presentation, let

$$f(s) = \frac{N(s)}{M(s)}. \tag{8.5.6}$$

The values of $f(s)$ and its first-order and second-order derivatives at the origin can be written as

$$f(0) = \frac{N(0)}{M(0)},$$
$$f'(0) = \frac{N'(0)M(0) - M'(0)N(0)}{M(0)^2},$$
$$f''(0) = \frac{\begin{array}{c} N''(0)M(0)^2 - M''(0)N(0)M(0) - \\ 2M'(0)N'(0)M(0) + 2M'(0)^2N(0). \end{array}}{M(0)^3}.$$

Consider two cases. First, assume that the plant is of first order:

$$G(s) = \frac{K}{(\tau s - 1)}e^{-\theta s}. \tag{8.5.7}$$

The quasi-H_∞ control and the H_2 control give the same closed-loop transfer function:

$$T(s) = \frac{(\beta s + 1)}{(\lambda s + 1)^2}e^{-\theta s}. \tag{8.5.8}$$

With the internal stability constraint in (8.4.4), the parameter β is obtained as follows:

$$\beta = \tau[(\lambda/\tau + 1)^2 e^{\theta/\tau} - 1].$$

Then

$$N(s) = \frac{(\tau s - 1)(\beta s + 1)}{K},$$
$$M(s) = \frac{(\lambda s + 1)^2 - (\beta s + 1)e^{-\theta s}}{s}.$$

This leads to

$$N(0) = -\frac{1}{K},$$
$$N'(0) = \frac{\tau - \beta}{K},$$
$$N''(0) = \frac{2\tau\beta}{K},$$
$$M(0) = 2\lambda + \theta - \beta,$$
$$M'(0) = \frac{2\lambda^2 - \theta^2 + 2\beta\theta}{2},$$
$$M''(0) = \frac{\theta^3 - 3\beta\theta^2}{3}.$$

The values of $f(s)$ and its first and second derivatives at the origin are

$$f(0) = -\frac{1}{K(2\lambda+\theta-\beta)},$$
$$f'(0) = \frac{(\tau-\beta)(2\lambda+\theta-\beta)+(\lambda^2-\theta^2/2+\beta\theta)}{K(2\lambda+\theta-\beta)^2},$$
$$f''(0) = \frac{\begin{array}{c}2\tau\beta(2\lambda+\theta-\beta)^2+(\theta^3/3-\beta\theta^2)(2\lambda+\theta-\beta)-\\ 2(\tau-\beta)(\lambda^2-\theta^2/2+\beta\theta)(2\lambda+\theta-\beta)-\\ 2(\lambda^2-\theta^2/2+\beta\theta)\end{array}}{K(2\lambda+\theta-\beta)^3}.$$

The controller parameters are as follows:

$$\begin{aligned} T_I &= -\tau+\beta-\frac{\lambda^2+\beta\theta-\theta^2/2}{2\lambda+\theta-\beta}, \\ K_C &= \frac{T_I}{-K(2\lambda+\theta-\beta)}, \\ T_D &= \frac{-\tau\beta-(\theta^3/6-\beta\theta^2/2)/(2\lambda+\theta-\beta)}{T_I}- \\ &\quad \frac{\lambda^2+\beta\theta-\theta^2/2}{2\lambda+\theta-\beta}. \end{aligned} \tag{8.5.9}$$

Next, assume that the plant is of second order:

$$G(s) = \frac{K}{(\tau_1 s-1)(\tau_2 s+1)}e^{-\theta s}. \tag{8.5.10}$$

In both the quasi-H_∞ control and H_2 control the closed-loop transfer function is

$$T(s) = \frac{\beta s+1}{(\lambda s+1)^3}e^{-\theta s}. \tag{8.5.11}$$

The internal stability constraint yields

$$\beta = \tau_1[(\lambda/\tau_1+1)^3 e^{\theta/\tau_1}-1].$$

Then

$$\begin{aligned} N(s) &= \frac{(\tau_1 s-1)(\tau_2 s+1)(\beta s+1)}{K}, \\ M(s) &= \frac{(\lambda s+1)^3-(\beta s+1)e^{-\theta s}}{s}. \end{aligned}$$

This leads to

$$
\begin{aligned}
N(0) &= -\frac{1}{K}, \\
N'(0) &= \frac{\tau_1 - \tau_2 - \beta}{K}, \\
N''(0) &= \frac{2\tau_1\tau_2 + 2\tau_1\beta_1 - 2\tau_2\beta}{K}, \\
M(0) &= 3\lambda + \theta - \beta, \\
M'(0) &= \frac{6\lambda^2 - \theta^2 + 2\beta\theta}{2}, \\
M''(0) &= \frac{6\lambda^3 - 3\beta\theta^2 + \theta^3}{3}.
\end{aligned}
$$

The values of $f(s)$ and its first and second derivatives at the origin are

$$
\begin{aligned}
f(0) &= -\frac{1}{K(3\lambda + \theta - \beta)}, \\
f'(0) &= \frac{(\tau_1 - \tau_2 - \beta)(3\lambda + \theta - \beta) + (3\lambda^2 - \theta^2/2 + \beta\theta)}{K(3\lambda + \theta - \beta)^2}, \\
f''(0) &= \frac{\begin{array}{l} 2(\tau_1\tau_2 + \tau_1\beta - \tau_2\beta)(3\lambda + \theta - \beta)^2 + \\ \quad (2\lambda^3 - \theta^2\beta + \theta^3/3)(3\lambda + \theta - \beta) - \\ \quad 2(\tau_1 - \tau_2 - \beta)(3\lambda^2 + \theta\beta - \theta^2/2)(3\lambda + \theta - \beta) - \\ \quad 2(3\lambda^2 + \theta\beta - \theta^2/2)^2 \end{array}}{K(3\lambda + \theta - \beta)^3}.
\end{aligned}
$$

Then, the controller parameters are

$$
\begin{aligned}
T_I &= -\frac{(\tau_1 - \tau_2 - \beta)(3\lambda + \theta - \beta) + (3\lambda^2 + \beta\theta - \theta^2/2)}{3\lambda + \theta - \beta}, \\
K_C &= \frac{T_I}{-K(3\lambda + \theta - \beta_1)}, \\
T_D &= -\frac{\begin{array}{l} 2(\tau_1\tau_2 + \tau_1\beta - \tau_2\beta)(3\lambda + \theta - \beta)^2 + \\ \quad (2\lambda^3 - \theta^2\beta + \theta^3/3)(3\lambda + \theta - \beta) - \\ \quad -(\tau_1 - \tau_2 - \beta)(6\lambda^2 + 2\theta\beta - \theta^2)(3\lambda + \theta - \beta) - \\ \quad 2(3\lambda^2 + \theta\beta - \theta^2/2)^2 \end{array}}{2T_I(3\lambda + \theta - \beta)^2}.
\end{aligned}
\tag{8.5.12}
$$

8.6 PID Design for the Best Achievable Performance

This section considers the design problem of the PID controller with the best achievable performance for unstable plants. Suppose that the plant is described by

$$G(s) = \frac{KN_+(s)N_-(s)}{M_+(s)M_-(s)}e^{-\theta s}. \tag{8.6.1}$$

According to the discussion in the last section, the controller can be expressed as

$$C(s) = \frac{f(s)}{s}. \tag{8.6.2}$$

The Maclaurin series expansion of $f(s)$ is

$$f(s) = f(0) + f'(0)s + \frac{f''(0)}{2!}s^2 + \frac{f^{(3)}(0)}{3!}s^3 + ... \tag{8.6.3}$$

A practical PID controller has the following expression:

$$C(s) = \frac{a_2 s^2 + a_1 s + a_0}{s(b_1 s + 1)}. \tag{8.6.4}$$

Let the Pade approximation of $f(s)$ be

$$\frac{a_2 s^2 + a_1 s + a_0}{b_1 s + 1}. \tag{8.6.5}$$

Then

$$\begin{bmatrix} a_0 \\ a_1 \\ a_2 \end{bmatrix} = \begin{bmatrix} f(0) & 0 \\ f'(0) & f(0) \\ f''(0)/2! & f'(0) \end{bmatrix} \begin{bmatrix} 1 \\ b_1 \end{bmatrix}, \tag{8.6.6}$$

$$b_1 f''(0)/2! = -f^{(3)}(0)/3!. \tag{8.6.7}$$

A little algebra yields that

$$\begin{aligned} a_0 &= f(0), \\ a_1 &= b_1 f(0) + f'(0), \\ a_2 &= b_1 f'(0) + f''(0)/2!, \\ b_1 &= -\frac{f^{(3)}(0)}{3f''(0)}. \end{aligned} \tag{8.6.8}$$

If the practical PID controller is in the form of

$$C = K_C\left(1 + \frac{1}{T_I s} + T_D s\right)\frac{1}{T_F s + 1},$$

the controller parameters are

$$K_C = a_1, \quad T_I = \frac{a_1}{a_0}, \quad T_D = \frac{a_2}{a_1}, \quad T_F = b_1. \tag{8.6.9}$$

All of these parameters should be positive.

Consider the first-order unstable plant:

$$G(s) = \frac{Ke^{-\theta s}}{\tau s - 1}. \tag{8.6.10}$$

The values of $f(s)$ and its first, second, and third derivatives at the origin are

$$\begin{aligned}
f(0) &= -\frac{1}{K(3\lambda + \theta - \beta)}, \\
f'(0) &= \frac{(\tau_1 - \tau_2 - \beta)(3\lambda + \theta - \beta) + (3\lambda^2 - \theta^2/2 + \beta\theta)}{K(3\lambda + \theta - \beta)^2}, \\
f''(0) &= \frac{\begin{array}{l} 2(\tau_1\tau_2 + \tau_1\beta - \tau_2\beta)(3\lambda + \theta - \beta)^2 + \\ \quad (2\lambda^3 - \theta^2\beta + \theta^3/3)(3\lambda + \theta - \beta) - \\ \quad 2(\tau_1 - \tau_2 - \beta)(3\lambda^2 + \theta\beta - \theta^2/2)(3\lambda + \theta - \beta) - \\ \quad 2(3\lambda^2 + \theta\beta - \theta^2/2)^2 \end{array}}{K(3\lambda + \theta - \beta)^3}, \\
f^{(3)}(0) &= \frac{\begin{array}{l} -6\beta\tau(\lambda^2 - \theta^2/2 + \beta\theta)(2\lambda + \theta - \beta)^2 + \\ \quad 6(\tau - \beta)(\lambda^2 - \theta^2/2 + \beta\theta)^2(2\lambda + \theta - \beta) - \\ \quad 3(\tau - \beta)(\theta^3/3 - \beta\theta^2)(2\lambda + \theta - \beta)^2 - \\ \quad (2\theta^3 - 6\beta\theta^2)(\lambda^2 - \theta^2/2 + \beta\theta)(2\lambda + \theta - \beta) + \\ \quad (\beta\theta^3 - \theta^4/4)(2\lambda + \theta - \beta)^2 + \\ \quad 6(\lambda^2 - \theta^2/2 + \beta\theta)^3 \end{array}}{K(2\lambda + \theta - \beta)^4}.
\end{aligned} \tag{8.6.11}$$

Clearly, the parameters of the PID controller are

$$\begin{aligned}
T_F &= -\frac{f^{(3)}(0)}{3f''(0)}, \\
K_C &= T_F f(0) + f'(0), \\
T_I &= \frac{K_C}{f(0)}, \\
T_D &= \frac{T_F f'(0) + f''(0)/2!}{K_C}.
\end{aligned} \tag{8.6.12}$$

The quantitative performance and robustness can also be obtained by increasing λ monotonically.

8.7 All Stabilizing PID Controllers for Unstable Plants

As we know, owing to the existence of θ, the closed-loop system is open-loop during $t < \theta$. If the plant is stable, there is no problem. The system output will ultimately reach a new equilibrium, even if no control action is added when $t \geq \theta$. Nevertheless, if the plant is unstable, the system output will continuously increase until the physical limitation is reached. This may cause such a problem. When $t < \theta$, the system output becomes very large. The PID controller does not act until $t = \theta$. Then can the controller pull this large system output back to a new equilibrium? In other words, is there any stabilizing PID controller? This section will discuss this problem.

Similar to the discussion for stable plants, the attention is concentrated on the first-order plant with time delay

$$G(s) = \frac{Ke^{-\theta s}}{\tau s - 1} \tag{8.7.1}$$

and the standard PID controller

$$C = K_C + \frac{K_I}{s} + K_D s, \tag{8.7.2}$$

where $K_I = K_C/T_I$ and $K_D = K_C T_D$.

Theorem 8.7.1. *If $\theta \geq 2\tau$, there exists no stabilizing PID controller for the first-order unstable plant with time delay.*

Proof. The following quasi-polynomial can be used to analyze the stability of the closed-loop system:

$$\delta^*(s) = -K(K_I + K_C s + K_D s^2) + (1 - \tau s)se^{\theta s}.$$

The imaginary part of $\delta^*(j\omega)$ is

$$\delta_i(\omega) = \omega[-KK_C + \cos(\theta\omega) + \tau\omega\sin(\theta\omega)].$$

Define the following function:

$$f(z, K_C) = \frac{-KK_C + \cos(z)}{\sin(z)},$$

where $z = \theta\omega$. To prove the theorem, it is sufficient to prove that the roots of $\delta_i(\omega)$ are not all real for $\theta \geq 2\tau$, or equivalently, $f(z, K_C)$ and the line $f(z) = -\tau z/\theta$ do not intersect in $(0, \pi)$. It was discussed in the proof of Theorem 4.6.1 that such a case implied instability.

Consider $K_{C1} < K_{C2}$. For any $z \in (0, \pi)$,

$$-KK_{C1} + \cos(z) > -KK_{C2} + \cos(z).$$

Since $\sin(z) > 0$,

$$f(z, K_{C1}) > f(z, K_{C2}).$$

In other words, for any fixed $z \in (0, \pi)$, $f(z, K_C)$ decreases monotonically with the increase of K_C. Hence, for $K_C > 1/K$ and any $z \in (0, \pi)$,

$$f(z, K_C) < f(z, \frac{1}{K}).$$

This implies that if $f(z) = -\tau z/\theta$ does not intersect the curve $f(z, 1/K)$ in $(0, \pi)$, it will not intersect any other curve $f(z, K_C)$ in the same interval.

It is observed that for any $z \in (0, \pi)$

$$f(z, \frac{1}{K}) = \frac{-1 + \cos(z)}{\sin(z)} = -\tan\left(\frac{z}{2}\right).$$

Define a continuous extension of $f(z, 1/K)$ over $[0, \pi)$ by

$$f_1(z, \frac{1}{K}) = -\tan\left(\frac{z}{2}\right).$$

Clearly, the curve $f_1(z, 1/K)$ intersects the line $-\tau z/\theta$ at $z = 0$ (Figure 8.7.1). Also, it is observed that the slope of the tangent to $f_1(z, 1/K)$ at $z = 0$ is given by

$$\frac{df_1}{dz} = -\frac{1}{2}\sec^2\left(\frac{z}{2}\right)\bigg|_{z=0} = -\frac{1}{2}.$$

If this slope is less than or equal to $-\tau/\theta$, then no further intersections will take place over $(0, \pi)$. Since $f(z, 1/K) = f_1(z, 1/K)$ in $(0, \pi)$, the curve $f(z, 1/K)$ will not intersect the line $-\tau z/\theta$ for $\theta \geq 2\tau$. This completes the proof. □

Now, assume that the condition in Theorem 8.7.1 is satisfied. When the closed-loop system is stable, within what range should the PID controller parameters be? The following theorem gives the answer.

Theorem 8.7.2. *If $\theta < 2\tau$, then the unstable plant can be stabilized by a PID controller if and only if*

$$\frac{1}{K} < K_C < K_T,$$

where

$$K_T = \frac{1}{K}\left[\frac{-\tau}{\theta}\alpha_1 \sin(\alpha_1) + \cos(\alpha_1)\right],$$

and α_1 is the solution to the equation

$$\tan(\alpha) = \frac{\tau}{\tau + \theta}\alpha$$

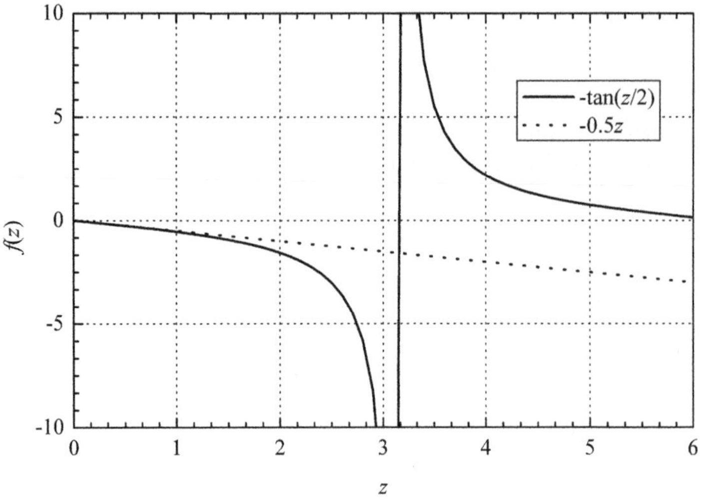

FIGURE 8.7.1
Plots of the curve $f_1(z, 1/K)$ and the line $f(z) = -\tau z/\theta$.
(From Silva et al., 2002. Reprinted by permission of the IEEE)

in the interval $(0, \pi)$*. In particular, when* $\theta = \tau$, $\alpha_1 = \pi/2$.

Furthermore, for each $K_C \in (1/K, K_T)$*, the stabilizing region of the integral constant and the derivative constant is the quadrilateral in Figure 8.7.2. Here*

$$
\begin{aligned}
m(z) &= \frac{\theta^2}{z^2}, \\
b(z) &= \frac{\theta}{Kz}\left[\sin(z) - \frac{\tau}{\theta}z\cos(z)\right], \\
w(z) &= -\frac{z}{K\theta}\left\{\sin(z) - \frac{\tau}{\theta}z[\cos(z)+1]\right\},
\end{aligned}
$$

and $z_j (j = 1, 2, ...)$ *are the positive real roots of*

$$-KK_C + \cos(z) + \frac{\tau}{\theta}z\sin(z) = 0.$$

These roots are arranged in an increasing order of magnitude.

Proof. The proof is similar to that for Theorem 4.6.1 and thus is omitted here. □

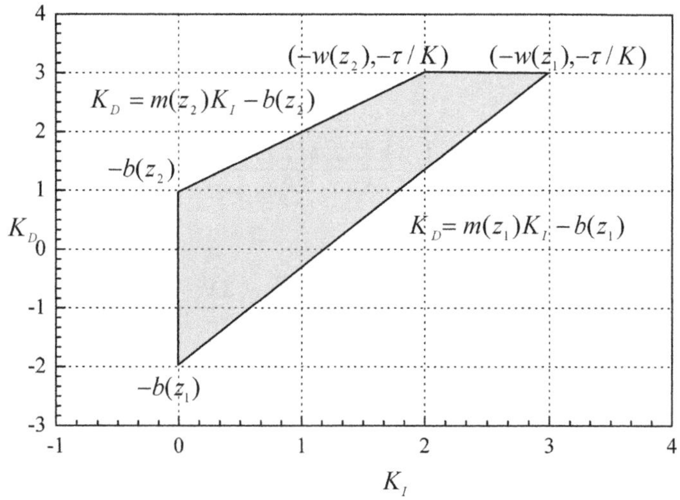

FIGURE 8.7.2
Stabilizing region of the integral and the derivative constants.
(From Silva et al., 2002. Reprinted by permission of the IEEE)

8.8 Summary

At the beginning of this chapter, a new parameterization is proposed. The main features are that it reflects the IMC structure and no coprime factorization is needed. The new parameterization is then applied to the controller design of the unstable plant with time delay.

Due to the constraint of internal stability, the controller in the system with an unstable plant can only be implemented in the unity feedback loop. Consider the following first-order plant:

$$G(s) = \frac{K}{\tau s - 1} e^{-\theta s}.$$

The H_∞ PID controller obtained in this chapter is

$$C(s) = \frac{\alpha}{K}(1 + \frac{1}{\beta s}),$$

where

$$\alpha = \frac{\lambda^2 + 2\lambda\tau + \theta\tau}{(\lambda + \theta)^2},$$

$$\beta = \frac{\lambda^2 + 2\lambda\tau + \theta\tau}{\tau - \theta}.$$

The H_2 PID controller is

$$C(s) = \frac{1}{Ks}\frac{(1 + \frac{\tau\theta}{\tau-\theta}s)(\beta s + 1)}{(\frac{\lambda^3}{\tau^2} + \frac{\theta^2\beta}{\tau-\theta})s + \alpha},$$

where

$$\begin{aligned}
\alpha &= \frac{\theta^2}{\tau - \theta} + \frac{\lambda^3}{\tau^2} + \frac{3\lambda^2}{\tau}, \\
\beta &= \frac{\lambda^3}{\tau^2} + \frac{3\lambda^2}{\tau} + 3\lambda.
\end{aligned}$$

If there is a second-order plant:

$$G(s) = \frac{K}{(\tau_1 s - 1)(\tau_2 s + 1)}e^{-\theta s},$$

the H_∞ PID controller is

$$C(s) = \frac{1}{K}\frac{(\tau_2 s + 1)(\beta s + 1)}{s(\lambda^3 s/\tau_1 + \alpha)},$$

where

$$\begin{aligned}
\alpha &= \frac{\lambda^3 + 3\lambda^2\tau_1 + 3\lambda\tau_1\theta + \theta^2\tau_1}{\tau_1(\tau_1 - \theta)}, \\
\beta &= \frac{\lambda^3 + 3\lambda^2\tau_1 + 3\lambda\tau_1^2 + \theta\tau_1^2}{\tau_1(\tau_1 - \theta)}.
\end{aligned}$$

The H_2 PID controller is

$$C(s) = \frac{1}{K}\frac{(\tau_2 s + 1)(\beta s + 1)}{s(\lambda^2\theta s/\tau_1 + \alpha)},$$

where

$$\begin{aligned}
\alpha &= \frac{\lambda^2(\tau_1 + \theta) + 4\lambda\tau_1\theta + 2\theta^2\tau_1}{\tau_1(\tau_1 - \theta)}, \\
\beta &= \frac{(\lambda^2 + 2\lambda\tau_1)(\tau_1 + \theta) + 2\tau_1^2\theta}{\tau_1(\tau_1 - \theta)}.
\end{aligned}$$

The tuning of the system with an unstable plant is similar to that of the system with a stable plant.

The performance and robustness of the closed-loop system are also discussed in this chapter. The unstructured uncertainty profile of the plant with parameter uncertainty is derived, the performance limitation of the system with an unstable plant is given, the existence problem of a stabilizing PID controller is solved, and the parameter scope is determined.

Exercises

1. The control of an inverted pendulum is a classical problem in control theory and widely used as a benchmark for testing control algorithms. Consider the control problem of an inverted pendulum on a moving base (Figure E8.1). The design objective is to keep the position of the pendulum top (that is, y) in equilibrium (that is, $\alpha = 0$) by adjusting the position of the moving base (that is, x) in the presence of disturbance. y can be inferred from the knowledge of x and α. The transfer function relating y to x is

$$G(s) = \frac{-1/M_bL_b}{s^2 - (M_b + M_s)g/(M_bL_b)}.$$

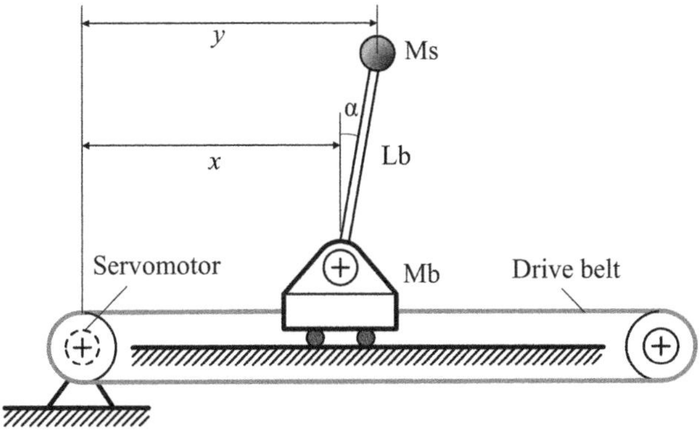

FIGURE E8.1
An inverted pendulum.

Let M_s=10kg, M_b=100kg, L_b=1m, and g=9.8m/s^2. The design specifications, for a unit step input, are that the settling time is less than 10s and the overshoot is less than 40%. Design a controller to satisfy these specifications.

2. Assume that a plant has simple poles $p_i (i = 1, 2, ..., r_p)$ in the open RHP. The filter is in the form of

$$J(s) = \frac{\beta_{r_p-1}s^{r_p-1} + ... + \beta_1 s + \beta_0}{(\lambda s + 1)^{m+r_p-1}},$$

where m is a positive integer and λ is the performance degree. The requirements on the filter are $J(p_i) = 1, i = 1, 2, ..., r_p$, Prove that

$$J(s) = \frac{1}{(\lambda s + 1)^{m+r_p-1}} \sum_{j=1}^{r_p} (\lambda p_j + 1)^{m+r_p-1} \prod_{i=1, i \neq j}^{r_p} \frac{s - p_i}{p_j - p_i}.$$

3. It is assumed that the real plant $\widetilde{G}(s)$ and the nominal plant $G(s)$ are rational and have the same number of RHP poles. $Q(s)$ is a stabilizing IMC controller for $G(s)$. Prove that the closed-loop system is internally stable with respect to the uncertainty profile $|\Delta_m(j\omega)|$ if and only if the filter $J(s)$ satisfies

$$|J(j\omega)| < |G(j\omega)Q(j\omega)\Delta_m(j\omega)|^{-1}.$$

for all frequencies.

4. Consider the plant with the transfer function

$$G(s) = \frac{Ke^{-\theta s}}{(\tau_1 s - 1)(\tau_2 s - 1)}.$$

Design an H_∞ PID controller.

5. Design an H_2 controller for the following plant:

$$G(s) = \frac{-s + \alpha}{-s + \beta}, \alpha > 0, \beta > 0, \alpha \neq \beta.$$

6. The problem considered here is to design a single controller that simultaneously stabilizes two plants. The problem of this type is referred to as the simultaneous stabilization problem. Two plants $G_1(s)$ and $G_2(s)$ can be stabilized by a single controller if and only if the "difference" plant $G(s) = G_1(s) - G_2(s)$ has an even number of real poles between every pair of real zeros in the closed RHP. Consider two plants

$$G_1(s) = \frac{1}{s + 1}, \quad G_2(s) = \frac{as + b}{(s + 1)(s - 1)},$$

where a and b are real constants, $a \neq 1$, and $b/a \neq -1$.

(a) Analyze the condition of simultaneous stabilization for the given plants.

(b) Design a stabilizing controller, if it exists.

Notes and References

The parameterization in Section 8.1 is drawn from Zhang et al. (2002) and Zhang et al. (2006) (Zhang W.D., F. Allgöwer and T. Liu. Controller parameterization for SISO and MIMO plants with time delay, *System Control Letters*, 2006, 55(10), 794–802. ©Elsevier). It can be viewed as a modified version of Youla parameterization.

Example 8.1.1 is from Zhang and Xu (2000).

Section 8.2 is adapted from Zhang (1998) and Zhang and Xu (2002a). The unstable plant with time delay can be used to describe different physical systems, for example, a reactor (Luyben and Melcic, 1978) or a plane (Enns et al., 1992).

Section 8.3 is from Zhang (1998) and Zhang and Xu (2002b). Discussion on the unstable reactor can be found in, for example, Luyben and Melcic (1978).

Section 8.4 is drawn from Zhang (1998). With regard to integrating plants and unstable plants, there exists difference between the design in this book and that in Morari and Zafiriou (1989, Chapter 5).

The analysis about the waterbed effect in Section 8.4 is based on Doyle et al. (1992). The unstructured uncertainty profile of parameter uncertainty was studied by Laughlin et al. (1987) (Laughlin D. L., D. E. Rivera, and M. Morari. Smith predictor design for robust performance, *Int. J. Control*, 1987, 46(2), 477–504. ©Taylor & Francis Ltd.).

The plant in Example 8.4.1 is based on Dorf and Bishop (2001, p. 551).

The Maclaurin PID controller in Section 8.5 is based on Lee et al. (2000) (Lee Y., J. Lee and S. Park. PID controller tuning for integrating and unstable processes with time delay, *Chem. Eng. Sci.*, 2000, 55(17), 3481–3493. ©Elsevier).

The parameter scope of the stabilizing PID controller in Section 8.7 was given by Bhattacharyya et al. (2009, Section 3.6) and Silva et al. (2002) (Silva G. J., A. Datta, and S. P. Bhattachcharyya. New results on the synthesis of PID controllers, *IEEE Trans. Auto. Control*, 2002, 47(2), 241–252. ©IEEE). A general result was provided by Ou et al. (2009).

The discussion about the control of unstable plants can also be found, for example, in Kwak et al. (1999), Paraskevopoulos et al. (2006), Xiang et al. (2007), and Thirunavukkarasu et al. (2009).

Exercise 1 is based on Dorf and Bishop (2001, p. 467).

Exercise 2 is adapted from Morari and Zafiriou (1989, Section 5.3).

Exercise 3 is adapted from Morari and Zafiriou (1989, Section 5.4).

The plant in Exercise 5 is from Morari and Zafiriou (1989, Section 5.7.2).

Exercise 6 is drawn from Doyle et al. (1992, Section 5.6). Discussion on the simultaneous stabilization problem can also be found in Dorato et al. (1992).

9

Complex Control Strategies

CONTENTS

In the preceding chapters, several basic control strategies for SISO control systems were discussed, including the unity feedback control and the Smith predictor. This chapter studies complex control strategies, including 2 DOF control, cascade control, anti-windup control, and feedforward control.

In 2 DOF control, the response to the reference is isolated from that to the disturbance. The controllers for the reference response and the disturbance response can be independently designed. The cascade control and the feedforward control are proposed, from different angles, to reject the disturbance effect on control system, while the purpose of anti-windup control is improving the performance when the system is subject to constraints on control variables.

In addition, the optimal rejection of input disturbance and the control of the plant with multiple time delays are also discussed in this chapter.

9.1 The 2 DOF Structure for Stable Plants

In Section 3.2, it was shown that the effect of the reference on the error was the same as that of the output disturbance on the system output:

$$\frac{e(s)}{r(s)} = \frac{y(s)}{d(s)} = \frac{1}{1+G(s)C(s)}. \tag{9.1.1}$$

Such a system has merely one degree of freedom and thus is called the single degree-of-freedom (1 DOF) system. When the reference and the disturbance have similar dynamic characteristics (for example, both of them are steps), a 1 DOF controller can simultaneously satisfy the requirements on the reference response and the disturbance response in many cases.

Sometimes, the dynamic characteristics of reference and disturbance are different. For example, the reference is a step while the disturbance at the plant output is a ramp. If both good reference response and good disturbance response are desired, the controller that reaches the two goals may not exist. In this case, an additional controller may have to be introduced so as to adjust the reference response and disturbance response independently. Thus there are two loops in this system. One is the reference loop, which is from the reference to the system output. The other is the disturbance loop from the disturbance at the plant output to the system output. Since the system has two degrees of freedom, it is referred to as the 2 DOF system.

A typical 2 DOF system is shown in Figure 9.1.1, where $C_1(s)$ is the controller of the disturbance loop and $C_2(s)$ is the controller of the reference loop. $C_2(s)$ is always stable. For convenience of presentation, the structure is named "Structure I." Structure I has many equivalents, as shown in Figure 9.1.2 and Figure 9.1.3. It should be pointed out that the $C_1(s)$s in Figure 9.1.2 and Figure 9.1.3 are not identical to the $C_1(s)$ in Figure 9.1.1.

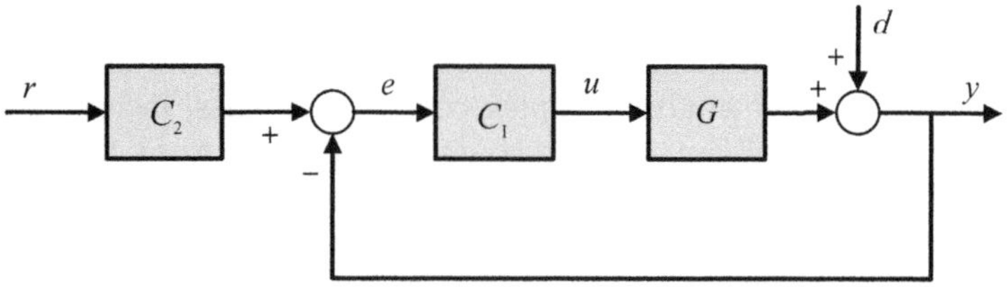

FIGURE 9.1.1
A typical 2 DOF system.

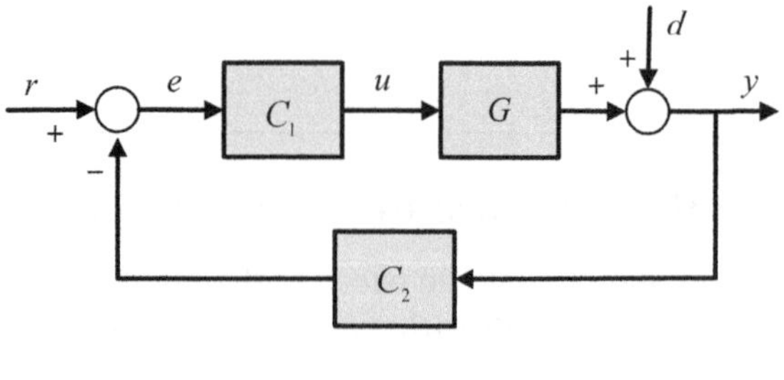

FIGURE 9.1.2
An equivalent of the typical 2 DOF system.

Consider the system shown in Figure 9.1.1. The input-output relationship

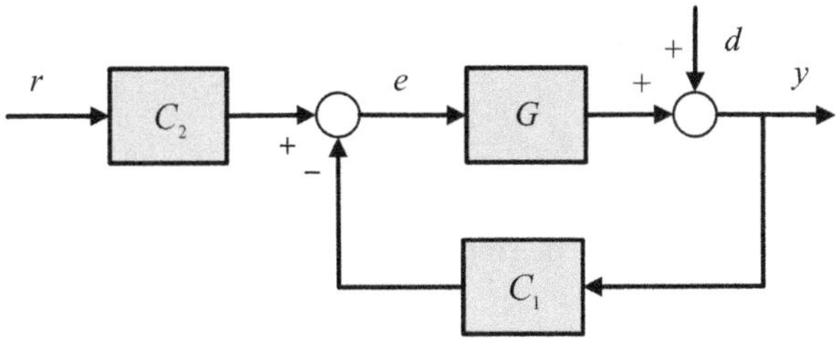

FIGURE 9.1.3
Another equivalent of the typical 2 DOF system.

is as follows:

$$\frac{e(s)}{r(s)} = \frac{C_2(s)}{1+G(s)C_1(s)}, \tag{9.1.2}$$

$$\frac{y(s)}{d(s)} = \frac{1}{1+G(s)C_1(s)}. \tag{9.1.3}$$

It can be seen that the internal stability of the closed-loop system is only determined by $C_1(s)$. The analysis for the internal stability is similar to that in a 1 DOF system. The design of a 2 DOF system involves two steps:

1. Designing $C_1(s)$ for the required disturbance response.
2. Designing $C_2(s)$ for the required reference response.

The design procedure of $C_1(s)$ is the same as that for the controller in a 1 DOF system. After $C_1(s)$ is designed, the loop consisting of $C_1(s)$ and $G(s)$ is viewed as an augmented plant, of which the transfer function is denoted by $T(s)$. The system consisting of $C_2(s)$ and $T(s)$ forms the IMC structure with an exact model. Accordingly, $C_2(s)$ can be directly designed. Since $C_2(s)$ is not involved in the feedback loop, it will not affect the disturbance loop.

To illustrate the design procedure, consider the following plant:

$$G(s) = \frac{K}{\tau s+1}e^{-\theta s}. \tag{9.1.4}$$

The system is required to track a step reference, and at the same time to reject the disturbance in the form of a ramp at the plant output.

First of all, design $C_1(s)$ for the required disturbance response. By utilizing the IMC controller $Q(s)$, $C_1(s)$ can be expressed as

$$C_1(s) = \frac{Q(s)}{1-G(s)Q(s)}. \tag{9.1.5}$$

The rational part of the plant is MP. Utilizing (6.2.5), we have

$$Q_{opt}(s) = \frac{\tau s+1}{K}. \tag{9.1.6}$$

Since the disturbance is a ramp, a Type 2 filter is demanded. In light of (5.7.5),

$$J(s) = \frac{2\lambda_1 s + 1}{(\lambda_1 s + 1)^2},$$

where λ_1 is the performance degree for adjusting the disturbance response. Simple computations give

$$Q(s) = Q_{opt}(s)J(s) = \frac{(\tau s + 1)(2\lambda_1 s + 1)}{K(\lambda_1 s + 1)^2}. \tag{9.1.7}$$

It is easy to compute $C_1(s)$ with (9.1.5). As the plant is stable, $C_1(s)$ can be implemented in the IMC structure.

Next, design $C_2(s)$ for the required reference response. The loop consisting of $C_1(s)$ and $G(s)$ is regarded as an augmented plant, whose transfer function is

$$T(s) = G(s)Q(s) = \frac{2\lambda_1 s + 1}{(\lambda_1 s + 1)^2} e^{-\theta s}. \tag{9.1.8}$$

Obviously, $T(s)$ is stable. According to the design procedure for the H_2 controller, the optimal $C_2(s)$ is the inverse of the rational part of $T(s)$. For a step reference, the suboptimal controller is

$$C_2(s) = \frac{(\lambda_1 s + 1)^2}{(2\lambda_1 s + 1)(\lambda_2 s + 1)}, \tag{9.1.9}$$

where λ_2 is the performance degree for adjusting the reference response.

Let $T_r(s)$ denote the transfer function from the reference to the system output, and $T_{d'}$ denote the transfer function from the disturbance at the plant input to the system output. It is easy to verify that the transfer function of the reference loop is

$$T_r(s) = \frac{1}{\lambda_2 s + 1} e^{-\theta s}. \tag{9.1.10}$$

The reference response to the unit step is

$$y(t) = \begin{cases} 0 & 0 < t < \theta \\ 1 - e^{-(t-\theta)/\lambda_2} & t \geq \theta \end{cases}. \tag{9.1.11}$$

The reference response can be adjusted by the performance degree λ_2 independently.

The transfer function with regard to the load response can be written as

$$T_{d'}(s) = \frac{K}{\tau s + 1} e^{-\theta s} \left[1 - \frac{2\lambda_1 s + 1}{(\lambda_1 s + 1)^2} e^{-\theta s}\right]. \tag{9.1.12}$$

The load response to the unit step disturbance is

$$y(t) = \begin{cases} 0 & 0 < t < \theta \\ K(1 - e^{-(t-\theta)/\tau}) & \theta \leq t < 2\theta \\ K\left[\begin{array}{l} \frac{\lambda_1}{\lambda_1 - \tau} e^{-(t-2\theta)/\lambda_1} - \\ \frac{\tau}{\lambda_1 - \tau} e^{-(t-2\theta)/\tau} - e^{-(t-\theta)/\tau} \end{array}\right] & t \geq 2\theta \end{cases}. \quad (9.1.13)$$

The disturbance response can be adjusted by employing the performance degree λ_1 independently.

Note that the disturbance loop of a 2 DOF system cannot provide better disturbance rejection performance than a 1 DOF system. Nevertheless, since the disturbance response and the reference response can be adjusted independently, better disturbance response and reference response can be reached simultaneously in a 2 DOF system.

When there exists uncertainty, the reference response cannot be thoroughly isolated from the disturbance response. In this case, the robust stability and the disturbance response is only determined by λ_1, while the reference response is mainly determined by λ_2. For robustness tuning, one can monotonically increase the performance degrees until the required response is obtained.

In Figure 9.1.4, a new 2 DOF structure is given. To distinguish it from Structure I, it is named "Structure II." In Structure II, $C_3(s)$ is the controller of the disturbance loop and $C_4(s)$ is the controller of the reference loop. Let

$$C_1(s) = C_3(s), \quad (9.1.14)$$

$$C_2(s) = \frac{C_4(s) + G(s)C_3(s)}{C_3(s)}. \quad (9.1.15)$$

Then Structure I and Structure II are equivalent to each other.

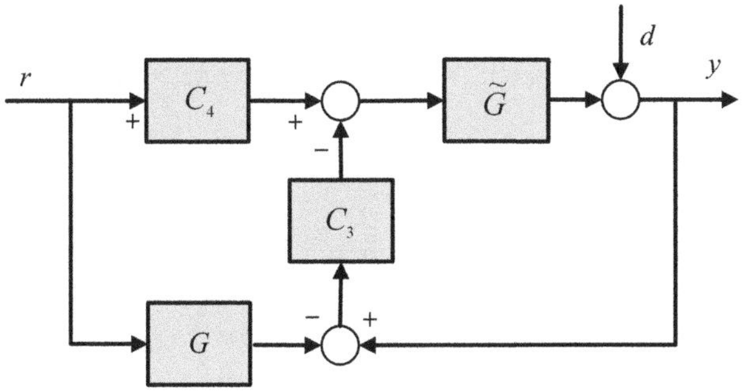

FIGURE 9.1.4
A New 2 DOF system.

The feature of Structure II is that the design is simpler than that for

Structure I. The design of $C_3(s)$ is similar to that for the unity feedback loop controller. Since the reference loop is an open one for the nominal plant, $C_4(s)$ can be designed as the inverse of the plant rational part.

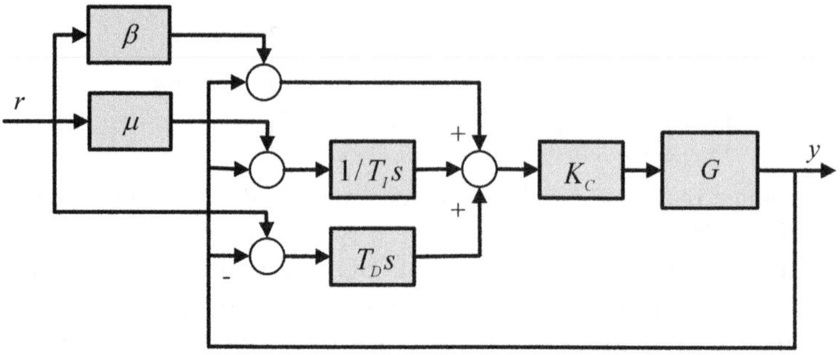

FIGURE 9.1.5
Structure of the RZN PID controller.

In Section 4.1, the RZN PID controller depicted in Figure 9.1.5 was introduced. By utilizing two parameters β and μ, an improved reference response is obtained. It is easy to verify that

$$\begin{aligned} T(s) &= \frac{G(s)K_C\left(\beta + \dfrac{1}{\mu T_I s} + T_D s\right)}{1 + G(s)K_C\left(1 + \dfrac{1}{T_I s} + T_D s\right)} \\ &= \frac{K_C\left(\beta + \dfrac{1}{\mu T_I s} + T_D s\right)}{K_C\left(1 + \dfrac{1}{T_I s} + T_D s\right)} \frac{G(s)K_C\left(1 + \dfrac{1}{T_I s} + T_D s\right)}{1 + G(s)K_C\left(1 + \dfrac{1}{T_I s} + T_D s\right)}. \end{aligned} \tag{9.1.16}$$

Let

$$\begin{aligned} C_1(s) &= K_C\left(1 + \frac{1}{T_I s} + T_D s\right), \\ C_2(s) &= \frac{K_C\left(\beta + \dfrac{1}{\mu T_I s} + T_D s\right)}{K_C\left(1 + \dfrac{1}{T_I s} + T_D s\right)}. \end{aligned} \tag{9.1.17}$$

Evidently, the introduction of β and μ is equivalent to providing an additional degree of freedom for the original system. In fact, the resulting system is a 2 DOF system.

9.2 The 2 DOF Structure for Unstable Plants

Because of the effects of RHP poles and time delay, the reference response of the 1 DOF system with an unstable plant usually exhibits excessive overshoot. A 2 DOF system can be used to overcome this disadvantage. This is because the 2 DOF system can roll off high frequency signals in the reference independently.

The design procedure of the 2 DOF system for unstable plants is similar to that for stable plants. However, to guarantee the internal stability, the controller of the disturbance loop can only be implemented in the unity feedback loop. The procedure will be illustrated by utilizing the first-order unstable plant. The design procedure for high-order plants is similar.

The first-order unstable plant can be expressed as

$$G(s) = \frac{K}{\tau s - 1} e^{-\theta s}. \tag{9.2.1}$$

As the main goal of this section is to discuss how to depress the excessive overshoot, it is assumed that the system is required to track a step reference and at the same time reject the effect of a step disturbance at the plant output.

Consider Structure I. First, design $C_1(s)$. The rational part of the plant is MP. Based on (8.4.8) we have

$$Q_{opt}(s) = \frac{\tau s - 1}{K}. \tag{9.2.2}$$

When the disturbance at the plant output is a step and the plant has only one RHP pole, the filter can easily be determined by (8.4.9). Then the H_2 suboptimal controller with the filter is as follows:

$$Q(s) = \frac{(\tau s - 1)\{\tau[(\lambda_1/\tau + 1)^2 e^{\theta/\tau} - 1]s + 1\}}{K(\lambda_1 s + 1)^2}. \tag{9.2.3}$$

The controller of disturbance loop is

$$\begin{aligned} C_1(s) &= \frac{Q(s)}{1 - G(s)Q(s)} \\ &= \frac{1}{K} \frac{(\tau s - 1)\{\tau[(\lambda_1/\tau + 1)^2 e^{\theta/\tau} - 1]s + 1\}}{(\lambda_1 s + 1)^2 - \{\tau[(\lambda_1/\tau + 1)^2 e^{\theta/\tau} - 1]s + 1\}e^{-\theta s}}. \end{aligned} \tag{9.2.4}$$

Since

$$\lim_{s \to 1/\tau} \{(\lambda_1 s + 1)^2 - \{\tau[(\lambda_1/\tau + 1)^2 e^{\theta/\tau} - 1]s + 1\}e^{-\theta s}\} = 0,$$

there exists a RHP zero-pole cancellation in $C_1(s)$. A rational approximation has to be used to remove it. This can be achieved in many ways. For example, the controller can be chosen as a PID controller in the form of

$$C_1(s) = K_C\left(1 + \frac{1}{T_I s} + T_D s\right). \tag{9.2.5}$$

With the Maclaurin series expansion, the following result was obtained in Section 8.5:

$$\begin{aligned} T_I &= -\tau + \beta_1 - \frac{\lambda^2 + \beta_1\theta - \theta^2/2}{2\lambda + \theta - \beta_1}, \\ K_C &= \frac{T_I}{-K(2\lambda + \theta - \beta_1)}, \\ T_D &= \frac{-\tau\beta_1 - (\theta^3/6 - \beta_1\theta^2/2)/(2\lambda + \theta - \beta_1)}{T_I} - \\ &\quad \frac{\lambda^2 + \beta_1\theta - \theta^2/2}{2\lambda + \theta - \beta_1}. \end{aligned} \tag{9.2.6}$$

Next, design $C_2(s)$. Regard the feedback loop consisting of $C_1(s)$ and $G(s)$ as an augmented plant. The transfer function of the augmented plant is

$$T(s) = \frac{G(s)C_1(s)}{1 + G(s)C_1(s)}. \tag{9.2.7}$$

The optimal $C_2(s)$ should be the inverse of $T(s)$ after the time delay in its numerator is removed. However, such a design procedure is tedious. Since $C_1(s)$ is an approximation of the ideal controller, $C_2(s)$ can be chosen as the inverse of the ideal $T(s)$ after the time delay in its numerator is removed. Then

$$C_2(s) = \frac{(\lambda_1 s + 1)^2}{\{\tau[(\lambda_1/\tau + 1)^2 e^{\theta/\tau} - 1]s + 1\}(\lambda_2 s + 1)}. \tag{9.2.8}$$

Now consider Structure II. In Structure II, $C_3(s)$ equals the $C_1(s)$ in Structure I. The optimal $C_4(s)$ can be chosen as the inverse of the plant rational part:

$$C_4(s) = \frac{\tau s - 1}{K(\lambda_2 s + 1)}. \tag{9.2.9}$$

$C_3(s)$ is an approximation of the ideal controller. Hence, the transfer function of the reference loop in the system with $C_3(s)$ is close to that in the system with the ideal controller:

$$T_r(s) = \frac{1}{\lambda_2 s + 1} e^{-\theta s}. \tag{9.2.10}$$

The reference response to the unit step is

$$y(t) = \begin{cases} 0 & 0 < t < \theta \\ 1 - e^{-(t-\theta)/\lambda_2} & t \geq \theta \end{cases}. \tag{9.2.11}$$

The reference response can be tuned by the performance degree λ_2 independently.

The transfer function corresponding to the load response is, approximately,

$$\begin{aligned} &T_{d'}(s) \\ &= \frac{K}{\tau s-1}e^{-\theta s}\left\{1-\frac{\tau[(\lambda_1/\tau+1)^2e^{\theta/\tau}-1]s+1}{(\lambda_1 s+1)^2}e^{-\theta s}\right\}. \end{aligned} \tag{9.2.12}$$

Let

$$\begin{aligned} a &= \frac{1}{\lambda_1}, \\ b &= -\frac{1}{\tau}, \\ c &= \frac{1}{\tau[(\lambda_1/\tau+1)^2e^{\theta/\tau}-1]}. \end{aligned}$$

The load response to the unit step disturbance is

$$y(t) = \begin{cases} 0 & 0 < t < \theta \\ K(1-e^{-b(t-\theta)}) & \theta \le t < 2\theta \\ K\Big(-e^{-b(t-\theta)} - \dfrac{a^2(b-c)}{(a-b)^2c}e^{-b(t-2\theta)} - & \\ \quad \dfrac{ab(c-a)+bc(a-b)}{(a-b)^2c}e^{-a(t-2\theta)} - & \\ \quad \dfrac{ab(c-a)}{(a-b)c}te^{-a(t-2\theta)}\Big) & t \ge 2\theta \end{cases}. \tag{9.2.13}$$

The disturbance response can be tuned by the performance degree λ_1 independently.

Two classes of control methods are frequently used for unstable plants in literature. In the first method, a controller is directly used to control the plant, for example, the 1 DOF control system or the 2 DOF control system. In the second method, an inner loop is introduced to stabilize the unstable plant, and then the controller is designed for the augmented stabilized plant.

The control system with an inner stabilizing loop is shown in Figure 9.2.1, where $C_s(s)$ is the stabilizer. $C_s(s)$ and $G(s)$ construct an augmented plant. $C_s(s)$ should be chosen so that the augmented plant is stable. $C(s)$ is then designed for the augmented stable plant.

The introduction of $C_s(s)$ makes the structure different from the unity feedback loop. As a result, the performance and robustness of the closed-loop system are difficult to analyze. It will be shown that the structure is in fact equivalent to the 2 DOF structure.

The closed-loop transfer function of the system is

$$T_r(s) = \frac{G(s)C(s)}{1+G(s)C_s(s)+G(s)C(s)}. \tag{9.2.14}$$

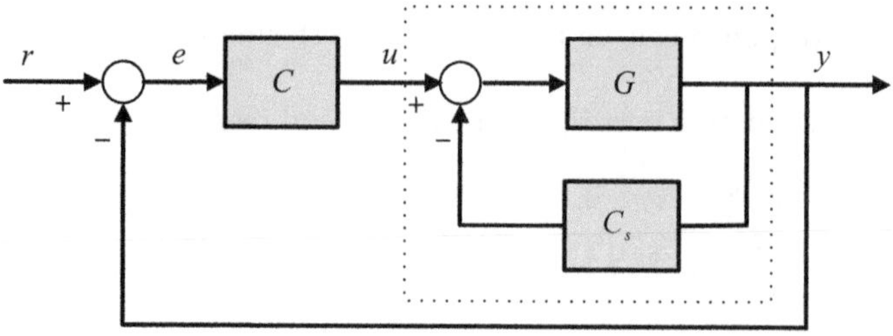

FIGURE 9.2.1
Control system with an inner stabilizing loop.

Rewrite it in the form of

$$T_r(s) = \frac{C(s)}{C_s(s) + C(s)} \frac{G(s)C(s) + G(s)C_s(s)}{1 + G(s)C_s(s) + G(s)C(s)}, \tag{9.2.15}$$

which is equivalent to the 2 DOF system with the following controllers:

$$C_1(s) = C_s(s) + C(s), \tag{9.2.16}$$

$$C_2(s) = \frac{C(s)}{C_s(s) + C(s)}. \tag{9.2.17}$$

This equivalent is given in Figure 9.2.2. Therefore,

1. The optimal method can be used to design the controller analytically in the system with an inner stabilizing loop.
2. The closed-loop response can be tuned quantitatively.
3. The robustness of the system can be analyzed with those methods developed for the unity feedback control system.

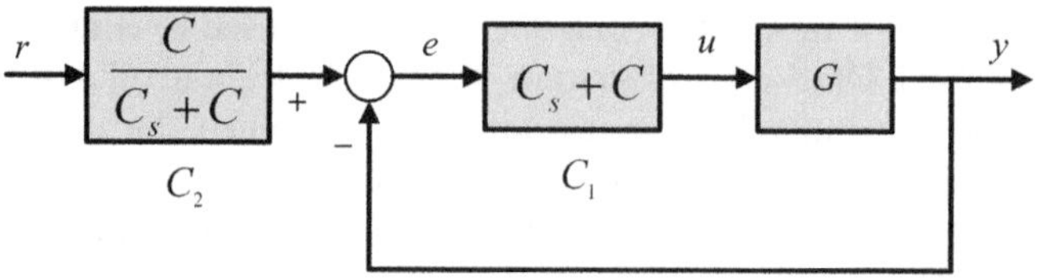

FIGURE 9.2.2
An equivalent of the control system with an inner stabilizing loop.

For the reason of simplicity, the stabilizer $C_s(s)$ is usually chosen to be a proportional controller. Assume that the $C(s)$ in Figure 9.2.1 is a PID controller:

$$C(s) = K_c \left(1 + \frac{1}{T_I s} + T_D s\right). \tag{9.2.18}$$

Then the controller of the disturbance loop in Figure 9.2.2 is also a PID controller:

$$\begin{aligned} C_1(s) &= C(s) + C_s(s) \\ &= K_c \frac{1 + (C_s/K_c + 1)T_I s + T_I T_D s^2}{T_I s}. \end{aligned} \quad (9.2.19)$$

Example 9.2.1. *A plane with fixed wings has the feature of self-regulating; that is, it is stable. However, a helicopter is generally unstable and thus is difficult to control. In a helicopter control system, the goal is to control the pitch angle by adjusting the rotor angle (Figure 9.2.3). The transfer function of the helicopter is*

$$\boxed{G(s) = \frac{25(s + 0.03)}{(s + 0.4)(s^2 - 0.36s + 0.16)}.}$$

The plant is MP. Following the discussion in Section 8.4, we have

$$Q_{opt}(s) = \frac{(s + 0.4)(s^2 - 0.36s + 0.16)}{25(s + 0.03)}.$$

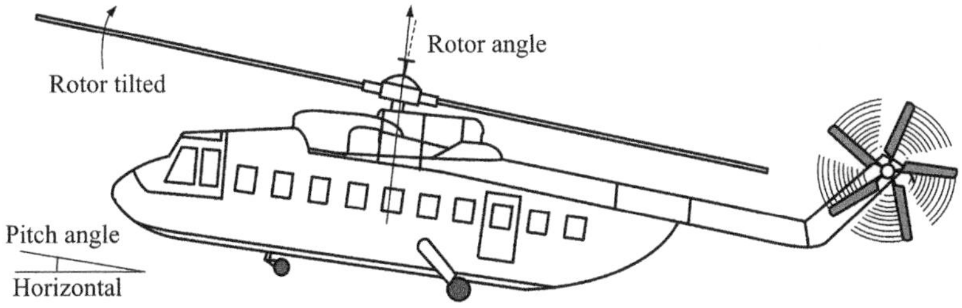

FIGURE 9.2.3
Helicopter control.

The plant has two RHP poles: $0.18 \pm 0.3572i$*. Choose the following filter:*

$$J(s) = \frac{\beta_2 s^2 + \beta_1 s + 1}{(\lambda_1 s + 1)^4}.$$

If one takes $\lambda_1 = 0.5$*, then* $\beta_2 = 1.6781$ *and* $\beta_1 = 1.9164$*. The controller of the disturbance loop in Structure I is*

$$\boxed{C_1(s) = \frac{(s + 0.4)(1.6781s^2 + 1.9164s + 1)}{25s(s + 0.03)(0.0625s + 0.5225)}.}$$

Since there is no time delay in the plant, the augmented plant consisting of

$C_1(s)$ and $T(s)$ is rational. Then, the controller of the reference loop is also a rational transfer function:

$$\boxed{C_2(s) = \frac{(\lambda_1 s+1)^4}{(1.6781s^2+1.9164s+1)(\lambda_2 s+1)^2}.}$$

Take $\lambda_2 = 0.5$. A unit step reference is added at $t = 0$ and a step load with the amplitude of -0.1 is added at $t = 20$. The closed-loop responses are shown in Figure 9.2.4. It can be seen that the disturbance responses are the same for the 1 DOF control system and the 2 DOF control system, while the reference response of the 2 DOF control system is improved.

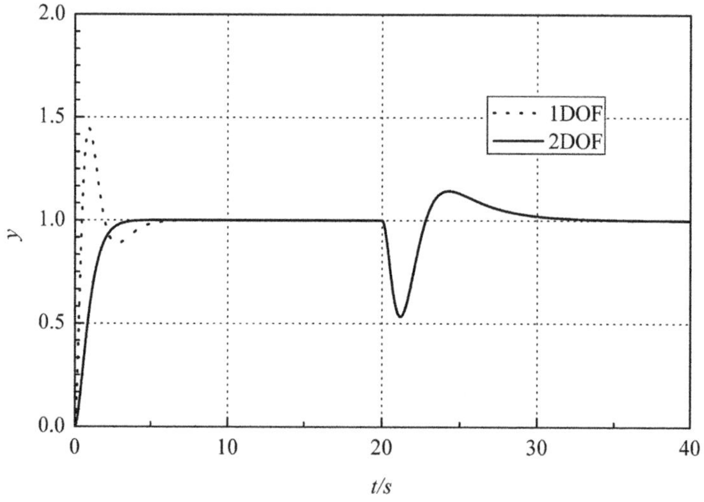

FIGURE 9.2.4
Responses of the 1 DOF system and 2 DOF system.

9.3 Cascade Control

The feature of the unity feedback configuration is that it has only one loop for one system output. There is another control configuration that uses more than one loop for one system output. This kind of configuration is referred to as the cascade control.

To see how the cascade control system works, consider the distillation column shown in Figure 9.3.1. Assume that the single loop structure is used;

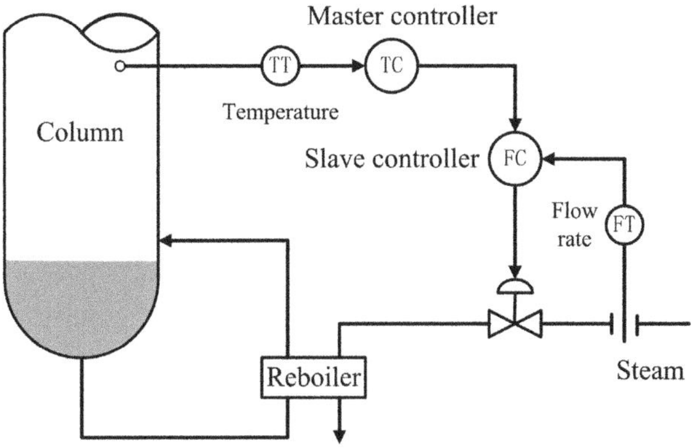

FIGURE 9.3.1
Temperature control system for a distillation column.

that is, only the master controller is used. The temperature at the bottom of the distillation column is controlled by adjusting the steam flow rate to the reboiler. With the increase of the steam pressure, the steam flow rate increases correspondingly. No correction will be made by the controller until the higher steam flow rate increases the vapor boilup and eventually raises the column temperature. Consequently, the whole system is disturbed by the pressure change of the supply steam.

The response can be improved by installing a slave controller in between the temperature controller and the controlled steam flow rate. This controller is used to control the flow rate. Such an arrangement constitutes a cascade control configuration. In this cascade control system, the slave controller will immediately detect the change of steam flow rate, and pinch the steam valve to pull the steam flow rate back to the desired value. As a result, the reboiler and the column are only slightly affected by the disturbance from the supply steam.

The features of cascade control are summarized as follows:

1. The controller, of which the reference is set by the operator, is the master controller. The other is the slave controller. The output signal of the master controller serves as the reference of the slave controller.

2. The two feedback loops are nested, with the slave loop located inside the master loop.

The diagram of the cascade control system is shown in Figure 9.3.2. In the example of distillation column, the input of $G_2(s)$ is the steam flow rate, and the output of $G_2(s)$ is the column temperature. $C_2(s)$ and $G_2(s)$ constitute the master loop. The output of $G_1(s)$ is the input of $G_2(s)$. $C_1(s)$ and $G_1(s)$

constitute the slave loop, whose reference is the output of $C_2(s)$. Normally, the slave controller is located close to a potential disturbance in order to improve the closed-loop response.

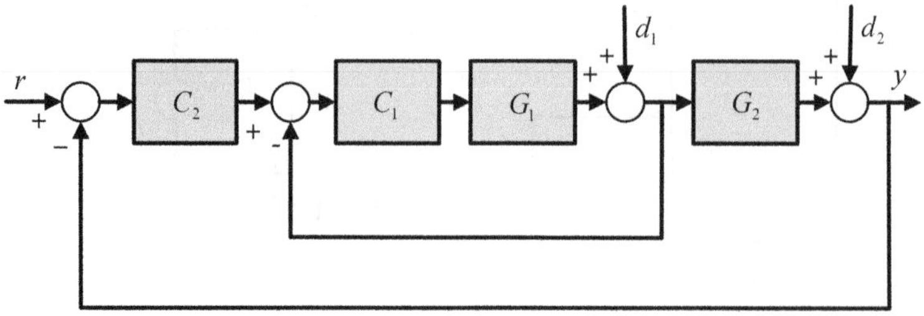

FIGURE 9.3.2
Diagram of the cascade control system.

A natural question is whether a cascade control system can always provide superior performance to the single loop system. This problem is discussed in what follows.

It is seen that there are two kinds of disturbances in a cascade control system. One is the disturbance $d_2(s)$ on the master loop. The other is the disturbance $d_1(s)$ on the slave loop.

Consider $d_2(s)$ first. Assume that only a single loop (that is, the master loop) is applied. The system response to $d_2(s)$ is

$$y(s) = [1 - G_1(s)G_2(s)Q(s)]d_2(s), \tag{9.3.1}$$

where $Q(s)$ is the IMC controller corresponding to the plant $G_1(s)G_2(s)$. When the cascade control structure is employed, the effect can be written as

$$y(s) = [1 - G_1(s)Q_1(s)G_2(s)Q_2(s)]d_2(s), \tag{9.3.2}$$

where $Q_1(s)$ is the IMC controller corresponding to $G_1(s)$ and $Q_2(s)$ is the IMC controller corresponding to the plant $G_1(s)Q_1(s)G_2(s)$. If $Q(s) = Q_1(s)Q_2(s)$ is taken, the two responses are the same.

Now consider $d_1(s)$. When only a single loop is used, the system response to $d_1(s)$ can be expressed as

$$y(s) = [1 - G_1(s)G_2(s)Q(s)]G_2(s)d_1(s). \tag{9.3.3}$$

If the cascade control structure is used, the system response to $d_1(s)$ is

$$y(s) = [1 - G_1(s)Q_1(s)]G_2(s)d_1(s). \tag{9.3.4}$$

The performance of cascade control is superior if $Q_1(s)$ can be designed so that

$$\min_{Q_1} \|[1 - G_1(s)Q_1(s)]G_2(s)\| \leq$$

$$\min_{Q} \|[1 - G_1(s)G_2(s)Q(s)]G_2(s)\|, \tag{9.3.5}$$

where $\|\cdot\|$ denotes some norm. (9.3.5) is an equality when $G_2(s)$ is MP and stable. Therefore, the cascade control is only useful when $G_2(s)$ has RHP zeros or a time delay.

Assume that $G_2(s)$ is NMP. The following two-step design procedure can be used to design the two controllers:

1. Assume that there is only the secondary loop. Design $C_1(s)$.
2. Regard the secondary loop and $G_2(s)$ as an augmented plant. Design $C_2(s)$.

The plants in the master loop and the slave loop are often described by the first-order model with time delay:

$$G_1(s) = \frac{K_1 e^{-\theta_1 s}}{\tau_1 s + 1}, \tag{9.3.6}$$

$$G_2(s) = \frac{K_2 e^{-\theta_2 s}}{\tau_2 s + 1}. \tag{9.3.7}$$

Use the two-step design procedure. Design $C_1(s)$ firstly:

$$Q_1(s) = \frac{\tau_1 s + 1}{K_1(\lambda_1 s + 1)}, \tag{9.3.8}$$

$$C_1(s) = \frac{Q_1(s)}{1 - G_1(s)Q_1(s)}. \tag{9.3.9}$$

The closed-loop transfer function is

$$T_1(s) = \frac{1}{\lambda_1 s + 1} e^{-\theta_1 s}. \tag{9.3.10}$$

Regard $T_1(s)$ and $G_2(s)$ as an augmented plant. Design $C_2(s)$:

$$Q_2(s) = \frac{(\tau_2 s + 1)(\lambda_1 s + 1)}{K_2(\lambda_2 s + 1)^2}, \tag{9.3.11}$$

$$C_2(s) = \frac{Q_2(s)}{1 - G_2(s)T_1(s)Q_2(s)}. \tag{9.3.12}$$

The overall closed-loop transfer function is

$$T_2(s) = \frac{1}{(\lambda_2 s + 1)^2} e^{-(\theta_1 + \theta_2)s}. \tag{9.3.13}$$

In the design, the nominal performance of the overall system is only determined by the performance degree λ_2. When there exists uncertainty, the robust performance relates to both λ_1 and λ_2. In general, the robustness is mainly determined by λ_2. This feature significantly reduces the complexity in tuning.

Example 9.3.1. *This example is given to illustrate how to design the cascade controller. Consider the temperature control system of the distillation column sketched in Figure 9.3.1. The dynamics of the steam flow rate can be expressed as*

$$G_1(s) = \frac{0.68}{0.39s+1}.$$

The dynamics of the column temperature is

$$G_2(s) = \frac{1.26e^{-0.5s}}{2.11s+1}.$$

Then

$$\begin{aligned} Q_1(s) &= \frac{0.39s+1}{0.68(\lambda_1 s+1)}, \\ C_1(s) &= \frac{0.39s+1}{0.68\lambda_1 s}. \end{aligned}$$

The closed-loop transfer function of the slave loop is

$$T_1(s) = \frac{1}{\lambda_1 s+1}.$$

It is readily obtained that

$$\begin{aligned} Q_2(s) &= \frac{(2.11s+1)(\lambda_1 s+1)}{1.26(\lambda_2 s+1)^2}, \\ C_2(s) &= \frac{(2.11s+1)(\lambda_1 s+1)}{1.26[(\lambda_2 s+1)^2 - e^{-0.5s}]}. \end{aligned}$$

9.4 An Anti-Windup Structure

It is necessary to consider at the beginning of design that the manipulated variable in a real system may be constrained by a physical limit. For example, in a paper-making process, the basis weight of paper is controlled by adjusting the flow rate of stock. The flow rate from a valve has a maximum value, which is determined by the full openness of the valve. If the controller output exceeds the maximum value, there will be something wrong with the feedback loop. In this situation, the valve remains fully open despite that the controller output may continue to change. If the controller includes integral action, the persistent error will be integrated. Thereby the output of the integrator becomes quite large. This phenomenon is known as the windup.

Windup puts the system in the state of saturation, which will affect the performance of the system. This issue has to be dealt with in an ad hoc fashion called anti-windup. Conceptually, windup can be prevented by stopping the integral action when the actuator saturates.

Assume that the saturation constraint on plant input is as follows:

$$\text{sat}[u(t)] = \begin{cases} u_{\min}(t) & u(t) < u_{\min}(t) \\ u(t) & u_{\min}(t) \leq u(t) \leq u_{\max}(t) \\ u_{\max}(t) & u(t) > u_{\ max}(t) \end{cases} . \tag{9.4.1}$$

A simple idea to avoid saturation is to adjust the controller parameter, so that the controller output is kept within its physical limit. Recall the control problem for the first-order plant with time delay in Section 6.4. The controller output is

$$\begin{aligned} u(s) &= Q(s)r(s) \\ &= \frac{\tau s + 1}{K(\lambda s + 1)} r(s). \end{aligned} \tag{9.4.2}$$

If the reference is the unit step, the corresponding time domain response can be written as

$$u(t) = K\left(\frac{\tau}{\lambda} e^{-t/\lambda} + 1 - e^{-t/\lambda}\right). \tag{9.4.3}$$

The controller output can be restricted by appropriately increasing the performance degree. The advantage of this method is that it is simple and the control structure is not changed; the disadvantage is that it does not sufficiently utilize $u(t)$ to improve the performance of the closed-loop system.

An optimization design method is introduced here, based on the IMC structure. Improved performance can be obtained by modifying the control structure.

Consider the IMC structure shown in Figure 9.4.1. Define

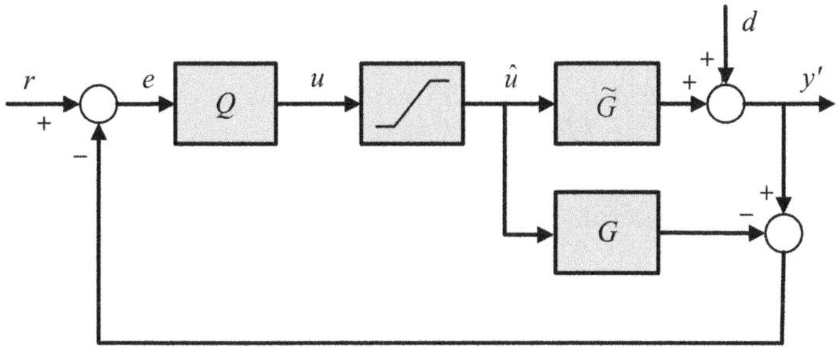

FIGURE 9.4.1
IMC structure in the presence of actuator constraint.

$$\begin{aligned} y'(t) &:= G(t) * \hat{u}(t) \\ &= \int_0^{\infty} G(t-\tau)\hat{u}(\tau)d\tau, \end{aligned} \tag{9.4.4}$$

where $y'(t)$ is the output of the constrained system. Because of the saturation constraints, $y'(t)$ necessarily differs from $y(t)$, the output of the unconstrained system. It is desirable to keep $y'(t)$ as close to $y(t)$ as possible. Mathematically, the problem is equivalent to solving the following optimization problem instantaneously at each time t:

$$\min_{\hat{u}(t)} |W(t) * y(t) - W(t) * y'(t)|, \tag{9.4.5}$$

where $W(t)$ is a weighting function that makes $W(s)G(s)$ bi-proper. The introduction of this weighting function is based on the following fact: if $G(s)$ is strictly proper, $\hat{u}(t)$ does not affect $y'(t)$ instantaneously and thus the minimization is meaningless. It should be noticed that the optimization is carried out continuously for $t \geq 0$, which differs from that over a horizon.

To deal with the problem, the modified IMC structure in Figure 9.4.2 is proposed. Here

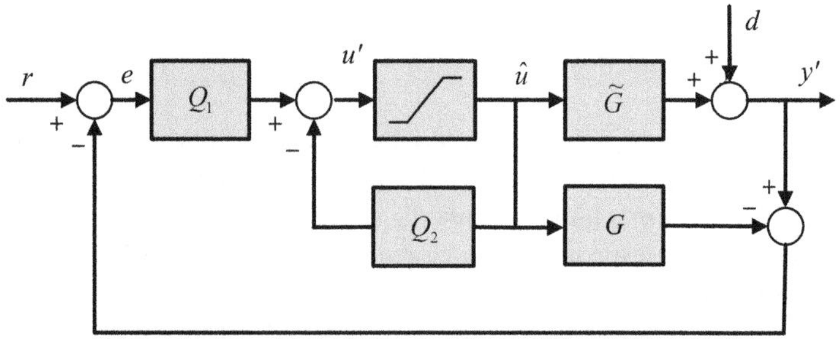

FIGURE 9.4.2
Modified IMC structure for anti-windup.

$$Q(s) = \frac{Q_1(s)}{1 + Q_2(s)}. \tag{9.4.6}$$

$Q(s)$ is bi-proper in general. Then

$$\begin{aligned} u'(s) &= Q_1(s)e(s) - Q_2(s)\hat{u}(s) \\ &= Q_1(s)e(s) - [Q_1(s)Q^{-1}(s) - 1]\hat{u}(s). \end{aligned} \tag{9.4.7}$$

Let

$$Q_3(s) = Q_1(s)Q^{-1}(s).$$

In the time domain,

$$u'(t) - \hat{u}(t) = Q_1(t) * e(t) - Q_3(t) * \hat{u}(t). \tag{9.4.8}$$

If $Q_1(s) = W(s)G(s)Q(s)$, then $\hat{u}(t)$ is the solution to the optimization problem. To show this, consider Figure 9.4.2. As $Q_3(s) = W(s)G(s)$, we have

$$\begin{aligned} u'(t) - \hat{u}(t) &= Q_1(t) * e(t) - Q_3(t) * \hat{u}(t) \\ &= W(t) * y(t) - W(t) * y'(t). \end{aligned}$$

Now let us see how the closed-loop response is optimized. When no saturation occurs, $\hat{u}(t) = u'(t) = \mathrm{sat}[u'(t)]$, $W(t) * y'(t) - W(t) * y(t) = 0$. Assume that saturation occurs. Since $\hat{u}(t)$ affects $W(t) * y'(t)$ linearly, $|W(t) * y'(t) - W(t) * y(t)|$ is a convex function of $\hat{u}(t)$. When $u'(t) = \hat{u}(t)$ is infeasible, the optimal solution must occur at the boundary, that is, $\hat{u}(t) = \mathrm{sat}[u'(t)]$.

In the modified IMC structure, $Q(s)$ is usually MP and always stable. If $Q(s)$ is MP and $Q_1(s)$ is NMP, then $[1 + Q_2(s)]^{-1}$ must be unstable. To guarantee the internal stability of the closed-loop system, $Q_1(s)$ must be MP and stable. $W(s)$ should be chosen so that $W(s)G(s)Q(s)$ is both MP and stable. When $W(s)$ is appropriately chosen, the input is kept saturated for the optimal amount of time until $|W(t) * y'(t) - W(t) * y(t)|$ becomes zero. Thus, the performance is greatly improved.

Different controller factorizations can be obtained by choosing different $W(s)$. Two special cases are discussed here.

1. $W(s) = G(s)^{-1}$. The optimization problem becomes

   $$\min_{\hat{u}(t)} |u(t) - \hat{u}(t)|. \tag{9.4.9}$$

   The solution corresponds to the conventional IMC structure, which "chops off" the control input resulting in performance deterioration. The stability of the closed-loop system is guaranteed.

2. $W(s)$ is chosen such that $Q_1(s)$ is a constant, for example, $Q_1(s) = Q(\infty)$. The optimization problem becomes

   $$\min_{\hat{u}(t)} |e(t) - e'(t)|, \tag{9.4.10}$$

   where

   $$e'(t) = Q^{-1}(t) * \hat{u}(t).$$

   The performance in this case is greatly improved, but the stability of the closed-loop system is not guaranteed. If the dynamics of $G(s)Q(s)$ is slow, minimizing the weighted error $[e(t) - e'(t)]$ may not be a good way to optimize the nonlinear performance. After the system comes out of the nonlinear region, the controller takes no action to compensate the effect of the error introduced during saturation.

In Item 1, $W(s)$ is chosen to guarantee the stability, while in Item 2 $W(s)$ is chosen to enhance the performance. Therefore, $W(s)$ can be tuned to tradeoff the performance and the stability of the constrained system.

For stable plants, the IMC structure in Figure 9.4.1 and the unity feedback loop in Figure 9.4.3 are equivalent. The modified IMC structure shown in Figure 9.4.2 can be directly extended to the unity feedback loop. The obtained anti-windup structure is shown in Figure 9.4.4. The controllers are defined as follows:

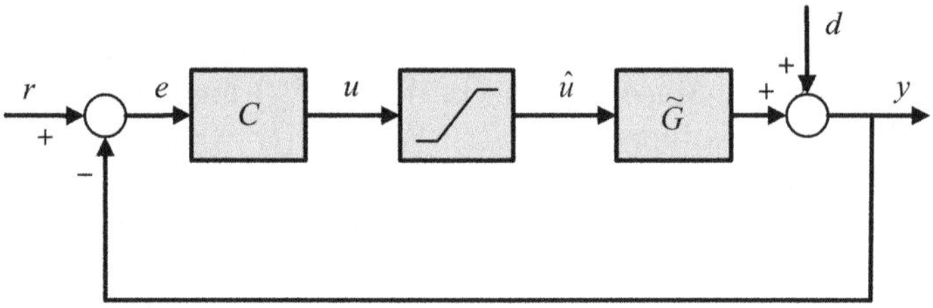

FIGURE 9.4.3
The unity feedback loop in the presence of actuator constraint.

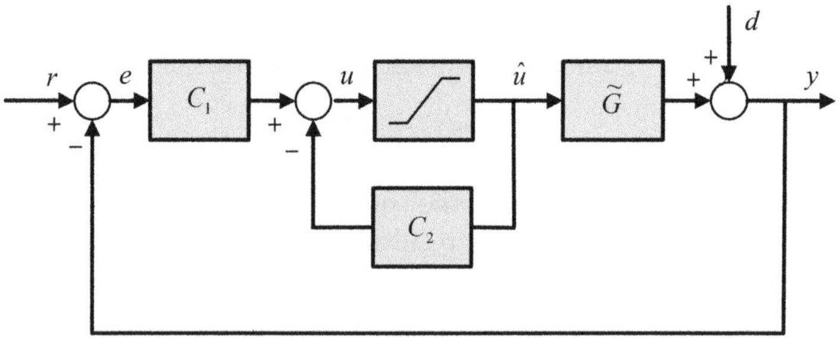

FIGURE 9.4.4
Modified unity feedback loop for anti-windup.

$$C_1(s) = Q_1(s), \tag{9.4.11}$$
$$C_2(s) = Q_2(s) - Q_1(s)G(s), \tag{9.4.12}$$

or

$$C_1(s) = C(\infty), \tag{9.4.13}$$
$$C_2(s) = \frac{C_1(s)}{C(s)} - 1, \tag{9.4.14}$$

where

$$C(s) = \frac{Q(s)}{1 - G(s)Q(s)}.$$

The latter controllers correspond to

$$W(s) = \frac{C_1(s)}{G(s)Q(s)}, \tag{9.4.15}$$

which implies

$$\min_{\hat{u}(t)} |C_1(t)[e(t) - e'(t)]|. \tag{9.4.16}$$

Example 9.4.1. *Consider the following plant:*

$$\boxed{G(s) = \frac{2}{100s + 1}.}$$

Case 1 *It is easy to obtain the following IMC controller:*

$$Q(s) = \frac{100s + 1}{2(\lambda s + 1)}.$$

It might as well take $\lambda = 20$.

Case 2 *Choosing $W(s) = 2.5(20s + 1)$ results in*

$$\boxed{Q_1(s) = 2.5, \;\; Q_2(s) = \frac{4}{100s + 1}.}$$

Case 3 *Choosing $W(s) = 50(s + 1)$ results in*

$$\boxed{Q_1(s) = \frac{50(s + 1)}{20s + 1}, \;\; Q_2(s) = \frac{99}{100s + 1}.}$$

Here $W(\infty)$ is chosen so that $Q_2(s)$ is strictly proper.

$Q(s)$ in Case 2 corresponds to minimizing $|e(t) - e'(t)|$, while $Q(s)$ in Case 3 approximately corresponds to minimizing $|y(t) - y'(t)|$.

Assume that the input is constrained within the saturation limits ± 1. The responses of the conventional IMC implementation and the modified IMC implementation to a unit step reference are shown in Figure 9.4.5 and Figure 9.4.6, respectively. The control input in Case 2 stays saturated until $e(t) = e'(t)$, while the control input in Case 3 stays saturated until $y(t) \approx y'(t)$.

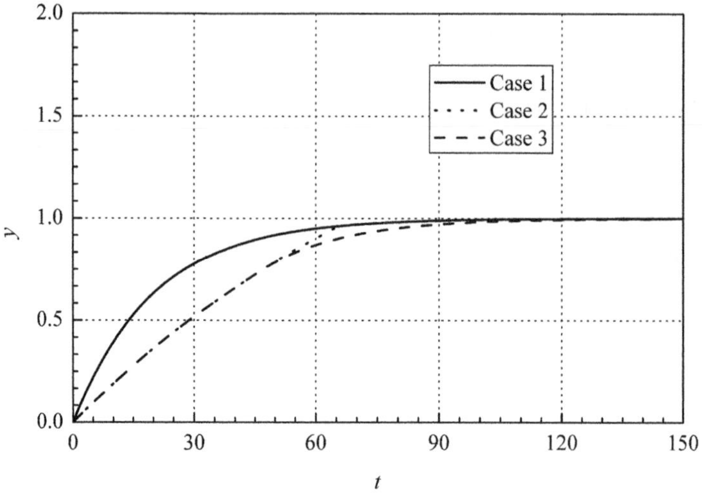

FIGURE 9.4.5
Responses of the system output.

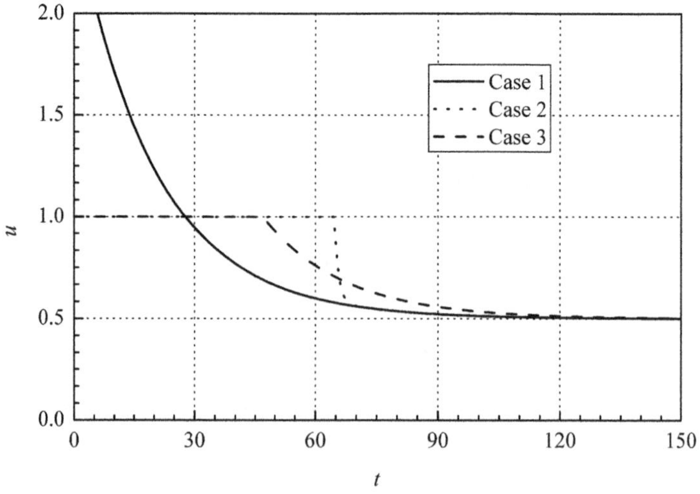

FIGURE 9.4.6
Responses of the plant input.

9.5 Feedforward Control

A control system should have good ability for disturbance rejection, so that the system output can be kept close to the reference in the presence of disturbances. In a feedback control loop, the disturbance can never be completely eliminated. The reason is simple. The feedback controller takes action only after the error caused by the disturbance happens.

When the disturbance entering a system is known, it can be compensated by introducing an additional loop. This is the basic idea of feedforward control. Theoretically, the feedforward controller has the potential to compensate disturbances exactly.

The feedforward structure is shown in Figure 9.5.1, where $C_f(s)$ is the feedforward controller, and $G_d(s)$ is the transfer function of the disturbance channel, through which the disturbance enters the system. The system output is

$$y(s) = G_d(s)d(s) - C_f(s)G(s)d(s). \tag{9.5.1}$$

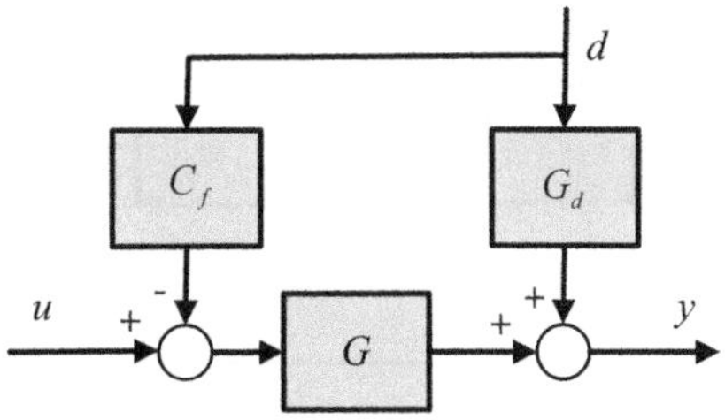

FIGURE 9.5.1
Feedforward control loop.

Assume that the model is exact. To compensate the disturbance completely, the controller should be

$$C_f(s) = \frac{G_d(s)}{G(s)}. \tag{9.5.2}$$

The exact compensation is achievable only when $C_f(s)$ is stable. This implies that $G(s)$ is MP or $G_d(s)$ has zeros wherever $G(s)$ has NMP zeros.

It is possible that $G(s)$ is NMP and has NMP zeros different from those of $G_d(s)$. In this case, $C_f(s)$ should be designed so that the effect of the disturbance on the system output is minimized. The transfer function from the disturbance to the system output is

$$\frac{y(s)}{d(s)} = G_d(s) - G(s)C_f(s).$$

Then the optimization problem can be expressed as

$$\min \|W(s)[G_d(s) - G(s)C_f(s)]\|, \tag{9.5.3}$$

where $\|\cdot\|$ denotes some norm and $W(s)$ is a weighting function. This problem can be converted into the quasi-H_∞ design problem or the H_2 design problem.

Feedforward control is an open-loop control strategy. If the model is not exact, the effect of the disturbance on the system output will not vanish. The system output will show a deviation from the reference. Consequently, feedforward control is seldom used alone. A frequently used scheme is the combined feedforward/feedback control, which is shown in Figure 9.5.2. In this structure, the effect of the disturbance on the system output is

$$y(s) = \frac{G_d(s)}{1 + G(s)C(s)}d(s) - \frac{C_f(s)G(s)}{1 + G(s)C(s)}d(s). \tag{9.5.4}$$

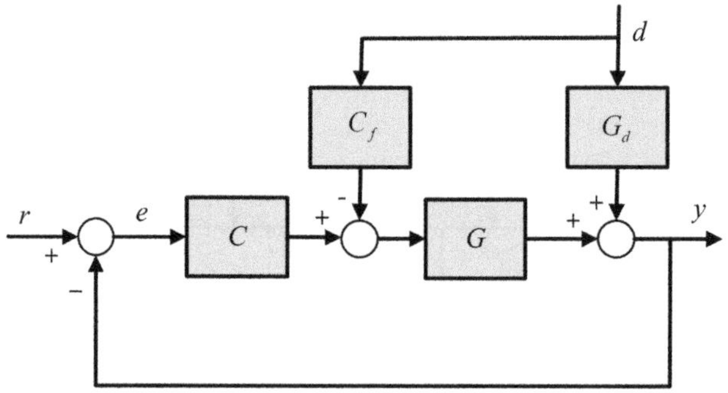

FIGURE 9.5.2
Combined feedback/feedforward control system.

The feedforward controller is still $C_f(s) = G_d(s)/G(s)$. The feedforward controller is used to compensate the disturbance, whereas the feedback action attempts to eliminate the effects of uncertainty. Since the characteristic equation of the system is not changed, the introduction of the feedforward controller will not affect the stability of the closed-loop system.

The feedforward controller can also be introduced to the IMC structure. In Figure 9.5.3, $Q(s)$ is the IMC controller and $Q_f(s)$ is the feedforward controller. They are related to the conventional controllers $C(s)$ and $C_f(s)$ through

$$C(s) = \frac{Q(s)}{1 - G(s)Q(s)}, \tag{9.5.5}$$

$$C_f(s) = \frac{Q_f(s) - G_d(s)Q(s)}{1 - G(s)Q(s)}. \tag{9.5.6}$$

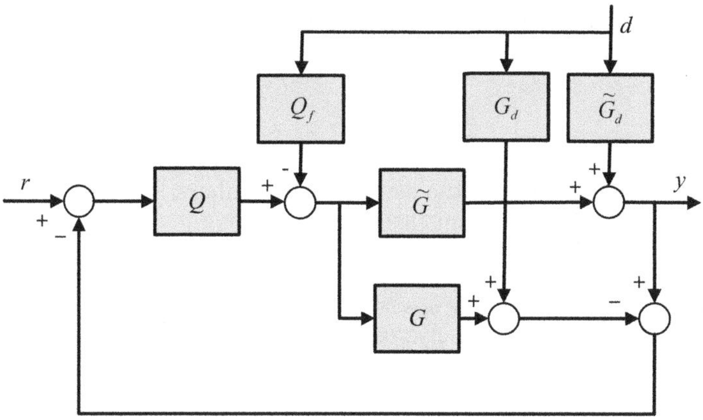

FIGURE 9.5.3
IMC feedback/feedforward control system.

When there is no model error,

$$y(s) = [G_d(s) - G(s)Q_f(s)]d(s). \tag{9.5.7}$$

To compensate the disturbance completely, the controller should be

$$Q_f(s) = \frac{G_d(s)}{G(s)}. \tag{9.5.8}$$

Note that the feedforward controller in the IMC structure is in the same form as that in the unity feedback loop.

Consider the design of feedforward controller for a system in which the plant and the transfer function of the disturbance channel can be adequately described by the first-order model with time delay; that is,

$$G(s) = \frac{Ke^{-\theta s}}{\tau s + 1}, \tag{9.5.9}$$

$$G_d(s) = \frac{K_d e^{-\theta_d s}}{\tau_d s + 1}. \tag{9.5.10}$$

The feedforward controller is as follows:

$$Q_f(s) = \frac{K_d}{K}\frac{\tau s + 1}{\tau_d s + 1}e^{(\theta - \theta_d)s}. \tag{9.5.11}$$

A typical case is $\theta \approx \theta_d$. Then the controller is simplified to

$$Q_f(s) = \frac{K_d}{K}\frac{\tau s + 1}{\tau_d s + 1}. \tag{9.5.12}$$

If the dynamic term of this controller is removed, the steady-state feedforward controller Q_{fs} will be obtained:

$$Q_{fs} = \frac{K_d}{K}. \tag{9.5.13}$$

When the model is exact, the feedforward/feedback loop can be viewed as the unity feedback loop without disturbance.

Example 9.5.1. *An important step in the paper-making process is drying. The dryer of a paper-making machine consists of many cylinders that are 1.5m in diameter and 2.1m in length. Each cylinder is fully filled with vapor. When the wet paper goes through the surface of these cylinders, most of the water evaporates and the final paper with proper moisture content is obtained (Figure 9.5.4).*

To reduce the fluctuation magnitude and frequency of the temperature change on the surface of cylinder, the feedforward/feedback scheme is applied. By mechanism analysis, the pressure inside the cylinder (which is used as the indicator of the cylinder temperature) is chosen as the system output, the steam flow rate is chosen as the control variable, and the steam temperature is regarded as the disturbance. The models of the cylinder are as follows:

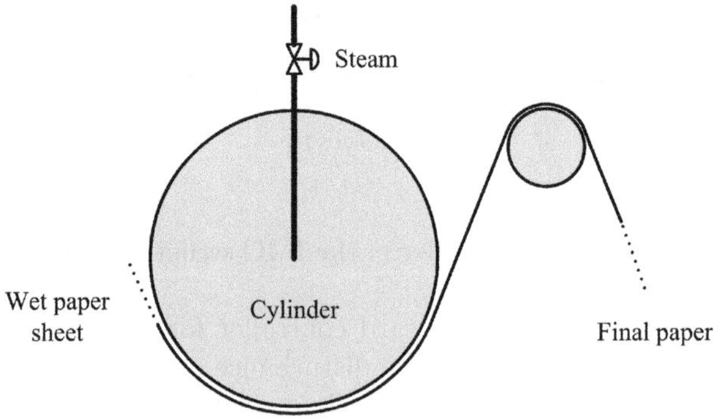

FIGURE 9.5.4
A cylinder.

$$\begin{aligned} G(s) &= \frac{1.65 \times 0.48}{(48s+1)(10s+1)}, \\ G_d(s) &= \frac{0.0636}{42.6s+1}. \end{aligned}$$

Then the feedforward controller is

$$C_f(s) = \frac{0.08(48s+1)(10s+1)}{42.6s+1}.$$

This is an improper transfer function. It can be physically realized by introducing a low-pass filter.

9.6 Optimal Input Disturbance Rejection

The discussion in the preceding chapters aims at the controller for output disturbances. The goal is to minimize $\|W(s)S(s)\|_2$ or $\|W(s)S(s)\|_\infty$. Design method of this kind has several merits:

1. The design procedure is simple and easy to understand.
2. The resulting controller is usually of low-order and easy to implement.
3. Attention is paid to both reference tracking and disturbance rejection.
4. The design for the regulator problem is identical to that for the servo problem.

The controller can also be designed for input disturbances. The problem has already been touched upon in Section 7.4. This section discusses the H_2 optimal control problem for input disturbance rejection.

Consider the unity feedback loop shown in Figure 2.3.1. The transfer function from the input disturbance to the system output is $G(s)S(s)$. Choose the performance index as min $\|W(s)G(s)S(s)\|_2$.

As introduced in Section 8.1, a general plant can be described by

$$G(s) = \frac{KN_+(s)N_-(s)}{M_+(s)M_-(s)}e^{-\theta s}. \tag{9.6.1}$$

The rational part of the plant is MP if $N_+(s) = 1$ and the plant is stable if $M_+(s) = 1$. For a step input, the $Q(s)$ that guarantees the internal stability and asymptotic tracking property can be written as

$$Q(s) = \frac{[1 + sQ_2(s)]M_+(s)}{K},$$

where $Q_2(s)$ is any stable transfer function that makes $Q(s)$ proper and satisfies

$$\lim_{s\to p_j}\frac{d^k}{ds^k}\left\{1 - \frac{[1 + sQ_2(s)]N_+(s)N_-(s)e^{-\theta s}}{M_-(s)}\right\} = 0, k = 0, 1, ..., l_j - 1.$$

Here $p_j(j = 1, 2, ..., r_p)$ are the l_j multiplicity unstable poles of $G(s)$. Take $W(s) = 1/s$. Then

$$\|W(s)G(s)S(s)\|_2^2$$

$$= \left\| W(s)G(s)\left\{1 - \frac{G(s)M_+(s)}{K}[1 + sQ_2(s)]\right\}\right\|_2^2$$

$$= \left\| \begin{array}{l} \dfrac{1}{s}\dfrac{KN_+(s)N_-(s)}{M_+(s)M_-(s)}e^{-\theta s} \\ \left[1 - \dfrac{N_+(s)N_-(s)}{M_-(s)}e^{-\theta s} - \dfrac{N_+(s)N_-(s)s}{M_-(s)}e^{-\theta s}Q_2(s)\right] \end{array} \right\|_2^2$$

$$= K^2 \left\| \begin{array}{l} \dfrac{N_-(s)N_+^2(-s)e^{\theta s} - N_+(s)}{sM_+(-s)M_-(s)N_+(s)} + \\ \dfrac{M_-(s) - N_-^2(s)N_+^2(-s)}{sM_+(-s)M_-^2(s)} - \dfrac{N_+^2(-s)N_-^2(s)}{M_+(-s)M_-^2(s)}Q_2(s) \end{array} \right\|_2^2. \quad (9.6.2)$$

Suppose that the plant is MP, that is, $\theta = 0$ and $N_+(s) = 1$. Then

$$\begin{aligned} &\|W(s)G(s)S(s)\|_2^2 \\ = \;& K^2 \left\| \frac{N_-(s) - 1}{sM_+(-s)M_-(s)} + \frac{M_-(s) - N_-^2(s)}{sM_+(-s)M_-^2(s)} - \frac{N_-^2(s)}{M_+(-s)M_-^2(s)}Q_2(s) \right\|_2^2 \\ = \;& K^2 \left\| \frac{M_-(s)N_-(s) - N_-^2(s)}{sM_+(-s)M_-^2(s)} - \frac{N_-^2(s)}{M_+(-s)M_-^2(s)}Q_2(s) \right\|_2^2. \end{aligned}$$

Since s is the factor of $M_-(s)N_-(s) - N_-^2(s)$, the optimal $Q_2(s)$ is

$$Q_{2opt}(s) = \frac{M_-(s)N_-(s) - N_-^2(s)}{sN_-^2(s)}.$$

The optimal $Q(s)$ can readily be obtained:

$$Q_{opt}(s) = \frac{M_-(s)M_+(s)}{KN_-(s)}. \quad (9.6.3)$$

Evidently, the controller is identical to the one designed for output disturbances. This implies that for MP plants, the design for input disturbances cannot provide any improved disturbance response. Nevertheless, the disturbance response can be improved for NMP plants.

To illustrate the problem, consider the simplest NMP plant described by the following transfer function:

$$G(s) = \frac{K(-z_r^{-1}s + 1)}{(\tau_1 s + 1)(\tau_2 s + 1)}, \quad (9.6.4)$$

where τ_1, τ_2, and z_r are positive numbers. We have

$$\|W(s)G(s)S(s)\|_2^2$$

$$= K^2 \left\| \frac{(z_r^{-1}s+1)^2 - (-z_r^{-1}s+1)}{s(\tau_1 s+1)(\tau_2 s+1)(-z_r^{-1}s+1)} + \frac{(\tau_1 s+1)(\tau_2 s+1) - (z_r^{-1}s+1)^2}{s(\tau_1 s+1)^2(\tau_2 s+1)^2} - \frac{(z_r^{-1}s+1)^2}{(\tau_1 s+1)^2(\tau_2 s+1)^2} Q_2(s) \right\|_2^2$$

$$= K^2 \left\| \frac{z_r^{-2}s + 3z_r^{-1}}{(\tau_1 s+1)(\tau_2 s+1)(-z_r^{-1}s+1)} + \frac{(\tau_1\tau_2 - z_r^{-2})s + (\tau_1+\tau_2-2z_r^{-1})}{(\tau_1 s+1)^2(\tau_2 s+1)^2} - \frac{(z_r^{-1}s+1)^2}{(\tau_1 s+1)^2(\tau_2 s+1)^2} Q_2(s) \right\|_2^2.$$

Let

$$\begin{aligned}
a_0 &= \frac{4z_r^{-2}}{\tau_1+\tau_2+z_r\tau_1\tau_2+z_r^{-1}}, \\
a_1 &= z_r^{-1} - a_0 + \tau_1 + \tau_2, \\
a_2 &= (z_r\tau_1\tau_2 - \tau_1 - \tau_2)a_0 + 3z_r^{-1}(\tau_1+\tau_2) + \tau_1\tau_2 - z_r^{-2}, \\
a_3 &= [z_r\tau_1\tau_2(\tau_1+\tau_2) - \tau_1\tau_2]a_0 + 3z_r^{-1}\tau_1\tau_2, \\
a_4 &= a_0 z_r \tau_1^2\tau_2^2.
\end{aligned}$$

It follows that

$$\begin{aligned}
&\|W(s)G(s)S(s)\|_2^2 \\
= \;& K^2 \left\| \frac{a_0}{-z_r^{-1}s+1} \right\|_2^2 + \\
& K^2 \left\| \frac{a_4 s^3 + a_3 s^2 + a_2 s + a_1}{(\tau_1 s+1)^2(\tau_2 s+1)^2} - \frac{(z_r^{-1}s+1)^2}{(\tau_1 s+1)^2(\tau_2 s+1)^2} Q_2(s) \right\|_2^2.
\end{aligned}$$

Minimize the right-hand side of the equation. By utilizing the optimal $Q_2(s)$, the optimal $Q(s)$ is obtained as follows:

$$Q_{opt}(s) = \frac{1}{K} + \frac{(a_4 s^3 + a_3 s^2 + a_2 s + a_1)s}{K(z_r^{-1}s+1)^2}. \tag{9.6.5}$$

The corresponding optimal performance is

$$\begin{aligned}
\text{ISE1} &= \min \|W(s)G(s)S(s)\|_2 \\
&= K \left\| \frac{a_0}{-z_r^{-1}s+1} \right\|_2 \\
&= K a_0 \sqrt{\frac{z_r}{2}}.
\end{aligned} \tag{9.6.6}$$

If instead of the input disturbance, the controller is designed for the output disturbance, one can obtain the following optimal controller based on (6.2.5):

$$Q_{opt}(s) = \frac{(\tau_1 s+1)(\tau_2 s+1)}{K(z_r^{-1}s+1)}. \tag{9.6.7}$$

The optimal performance corresponding to the input disturbance is

$$\begin{aligned}\text{ISE2} &= \min \|W(s)G(s)S(s)\|_2 \\ &= K\left\|\frac{2z_r^{-1}}{(\tau_1 s+1)(\tau_2 s+1)}\right\|_2 \\ &= Kz_r^{-1}\sqrt{\frac{2}{\tau_1+\tau_2}}.\end{aligned} \tag{9.6.8}$$

To compare the two performances, we calculate the ratio of them:

$$\begin{aligned}\frac{\text{ISE2}}{\text{ISE1}} &= \frac{z_r(\tau_1+\tau_2)+\tau_1\tau_2 z_r^2+1}{2\sqrt{z_r(\tau_1+\tau_2)}} \\ &= \frac{\sqrt{z_r(\tau_1+\tau_2)}}{2}+\frac{1}{2\sqrt{z_r(\tau_1+\tau_2)}}+\frac{\tau_1\tau_2 z_r^2}{2\sqrt{z_r(\tau_1+\tau_2)}}.\end{aligned} \tag{9.6.9}$$

It is concluded that the ratio is always greater than one. To see this, let $a=\sqrt{z_r(\tau_1+\tau_2)}$. Since $(a-1)^2 \geq 0$, $a+1/a \geq 2$. This implies that the sum of the first two terms in the right-hand side of (9.6.9) is greater than or equal to one. The third term is a positive number. The conclusion is that, for NMP plants, the disturbance response can be improved by designing the controller for input disturbances.

It is easy to verify that $S(s)$ tends to be zero when the controller tends to be optimal in the system with an MP plant. Both designs can approach the optimal rejection for input disturbances. Nevertheless, in the system with an NMP plant, when the controller tends to be optimal, instead of zero, $S(s)$ tends to be a constant.

It is seen from (9.6.9) that the effect of the NMP zero is evident. The larger the z_r, the larger the performance difference between the two designs. Now consider the effect of the plant poles. When the plant poles are close to the imaginary axis (that is, τ_1 or τ_2 is large), the third term in the right-hand side of (9.6.9) is large, which implies that the performance difference between the two designs is large.

In practice, the controller is seldom designed for input disturbances if the plant does not have any poles close to the imaginary axis. On one hand, the result is complex. On the other hand, the reference response usually worsens. This can be seen in the following example.

Example 9.6.1. *Anesthesia can be administered automatically by a control system. For certain operations, such as brain and eye surgery, involuntary muscle movements can be disastrous. To ensure adequate operating conditions for the surgeon, muscle relaxant drugs, which block involuntary muscle movements, are administered.*

A conventional method used by anesthesiologists for muscle relaxant administration is injecting a bolus dose, whose size is determined by experience, and to inject supplements as required. However, an anesthesiologist may sometimes fail to maintain a steady level of relaxation, resulting in a large amount of drug consumption or unexpected side effect. Significant improvements may be achieved by introducing the concept of automatic control.

As the level of relaxation cannot be directly measured, the arterial blood pressure is chosen as its proxy. When the blood pressure increases, the level of relaxation decreases. Assume that the body dynamics is

$$\boxed{G(s) = \frac{Ke^{-\theta s}}{\tau s + 1},}$$

where $K = 2$, $\tau = 0.2$, *and* $\theta = 0.1$. *If the controller is designed for output disturbances, the controller can be obtained from (6.2.6):*

$$\boxed{Q(s) = \frac{0.2s + 1}{2(\lambda s + 1)}.}$$

It might as well take $\lambda = 0.02$.

Consider the controller for input disturbances. Using (9.6.2), one obtains the following controller:

$$Q(s) = \frac{(\tau s + 1)[1 + \tau s(1 - e^{-\theta/\tau})]}{K(\lambda s + 1)^2}.$$

Substituting the plant parameters into the controller yields

$$\boxed{Q(s) = \frac{(0.2s + 1)(0.0787s + 1)}{2(\lambda s + 1)^2}.}$$

To obtain the same rise time, take $\lambda = 0.04$. *The closed-loop responses are shown in Figure 9.6.1. The design for the output disturbance has a better reference response, while the design for input disturbances has a better disturbance response. As the price, the controller for the input disturbance is complex and the reference response worsens.*

When the plant has poles close to the imaginary axis, three ways can be used to design the controller for input disturbances:

1. Introduce zeros by $S(s)$ to cancel the poles of $G(s)$ in the performance index $\|W(s)G(s)S(s)\|_2$.
2. Design the system as a Type 2 one.
3. Introduce zeros by $W(s)$ to cancel the poles of $G(s)$ in the performance index $\|W(s)S(s)\|_2$.

The first method has been discussed in the first half part of this section. The second method was, in fact, given in Chapter 7. The third method is mainly used in two cases: to simplify the design or to tradeoff between the design for the input disturbance and the one for the output disturbance.

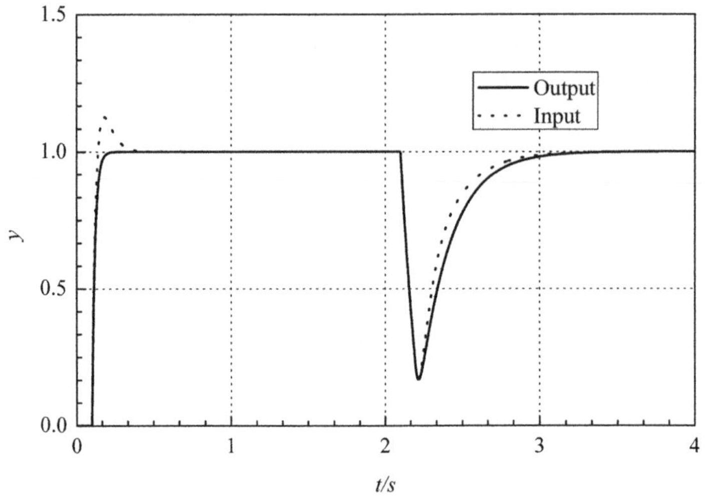

FIGURE 9.6.1
Responses of systems designed for different disturbances.

In the plant whose order is greater than one, normally not all poles are close to the imaginary axis. Assume that there is only one pole, τ_1, close to the imaginary axis in (9.6.4). The controller designed by the first method is complex. A simplified method is to choose a simple weighting function in the third method. For example, the following weighting function is taken:

$$W(s) = \frac{(\lambda s + 1)(\gamma s + 1)}{s(\tau_1 s + 1)}, \tag{9.6.10}$$

where $\gamma \in [\lambda, \tau_1]$. The rest of the design is similar to the first method.

It can be seen that this choice implies that only the slow pole, τ_1, is considered in the weighting function. The main design objective is the input disturbance when $\gamma = \lambda$, while it is the output disturbance when $\gamma = \tau_1$. By choosing an appropriate γ in between λ and τ_1, one can easily tradeoff between the design for the input disturbance and that for the output disturbance.

9.7 Control of Plants with Multiple Time Delays

Consider the plant

$$G(s) = \frac{KN_+(s)N_-(s)}{M_+(s)M_-(s)} e^{-\theta s}. \tag{9.7.1}$$

The control problem for this kind of plant is the most frequently encountered problem in linear control systems. Nevertheless, there is a more complex case; that is, there are multiple time delays in the nominator or denominator of plant. This corresponds to the situation where the plant has multiple state time delays and output time delays.

It is fairly difficult to rigorously treat such a plant in control system design. Therefore, it is usually reduced to the form of a rational transfer function with time delay. To explain why this treatment is necessary, consider a plant with dual time delays. The plant is described by

$$G(s) = \frac{e^{-\theta_1 s}}{\tau s + 1} - \frac{K e^{-\theta_2 s}}{\tau s + 1}, \tag{9.7.2}$$

which consists of two parallel stable plants with the same time constant.

Without loss of generality, let $K > 1,\ \theta_1 < \theta_2$ and $\theta = \theta_2 - \theta_1$. Then

$$G(s) = \frac{1 - K e^{-\theta s}}{\tau s + 1} e^{-\theta_1 s}.$$

If the design specification is to minimize $\|W(s)S(s)\|_2$, the optimal controller $Q(s)$ should be identical to that for the following plant:

$$G(s) = \frac{1 - K e^{-\theta s}}{\tau s + 1}. \tag{9.7.3}$$

For a step input, the $Q(s)$ satisfying the internal stability and asymptotic tracking property can be written as

$$Q(s) = \frac{1}{1 - K} + sQ_1(s), \tag{9.7.4}$$

where $Q_1(s)$ is stable. Take $W(s) = 1/s$, then

$$\begin{aligned}
&\|W(s)S(s)\|_2^2 \\
&= \left\| W(s) \left\{ 1 - G(s) \left[\frac{1}{1-K} + sQ_1(s) \right] \right\} \right\|_2^2 \\
&= \left\| \frac{1}{s} \left\{ 1 - \frac{1 - K e^{-\theta s}}{\tau s + 1} \left[\frac{1}{1-K} + sQ_1(s) \right] \right\} \right\|_2^2 \\
&= \left\| \frac{(\tau s + 1)(1 - K) - (1 - K e^{-\theta s})}{s(\tau s + 1)(1 - K)} - \frac{1 - K e^{-\theta s}}{\tau s + 1} Q_1(s) \right\|_2^2 .
\end{aligned}$$

Let $1 - K e^{-\theta s} = 0$. An infinite number of RHP zeros of $G(s)$ are obtained:

$$z_k = \frac{\ln K + 2k\pi j}{\theta}, \tag{9.7.5}$$

where k is any integer. An all-pass function can be constructed with the following poles:

$$p_k = \frac{-\ln K + 2k\pi j}{\theta}. \tag{9.7.6}$$

These poles are the roots of $-K + e^{-\theta s} = 0$. They are the mirror images of z_k. Consequently,

$$\begin{aligned}
&\|W(s)S(s)\|_2^2 \\
&= \left\| \frac{[(\tau s+1)(1-K)-(1-Ke^{-\theta s})](-K+e^{-\theta s})}{s(\tau s+1)(1-K)(1-Ke^{-\theta s})} - \frac{-K+e^{-\theta s}}{\tau s+1}Q_1(s) \right\|_2^2 \\
&= \left\| \frac{-K+e^{-\theta s}-1+Ke^{-\theta s}}{s(1-Ke^{-\theta s})} + \frac{(\tau s+1)(1-K)-(-K+e^{-\theta s})}{s(\tau s+1)(1-K)} - \frac{-K+e^{-\theta s}}{\tau s+1}Q_1(s) \right\|_2^2 \\
&= \left\| \frac{-K+e^{-\theta s}-1+Ke^{-\theta s}}{s(1-Ke^{-\theta s})} \right\|_2^2 + \\
&\quad \left\| \frac{(\tau s+1)(1-K)-(-K+e^{-\theta s})}{s(\tau s+1)(1-K)} - \frac{-K+e^{-\theta s}}{\tau s+1}Q_1(s) \right\|_2^2 .
\end{aligned}$$

Minimizing the right-hand side of the equality yields $Q_1(s)$. Then the optimal controller is

$$Q_{opt}(s) = \frac{\tau s+1}{-K+e^{-\theta s}}. \tag{9.7.7}$$

Introduce the following filter:

$$J(s) = \frac{1}{\lambda s+1}.$$

The unity feedback loop controller is

$$C(s) = \frac{\tau s+1}{(\lambda s+1)(-K+e^{-\theta s})-(1-Ke^{-\theta s})e^{-\theta_1 s}}. \tag{9.7.8}$$

To implement the controller in the unity feedback loop, the reduction technique has to be used.

For general plants, it is impossible to analytically design the controller when the time delay is rigorously treated. The difficulty lies in that the all-pass part of the plant cannot be constructed. For example, consider the following plant with dual time delays:

$$G(s) = \frac{e^{-\theta_1 s}}{\tau_1 s+1} + \frac{e^{-\theta_2 s}}{\tau_2 s-1}$$

$$= \frac{(\tau_2 s - 1)e^{-\theta_1 s} + (\tau_1 s + 1)e^{-\theta_2 s}}{(\tau_1 s + 1)(\tau_2 s - 1)}. \tag{9.7.9}$$

Let $(\tau_2 s - 1)e^{-\theta_1 s} + (\tau_1 s + 1)e^{-\theta_2 s} = 0$. The zeros of the plant are the solution to the following equation:

$$e^{-(\theta_1 - \theta_2)s} = \frac{\tau_1 s + 1}{-\tau_2 s + 1}. \tag{9.7.10}$$

It is a challenge to construct an all-pass transfer function with this equation.

9.8 Summary

When the dynamic characteristics of reference and disturbance are substantially different and both good reference tracking and good disturbance rejecting are desired, it is advantageous to use a 2 DOF control system. This chapter discusses the 2 DOF control problem and an analytical design procedure is proposed. It can be seen that Structure I is actually the unity feedback loop with a prefilter. Structure II provides an alternative way to implement the prefiltering.

One main function of feedback controllers is rejecting the disturbance. In some systems, the system response to the disturbance can be improved by introducing a secondary measurement point and a secondary feedback controller, that is, by using the cascade control. It is pointed out that the performance cannot always be improved by the cascade control. The cascade control is only useful when the master plant is NMP.

There is always saturation nonlinearity in a real actuator. The saturation nonlinearity will cause windup and prevent the controller from reducing the error signal efficiently. In this case, anti-windup schemes are proposed to improve the system performance. An optimization-based design method is presented in this chapter, which is very effective for anti-windup.

A disadvantage of feedback control is that the corrective action for disturbance does not begin until the system output deviates from the reference. Compared with feedback control, the advantage of feedforward control is that the disturbance can be compensated completely at the time it happens. The inherent weakness of the feedforward control is that it requires exact information about the plant and disturbance. In this chapter, an analytical design method is provided for feedforward/feedback control.

In this chapter, the optimal rejection problem of input disturbances and the control problem of the plant with multiple time delays are also studied. The two problems are solved with the design technique introduced in the preceding chapters. It is also pointed out that when the plant is MP, the design for the input disturbance cannot provide any improvement on disturbance response.

Exercises

1. Does the prefilter in a 2 DOF scheme affect the stability and robust stability of the closed-loop system? Why?
2. A parallel cascade control structure is shown in Figure E9.1. Compare the parallel cascade control system with the series cascade control system in Figure 9.3.2 and give some suggestions on the design of the parallel cascade controllers.

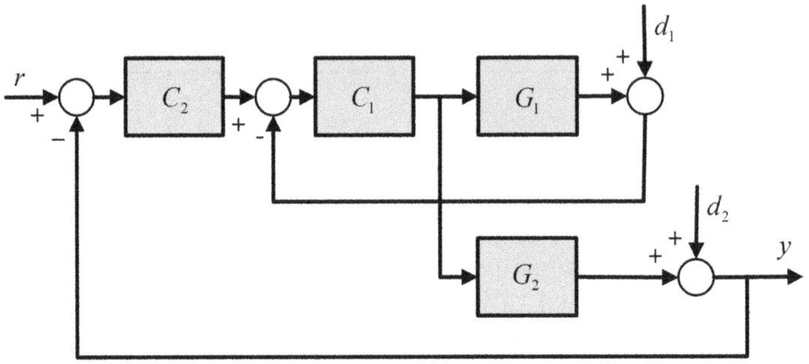

FIGURE E9.1
Parallel cascade control structure.

3. Consider the conventional feedback/feedforward control structure in Figure 9.5.2 and the IMC feedback/feedforward control structure in Figure 9.5.3. Is $C_f(s)$ equivalent to $Q_f(s)$?
4. Consider the nuclear reactor described in Example 6.4.1. The plant model is given by

$$G(s) = \frac{e^{-0.4s}}{0.2s+1}.$$

Assume that the transfer function of the disturbance channel is

$$G_d(s) = \frac{2e^{-0.1s}}{2s+1}.$$

Design a feedforward controller.

5. The solar-powered rover is a highly helpful tool for interstellar exploration. It can be used for a mission of several months in severe environment. A typical appearance of the rover is a car with solar panel and camera equipped on its back (Figure E5.1).

The steering of the car is controlled by a feedback control system.

The objective is to reach a step response to a steering command with zero steady-state error, overshoot less than 20%, peak time less than 0.3s, and $|u(t)| \leq 50$. The dynamics of the steering plant is given by

$$G(s) = \frac{2e^{-0.1s}}{0.2s + 1}.$$

Design an acceptable controller for the steering system.

6. For the plant in Example 9.6.1 and the input disturbance, prove

$$Q(s) = \frac{(\tau s + 1)[1 + \tau s(1 - e^{-\theta/\tau})]}{K(\lambda s + 1)^2}.$$

Notes and References

Section 9.1 and Section 9.2 are from Zhang et al. (2004) and Zhang (2006). An alternative 2 DOF control structure was discussed by Astrom et al. (1994), Zhang and Sun (1996b), and Zhang et al. (1998a). Kwok and Davison (2007) discussed an interesting problem regarding the complexity of control structure.

The plant in Example 9.2.1 is from Dorf and Bishop (2001, p. 386).

Most discussion in Section 9.3 is based on Morari and Zafiriou (1989). The cascade control system is also studied by Maffezzoni et al. (1990) and Kaya (2001).

The design method in Section 9.4 was proposed by Zheng et al. (1994) (Zheng A., M. V. Kothare and M. Morari. Anti-windup design for internal model control, *Int. J. Control*, 1994, 60(5), 1015–1024. ©Taylor & Francis Ltd.). The factorization in Item 2 corresponds to the model state feedback proposed in Coulibaly et al. (1990).

There is a lot of literature on the topic of anti-windup. The relationships among different anti-windup schemes were discussed by Edwards and Postlethwaite (1998) and Zaccarian and Teel (2002).

Section 9.5 is adapted from Morari and Zafiriou (1989, Section 6.3). Lewin and Scali (1988) analyzed the robustness of feedforward control system.

The plant in Example 9.5.1 is from Sun (1993, Section 11.2).

Section 9.6 and 9.7 are from Zhang et al. (2006).

The discussion about the input disturbance rejection problem can also be found in Doyle et al. (1992). The design method for rejecting the input disturbance by choosing appropriate weighting functions was discussed in detail in Alcantara et al. (2011).

The anesthesia control problem was discussed by Frei et al. (2000) and

Linkens (1993). The plant in Example 9.6.1 is from Dorf and Bishop (2001, p. 466).

The parallel cascade control structure in Exercise 2 is from Semino and Brambilla (1996).

Exercise 5 is adapted from Dorf and Bishop (2001, p. 728).

10

Analysis of MIMO Systems

CONTENTS

The control systems considered in the preceding chapters are SISO systems. In these systems, the plant has only a single output and a single manipulated input. Some systems are MIMO systems, in which the plant has more than one output and requires more than one manipulated input. From this chapter on, the analysis and design of MIMO systems will be the main subject. The discussion will concentrate on the square plant, that is, the plant with the same number of inputs and outputs.

The analysis and design of MIMO systems require more mathematical knowledge than the SISO case. As in SISO systems, an effort is made to treat MIMO systems with the least mathematical knowledge.

For convenience of understanding the subsequent chapters, some necessary terminologies and basic concepts are introduced in this chapter, including zero and pole, norm and system gain, internal stability, and performance index. These terminologies and concepts are clear in the SISO system, but their extension to the MIMO case should be properly defined.

For the sake of providing a panorama of the methods and results in this area, the uncertainty description and the test for robust stability and robust performance are introduced in Sections 10.5 and 10.6. The discussion in these two sections is very theoretical. Those application-oriented readers can skip them. For an in-depth study on the mathematical knowledge for MIMO systems, readers are referred to monographs about this topic.

10.1 Zeros and Poles of a MIMO Plant

Consider linear time invariant and causal plants. An $n \times n$ MIMO plant $\boldsymbol{G}(t)$ is causal if all of its elements $G_{ij}(t)(i = 1, 2, ..., n; j = 1, 2, ..., n)$ are causal. Such a MIMO plant can be described by a square transfer function matrix $\boldsymbol{G}(s)$, whose elements are in the form of a proper transfer functions with real coefficients. A transfer function matrix $\boldsymbol{G}(s)$ is proper if all its elements $G_{ij}(s)$ are proper; $\boldsymbol{G}(s)$ is strictly proper if all its elements $G_{ij}(s)$ are strictly proper. All transfer function matrices that are not proper are improper.

In SISO systems, it is seen that zeros and poles play an important role in the analysis and design of control systems. Normally, the designer has to compute zeros and poles at the beginning of design. At least it should be known whether the plant has zeros and poles in the RHP. In this section, the concept of SISO zero and pole is generalized to $n \times n$ MIMO systems.

The pole polynomial $\pi(s)$ of $\boldsymbol{G(s)}$ is the least common denominator of all non-identically zero minors of $\boldsymbol{G(s)}$. The pole is the root of the equation $\pi(s) = 0$. A system is stable if and only if all its poles are in the open LHP.

Example 10.1.1. *Consider the following plant:*

$$\boldsymbol{G}(s) = \begin{bmatrix} \frac{1}{s+3} & \frac{1}{s-2} \\ \frac{3}{s+3} & \frac{s+1}{s-2} \end{bmatrix}.$$

The minors of order 1 are the determinants of the elements of $\boldsymbol{G}(s)$. *The minor of order 2 is the determinant of the plant itself. In view of the minors of all orders, the least common denominator is*

$$\pi(s) = (s + 3)(s - 2).$$

Therefore, $\boldsymbol{G}(s)$ *has two poles: one is at* $s = -3$ *and the other is at* $s = 2$.

Let "det" denote the determinant of a matrix. The zero polynomial $\zeta(s)$ is the numerator of $\det[\boldsymbol{G}(s)]$, provided that $\det[\boldsymbol{G}(s)]$ has been adjusted to have the pole polynomial as its denominator. The zeros of $\boldsymbol{G}(s)$ are the roots of the equation $\zeta(s) = 0$.

If a point is the zero of $\boldsymbol{G}(s)$, the rank of $\boldsymbol{G}(s)$ at this point is less than its normal rank. The normal rank is defined as the rank of $\boldsymbol{G}(s)$ for every s in the set of complex numbers, except for a finite number of points.

Example 10.1.2. *Consider the plant in Example 10.1.1. Adjust* $\det \boldsymbol{G}(s)$ *so that its denominator is the pole polynomial:*

$$\det[\boldsymbol{G}(s)] = \frac{s - 2}{(s + 3)(s - 2)}.$$

Then

$$\zeta(s) = s - 2.$$

Hence, $\boldsymbol{G}(s)$ *has one zero at* $s = 2$.

Similar to SISO systems, when $\boldsymbol{G(s)}$ has only one pole at $s = p_0$, p_0 is said to be a simple pole. If $\boldsymbol{G(s)}$ has multiple poles at $s = p_0$, p_0 is a multiple pole. Zeros and their multiplicity can be defined similarly.

It is observed in Example 10.1.1 and Example 10.1.2 that MIMO plants may have zeros and poles at the same location. In general, it is impossible to find all plant zeros from the condition $\det[\boldsymbol{G}(s)] = 0$. When forming the determinant, zeros and poles at the same location cancel each other.

It is seen that the zero location of a MIMO system is no longer related to the zero location of the individual SISO transfer function constituting the MIMO system. Thus, it is possible for a MIMO system to be NMP even though all SISO transfer functions are MP. A plant $\boldsymbol{G(s)}$ is NMP if its transfer function matrix contains zeros in the closed RHP or contains a time delay. Otherwise, the plant is MP. With regard to MIMO plants, different definitions for time delay will result in different MP plants. This book gives a definition with respect to the decoupling control in Chapter 13.

Example 10.1.3. *The plant*

$$\boldsymbol{G}(s) = \frac{1}{s+1}\begin{bmatrix} s+3 & 2 \\ 3 & 1 \end{bmatrix}$$

has a zero at $s = 3$ and thus is NMP. However, all SISO transfer functions are MP.

The normal rank of $\boldsymbol{G}(s)$ is 2. Substituting $s = 3$ into $\boldsymbol{G}(s)$ yields

$$\boldsymbol{G}(3) = \frac{1}{4}\begin{bmatrix} 6 & 2 \\ 3 & 1 \end{bmatrix}.$$

It can be seen that the feature of the resulting matrix is that its two rows are not independent. The rank of $\boldsymbol{G}(3)$ is 1, which is less than the normal rank of $\boldsymbol{G}(s)$.

The zero multiplicity is closely related to the plant rank at the zero.

Theorem 10.1.1. *If the rank of $\boldsymbol{G}(z)$ is k, then $\boldsymbol{G}(s)$ has a zero at $s = z$ with multiplicity at least $(n - k)$.*

Proof. Since $\text{rank}[\boldsymbol{G}(z)] = k$, with appropriate elementary transformations $\boldsymbol{G}(s)$ can be transformed into the following form:

$$\boldsymbol{G}_1(s) = \begin{bmatrix} b_1(s) \\ \vdots \\ b_k(s) \\ b_{k+1}(s) \\ \vdots \\ b_n(s) \end{bmatrix},$$

where $b_1(z), b_2(z), ..., b_k(z)$ are linearly independent rows. $b_{k+i}(z)(i = 1, 2, ..., n-k)$ either are zero or can be written as

$$b_{k+i}(z) = \alpha_{i1}b_1(z) + \alpha_{i2}b_2(z) + ... + \alpha_{ik}b_k(z),$$

where $\alpha_{i1}, \alpha_{i2}, ..., \alpha_{ik}$ are constants that are not all zero.

Carry out elementary transformations with regard to $\boldsymbol{G}_1(s)$, so that

$$\boldsymbol{G}_2(s) = \begin{bmatrix} b_1(s) \\ \vdots \\ b_k(s) \\ b'_{k+1}(s) \\ \vdots \\ b'_n(s) \end{bmatrix},$$

where $b'_{k+i}(s) = b_{k+i}(s) - \alpha_{i1}b_1(s) - \alpha_{i2}b_2(s) - ... - \alpha_{ik}b_k(s)$ if $b_{k+i}(z) \neq 0$ and $b'_{k+i}(s) = b_{k+i}(s)$ if $b_{k+i}(z) = 0$. Then $b'_{k+i}(s)(i = 1, 2, ..., n-k)$ must contain the factor $s - z$.

As elementary transformations do not change the value of the determinant, $\boldsymbol{G}(s)$ has a zero at z with multiplicity least $(n-k)$. □

10.2 Singular Values

Consider a fixed frequency ω. $T(s)$ is a complex number at $s = j\omega$.

For a SISO system, $y(s) = T(s)r(s)$. The gain at a given frequency can be simply expressed as

$$\frac{|y(j\omega)|}{|r(j\omega)|} = \frac{|T(j\omega)r(j\omega)|}{|r(j\omega)|} = |T(j\omega)|. \tag{10.2.1}$$

The gain depends on the frequency, but is independent of the input magnitude.

In an $n \times n$ system $\boldsymbol{T}(s)$, the case is different because both the input $\boldsymbol{r}(s)$ and the output $\boldsymbol{y}(s) = \boldsymbol{T}(s)\boldsymbol{r}(s)$ are vectors. To investigate the gain at a given frequency, we need to "sum up" the magnitudes of the input signal and output signal in each vector by utilizing norms. If the 2-norm, a frequently used measure, is selected, the "size" of the input signal is

$$\|\boldsymbol{r}(j\omega)\|_2 := \left[\sum_{j=1}^{n} |r_j(j\omega)|^2\right]^{1/2}, \tag{10.2.2}$$

and the "size" of the output signal is

$$\|\boldsymbol{y}(j\omega)\|_2 := \left[\sum_{i=1}^{n} |y_i(j\omega)|^2\right]^{1/2}. \tag{10.2.3}$$

The gain of the system at a given frequency can then be described by the ratio

$$\begin{aligned}\frac{\|\boldsymbol{y}(j\omega)\|_2}{\|\boldsymbol{r}(j\omega)\|_2} &= \frac{\|\boldsymbol{T}(j\omega)\boldsymbol{r}(j\omega)\|_2}{\|\boldsymbol{r}(j\omega)\|_2} \\ &= \left[\frac{\sum_i |y_i(j\omega)|^2}{\sum_j |r_j(j\omega)|^2}\right]^{1/2}.\end{aligned} \tag{10.2.4}$$

In addition to the frequency, the gain depends on the input direction. Hence, the above definition for the gain is not an ideal one.

Example 10.2.1. *Consider a* 2×2 *system*

$$\boldsymbol{T} = \begin{bmatrix} 1 & 2 \\ 3 & 4 \end{bmatrix}.$$

The inputs are

$$\boldsymbol{r_1} = \begin{bmatrix} 1 \\ 0 \end{bmatrix}, \boldsymbol{r_2} = \begin{bmatrix} 0 \\ 1 \end{bmatrix}, \boldsymbol{r_3} = \begin{bmatrix} 0.707 \\ 0.707 \end{bmatrix},$$
$$\boldsymbol{r_4} = \begin{bmatrix} 0.707 \\ -0.707 \end{bmatrix}, \boldsymbol{r_5} = \begin{bmatrix} 0.8 \\ -0.6 \end{bmatrix},$$

These inputs have the same magnitude $\|\boldsymbol{r}\|_2 = 1$, *but they are in different directions. Compute the system outputs for the five inputs:*

$$\boldsymbol{y_1} = \begin{bmatrix} 1 \\ 3 \end{bmatrix}, \boldsymbol{y_2} = \begin{bmatrix} 4 \\ 2 \end{bmatrix}, \boldsymbol{y_3} = \begin{bmatrix} 2.12 \\ 4.95 \end{bmatrix},$$
$$\boldsymbol{y_4} = \begin{bmatrix} -0.707 \\ -0.707 \end{bmatrix}, \boldsymbol{y_5} = \begin{bmatrix} -0.40 \\ 0 \end{bmatrix}.$$

The 2-norms of these outputs are

$$\|\boldsymbol{y_1}\|_2 = 3.16, \|\boldsymbol{y_2}\|_2 = 4.47, \|\boldsymbol{y_3}\|_2 = 5.38,$$
$$\|\boldsymbol{y_4}\|_2 = 1.00, \|\boldsymbol{y_5}\|_2 = 0.40.$$

It is observed that the inputs with the same magnitude and different directions relate to different system gains.

As it is known, the eigenvalues of a MIMO system reflect its gain characteristic. Let $\lambda_{ei}[\boldsymbol{T}(j\omega)](i = 1, 2, ..., n)$ denote the eigenvalues of $\boldsymbol{T}(j\omega)$. The sum of the eigenvalues of $\boldsymbol{T}(j\omega)$ is equal to the trace of $\boldsymbol{T}(j\omega)$ (that is, the sum of the diagonal elements):

$$\text{Trace}[\boldsymbol{T}(j\omega)] = \sum_i \lambda_{ei}[\boldsymbol{T}(j\omega)].$$

The largest eigenvalue is called the spectral radius, which is denoted by

$$\rho[\boldsymbol{T}(j\omega)] = \max_i |\lambda_{ei}[\boldsymbol{T}(j\omega)]|.$$

Let the complex number t_{ij} be the elements of $\boldsymbol{T}(j\omega)$. Define the norm of a complex matrix as

$$\|\boldsymbol{T}(j\omega)\|_\infty = \max_i \sum_j |t_{ij}|.$$

If $\lambda_{ei}[\boldsymbol{T}(j\omega)]$ is an eigenvalue of $\boldsymbol{T}(j\omega)$ and $\boldsymbol{r}(j\omega)$ is the corresponding eigenvector, then

$$|\boldsymbol{T}(j\omega)\boldsymbol{r}(j\omega)| = |\lambda_{ei}[\boldsymbol{T}(j\omega)]||\boldsymbol{r}(j\omega)|.$$

As

$$|\boldsymbol{T}(j\omega)\boldsymbol{r}(j\omega)| \leq \|\boldsymbol{T}(j\omega)\|_\infty |\boldsymbol{r}(j\omega)|,$$

we have

$$\rho[\boldsymbol{T}(j\omega)] \leq \|\boldsymbol{T}(j\omega)\|_\infty.$$

The eigenvalues of a system, however, do not provide a useful means for generalizing the SISO gain, because they only measure the gain in the special case where the input and the output are in the same direction.

Example 10.2.2. *Consider the system* $\boldsymbol{y} = \boldsymbol{T}\boldsymbol{r}$ *with*

$$\boldsymbol{T} = \begin{bmatrix} 0 & 1 \\ 0 & 0 \end{bmatrix}.$$

There are two eigenvalues

$$\lambda_{e1}[\boldsymbol{T}] = \lambda_{e2}[\boldsymbol{T}] = 0.$$

It is clearly misleading to conclude that the gain of the system is zero. For example, with the input

$$\boldsymbol{r} = [0\ 1]^T,$$

we have the output

$$\boldsymbol{y} = [1\ 0]^T.$$

A good measure for the MIMO gain at a given frequency ω is the singular value. The singular values of the complex matrix $\boldsymbol{T}(j\omega)$, denoted by $\sigma_i[\boldsymbol{T}(j\omega)]$, are the n square roots of the eigenvalues of $\boldsymbol{T}^H(j\omega)\boldsymbol{T}(j\omega)$, that is,

$$\sigma_i[\boldsymbol{T}(j\omega)] = \left\{\lambda_{ei}[\boldsymbol{T}^H(j\omega)\boldsymbol{T}(j\omega)]\right\}^{1/2}, i = 1, 2, ..., n, \quad (10.2.5)$$

where the superscript H denotes the complex conjugate transpose of a matrix: $\boldsymbol{T}^H(j\omega) = \bar{\boldsymbol{T}}^T(j\omega)$. For convenience, the ordering $\sigma_1 \geq \sigma_2 \geq ... \geq \sigma_n$ is adopted. In general, the singular values must be computed numerically. However, for 2×2 matrices, analytical expressions can be obtained.

Instead of a single gain, there are a group of gains in MIMO systems. The largest gain in all input directions is equal to the maximum singular value:

$$\bar{\sigma}[\boldsymbol{T}(j\omega)] = \max_{\boldsymbol{r}(j\omega)\neq 0} \frac{\|\boldsymbol{T}(j\omega)\boldsymbol{r}(j\omega)\|_2}{\|\boldsymbol{r}(j\omega)\|_2}, \quad (10.2.6)$$

and the smallest gain in all input directions is equal to the minimum singular value:

$$\underline{\sigma}[\boldsymbol{T}(j\omega)] = \min_{\boldsymbol{r}(j\omega)\neq 0} \frac{\|\boldsymbol{T}(j\omega)\boldsymbol{r}(j\omega)\|_2}{\|\boldsymbol{r}(j\omega)\|_2}. \quad (10.2.7)$$

A convenient way to represent a matrix and expose its internal structure is using singular value decomposition (SVD). The SVD of $\boldsymbol{T}(j\omega)$ is given by

$$\begin{aligned}\boldsymbol{T}(j\omega) &= \boldsymbol{U}(j\omega)\boldsymbol{\Sigma}(j\omega)\boldsymbol{V}^{\boldsymbol{H}}(j\omega) \\ &= \sum_{i=1}^{n} \sigma_i[\boldsymbol{T}(j\omega)]\boldsymbol{u_i}(j\omega)\boldsymbol{v_i}^H(j\omega), \quad (10.2.8)\end{aligned}$$

where

$$\boldsymbol{U}(j\omega) = [\ \boldsymbol{u_1}(j\omega) \quad \boldsymbol{u_2}(j\omega) \quad ... \quad \boldsymbol{u_n}(j\omega)\ ], \quad (10.2.9)$$

and $\boldsymbol{U}^H(j\omega) = \boldsymbol{U}^{-1}(j\omega)$,

$$\boldsymbol{V}(j\omega) = [\ \boldsymbol{v_1}(j\omega) \quad \boldsymbol{v_2}(j\omega) \quad ... \quad \boldsymbol{v_n}(j\omega)\ ], \quad (10.2.10)$$

and $\boldsymbol{V}^H(j\omega) = \boldsymbol{V}^{-1}(j\omega)$. $\boldsymbol{\Sigma}(j\omega)$ can be written as

$$\boldsymbol{\Sigma}(j\omega) = \text{diag}\{\sigma_1[\boldsymbol{T}(j\omega)], \sigma_2[\boldsymbol{T}(j\omega)], ..., \sigma_n[\boldsymbol{T}(j\omega)]\}. \quad (10.2.11)$$

Here "diag" denotes a diagonal matrix.

It is easy to check that the columns of $\boldsymbol{V}(j\omega)$ and $\boldsymbol{U}(j\omega)$ are unit eigenvectors of $\boldsymbol{T}^H(j\omega)\boldsymbol{T}(j\omega)$ and $\boldsymbol{T}(j\omega)\boldsymbol{T}^H(j\omega)$, respectively. They are known as the right singular vectors and the left singular vectors of the matrix. The right singular vectors represent the input direction, while the left singular vectors represent the output direction. By SVD an arbitrary matrix can be decomposed into a "rotation" ($\boldsymbol{V}^H(j\omega)$), followed by scaling ($\boldsymbol{\Sigma}(j\omega)$), and then followed by a "rotation" ($\boldsymbol{U}(j\omega)$).

Example 10.2.3. *Consider the following system again:*

$$\boldsymbol{T} = \begin{bmatrix} 1 & 2 \\ 3 & 4 \end{bmatrix}.$$

The singular value decomposition is

$$\boldsymbol{T} = \begin{bmatrix} -0.4046 & -0.9145 \\ -0.9145 & 0.4046 \end{bmatrix} \begin{bmatrix} 5.4650 & 0 \\ 0 & 0.3660 \end{bmatrix} \begin{bmatrix} -0.5760 & 0.8174 \\ -0.8174 & -0.5760 \end{bmatrix}^H.$$

The largest gain of 5.4650 relates to the input in the direction

$$\begin{bmatrix} -0.5760 \\ -0.8174 \end{bmatrix},$$

and the smallest gain of 0.3660 relates to the input in the direction

$$\begin{bmatrix} 0.8174 \\ -0.5760 \end{bmatrix}.$$

10.3 Norms for Signals and Systems

In the last section, the MIMO gain was considered only at individual frequencies. In system design, it is very useful to estimate the MIMO gain over the whole frequency range. This can be achieved by defining norms for vector functions and matrix-valued functions.

Consider the vector function $\boldsymbol{r}(t)$ of dimension n. The 2-norm of $\boldsymbol{r}(t)$ is defined as

$$\|\boldsymbol{r}(t)\|_2 = \left[\int_{-\infty}^{\infty} \boldsymbol{r}^T(t)\boldsymbol{r}(t)dt\right]^{1/2}. \tag{10.3.1}$$

For the $\boldsymbol{r}(t)$ whose energy is bounded, Parseval's theorem yields an equivalent expression:

$$\|\boldsymbol{r}(s)\|_2 := \left[\frac{1}{2\pi}\int_{-\infty}^{\infty} \boldsymbol{r}^H(j\omega)\boldsymbol{r}(j\omega)d\omega\right]^{1/2}. \tag{10.3.2}$$

Assume that the matrix-valued function $\boldsymbol{T}(s)$ of dimension $n \times n$ is a strictly proper transfer function matrix without poles on the imaginary axis. The definition of the 2-norm of $\boldsymbol{T}(s)$ is

$$\|\boldsymbol{T}(s)\|_2 := \left\{\frac{1}{2\pi}\int_{-\infty}^{\infty} \text{Trace}\left[\boldsymbol{T}^H(j\omega)\boldsymbol{T}(j\omega)\right] d\omega\right\}^{1/2}. \tag{10.3.3}$$

If $\boldsymbol{T}(s)$ has no poles in the RHP, by Parseval's theorem we have

$$\begin{aligned} \|\boldsymbol{T}(s)\|_2 &= \|\boldsymbol{T}(t)\|_2 \\ &= \left\{ \int_{-\infty}^{\infty} \text{Trace} \left[\boldsymbol{T}^T(t) \boldsymbol{T}(t) \right] dt \right\}^{1/2} . \end{aligned} \tag{10.3.4}$$

Assume that $\boldsymbol{T}(s)$ of dimension $n \times n$ is a proper transfer function matrix without poles on the imaginary axis. The definition of the ∞-norm of $\boldsymbol{T}(s)$ is

$$\|\boldsymbol{T}(s)\|_\infty := \sup_\omega \bar{\sigma} \left[\boldsymbol{T}(j\omega) \right]. \tag{10.3.5}$$

The ∞-norm is sub-multiplicative:

$$\|\boldsymbol{T}_1(s)\boldsymbol{T}_2(s)\|_\infty \leq \|\boldsymbol{T}_1(s)\|_\infty \|\boldsymbol{T}_2(s)\|_\infty . \tag{10.3.6}$$

For beginners, it is easy to confuse the norm concept in this section with that in the last section where the norm was defined for vectors and matrices whose elements were complex numbers. The norm in this section is defined for vectors and matrices whose elements are functions.

Consider the linear system

$$\boldsymbol{y}(s) = \boldsymbol{T}(s)\boldsymbol{r}(s), \tag{10.3.7}$$

where $\boldsymbol{T}(s)$ is stable. An interesting problem is how to quantify the least upper bound of the system output $y(s)$ for a known input $\boldsymbol{r}(s)$, or equivalently, how large the system gain is. The system gains are shown in Table 10.3.1.

TABLE 10.3.1
System gains for MIMO systems.

	$\boldsymbol{r}(t) = \delta(t)\boldsymbol{I}$	$\|\boldsymbol{r}(t)\|_2$
$\|\boldsymbol{y}(t)\|_2$	$\|\boldsymbol{T}(s)\|_2$	$\|\boldsymbol{T}(s)\|_\infty$

Consider Entry (1, 1) of Table 10.3.1. Assume that the impulse is applied to each input in due order: $\boldsymbol{r}(t) = \delta(t)\boldsymbol{I}$. We have

$$\|\boldsymbol{y}(t)\|_2 = \|\boldsymbol{T}(t)\|_2 = \|\boldsymbol{T}(s)\|_2 . \tag{10.3.8}$$

This implies that for this specific input, the energy of the system output is the square of the 2-norm of the system transfer function matrix.

Consider Entry (1, 2). Assume that the input is bounded: $\|\boldsymbol{r}(s)\|_2 \leq 1$. The energy of the system output is bounded by the square of the ∞-norm of the system transfer function. This is shown as follows:

$$\|\boldsymbol{y}(s)\|_2^2 = \frac{1}{2\pi} \int_{-\infty}^{\infty} \boldsymbol{r}^H(j\omega)\boldsymbol{T}^H(j\omega)\boldsymbol{T}(j\omega)\boldsymbol{r}(j\omega)d\omega$$

$$\leq \sup_{\omega} |\boldsymbol{T}(j\omega)|^2 \frac{1}{2\pi} \int_{-\infty}^{\infty} \boldsymbol{r}^H(j\omega)\boldsymbol{r}(j\omega)d\omega,$$

or

$$\|\boldsymbol{y}(t)\|_2^2 \leq \|\boldsymbol{T}(s)\|_\infty^2 \|\boldsymbol{r}(t)\|_2^2. \tag{10.3.9}$$

Thus, $\|\boldsymbol{T}(s)\|_\infty$ is an upper bound for the energy of the system output. To prove that it is the least upper bound, it is enough to show that the bound can be reached for a specific input. The specific input is a constructed frequency domain impulse occurring at the frequency where $|\boldsymbol{T}(j\omega)|$ is maximum. The constructing procedure, which can be found in Section 3.1, will not be repeated here.

10.4 Nominal Stability and Performance

In order to work well in a real system, the controller has to meet the following objectives:

- Nominal stability (NS).
- Nominal performance (NP).
- Robust stability (RS).
- Robust performance (RP).

The nominal stability is mandatory, while the robust performance is normally the ultimate design objective (Figure 10.4.1). Nominal stability and nominal performance are addressed in this section. The other two objectives are going to be studied in the next two sections.

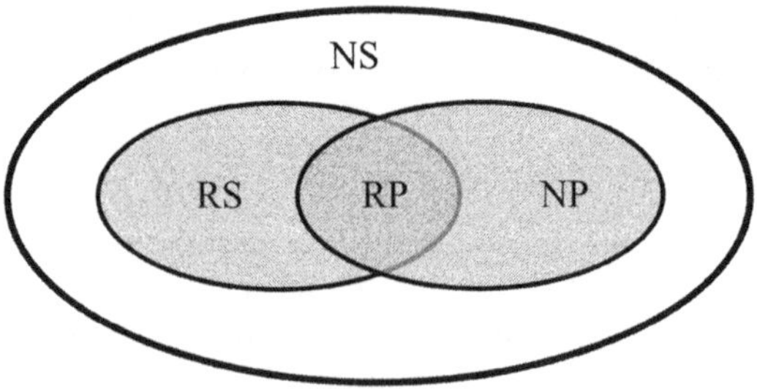

FIGURE 10.4.1
Design objectives of a control system.

Consider the control system consisting of an $n \times n$ plant $\boldsymbol{G}(s)$ and an $n \times n$ controller $\boldsymbol{C}(s)$ (Figure10.4.2). Assume that there is not any unstable hidden mode in $\boldsymbol{G}(s)$. This assumption implies that the plant can be stabilized with feedback control. It is satisfied if there is no RHP zero-pole cancellation in $\boldsymbol{G}(s)$.

The transfer function matrix of the open-loop system is given by

$$\boldsymbol{L}(s) = \boldsymbol{G}(s)\boldsymbol{C}(s). \tag{10.4.1}$$

The objective of the system is to keep the output $\boldsymbol{y}(s)$ close to the reference $\boldsymbol{r}(s)$. The first step is to analyze the stability.

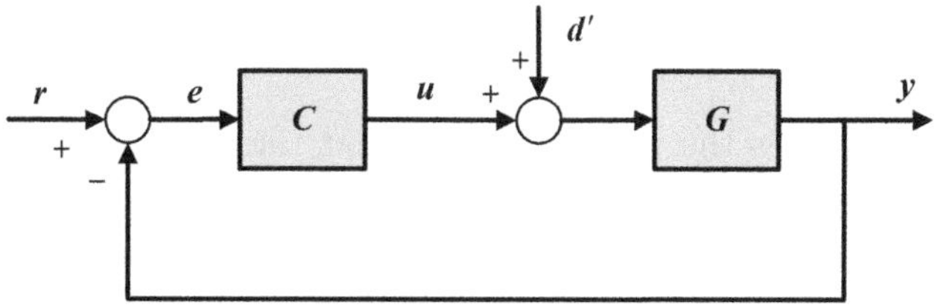

FIGURE 10.4.2
MIMO unity feedback control loop.

Theorem 10.4.1 (MIMO Nyquist Stability Criterion)**.** *Let n be the number of unstable poles of* $\boldsymbol{L}(s)$*. The closed-loop system is stable if and only if the Nyquist plot of* $\det[\boldsymbol{I} + \boldsymbol{L}(s)]$ *does not pass through the origin, and encircles it n times counterclockwise.*

The test for internal stability introduced in Chapter 3 is also applicable to MIMO systems. The unity feedback control system shown in Figure 10.4.2 is internally stable if and only if all the elements in the transfer function matrix $\boldsymbol{H}(s)$ are stable.

$$\begin{bmatrix} \boldsymbol{y}(s) \\ \boldsymbol{u}(s) \end{bmatrix} = \boldsymbol{H}(s) \begin{bmatrix} \boldsymbol{r}(s) \\ \boldsymbol{d}'(s) \end{bmatrix},$$

where

$$\boldsymbol{H}(s) = \begin{bmatrix} \boldsymbol{G}(s)\boldsymbol{C}(s)[\boldsymbol{I} + \boldsymbol{G}(s)\boldsymbol{C}(s)]^{-1} & [\boldsymbol{I} + \boldsymbol{G}(s)\boldsymbol{C}(s)]^{-1}\boldsymbol{G}(s) \\ \boldsymbol{C}(s)[\boldsymbol{I} + \boldsymbol{G}(s)\boldsymbol{C}(s)]^{-1} & -\boldsymbol{C}(s)[\boldsymbol{I} + \boldsymbol{G}(s)\boldsymbol{C}(s)]^{-1}\boldsymbol{G}(s) \end{bmatrix}. \tag{10.4.2}$$

The performance analysis for MIMO systems is similar to that for SISO systems. Consider the IMC structure shown in Figure 10.4.3, where $\tilde{\boldsymbol{G}}(s)$ is

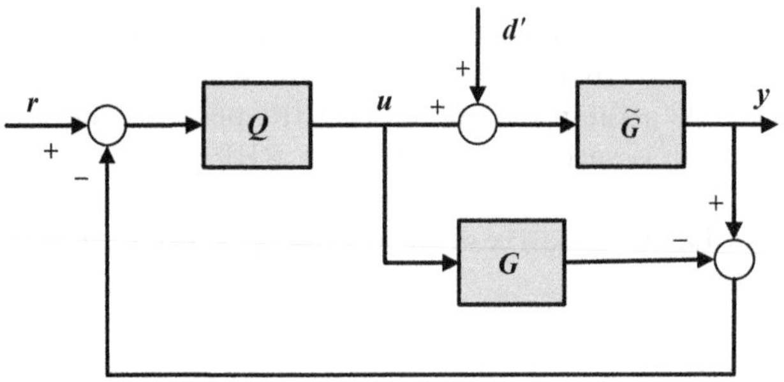

FIGURE 10.4.3
MIMO IMC structure.

the plant and $\boldsymbol{G}(s)$ is the model. Assume that the model is exact (that is, $\tilde{\boldsymbol{G}}(s) = \boldsymbol{G}(s)$). The unit feedback loop controller can be written as

$$\boldsymbol{C}(s) = \boldsymbol{Q}(s)[\boldsymbol{I} - \boldsymbol{G}(s)\boldsymbol{Q}(s)]^{-1}. \tag{10.4.3}$$

Define the sensitivity function as

$$\begin{aligned}\boldsymbol{S}(s) &= [\boldsymbol{I} + \boldsymbol{G}(s)\boldsymbol{C}(s)]^{-1} \\ &= \boldsymbol{I} - \boldsymbol{G}(s)\boldsymbol{Q}(s).\end{aligned} \tag{10.4.4}$$

The complementary sensitivity function is

$$\begin{aligned}\boldsymbol{T}(s) &= \boldsymbol{I} - \boldsymbol{S}(s) \\ &= \boldsymbol{G}(s)\boldsymbol{C}(s)[\boldsymbol{I} + \boldsymbol{G}(s)\boldsymbol{C}(s)]^{-1} \\ &= \boldsymbol{G}(s)\boldsymbol{Q}(s).\end{aligned} \tag{10.4.5}$$

First of all, the steady-state performance of the closed-loop system is characterized. Let m be the largest integer satisfying

$$\text{rank}\{\lim_{s\to 0}[s^m \boldsymbol{L}(s)]\} = n. \tag{10.4.6}$$

$\boldsymbol{L}(s)$ is said to be of Type m. It is seen that $\boldsymbol{L}(s)$ has at least $n \times m$ poles at the origin. The corresponding sensitivity function matrix satisfies

$$\lim_{s\to 0}[s^{-k}\boldsymbol{S}(s)] = \boldsymbol{0}, k = 1, 2, ..., m-1. \tag{10.4.7}$$

If the closed-loop system is stable, as $t \to \infty$ the closed-loop system perfectly tracks references of the form $\sum_{k=0}^{m} \boldsymbol{a_k} s^{-k}$, where $\boldsymbol{a_k}$ are real constant vectors.

In particular, a Type 1 system requires

$$\lim_{s\to 0}[\boldsymbol{G}(s)\boldsymbol{Q}(s)] = \boldsymbol{I}, \tag{10.4.8}$$

and a Type 2 system requires

$$\lim_{s\to 0}[\boldsymbol{G}(s)\boldsymbol{Q}(s)] = \boldsymbol{I}, \tag{10.4.9}$$

$$\lim_{s\to 0}\frac{d}{ds}[\boldsymbol{G}(s)\boldsymbol{Q}(s)] = \boldsymbol{0}. \tag{10.4.10}$$

Then, consider the dynamic performance of MIMO systems. The performance indices include engineering-oriented indices and optimal indices.

For the main channels, the most common engineering-oriented indices are the same as those in SISO systems, including overshoot, rise time, and so on. With regard to the coupled channels, the engineering-oriented indices are frequently specified in terms of the amplitudes of coupled responses. This is illustrated in the following example.

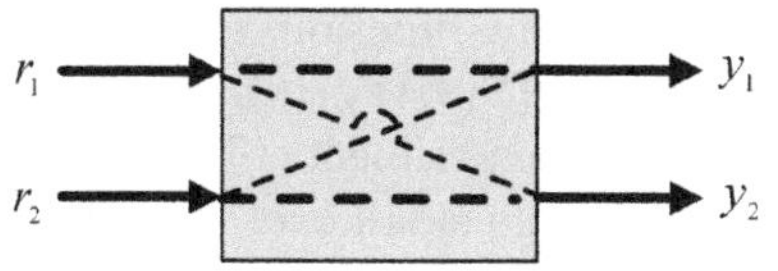

FIGURE 10.4.4
A 2×2 MIMO system.

Example 10.4.1. *In a 2×2 system, the inputs are $r_1(s)$ and $r_2(s)$ and the outputs are $y_1(s)$ and $y_2(s)$, respectively (Figure 10.4.4). The main channels are $r_1(s) - y_1(s)$ and $r_2(s) - y_2(s)$; that is, $y_1(s)$ is mainly controlled by $r_1(s)$ and $y_2(s)$ is mainly controlled by $r_2(s)$. The coupled channels are $r_1(s) - y_2(s)$ and $r_2(s) - y_1(s)$. The design requirements are as follows:*

1. *The overshoots of the main channels are less than 10%.*
2. *$y_1(t) < 0.2$ for $r_1(s) = 0$ and $r_2(s) = 1/s$.*
3. *$y_2(t) < 0.4$ for $r_1(s) = 1/s$ and $r_2(s) = 0$.*

The overarching characteristic that distinguishes QPCT from others is that QPCT can be used to design controllers satisfying the quantitative performance requirements in the above example in an easy way.

The optimal indices are normally specified in terms of the norm of the weighted sensitivity transfer function matrix. Let $\boldsymbol{W_{p1}}(s)$ and $\boldsymbol{W_{p2}}(s)$ be two weighting functions. The H_2 optimal control of MIMO system is defined as

$$\begin{aligned} &\min \|\boldsymbol{W_{p2}}(s)\boldsymbol{S}(s)\boldsymbol{W_{p1}}(s)\|_2^2 \\ = &\min \frac{1}{2\pi}\int_{-\infty}^{\infty} \text{Trace}\left\{ \begin{array}{c} [\boldsymbol{W_{p2}}(j\omega)\boldsymbol{S}(j\omega)\boldsymbol{W_{p1}}(j\omega)]^H \cdot \\ [\boldsymbol{W_{p2}}(j\omega)\boldsymbol{S}(j\omega)\boldsymbol{W_{p1}}(j\omega)] \end{array}\right\} d\omega. \end{aligned} \tag{10.4.11}$$

The index for the H_∞ optimal control is expressed as

$$\begin{aligned} & \min \|\boldsymbol{W_{p2}}(s)\boldsymbol{S}(s)\boldsymbol{W_{p1}}(s)\|_\infty \\ = & \min \sup_\omega \bar{\sigma}\left[\boldsymbol{W_{p2}}(j\omega)\boldsymbol{S}(j\omega)\boldsymbol{W_{p1}}(j\omega)\right]. \end{aligned} \tag{10.4.12}$$

Generally speaking, $\boldsymbol{W_{p1}}(s)$ is more important than $\boldsymbol{W_{p2}}(s)$ in the above indices, because $\boldsymbol{W_{p1}}(s)$ is needed in all cases, while $\boldsymbol{W_{p2}}(s)$ is not always necessary.

$\boldsymbol{W_{p1}}(s)$ is the input weighting function. Excite the system in separate experiments with n different linearly independent inputs $\boldsymbol{r_i}(s)(i = 1, 2, ..., n)$. For one experiment the error is $\boldsymbol{e_i}(s) = \boldsymbol{S}(s)\boldsymbol{r_i}(s)$. Define $\boldsymbol{W_{p1}}(s) = [\boldsymbol{r_1}(s), \boldsymbol{r_2}(s), ..., \boldsymbol{r_n}(s)]$. The columns of $\boldsymbol{S}(s)\boldsymbol{W_{p1}}(s)$ are the errors from the n experiments. In practice, step inputs are of primary importance. In this case, one can take $\boldsymbol{W_{p1}}(s) = s^{-1}\boldsymbol{I}(s)$.

$\boldsymbol{W_{p2}}(s)$ is the output weighting function. Premultiplication by the output weight $\boldsymbol{W_{p2}}(s)$ generates $\boldsymbol{W_{p2}}(s)\boldsymbol{S}(s)\boldsymbol{W_{p1}}(s)$. The columns of the matrix are the weighted errors from the n experiments. The output weight is used since it may be desirable to make errors small over some frequency ranges.

In the method of this book, a filter is introduced to reach the same goal. Hence, $\boldsymbol{W_{p2}}(s) = \boldsymbol{I}$ is taken. Comparatively, the merits of penalizing errors by a filter are that the design procedure is simple and the degree of penalizing can be tuned easily.

Now only one weighting function needs to be considered. Hence, $\boldsymbol{W_{p1}}(s)$ is denoted by $\boldsymbol{W}(s)$ for simplicity.

10.5 Robust Stability of MIMO Systems

The description of uncertainty and the test of robustness for MIMO plants are very complicated. The result is not a simple generalization of the SISO case.

In the SISO case, the uncertainty is described by an uncertain plant family. This family corresponds to a Nyquist band consisting of a union of disks with certain radius. Similar uncertainty description can be developed for MIMO systems. Since the commutative property does not hold for matrix multiplication in general, one has to distinguish the uncertainty occurring at the plant input and that at the plant output (Figure 10.5.1).

Assume that the uncertain plants have the same number of RHP poles as the nominal plant. Let the subscript "I" stand for "Input" and the subscript "O" stand for "Output." The uncertain plant can be described in the following form:

Plant with multiplicative output uncertainty:

$$\tilde{\boldsymbol{G}}(s) = [\boldsymbol{I} + \boldsymbol{\delta_O}(s)]\boldsymbol{G}(s). \tag{10.5.1}$$

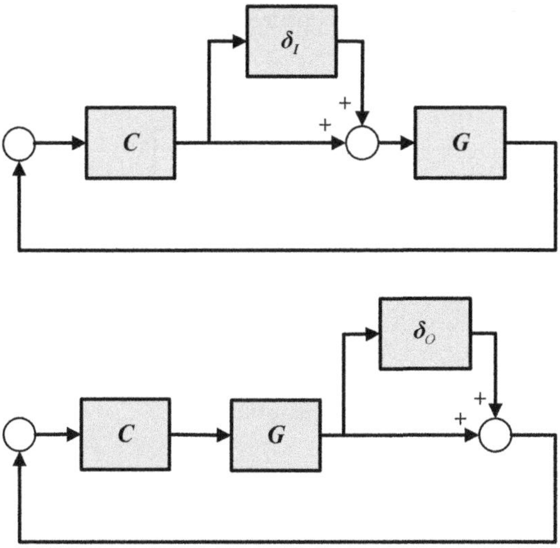

FIGURE 10.5.1
Input uncertainty $\delta_I(s)$ and output uncertainty $\delta_O(s)$.

Plant with multiplicative input uncertainty:

$$\tilde{\boldsymbol{G}}(s) = \boldsymbol{G}(s)[\boldsymbol{I} + \boldsymbol{\delta_I}(s)]. \tag{10.5.2}$$

Both of the uncertainties can be described in a unified form:

$$\boldsymbol{\delta}(s) = \boldsymbol{W_2}(s)\boldsymbol{\Delta}(s)\boldsymbol{W_1}(s), \tag{10.5.3}$$

where $\boldsymbol{W_1}(s)$ and $\boldsymbol{W_2}(s)$ are stable weighting function matrices. $\boldsymbol{\Delta}(s)$ is a stable transfer function matrix denoting the normalized uncertainty:

$$\bar{\sigma}[\boldsymbol{\Delta}(j\omega)] \leq 1, \quad \forall\omega, \tag{10.5.4}$$

or equivalently,

$$\|\boldsymbol{\Delta}(s)\|_\infty \leq 1.$$

The unstructured uncertainty is constructed by lumping different sources of uncertainties into a single uncertainty. Let $\Delta_m(s)$ be a stable scalar weighting function. The unstructured uncertainty is usually interpreted as follows:

$$\boldsymbol{\delta}(s) = \Delta_m(s)\boldsymbol{\Delta}(s), \tag{10.5.5}$$

that is, in the unified form of the uncertainty description, $\boldsymbol{W_1}(s) = \Delta_m(s)$ and $\boldsymbol{W_2}(s) = \boldsymbol{I}$, or $\boldsymbol{W_2}(s) = \Delta_m(s)$ and $\boldsymbol{W_1}(s) = \boldsymbol{I}$. $\Delta_m(s)$ gives the magnitude profile of the uncertainty $\boldsymbol{\delta}(s)$:

$$\bar{\sigma}\left[\boldsymbol{\delta}(j\omega)\right] \leq |\Delta_m(j\omega)|, \quad \forall\omega. \tag{10.5.6}$$

$\tilde{\boldsymbol{G}}(j\omega)$ describes a disk with the center $\boldsymbol{G}(j\omega)$ and the radius $|\Delta_m(j\omega)|$ at each frequency ω.

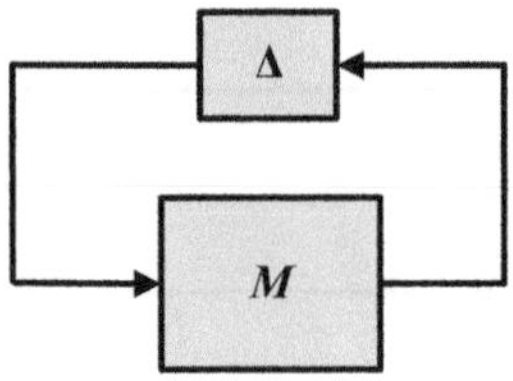

FIGURE 10.5.2
General $\boldsymbol{M\Delta}$ structure for robustness analysis.

To analyze the robust stability of the closed-loop system in a unified framework, the systems shown in Figure 10.5.1 are usually redrawn in the $\boldsymbol{M\Delta}$ form shown in Figure 10.5.2. Let

$$\begin{aligned}\boldsymbol{S_I}(s) &= [\boldsymbol{I}+\boldsymbol{C}(s)\boldsymbol{G}(s)]^{-1},\\ \boldsymbol{T_I}(s) &= [\boldsymbol{I}+\boldsymbol{C}(s)\boldsymbol{G}(s)]^{-1}\boldsymbol{C}(s)\boldsymbol{G}(s),\end{aligned}$$

and

$$\begin{aligned}\boldsymbol{S_O}(s) &= [\boldsymbol{I}+\boldsymbol{G}(s)\boldsymbol{C}(s)]^{-1},\\ \boldsymbol{T_O}(s) &= \boldsymbol{G}(s)\boldsymbol{C}(s)[\boldsymbol{I}+\boldsymbol{G}(s)\boldsymbol{C}(s)]^{-1}.\end{aligned}$$

It is easy to verify that, for the input uncertainty $\boldsymbol{\delta}_I(s) = \Delta_{Im}(s)\boldsymbol{\Delta}_I(s)$,

$$\boldsymbol{M}(s) = -\boldsymbol{T_I}(s)\Delta_{Im}(s), \boldsymbol{\Delta}(s) = \boldsymbol{\Delta}_I(s), \tag{10.5.7}$$

and for the output uncertainty $\boldsymbol{\delta}_O(s) = \Delta_{Om}(s)\boldsymbol{\Delta}_O(s)$,

$$\boldsymbol{M}(s) = -\boldsymbol{T_O}(s)\Delta_{Om}(s), \boldsymbol{\Delta}(s) = \boldsymbol{\Delta}_O(s). \tag{10.5.8}$$

When the nominal system is internally stable, $\boldsymbol{M}(s)$ is evidently stable. The following theorem gives the condition for the robust stability of the closed-loop system.

Theorem 10.5.1. *The closed-loop system shown in Figure 10.5.2 is stable for all* $\boldsymbol{\Delta}(s)$*s if and only if one of the following two equivalent conditions is satisfied:*

1. $\det[\boldsymbol{I} - \boldsymbol{M}(j\omega)\boldsymbol{\Delta}(j\omega)] \neq 0, \quad \forall\omega, \forall\boldsymbol{\Delta}(j\omega).$
2. $\|\boldsymbol{M}(s)\|_\infty < 1.$

Proof. 1. By assumption, the nominal system is internally stable; the uncertain plant and the nominal plant have the same number of RHP poles. The closed-loop system shown in Figure 10.5.2 is stable for all $\boldsymbol{\Delta}(s)$s, if and only if $\det[\boldsymbol{I}+\boldsymbol{M}(s)\boldsymbol{\Delta}(s)]$ encircle the origin as many times as the nominal system.

If the Nyquist plot of $\det[\boldsymbol{I}+\boldsymbol{M}(s)\boldsymbol{\Delta}(s)]$ does not pass through the origin, the number of encirclements will not change. This is equivalent to

$$\det[\boldsymbol{I}-\boldsymbol{M}(j\omega)\boldsymbol{\Delta}(j\omega)] \neq 0, \quad \forall\omega, \forall\boldsymbol{\Delta}(j\omega).$$

2. The result can be proved by contradiction.

First, it is shown that $\rho[\boldsymbol{M}(j\omega)\boldsymbol{\Delta}(j\omega)] < 1$ is sufficient. Assume that there exist a frequency ω' and an uncertainty $\boldsymbol{\Delta}'(j\omega')$ such that $\rho[\boldsymbol{M}(j\omega')\boldsymbol{\Delta}'(j\omega')] < 1$, but

$$\det[\boldsymbol{I}-\boldsymbol{M}(j\omega')\boldsymbol{\Delta}'(j\omega')] = 0,$$

which is equivalent to

$$\prod_i \lambda_{ei}[\boldsymbol{I}-\boldsymbol{M}(j\omega')\boldsymbol{\Delta}'(j\omega')] = 0.$$

This implies that for some i

$$1-\lambda_{ei}[\boldsymbol{M}(j\omega')\boldsymbol{\Delta}'(j\omega')] = 0.$$

Consequently,

$$\rho[\boldsymbol{M}(j\omega')\boldsymbol{\Delta}'(j\omega')] \geq 1,$$

which is a contradiction. Therefore, $\rho[\boldsymbol{M}(j\omega)\boldsymbol{\Delta}(j\omega)] < 1$ is sufficient for robust stability.

Because $\rho[\boldsymbol{M}(j\omega)\boldsymbol{\Delta}'(j\omega)] \leq \|\boldsymbol{M}(j\omega)\boldsymbol{\Delta}'(j\omega)\|_\infty \leq \|\boldsymbol{M}(s)\|_\infty$, $\|\boldsymbol{M}(s)\|_\infty < 1$ is also sufficient.

Next, to prove the necessity of $\|\boldsymbol{M}(s)\|_\infty < 1$, assume that the closed-loop system is stable, but $\|\boldsymbol{M}(s)\|_\infty \geq 1$. Then for some frequency ω' we have $\sigma_1[\boldsymbol{M}(j\omega')] \geq 1$. It will be shown that there exists a $\boldsymbol{\Delta}'(s)$ with $\|\boldsymbol{\Delta}'(s)\|_\infty \leq 1$ such that the closed-loop system is unstable.

Let the SVD of $\boldsymbol{M}(j\omega')$ be

$$\boldsymbol{M}(j\omega') = \boldsymbol{U}(j\omega')\boldsymbol{\Sigma}(j\omega')\boldsymbol{V}^H(j\omega').$$

Define

$$\boldsymbol{D}(j\omega') = \mathrm{diag}\{1/\sigma_1[\boldsymbol{M}(j\omega')], 0, ..., 0\},$$

and

$$\boldsymbol{\Delta}'(s) = \boldsymbol{V}(s)\boldsymbol{D}(j\omega')\boldsymbol{U}^H(s).$$

$\boldsymbol{V}(s)$ and $\boldsymbol{U}(s)$ can be easily constructed from the complex matrices

$\boldsymbol{V}(j\omega')$ and $\boldsymbol{U}(j\omega')$. The first vector of $\boldsymbol{V}(j\omega')$, $\boldsymbol{v}_1(j\omega')$, is used to illustrate the procedure. Write it in the following form:

$$\boldsymbol{v}_1^T(j\omega') = [v_{11}e^{j\phi_1}, v_{12}e^{j\phi_2}, ..., v_{1n}e^{j\phi_n}],$$

where v_{1j} are real numbers, and are chosen so that $\phi_j \in [-\pi, 0)$, $j = 1, 2, ..., n$. Choose $\alpha_j \geq 0$ so that

$$\angle\left(\frac{\alpha_j - j\omega'}{\alpha_j + j\omega'}\right) = \phi_j,$$

$\boldsymbol{v}_1(s)$ can be taken as

$$\boldsymbol{v}_1^T(s) = \left[v_{11}\frac{\alpha_1 - s}{\alpha_1 + s}, v_{12}\frac{\alpha_2 - s}{\alpha_2 + s}, \ldots, v_{1n}\frac{\alpha_n - s}{\alpha_n + s}\right].$$

Clearly,

$$\left\|\boldsymbol{\Delta}'(s)\right\|_\infty = 1/\sigma_1[\boldsymbol{M}(j\omega')] \leq 1$$

and

$$\begin{aligned} & \det[\boldsymbol{I} - \boldsymbol{M}(j\omega')\boldsymbol{\Delta}'(j\omega')] \\ = \ & \det[\boldsymbol{I} - \boldsymbol{U}(j\omega')\boldsymbol{\Sigma}(j\omega')\boldsymbol{V}^H(j\omega')\boldsymbol{V}(j\omega')\boldsymbol{D}(j\omega')\boldsymbol{U}^H(j\omega')] \\ = \ & \det[\boldsymbol{I} - \boldsymbol{U}(j\omega')\boldsymbol{\Sigma}(j\omega')\boldsymbol{D}(j\omega')\boldsymbol{U}^H(j\omega')] \\ = \ & 0, \end{aligned}$$

which implies the closed-loop system is unstable. □

In particular, for the input uncertainty and output uncertainty (Figure 10.5.1), there are the following results.

Corollary 10.5.2. *The closed-loop system is stable for the multiplicative input uncertainty if and only if*

$$\|\boldsymbol{T_I}(s)\Delta_{Im}(s)\|_\infty < 1.$$

Corollary 10.5.3. *The closed-loop system is stable for the multiplicative output uncertainty if and only if*

$$\|\boldsymbol{T_O}(s)\Delta_{Om}(s)\|_\infty < 1.$$

Sometimes, the unstructured uncertainty description is conservative. In this case, it is desirable to use the structured uncertainty description. Unfortunately, in most cases simple and meaningful conditions cannot be obtained for a rigorous structured uncertainty description. A compromise is that some uncertainties are described in a structured manner, while the rest are lumped into a single unstructured uncertainty. The combined unstructured/structured uncertainty is usually expressed in the form of a block diagonal matrix:

$$\delta(s) = \boldsymbol{W_2}(s)\boldsymbol{\Delta}(s)\boldsymbol{W_1}(s), \tag{10.5.9}$$

with

$$\begin{aligned}
\boldsymbol{\Delta}(s) &= \text{diag}\{\boldsymbol{\Delta_1}(s), \boldsymbol{\Delta_2}(s), ..., \boldsymbol{\Delta_m}(s)\}, \\
\|\boldsymbol{\Delta}_i(s)\|_\infty &\le 1, i = 1, 2, ..., m, \\
\boldsymbol{W_1}(s) &= \text{diag}\{\boldsymbol{W_{11}}(s), \boldsymbol{W_{12}}(s), ..., \boldsymbol{W_{1m}}(s)\}, \\
\boldsymbol{W_2}(s) &= \text{diag}\{\boldsymbol{W_{21}}(s), \boldsymbol{W_{22}}(s), ..., \boldsymbol{W_{2m}}(s)\}.
\end{aligned}$$

$\boldsymbol{W_1}(s)$ and $\boldsymbol{W_2}(s)$ are stable transfer function matrices.

By following the steps in the proof of Theorem 10.5.1, it can be proved that the robust stability is guaranteed if and only if

$$\det[\boldsymbol{I} - \boldsymbol{M}(j\omega)\boldsymbol{\Delta}(j\omega)] \neq 0, \quad \forall\omega, \forall\boldsymbol{\Delta}(j\omega).$$

Nevertheless, the second condition, namely $\|\boldsymbol{M}(s)\|_\infty < 1$, is only sufficient for robust stability. In the proof of Theorem 10.5.1, all uncertainties that satisfy $\|\boldsymbol{\Delta}(s)\|_\infty \le 1$ are permissible. Now, only those having the specific block diagonal structure are permissible. The condition can be conservative. To deal with this problem, the structured singular value (SSV) is proposed. It can be regarded as a generalization of the singular value.

Definition 10.5.1. *Find the smallest* $\bar{\sigma}[\boldsymbol{\Delta}(j\omega)]$ *that makes* $\boldsymbol{I} - \boldsymbol{M}(j\omega)\boldsymbol{\Delta}(j\omega)$ *singular. SSV is defined as* $\mu[\boldsymbol{M}(j\omega)] = 1/\bar{\sigma}[\boldsymbol{\Delta}(j\omega)]$. *If no* $\boldsymbol{\Delta}(j\omega)$ *exists such that* $\det[\boldsymbol{I} - \boldsymbol{M}(j\omega)\boldsymbol{\Delta}(j\omega)] = 0$, *then* $\mu[\boldsymbol{M}(j\omega)] = 0$.

The singularity of a complex matrix means that its determinant is zero. It is noted that SSV depends on not only $\boldsymbol{M}(j\omega)$ but also the structure of $\boldsymbol{\Delta}(j\omega)$. At present, SSV can only be computed numerically.

In the case where $\boldsymbol{\Delta}(j\omega)$ is unstructured (that is, it is a full matrix), $\mu[\boldsymbol{M}(j\omega)] = \bar{\sigma}[\boldsymbol{M}(j\omega)]$.

Theorem 10.5.4. *The closed-loop system is stable for all* $\boldsymbol{\Delta}(s)s(\|\boldsymbol{\Delta}(s)\|_\infty \le 1)$ *if and only if*

$$\mu[\boldsymbol{M}(j\omega)] < 1, \quad \forall\omega.$$

Proof. If $\mu[\boldsymbol{M}(j\omega)] < 1$ at all frequencies, then $\bar{\sigma}[\boldsymbol{\Delta}(j\omega)] > 1$, which implies that no permissible $\boldsymbol{\Delta}(j\omega)$ exists so that $\det[\boldsymbol{I} - \boldsymbol{M}(j\omega)\boldsymbol{\Delta}(j\omega)] = 0$. Hence, the system is stable.

Assume that the system is stable, but $\mu[\boldsymbol{M}(j\omega')] \ge 1$ at some frequency ω'. From the definition of SSV, there must exist an uncertainty $\bar{\sigma}[\boldsymbol{\Delta}(j\omega')] \le 1$ such that $\det[\boldsymbol{I} - \boldsymbol{M}(j\omega')\boldsymbol{\Delta}(\omega')] = 0$. Then the system is unstable. This contradicts the assumption. □

10.6 Robust Performance of MIMO Systems

In some design methods for MIMO control systems, the general control configuration shown in Figure 10.6.1 is used. The block $\boldsymbol{N}(s)$ in the configuration has two sets of inputs and outputs. In the first set, the inputs w consists of all exogenous signals (such as reference or disturbance), while the outputs z consists of those outputs whose behavior is of interest (such as plant output or error signal). The second set of inputs and outputs are the outputs and inputs of the plant uncertainty, respectively.

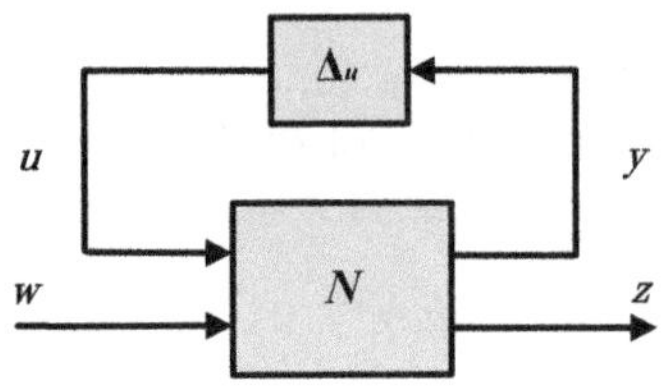

FIGURE 10.6.1
General control configuration.

Assume that the uncertainty is expressed as a block diagonal matrix

$$\boldsymbol{\Delta_u}(s) = \mathrm{diag}\{\boldsymbol{\Delta_1}(s), \boldsymbol{\Delta_2}(s), ..., \boldsymbol{\Delta_m}(s)\} \qquad (10.6.1)$$

with

$$\|\boldsymbol{\Delta_u}(s)\|_\infty \leq 1.$$

The subscript "u" stands for "uncertainty." $\boldsymbol{N}(s)$ is stable when the nominal system is internally stable. Partition $\boldsymbol{N}(s)$ as

$$\boldsymbol{N}(s) = \begin{bmatrix} \boldsymbol{N_{11}}(s) & \boldsymbol{N_{12}}(s) \\ \boldsymbol{N_{21}}(s) & \boldsymbol{N_{22}}(s) \end{bmatrix} \qquad (10.6.2)$$

with the dimensions of its parts compatible with the input and output signals. The transfer function matrix from $\boldsymbol{w}(s)$ to $\boldsymbol{z}(s)$, $\boldsymbol{F}(\boldsymbol{N}, \boldsymbol{\Delta_u})$, can be expressed in the form of a linear fractional transformation (LFT):

$$\boldsymbol{F}(\boldsymbol{N}, \boldsymbol{\Delta_u}) = \boldsymbol{N_{22}}(s) + \boldsymbol{N_{21}}(s)\boldsymbol{\Delta}(s)[\boldsymbol{I} - \boldsymbol{N_{11}}(s)\boldsymbol{\Delta}(s)]^{-1}\boldsymbol{N_{12}}(s). \quad (10.6.3)$$

This can be obtained by eliminating $\boldsymbol{y}(s)$ and $\boldsymbol{u}(s)$ from the following equations:

$$\begin{bmatrix} \boldsymbol{y}(s) \\ \boldsymbol{z}(s) \end{bmatrix} = \begin{bmatrix} \boldsymbol{N_{11}}(s) & \boldsymbol{N_{12}}(s) \\ \boldsymbol{N_{21}}(s) & \boldsymbol{N_{22}}(s) \end{bmatrix} \begin{bmatrix} \boldsymbol{u}(s) \\ \boldsymbol{w}(s) \end{bmatrix},$$

$$u(s) = \Delta_u(s)y(s).$$

In particular, when there is no uncertainty,

$$F(N, \Delta_u) = N_{22}(s), \tag{10.6.4}$$

that is, the nominal transfer function matrix from $w(s)$ to $z(s)$ is $N_{22}(s)$.

It is observed from the expression of $F(N, \Delta_u)$ that the only possible source of instability is the term $[I - N_{11}(s)\Delta(s)]^{-1}$ when the nominal system is stable. By identifying $N_{11}(s)$ with $M(s)$, the stability of the system in Figure 10.6.1 can be tested with Figure 10.5.2.

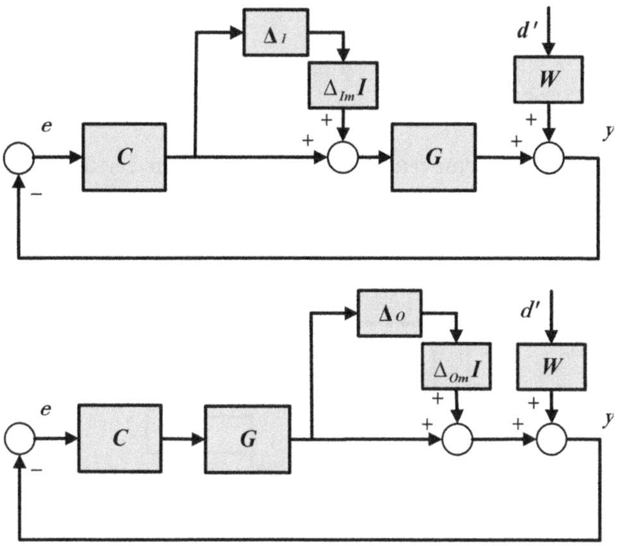

FIGURE 10.6.2
Systems with input or output uncertainty.

Consider the uncertain systems shown in Figure 10.6.2, where $W(s)$ is the input weighting function. Define $z(s) = y(s)$ and $w(s) = d'(s)$. $F(N, \Delta_u)$ is the perturbed weighting sensitivity function. For the input uncertainty, it is not difficult to convert the diagram into the form shown in Figure 10.6.1:

$$N(s) = \begin{bmatrix} C(s)S_I(s)G(s)\Delta_{Im}(s) & -C(s)S_I(s)W(s) \\ S_I(s)G(s)\Delta_{Im}(s) & S_I(s)W(s) \end{bmatrix}. \tag{10.6.5}$$

As to the output uncertainty, we have

$$N(s) = \begin{bmatrix} T_O(s)\Delta_{Om}(s) & -T_O(s)W(s) \\ S_O(s)\Delta_{Om}(s) & S_O(s)W(s) \end{bmatrix}. \tag{10.6.6}$$

Robust performance means that the performance objective is satisfied even when there exists uncertainty. Suppose the robust performance is measured in terms of ∞ norm. The robust performance can be expressed as

$$\|\boldsymbol{F}(\boldsymbol{N}, \boldsymbol{\Delta_u})\|_\infty < 1. \tag{10.6.7}$$

To analyze the robust performance, an uncertainty $\boldsymbol{\Delta_p}(s)$ ($\bar{\sigma}[\boldsymbol{\Delta_p}(j\omega)] \leq 1$) is introduced, as shown in Figure 10.6.3. Here the subscript p stands for "performance." $\boldsymbol{\Delta_p}(s)$ is a fictitious uncertainty. The reason for introducing such an uncertainty is to build a relationship between the $\boldsymbol{N\Delta}$ structure and the $\boldsymbol{M\Delta}$ structure. In this way, the preceding result can be utilized to derive the condition for testing robust performance. Now the new uncertainty can be written as

$$\begin{aligned} \boldsymbol{\Delta}(s) &= \text{diag}\{\boldsymbol{\Delta_u}(s), \boldsymbol{\Delta_p}(s)\}, \\ \|\boldsymbol{\Delta}(s)\|_\infty &\leq 1. \end{aligned} \tag{10.6.8}$$

The following theorem can readily be obtained from Theorem 10.5.4.

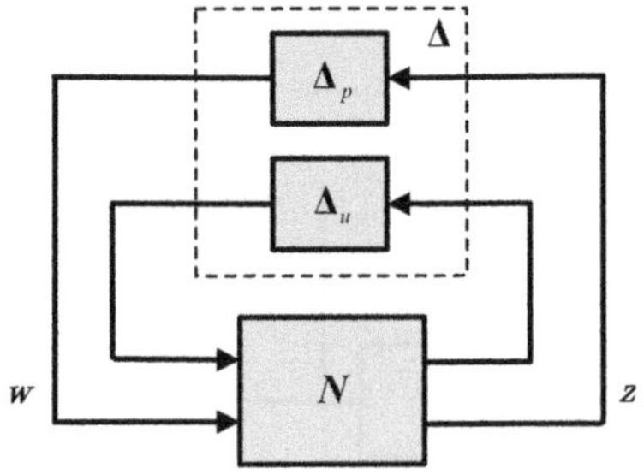

FIGURE 10.6.3
$\boldsymbol{N\Delta}$ structure for checking robust performance.

Theorem 10.6.1. *A stable system $\boldsymbol{N}(s)$ satisfies the robust performance condition $\|\boldsymbol{F}(\boldsymbol{N}, \boldsymbol{\Delta_u})\|_\infty < 1$ if and only if*

$$\mu_\Delta[\boldsymbol{N}(j\omega)] < 1, \forall\omega,$$

where μ is computed with respect to the block diagonal uncertainty $\boldsymbol{\Delta}(s)$.

10.7 Summary

This chapter reviews the basic concepts and analyzing methods of MIMO systems. The stability of MIMO systems can be checked by the MIMO Nyquist stability criterion. Similar to the SISO case, the quantitative engineering performance index cannot be used for optimal controller design. The optimal controller can be obtained through minimizing the 2-norm or ∞-norm of the weighted sensitivity function matrix.

In MIMO systems, the uncertainty is expressed in the form of an input uncertainty description or an output uncertainty description:

$$\tilde{\boldsymbol{G}}(s) = [\boldsymbol{I} + \boldsymbol{\delta_O}(s)]\boldsymbol{G}(s) \text{ or } \tilde{\boldsymbol{G}}(s) = \boldsymbol{G}(s)[\boldsymbol{I} + \boldsymbol{\delta_I}(s)].$$

The uncertainty can be described as a norm-bounded matrix with the same dimension as the plant (that is, the unstructured uncertainty). The advantage of the description is that a simple necessary and sufficient condition for testing robust stability can be obtained:

$$\|\boldsymbol{M}(s)\|_\infty < 1.$$

When this description is conservative, an uncertainty description involving multiple norm-bounded uncertainties can be introduced (that is, the structured uncertainty):

$$\boldsymbol{\Delta}(s) = \text{diag}\{\boldsymbol{\Delta_1}(s), \boldsymbol{\Delta_2}(s), ..., \boldsymbol{\Delta_m}(s)\}.$$

A necessary and sufficient condition for robust stability can be established via SSV:

$$\mu[\boldsymbol{M}(j\omega)] < 1, \quad \forall\omega.$$

The $\boldsymbol{N\Delta}$ structure is introduced in this chapter. The major advantage of this structure is that it allows us to express the robust performance test as a robust stability problem in the $\boldsymbol{M\Delta}$ structure. The nominally stable system $\boldsymbol{N}(s)$ subject to the block diagonal uncertainty $\boldsymbol{\Delta}(s)$ satisfies the robust performance condition if and only if

$$\mu_\Delta[\boldsymbol{N}(j\omega)] < 1, \quad \forall\omega.$$

Exercises

1. Consider the 2×2 plant

$$\boldsymbol{G}(s) = \frac{1}{s+3}\begin{bmatrix} s-1 & 4 \\ 4.5 & 2(s-1) \end{bmatrix}.$$

Compute its poles and zeros.

2. Four-wheel steering (4 WS) is a system that allows the rear wheels to turn for maneuver, rather than merely follow the front wheels. Most 4 WS systems can control the rear wheels in the following fundamental ways (Figure E10.1):

    - At low speeds, the rear wheels are turned in the opposite direction from the front wheels. This can lessen the turning radius by 20% approximately.
    - At higher speeds on highway, the rear wheels are turned in the same direction as the front wheels. This improves the lane-changing maneuverability.

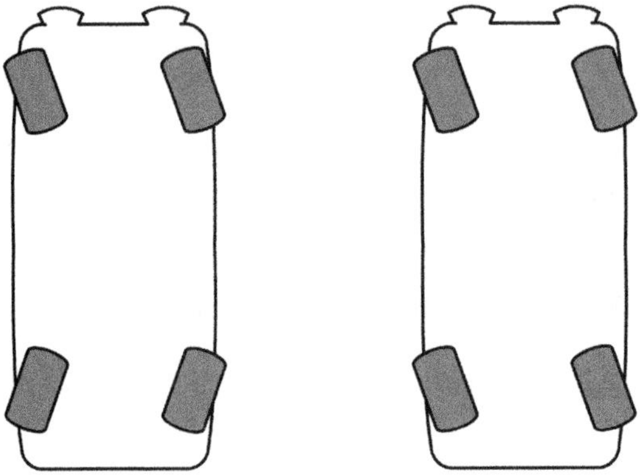

FIGURE E10.1
Control of the rear wheels in a 4 WS system.

The model of a 4 WS car is given as follows:

$$G(s) = \frac{1}{s^2 + 25.6s + 173.6} \begin{bmatrix} 4.9s + 23.4 & 61.3s + 722.8 \\ 5.3s + 150.3 & -80.2s - 722.8 \end{bmatrix}.$$

(a) For a 2 × 2 matrix $G(s)$, derive the analytical expression for computing the singular value.

(b) Utilize the obtained formula to compute the singular value of the 4WS model.

3. Prove

$$\boldsymbol{Q}(s)[\boldsymbol{I} - \boldsymbol{G}(s)\boldsymbol{Q}(s)]^{-1} = [\boldsymbol{I} - \boldsymbol{Q}(s)\boldsymbol{G}(s)]^{-1}\boldsymbol{Q}(s).$$

4. The uncertainty can be described in the form of an inverse multiplicative uncertainty description (Figure E10.2). Assume that the uncertainty is an unstructured one:

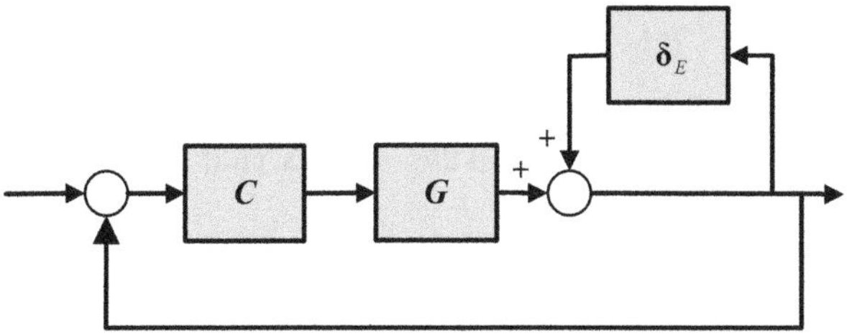

FIGURE E10.2
The inverse multiplicative uncertainty.

$$\begin{aligned}\boldsymbol{\delta_E}(s) &= \Delta_{Em}(s)\boldsymbol{\Delta_E}(s),\\ \|\boldsymbol{\Delta_E}(s)\|_\infty &\leq 1.\end{aligned}$$

Compute $\boldsymbol{M}(s)$ of the $\boldsymbol{M\Delta}$ structure for the inverse multiplicative uncertainty.

5. Consider a system with simultaneous multiplicative input and output uncertainties (Figure E10.3). The uncertain plants are given by

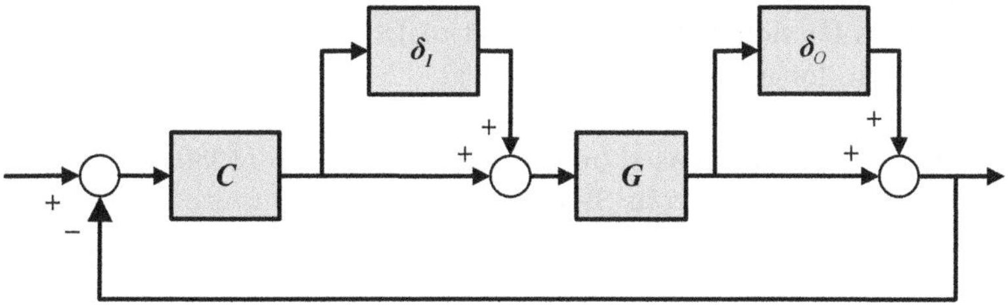

FIGURE E10.3
Simultaneous multiplicative input and output uncertainties.

$$\tilde{\boldsymbol{G}}(s) = [\boldsymbol{I} + \boldsymbol{\delta_O}(s)]\boldsymbol{G}(s)[\boldsymbol{I} + \boldsymbol{\delta_I}(s)],$$
$$\boldsymbol{\delta_I}(s) = \boldsymbol{W_{2I}}(s)\boldsymbol{\Delta_I}(s)\boldsymbol{W_{1I}}(s), \|\boldsymbol{\Delta_I}(s)\|_\infty \leq 1,$$
$$\boldsymbol{\delta_O}(s) = \boldsymbol{W_{2O}}(s)\boldsymbol{\Delta_O}(s)\boldsymbol{W_{1O}}(s), \|\boldsymbol{\Delta_O}(s)\|_\infty \leq 1.$$

Compute the $\boldsymbol{M}(s)$ of $\boldsymbol{M\Delta}$ structure for the simultaneous multiplicative input and output uncertainties.

6. Let $\boldsymbol{M}$ be a complex 2×2 matrix:

$$\boldsymbol{M} = \begin{bmatrix} a & a \\ b & b \end{bmatrix}.$$

Prove

$$\mu(\boldsymbol{M}) = \begin{cases} \rho(\boldsymbol{M}) = |a+b| & \text{for } \boldsymbol{\Delta} = \delta \boldsymbol{I} \\ \bar{\sigma}(\boldsymbol{M}) = \sqrt{2a^2+2b^2} & \text{for } \boldsymbol{\Delta} \text{ that is a full matrix} \end{cases}.$$

Notes and References

This chapter closely follows Morari and Zafiriou (1989) and Skogestad and Postlethwaite (2005). The two books provide detailed discussion on the analyzing problem of the uncertain MIMO system.

There are different definitions for zero and pole. Definitions given in Section 10.1 are from Morari and Zafiriou (1989). The original version can be found in Postlethwaite and MacFarlane (1979). Laub and Moore (1978) discussed the numerical computation of zero.

Sections 10.2 and 10.3 are written based on Skogestad and Postlethwaite (2005) and Morari and Zafiriou (1989).

Sections 10.4–10.6 are adapted from Morari and Zafiriou (1989). The proof for the necessity of the second condition in Theorem 10.5.1 is based on the idea in Theorem 8.1 of Zhou and Doyle (1997).

The SSV in Section 10.5 is attributed to Doyle (1982). The computation can be found in, for example, Doyle (1983).

The plant in Exercise 2 is from Liu et al. (2002).

Exercises 4 and 5 are based on Morari and Zafiriou (1989, Chapter 11).

Exercise 6 is adapted from Skogestad and Postlethwaite (2005, Example 8.7).

11

Classical Design Methods for MIMO Systems

CONTENTS

In Chapter 10, it was seen that the analysis of a MIMO system was complicated when compared to that of a SISO system. For a MIMO system it is much more difficult to give a simple and meaningful explanation of the result than for a SISO system, since there is a control action tradeoff among different inputs, as well as a performance tradeoff among different outputs.

This chapter studies the design problem of MIMO control systems. An important feature of MIMO plants is the interaction (or coupling) between inputs and outputs, that is, each input may affect more than one output. This feature makes the design of a MIMO system very challenging.

Many design methods have been proposed for MIMO systems. This chapter introduces those classical design methods that are frequently used in industry. The advantage of classical design methods is that they are easy to understand and use.

11.1 Interaction Analysis

The first step in many design methods for MIMO systems is analyzing the extent of interaction. The purpose is to determine proper pairings between plant inputs and outputs so that the plant input and the plant output that have the largest effect on each other are matched up.

An extensively adopted method for interaction analysis is relative gain array (RGA). It provides a measure for the steady-state gain between a given input-output pairing. By selecting sensitive input-output connections, interaction can be reduced.

Consider a system with two inputs, u_1 and u_2, and two outputs, y_1 and y_2,

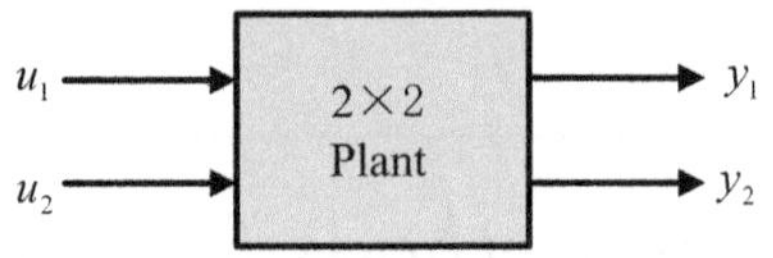

FIGURE 11.1.1
A 2×2 MIMO plant.

where each output is affected by both inputs (Figure 11.1.1). RGA is defined as

$$\mathbf{\Lambda} = \begin{bmatrix} \lambda_{11} & \lambda_{12} \\ \lambda_{21} & \lambda_{22} \end{bmatrix}. \tag{11.1.1}$$

where

$$\lambda_{ij} = \frac{\text{Open-loop gain } k_{ij} \text{ between } y_i \text{ and } u_j}{\text{Closed-loop gain } a_{ij} \text{ between } y_i \text{ and } u_j}, i, j = 1, 2$$

is called the relative gain between y_i and u_j. It should be emphasized that the relative gain λ_{ij} is entirely unrelated to the performance degree, for which a similar symbol is used.

The open-loop gains are shown in Table 11.1.1, which are derived from the steady-state open-loop relationship between the input and output:

$$\Delta y_1 = k_{11}\Delta u_1 + k_{12}\Delta u_2, \tag{11.1.2}$$
$$\Delta y_2 = k_{21}\Delta u_1 + k_{22}\Delta u_2. \tag{11.1.3}$$

TABLE 11.1.1
Open-loop gains.

	u_1	u_2
y_1	$k_{11} = \left.\frac{\Delta y_1}{\Delta u_1}\right\vert_{u_2}$	$k_{12} = \left.\frac{\Delta y_1}{\Delta u_2}\right\vert_{u_1}$
y_2	$k_{21} = \left.\frac{\Delta y_2}{\Delta u_1}\right\vert_{u_2}$	$k_{22} = \left.\frac{\Delta y_2}{\Delta u_2}\right\vert_{u_1}$

The open-loop gain can be determined by utilizing experimental tests. For example, to evaluate k_{11}, when the plant is operated at the steady state, one can make a small change Δu_1 in u_1 and u_2 is kept constant (that is, $\Delta u_2 = 0$). Let Δy_1 be the output offset. The open-loop gain between y_1 and u_1 is given by

$$k_{11} = \left.\frac{\Delta y_1}{\Delta u_1}\right|_{u_2}. \tag{11.1.4}$$

As the closed-loop gain a_{ij} is not independent of k_{ij}, λ_{ij} can be computed with only k_{ij}. According to the definition of the closed-loop gain, $a_{11} = \Delta y_1/\Delta u_1$ when $\Delta y_2 = 0$. Then

$$0 = k_{21}\Delta u_1 + k_{22}\Delta u_2.$$

Solving for Δu_2 results in

$$\Delta u_2 = -\frac{k_{21}}{k_{22}}\Delta u_1.$$

It follows that

$$\Delta y_1 = k_{11}\Delta u_1 - \frac{k_{12}k_{21}}{k_{22}}\Delta u_1.$$

Therefore,

$$\begin{aligned} a_{11} &= \left.\frac{\Delta y_1}{\Delta u_1}\right|_{\Delta y_2=0} \\ &= k_{11} - \frac{k_{12}k_{21}}{k_{22}}. \end{aligned} \tag{11.1.5}$$

The relative gain is

$$\begin{aligned} \lambda_{11} &= \frac{k_{11}}{a_{11}} \\ &= \frac{k_{11}k_{22}}{k_{11}k_{22} - k_{12}k_{21}}. \end{aligned} \tag{11.1.6}$$

A useful property of the relative gain matrix is that each row and each column sums to 1. Thus, in a 2×2 system, only one of the four elements needs to be computed explicitly.

The relative gain provides a useful measure for interaction. In particular:

1. If $\lambda_{ij} = 0$, the open-loop gain is zero. u_j has no effect on y_i.
2. If $\lambda_{ij} = 1$, the loop consisting of u_j and y_i is not affected by other loops.
3. If $0 < \lambda_{ij} < 1$, there exists interaction among different loops. The worst case is $\lambda_{ij} = 0.5$.
4. If $\lambda_{ij} < 0$, the open-loop gain is in the opposite direction from the closed-loop gain. This case should be avoided.

The rule to reduce interaction by pairing plant inputs and outputs is that the control loops should be selected in such a way that the relative gains are positive, and as close as possible to unity. Since only steady-state responses are considered, the rule does not guarantee the minimum dynamic interaction.

The definition of RGA and its application in selecting control loops are not confined to systems with two inputs and two outputs. The extension to $n \times n$ systems is straightforward. The RGA can be computed for an $n \times n$ plant with the following procedure.

First, arrange the k_{ij}s in a matrix:

$$\boldsymbol{K} = \begin{bmatrix} k_{11} & \cdots & k_{1n} \\ \vdots & \ddots & \vdots \\ k_{n1} & \cdots & k_{nn} \end{bmatrix}. \tag{11.1.7}$$

Next, compute a new matrix, by first inverting, then transposing the matrix $\boldsymbol{K}$:

$$(\boldsymbol{K}^{-1})^T = \begin{bmatrix} c_{11} & \cdots & c_{1n} \\ \vdots & \ddots & \vdots \\ c_{n1} & \cdots & c_{nn} \end{bmatrix}. \tag{11.1.8}$$

The element in the ith row and jth column of $(\boldsymbol{K}^{-1})^T$ is the reciprocal of a_{ij}, that is,

$$c_{ij} = \frac{1}{a_{ij}}. \tag{11.1.9}$$

The RGA is given by

$$\boldsymbol{\Lambda} = \boldsymbol{K} \otimes (\boldsymbol{K}^{-1})^T, \tag{11.1.10}$$

where "$\otimes$" denotes the element-by-element product.

Example 11.1.1. *Blending is a frequently encountered process in industry. For example, in a paper-making process, the thick pulp from the stock preparation system is blended with the recycled water, and then delivered to the head box. Choose the flow rate of the thick pulp, u_1, and the flow rate of the recycled water, u_2, as the plant inputs. The system outputs are the flow rate of the thin pulp, y_1, and its consistency, y_2. When the condenser paper is produced, the consistencies of the thick pulp and the recycled water are 0.66% and 0.03%, respectively. The desired flow rate and consistency of the thin pulp are 152 kg/min and 0.25%, respectively. The mass balance yields*

$$\begin{aligned} 152 &= u_1 + u_2, \\ 152 \times 0.25\% &= u_1 \times 0.66\% + u_2 \times 0.03\%. \end{aligned}$$

The solution to the above equations is

$$u_1 = 53.08, u_2 = 98.92.$$

Change u_1 by one unit (that is, u_1 changes from 53.08 to 54.08) while keep u_2 constant. The following steady-state outputs are obtained:

$$y_1 = 153, y_2 = 0.2527\%.$$

Therefore,

$$k_{11} = \left.\frac{\Delta y_1}{\Delta u_1}\right|_{u_2} = \frac{1}{1} = 1.$$

Change u_1 by one unit (that is, u_1 changes from 53.08 to 54.08) while keep y_2 the same. We have

$$y_1 = 154.87, u_2 = 100.79.$$

Then

$$a_{11} = \left.\frac{\Delta y_1}{\Delta u_1}\right|_{y_2} = \frac{1.87}{1} = 1.87.$$

Consequently, the RGA is

$$\mathbf{\Lambda} = \begin{bmatrix} 0.535 & 0.465 \\ 0.465 & 0.535 \end{bmatrix}.$$

The two loops with the minimum interaction are formed when y_1 is controlled by u_1 and y_2 is controlled by u_2.

11.2 Decentralized Controller Design

Once appropriate loop pairings are determined, the next step is to design a controller for the MIMO system. This section introduces decentralized control. In decentralized control, multiple independent SISO controllers are designed. Each controller uses one plant input to control a preassigned output. Feedback is utilized to overcome the interaction. Hence, the pairing problem is much more important in decentralized control than in other methods.

The advantage of decentralized control is as follows:

1. The approach is easy to understand.
2. Good control can be reached in many cases.
3. The complexity and the cost of hardware are low.

Compared to the system with a full controller matrix, the constraint imposed by the decentralized control on the controller structure leads to performance deterioration. The designer must weigh which aspect is more important, the performance or the simplicity.

Assume that the plant is denoted by an $n \times n$ transfer function matrix $\boldsymbol{G}(s)$:

$$\boldsymbol{G}(s) = \begin{bmatrix} G_{11}(s) & \cdots & G_{1n}(s) \\ \vdots & \ddots & \vdots \\ G_{n1}(s) & \cdots & G_{nn}(s) \end{bmatrix},$$

and the controller is denoted by an $n \times n$ transfer function matrix $\boldsymbol{C}(s)$. Without loss of generality, in a decentralized control system $\boldsymbol{C}(s)$ is diagonal:

$$\boldsymbol{C}(s) = \operatorname{diag}\left\{C_{11}(s), ..., C_{nn}(s)\right\}.$$

The reference $\boldsymbol{r}(s)$ is a vector of $n \times 1$ dimension:

$$\boldsymbol{r}(s) = \begin{bmatrix} r_1(s) & r_2(s) & ... & r_n(s) \end{bmatrix}^T,$$

and the system output $\boldsymbol{y}(s)$ is a vector of $n \times 1$ dimension as well:

$$\boldsymbol{y}(s) = \begin{bmatrix} y_1(s) & y_2(s) & ... & y_n(s) \end{bmatrix}^T.$$

Then we have

$$\boldsymbol{y}(s) = \boldsymbol{G}(s)\boldsymbol{C}(s)[\boldsymbol{I} + \boldsymbol{G}(s)\boldsymbol{C}(s)]^{-1}\boldsymbol{r}(s). \tag{11.2.1}$$

The control structure for the simplest case (that is, the 2×2 case) is shown in Figure 11.2.1.

The characteristic equation of this system is

$$\det[\boldsymbol{I} + \boldsymbol{G}(s)\boldsymbol{C}(s)] = 0. \tag{11.2.2}$$

This equation can be used to test the closed-loop stability. For a stable plant, if the Nyquist plot of $\det[\boldsymbol{I} + \boldsymbol{G}(s)\boldsymbol{C}(s)]$ passes through or encircles the origin, the closed-loop system is unstable. Define a scalar function $W_c(s)$:

$$W_c(s) = -1 + \det\left[\boldsymbol{I} + \boldsymbol{G}(s)\boldsymbol{C}(s)\right]. \tag{11.2.3}$$

The closer the Nyquist plot of $W_c(s)$ is to the point $(-1, 0)$, the closer the closed-loop system is to instability. Define the closed-loop logarithm modulus for a MIMO system as follows:

$$L_c = 20 \lg \left| \frac{W_c(j\omega)}{1 + W_c(j\omega)} \right|. \tag{11.2.4}$$

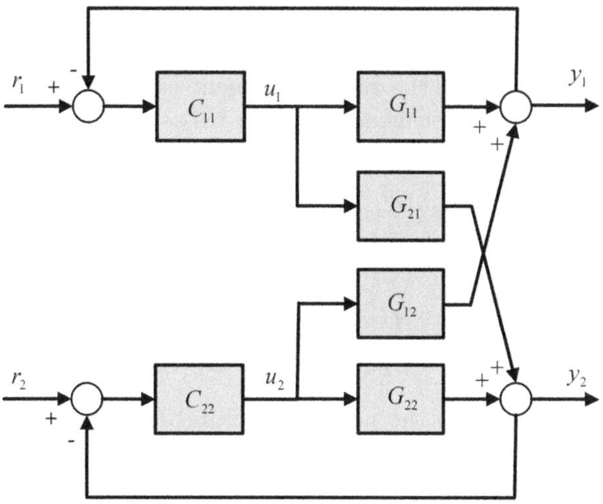

FIGURE 11.2.1
A 2 × 2 decentralized control system.

By testing on a large number of MIMO plants, it is found that the following choice can provide a reasonable tradeoff between stability and performance:

$$\max(L_c) = 2n.$$

In a SISO system, $\max(L_c)$ is the resonance peak.

Different controller parameters may reach $\max(L_c) = 2n$ for the same plant. When better response is desired, further tuning has to be carried out.

The SISO H_∞ or H_2 design methods introduced in the preceding chapters can be used to design a decentralized controller. The design procedure is as follows:

1. Calculate SISO controllers for each individual loop.
2. Close all loops. Take the same performance degrees temporarily: $\lambda_1 = \lambda_2 = ... = \lambda_n$. Increase the performance degrees from small to large so that the closed-loop system is stable.
3. Tune each performance degree to reach the required closed-loop response (for example, $\max(L_c) = 2n$).

The advantage of the above decentralized H_∞ or decentralized H_2 design is that each channel can be tuned easily for the required response; the tuning does not require the exact information about the uncertainty.

Now let us analyze the stability of a decentralized control system. For the original system, the sensitivity transfer function matrix and complementary sensitivity transfer function matrix are, respectively, as follows:

$$\boldsymbol{S}(s) \quad = \quad [\boldsymbol{I} + \boldsymbol{G}(s)\boldsymbol{C}(s)]^{-1}, \tag{11.2.5}$$

$$T(s) = G(s)C(s)[I + G(s)C(s)]^{-1}. \quad (11.2.6)$$

In decentralized control, the diagonal elements are regarded as the nominal plant and the nondiagonal elements are regarded as the uncertainty. Write the nominal plant as follows:

$$G_a(s) = \text{diag}\{G_{11}(s), G_{22}(s), ..., G_{nn}(s)\}. \quad (11.2.7)$$

The nominal plant and the controller constitute a new system:

$$S_a(s) = [I + G_a(s)C(s)]^{-1}, \quad (11.2.8)$$
$$T_a(s) = G_a(s)C(s)[I + G_a(s)C(s)]^{-1}. \quad (11.2.9)$$

Regard the new system as the nominal one and the original system as the uncertain one. The uncertainty is described by the multiplicative output uncertainty $\delta_a(s)$. Assume that $G(s)$ and $G_a(s)$ have the same RHP poles. The following theorem provides a test for the stability of the closed-loop system.

Theorem 11.2.1. *Assume that $T_a(s)$ is stable. The closed-loop system $T(s)$ is stable if and only if the Nyquist plot of $\det[I + \delta_a(s)T_a(s)]$ does not pass through or encircle the origin.*

Proof. For the original system, we have the following identity:

$$\begin{aligned} I + G(s)C(s) &= I + [I + \delta_a(s)]G_a(s)C(s) \\ &= [I + \delta_a(s)T_a(s)][I + G_a(s)C(s)]. \end{aligned}$$

Let the number of the unstable poles of $G(s)$ and $G_a(s)$ be k. $T(s)$ is stable if and only if

$$\det[I + G(s)C(s)] = \det[I + G_a(s)C(s)]\det[I + \delta_a(s)T_a(s)]$$

does not pass through the origin and encircles the origin k times counterclockwise. Because $T_a(s)$ is stable by assumption, $\det[I + G_a(s)C(s)]$ does not pass through the origin and encircles the origin k times counterclockwise. It is immediately known that $\det[I + \delta_a(s)T_a(s)]$ should not pass through or encircle the origin. □

Example 11.2.1. *Basis weight and moisture content are two primary controlled variables in the paper-making process. In general, the basis weight is controlled by adjusting the flow rate of stock. The higher the flow rate, the larger the basis weight. The moisture content can be controlled by adjusting the steam pressure of dryer. The larger the steam pressure, the higher the temperature of the dryer and the less the moisture content (Figure 11.2.2). For producing the paper of 78g/m^2, the model of a paper machine is*

$$\boxed{G(s) = \begin{bmatrix} \frac{5.15e^{-2.8s}}{1.8s+1} & \frac{-0.20e^{-1.2s}}{2.23s+1} \\ \frac{0.44e^{-2.8s}}{1.8s+1} & \frac{-1.26e^{-1.2s}}{2.23s+1} \end{bmatrix}}.$$

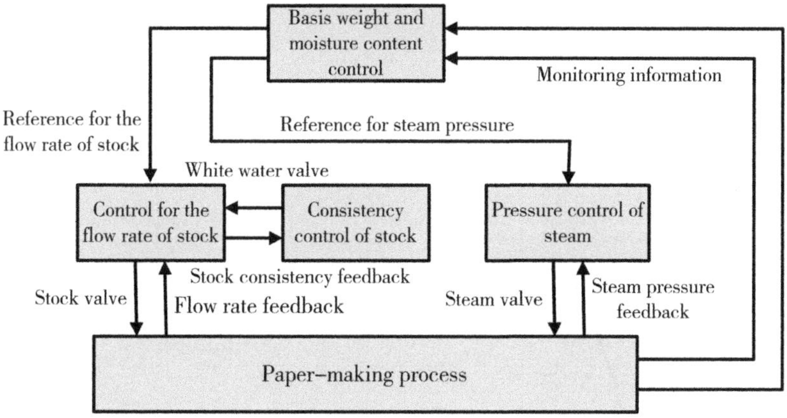

FIGURE 11.2.2
Control strategy of a paper machine.

Suppose that the H_∞ PID controller given by (4.2.11) is used:

$$\boxed{C(s) = \begin{bmatrix} C_{11}(s) & 0 \\ 0 & C_{22}(s) \end{bmatrix},}$$

where

$$\begin{aligned} C_{11}(s) &= \frac{1}{5.15}\frac{(1.8s+1)(1+1.4s)}{\lambda^2 s^2 + (2\lambda + 1.4)s}, \\ C_{22}(s) &= \frac{1}{1.26}\frac{(2.23s+1)(1+0.6s)}{\lambda^2 s^2 + (2\lambda + 0.6)s}. \end{aligned}$$

Assume that the quantitative design requirement is that $y_1(t) < 0.15$ for $r_1(s) = 0$ and $r_2(s) = 1/s$, and $y_2(t) < 0.05$ for $r_1(s) = 1/s$ and $r_2(s) = 0$. The controller parameters are taken as $\lambda_1 = 0.9\theta_{11}$ and $\lambda_2 = 0.6\theta_{22}$ respectively, where θ_{11} and θ_{22} are time delays of the diagonal elements respectively. The responses of the closed-loop system are fast and steady (Figure 11.2.3).

11.3 Decoupler Design

When the interaction among control loops is weak, the decentralized control method works well. For the control loops with severe interaction, decoupling control is a better method. In decoupling control, an additional compensation

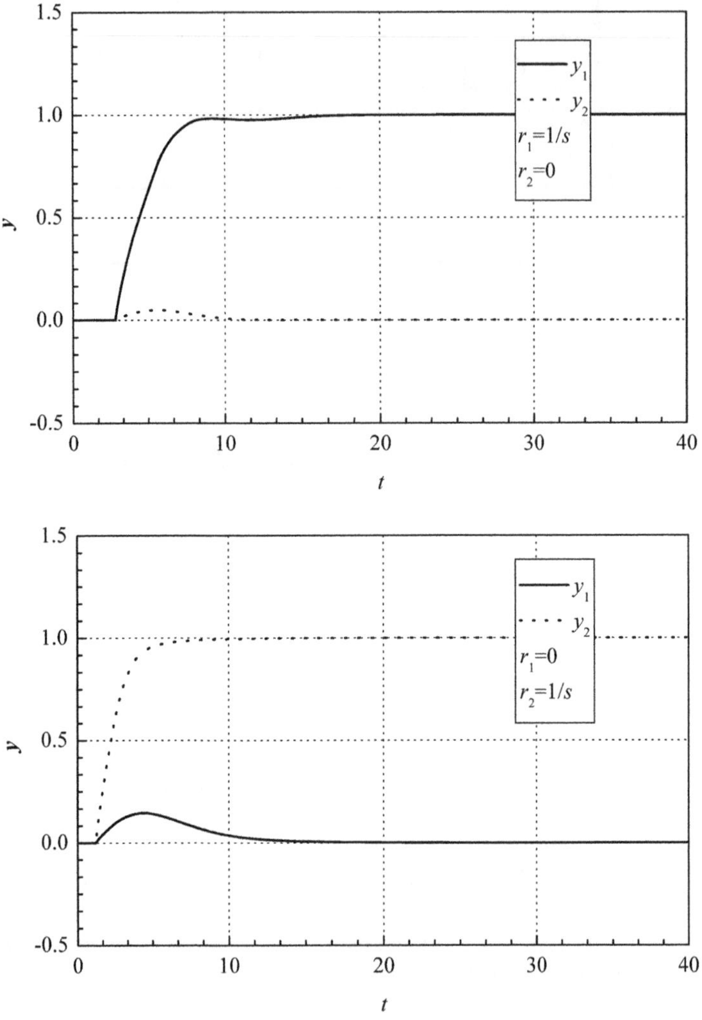

FIGURE 11.2.3
Responses of the decentralized H_∞ controller.

structure called a decoupler is introduced to reduce the interaction. The function of a decoupler is decomposing a MIMO plant into a series of independent SISO plants. Then the SISO design methods (for example, the H_∞ method or H_2 method introduced in the preceding chapters) can be utilized to design MIMO control systems. Figure 11.3.1 shows the general decoupling structure.

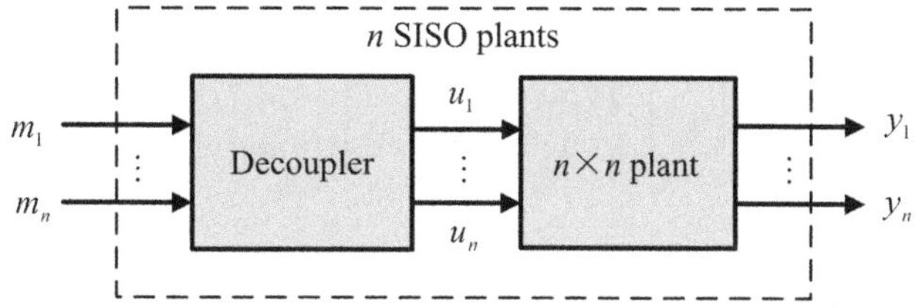

FIGURE 11.3.1
MIMO plant with a decoupler.

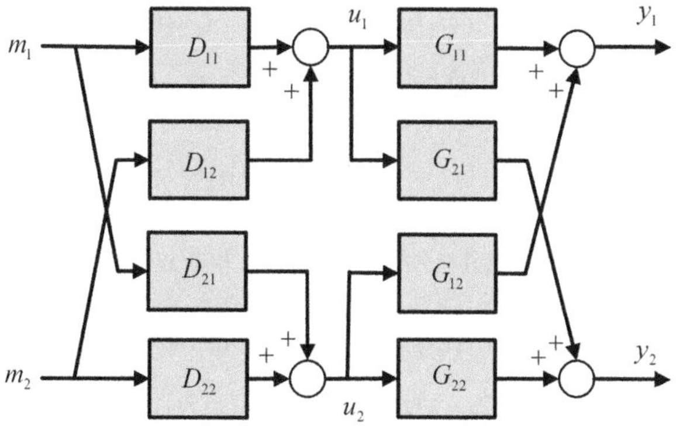

FIGURE 11.3.2
Decoupling for a 2 × 2 plant.

For ease of understanding, consider a 2 × 2 MP plant first. The decoupling system is shown in Figure 11.3.2, where the decoupler inputs are two new manipulated variables $m_1(s)$ and $m_2(s)$, and its outputs are the original manipulated variables $u_1(s)$ and $u_2(s)$. For the original plant we have

$$\begin{bmatrix} y_1(s) \\ y_2(s) \end{bmatrix} = \begin{bmatrix} G_{11}(s) & G_{12}(s) \\ G_{21}(s) & G_{22}(s) \end{bmatrix} \begin{bmatrix} u_1(s) \\ u_2(s) \end{bmatrix}. \tag{11.3.1}$$

The decoupler equation can be written as

$$\begin{bmatrix} u_1(s) \\ u_2(s) \end{bmatrix} = \begin{bmatrix} D_{11}(s) & D_{12}(s) \\ D_{21}(s) & D_{22}(s) \end{bmatrix} \begin{bmatrix} m_1(s) \\ m_2(s) \end{bmatrix}. \tag{11.3.2}$$

Then the equations for the plant-decoupler combination are

$$\begin{bmatrix} y_1(s) \\ y_2(s) \end{bmatrix} = \begin{bmatrix} G_{11}(s) & G_{12}(s) \\ G_{21}(s) & G_{22}(s) \end{bmatrix} \begin{bmatrix} D_{11}(s) & D_{12}(s) \\ D_{21}(s) & D_{22}(s) \end{bmatrix} \begin{bmatrix} m_1(s) \\ m_2(s) \end{bmatrix}. \tag{11.3.3}$$

The decoupler should be designed so that the non-diagonal elements are zero. In this case, $y_1(s)$ is only affected by $u_1(s)$ and $y_2(s)$ is only affected by $u_2(s)$. Let

$$\begin{bmatrix} y_1(s) \\ y_2(s) \end{bmatrix} = \begin{bmatrix} P_{11}(s) & 0 \\ 0 & P_{22}(s) \end{bmatrix} \begin{bmatrix} m_1(s) \\ m_2(s) \end{bmatrix}. \tag{11.3.4}$$

We have

$$\begin{bmatrix} G_{11}(s) & G_{12}(s) \\ G_{21}(s) & G_{22}(s) \end{bmatrix} \begin{bmatrix} D_{11}(s) & D_{12}(s) \\ D_{21}(s) & D_{22}(s) \end{bmatrix} = \begin{bmatrix} P_{11}(s) & 0 \\ 0 & P_{22}(s) \end{bmatrix}. \tag{11.3.5}$$

This equality involves four equations and six unknowns. The solution is not unique. For example, the following are possible solutions:

$$D_{11}(s) = 1, D_{12}(s) = -\frac{G_{12}(s)}{G_{11}(s)}, D_{21}(s) = -\frac{G_{21}(s)}{G_{22}(s)}, D_{22}(s) = 1,$$

$$D_{11}(s) = 1, D_{12}(s) = 1, D_{21}(s) = -\frac{G_{21}(s)}{G_{22}(s)}, D_{22}(s) = -\frac{G_{11}(s)}{G_{12}(s)},$$

$$D_{11}(s) = -\frac{G_{22}(s)}{G_{21}(s)}, D_{12}(s) = -\frac{G_{12}(s)}{G_{11}(s)}, D_{21}(s) = 1, D_{22}(s) = 1,$$

$$D_{11}(s) = -\frac{G_{22}(s)}{G_{21}(s)}, D_{12}(s) = 1, D_{21}(s) = 1, D_{22}(s) = -\frac{G_{11}(s)}{G_{12}(s)}.$$

With the above solutions, $P_{11}(s)$ and $P_{22}(s)$ can be determined as follows:

$$P_{11}(s) = G_{11}(s)D_{11}(s) + G_{12}(s)D_{21}(s)$$
$$P_{22}(s) = G_{21}(s)D_{12}(s) + G_{22}(s)D_{22}(s)$$

To obtain a unique solution, three typical methods are frequently used. The first is to take the inverse of the plant as the decoupler, that is,

$$\begin{bmatrix} P_{11}(s) & 0 \\ 0 & P_{22}(s) \end{bmatrix} = \boldsymbol{I}. \tag{11.3.6}$$

The second is to take 1 as the diagonal elements of the decoupler:

$$\begin{bmatrix} D_{11}(s) & D_{12}(s) \\ D_{21}(s) & D_{22}(s) \end{bmatrix} = \begin{bmatrix} 1 & D_{12}(s) \\ D_{21}(s) & 1 \end{bmatrix}. \tag{11.3.7}$$

And the third is to take 0 as the diagonal elements of the decoupler:

$$\begin{bmatrix} D_{11}(s) & D_{12}(s) \\ D_{21}(s) & D_{22}(s) \end{bmatrix} = \begin{bmatrix} 0 & D_{12}(s) \\ D_{21}(s) & 0 \end{bmatrix}. \tag{11.3.8}$$

A natural question is which decoupling scheme the designer should choose among the three. Two aspects should be considered with regard to the question. The first is whether the obtained decoupler is realizable. Some methods cannot give a decoupler that is physically realizable, in particular, when the plant involves RHP zeros or a time delay. The second is whether a simple decoupler can be obtained. The ultimate objective of the decoupler design is to eliminate the effect of interaction. If the decoupler is realizable, the simpler, certainly the better.

Now consider the decoupler design problem for a general plant. The design equations can be conveniently summarized with matrix notations. Let the plant be an $n \times n$ transfer function matrix $\boldsymbol{G}(s)$ and the decoupler be an $n \times n$ transfer function matrix $\boldsymbol{D}(s)$. We have

$$\boldsymbol{y}(s) = \boldsymbol{G}(s)\boldsymbol{D}(s)\boldsymbol{m}(s). \tag{11.3.9}$$

Assume that the decoupled plant is an $n \times n$ diagonal matrix $\boldsymbol{P}(s)$. It is desirable that

$$\boldsymbol{y}(s) = \boldsymbol{P}(s)\boldsymbol{m}(s). \tag{11.3.10}$$

By comparison, the decoupler is given by

$$\boldsymbol{D}(s) = \boldsymbol{G}(s)^{-1}\boldsymbol{P}(s). \tag{11.3.11}$$

Mathematically, the calculation of the decoupler is trivial. If $\boldsymbol{P}(s)$ is given (for example, it is taken to be a unity matrix), $\boldsymbol{D}(s)$ can be determined uniquely as long as $\boldsymbol{G}(s)$ is not singular. The non-singularity of a transfer function matrix means that its determinant is not identically zero, or equivalently, the transfer function matrix is not singular for every s in the set of complex numbers, except for a finite number of points. The inverse of a non-singular $\boldsymbol{G}(s)$ can be expressed as

$$\boldsymbol{G}^{-1}(s) = \frac{\text{adj}[\boldsymbol{G}(s)]}{\det[\boldsymbol{G}(s)]}, \tag{11.3.12}$$

where adj$(\cdot)$ denotes the adjoint. The determinant of $\boldsymbol{G}(s)$ can be calculated as a signed sum of the permutations taking one and only one element from every row and column of $\boldsymbol{G}(s)$:

$$\det[\boldsymbol{G}(s)] = \sum_{j_1 j_2 \ldots j_n} [\text{sgn}(j_1 j_2 \ldots j_n)] G_{1j_1} G_{2j_2} \ldots G_{nj_n}, \tag{11.3.13}$$

where $j_1 j_2 \ldots j_n$ denotes a permutation of the number $1, 2, \ldots, n$. The value of $\text{sgn}(j_1 j_2 \ldots j_n)$ can be found by using the number of permutation inversion:

$$\text{sgn}(j_1 j_2 \ldots j_n) = \left\{ \begin{array}{ll} 1 & \text{if the number of permutation inversion is even} \\ -1 & \text{if the number of permutation inversion is odd} \end{array} \right\}.$$

The element of the adjoint matrix, adj$[\boldsymbol{G}(s)]$, is the cofactor of $G_{ij}(s)$, which is the signed determinant of $\boldsymbol{G}(s)$ with the row i and column j removed.

Nevertheless, the problem is not so easy from a control theory point of view. On one hand, the obtained decoupler may not be physically realizable. On the other hand, the decoupled system may be internally unstable when the plant is NMP or unstable. These problems can be solved well with the methods in the next two chapters.

11.4 Summary

The first step in many design methods for MIMO systems is determining the match between plant inputs and outputs. This problem is particularly important in decentralized control. It can be dealt with by employing RGA.

Two classical design methods for MIMO systems have been reviewed in this chapter. One is the decentralized control and the other is the decoupling control. The decentralized control emphasizes the simplicity of control configuration, while the decoupling control provides better performance with a relatively complex structure. The common feature of the two methods is that they are easy to understand.

A decentralized control method is given in this chapter based on the SISO H_2/H_∞ design method, from which it can be seen that the design methods developed in the preceding chapters can be easily integrated into the decentralized control and the decoupling control. One main merit of such a design is that the closed-loop response can be easily tuned for quantitative requirements.

Exercises

1. Consider a 3 × 3 plant:

$$\boldsymbol{G}(0) = \begin{bmatrix} 16.8 & 30.5 & 4.30 \\ -16.7 & 31.0 & -1.41 \\ 1.27 & 54.1 & 5.40 \end{bmatrix}.$$

   Compute the RGA.

2. Consider the system shown in Figure E11.1, where $\boldsymbol{G}(s)$ is an $n \times n$ proper rational plant and $\boldsymbol{C} = \text{diag}\{c_i\}, c_i > 0, i = 1, 2, ..., n$. $\boldsymbol{G}(s)$ is integral stabilizable if there exists a $c > 0$ such that the closed-loop system is stable for $\boldsymbol{C} = c\boldsymbol{I}$. Prove that $\boldsymbol{G}(s)$ is integral stabilizable only if $\det \boldsymbol{G}(0) > 0$.

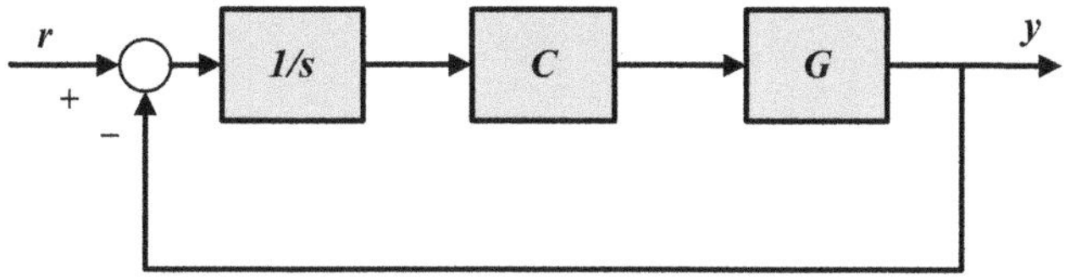

FIGURE E11.1
System with an integrator and a diagonal controller.

3. Consider a stable MP plant $\boldsymbol{G}(s)$. Discuss the feasibility of designing a controller to make the sensitivity function matrix $\boldsymbol{S}(s)$ diagonal.
4. Consider the following plant:

$$\boldsymbol{G}(s) = \begin{bmatrix} \frac{0.5e^{-1.5s}}{s+1} & \frac{e^{-0.5s}}{2s+1} \\ \frac{2e^{-1.0s}}{0.5s+1} & \frac{1}{s+1} \end{bmatrix}.$$

Give two different decouplers and compare them.

5*. Consider the following plant:

$$\boldsymbol{G}(s) = \frac{s-1}{(s-1)^2(s-4)} \begin{bmatrix} 2 & s-4 \\ s-4 & 0 \end{bmatrix}.$$

Is this plant stabilizable?

Notes and References

Section 11.1 is adapted from Deshpande and Ash (1981). The RGA was proposed by Bristol (1966).

The plant in Example 11.1.1 is from Sun (1993, p. 329).

The method in Section 11.2 was proposed by Zhang (1996). The maximum logarithm modulus was presented by Luyben (1986). Theorem 11.2.1 was given by Morari and Zafiriou (1989, Section14.4.1).

The plant in Example 11.2.1 is from Zhang et al. (2001). Figure 11.2.2 is from Zhang (1996).

The classical decoupling design was discussed in detail in Liu (1983). The calculation of the determinant of a matrix can be found in, for example, Lipschutz (1968).

The plant in Exercise 1 is from Skogestad and Postlethwaite (2005, Section 3.4.2).

Exercise 2 is adapted from Morari and Zafiriou (1989, Section 14.3).

The plant in Exercise 4 is from Stephanopoulos (1984, Section 24.3).

12

Quasi-H_∞ Decoupling Control

CONTENTS

The classical method for the design of decoupling control systems involves two steps. The first step is designing a decoupler such that the MIMO plant is decomposed into a series of independent SISO plants. In the second step, a SISO method is used to design controllers for these SISO plants. There are some unsolved problems in this method:

- It is not applicable to NMP plants and plants with RHP poles.
- It is difficult to analyze the effect of the decoupler on the closed-loop performance and robustness.

To solve these problems, an alternative design method is developed in this chapter. The new method is applicable to general plants. In this method, the decoupler and controller are designed in one step. The performance and robustness of the resulting system can be analyzed by employing those developed techniques.

12.1 Diagonal Factorization for Quasi-H_∞ Control

Consider the IMC structure shown in Figure 12.1.1, where $\tilde{\boldsymbol{G}}(s)$ is an $n \times n$ plant, $\boldsymbol{G}(s)$ is the model, and $\boldsymbol{Q}(s)$ is an $n \times n$ controller. The closed-loop transfer function matrix is

$$\boldsymbol{T}(s) = \boldsymbol{G}(s)\boldsymbol{Q}(s). \tag{12.1.1}$$

In quasi-H_∞ decoupling control, the plant can be proper, have a time delay, have poles on the imaginary axis, or have poles and zeros in the open RHP. If the plant has poles in the closed RHP, the controller has to be implemented in the unity feedback loop shown in Figure 12.1.2, where $\boldsymbol{C}(s)$ is an $n \times n$ controller:

$$\boldsymbol{C}(s) = \boldsymbol{Q}(s)[\boldsymbol{I} - \boldsymbol{G}(s)\boldsymbol{Q}(s)]^{-1}. \quad (12.1.2)$$

The closed-loop transfer function matrix is

$$\boldsymbol{T}(s) = \boldsymbol{G}(s)\boldsymbol{C}(s)[\boldsymbol{I} + \boldsymbol{G}(s)\boldsymbol{C}(s)]^{-1}. \quad (12.1.3)$$

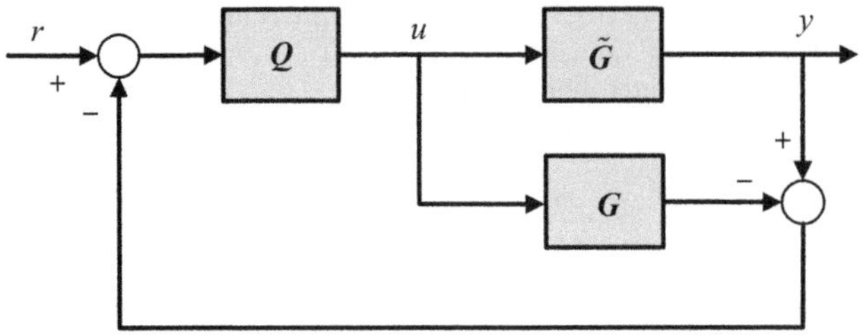

FIGURE 12.1.1
IMC structure with an $n \times n$ plant.

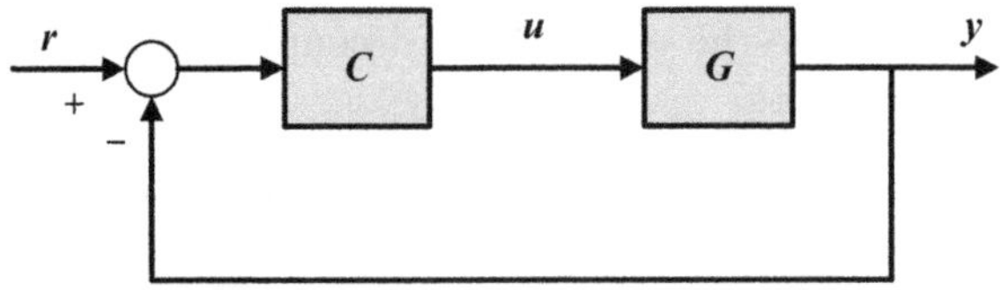

FIGURE 12.1.2
Unity feedback loop with an $n \times n$ plant.

As stated in Section 10.4, the following assumptions are made for the plant:

1. There is no unstable hidden mode in $\boldsymbol{G}(s)$.

For quasi-H_∞ control, it is further assumed that

2. $\boldsymbol{G}(s)$ is of full normal rank, that is, $\text{rank}[\boldsymbol{G}(s)] = n$.

If the second assumption is not satisfied, the closed-loop transfer function matrix $\boldsymbol{T}(s)$ must be identically singular. When $\boldsymbol{G}(s)$ is not of full normal rank, a slight perturbation can be introduced in the coefficients of the plant so as to satisfy the second condition.

Unlike in the classical decoupling control, in the quasi-H_∞ control there is no independent step for decoupler design. The decoupler and controller are expressed in the form of one transfer function matrix and are designed in one step.

The key of the design is defining a diagonal factorization, in which the closed RHP zero and time delay are separated from the plant. In the next section, it will be seen that the decoupled response is obtained on the basis of such a factorization.

The factorization for the closed RHP zero and time delay can be obtained easily for a SISO plant, but the extension to the MIMO case should be properly defined. Write the plant $\boldsymbol{G}(s)$ with multiple time delays in the form of

$$\boldsymbol{G}(s) = \begin{bmatrix} G_{11}(s)e^{-\theta_{11}s} & \cdots & G_{1n}(s)e^{-\theta_{1n}s} \\ \vdots & \ddots & \vdots \\ G_{n1}(s)e^{-\theta_{n1}s} & \cdots & G_{nn}(s)e^{-\theta_{nn}s} \end{bmatrix}, \tag{12.1.4}$$

where $G_{ij}(s)(i,j = 1,2,...,n)$ are scalar rational transfer functions and $\theta_{ij}(\theta_{ij} \geq 0)$ are time delays. Let the inverse of the plant be

$$\boldsymbol{G}^{-1}(s) = \begin{bmatrix} G^{11}(s)e^{-\theta^{11}s} & \cdots & G^{1n}(s)e^{-\theta^{1n}s} \\ \vdots & \ddots & \vdots \\ G^{n1}(s)e^{-\theta^{n1}s} & \cdots & G^{nn}(s)e^{-\theta^{nn}s} \end{bmatrix}, \tag{12.1.5}$$

where $G^{ji}(s)e^{-\theta^{ji}s}(j,i = 1,2,...,n)$ are the elements of $\boldsymbol{G}^{-1}(s)$; θ^{ji} are the real numbers denoting the maximum time delay that can be separated from each element. After θ^{ji} are removed, the remainder should be physically realizable. The following example is used to illustrate how to compute θ^{ji}.

Example 12.1.1. *Consider the following element:*

$$\frac{e^{-3s}}{2s+1} + \frac{e^{-2s}}{3s+1} = \left(\frac{e^{-s}}{2s+1} + \frac{1}{3s+1}\right)e^{-2s}.$$

The maximum time delay that can be separated is 2. In the following element

$$\frac{e^{-3s}}{2s+1} + \frac{e^{2s}}{3s+1} = \left(\frac{e^{-5s}}{2s+1} + \frac{1}{3s+1}\right)e^{2s},$$

the maximum time delay that can be separated is −2.

Consider the factorization for time delay first. Some elements of $\boldsymbol{G}^{-1}(s)$ may contain predictions (that is, $\theta^{ji} < 0$). This implies that the resulting control system is physically irrealizable. To avoid this, these predictions have to be removed. This can be reached by postmultiplying $\boldsymbol{G}^{-1}(s)$ with a diagonal matrix $\boldsymbol{G_D}(s)$. Let the matrix without predictions be $\boldsymbol{G_O}^{-1}(s)$. We have $\boldsymbol{G_O}(s) = \boldsymbol{G_D^{-1}}(s)\boldsymbol{G}(s)$. $\boldsymbol{G_D}(s)$ should be chosen so that it counteracts those predictions and at the same time no additional time delays are introduced.

Definition 12.1.1. *Let $\theta_{li}(i = 1, 2, ..., n)$ be the largest prediction of the ith column of $\boldsymbol{G}^{-1}(s)$, that is, $\theta_{li} = \max_j \theta^{ji}, j = 1, 2, ..., n$. The H_∞ diagonal factorization for time delay is defined as*

$$\boldsymbol{G_D}(s) = \begin{bmatrix} e^{-\theta_{l1}s} & \cdots & 0 \\ \vdots & \ddots & \vdots \\ 0 & \cdots & e^{-\theta_{ln}s} \end{bmatrix}.$$

In particular, for rational plants $\boldsymbol{G_D}(s) = \boldsymbol{I}$.

Since the elements of $\boldsymbol{G_D}(s)$ are time delays, no closed RHP zeros and poles are cancelled when forming $\boldsymbol{G_O}(s)$.

Example 12.1.2. *Assume that*

$$\boldsymbol{G}^{-1}(s) = \begin{bmatrix} * & *e^{4s} & * \\ * & *e^{-3s} & * \\ * & *e^{6s} & * \end{bmatrix}.$$

Here $*$ denotes an arbitrary entry. The time delays of the elements in the second column are 4, -3, and 6, respectively. Then $\theta_{l2} = 6$.

Next, consider the factorization for closed RHP zeros. The plant may have closed RHP zeros in addition to a time delay. This implies that ${\boldsymbol{G_O}}^{-1}(s)$ is unstable. Then, an internally unstable system is obtained. To make ${\boldsymbol{G_O}}^{-1}(s)$ stable, the unstable poles in each element must be removed. This can be reached by postmultiplying ${\boldsymbol{G_O}}^{-1}(s)$ with a diagonal matrix $\boldsymbol{G_N}(s)$. Let the obtained matrix be ${\boldsymbol{G_{MP}}}^{-1}(s)$. One readily obtains that $\boldsymbol{G_{MP}}(s) = {\boldsymbol{G_N}}^{-1}(s)\boldsymbol{G_O}(s)$. According to the definition in Section 10.1, $\boldsymbol{G_{MP}}(s)$ is MP.

Assume that $z_j(\mathrm{Re}(z_j) \geq 0, j = 1, 2, ..., r_z)$ are unstable poles of ${\boldsymbol{G_O}}^{-1}(s)$.

Definition 12.1.2. *Let $k_{ij}(i = 1, 2, ..., n)$ be the largest multiplicity of the unstable pole $z_j(j = 1, 2, ..., r_z)$ in the ith column of ${\boldsymbol{G_O}}^{-1}(s)$. The H_∞ diagonal factorization for closed RHP zeros is defined as*

$$\boldsymbol{G_N}(s) = \begin{bmatrix} \prod_{j=1}^{r_z} (-s/z_j + 1)^{k_{1j}} & \cdots & 0 \\ \vdots & \ddots & \vdots \\ 0 & \cdots & \prod_{j=1}^{r_z} (-s/z_j + 1)^{k_{nj}} \end{bmatrix}.$$

In particular, for MP plants $\boldsymbol{G_N}(s) = \boldsymbol{I}$.

Example 12.1.3. *Assume that the second column of ${\boldsymbol{G_O}}^{-1}(s)$ involves only one unstable pole at $s = 1$ (that is, $z_1 = 1$). The multiplicities in its elements are shown as follows:*

$$\boldsymbol{G_O}^{-1}(s) = \begin{bmatrix} * & \frac{*}{*(s-1)^4} & * \\ * & \frac{*}{*(s-1)^3} & * \\ * & \frac{*}{*(s-1)^6} & * \end{bmatrix}.$$

The multiplicities of this RHP pole in the elements of the second column are 4, 3, and 6, respectively. Then $k_{21} = 6$.

The factorization of $\boldsymbol{G}(s)$ for both the closed RHP zeros and the time delay can be written as follows:

$$\boldsymbol{G}(s) = \boldsymbol{G_D}(s)\boldsymbol{G_N}(s)\boldsymbol{G_{MP}}(s). \qquad (12.1.6)$$

In the factorization, $\boldsymbol{G_D}(s)$ denotes the time delay part of the plant, $\boldsymbol{G_N}(s)$ is related to the closed RHP zeros of the plant, and $\boldsymbol{G_{MP}}(s)$ is the MP part of the plant. $\boldsymbol{G_O}(s) = \boldsymbol{G_N}(s)\boldsymbol{G_{MP}}(s)$ is the "rational" part of the plant. It should be emphasized that $\boldsymbol{G_O}(s)$ may not be rational in MIMO systems, since the element of $\boldsymbol{G_O}(s)$ may have a time delay.

When the element in the transfer function matrix of a plant involves a time delay, the plant may have infinite RHP zeros, that is, $r_z \to \infty$ or $k_{ij} \to \infty$. The following example illustrates such a case.

Example 12.1.4. *Consider the plant described by the following transfer function matrix:*

$$\boldsymbol{G} = \begin{bmatrix} 1 & 1 \\ 1 & 2e^{-s} \end{bmatrix}.$$

$\boldsymbol{G}(s)$ *has zeros at* $s = \ln 2 + j2k\pi, k = 0, \pm 1, \pm 2, \ldots$. *Following the factorization given in this section, we have*

$$\begin{aligned} \boldsymbol{G_D}(s) &= \boldsymbol{I}, \\ \boldsymbol{G_N}(s) &= (-2e^{-s}+1)\boldsymbol{I}, \\ \boldsymbol{G_{MP}}(s) &= \frac{1}{-2e^{-s}+1}\begin{bmatrix} 1 & 1 \\ 1 & 2e^{-s} \end{bmatrix}. \end{aligned}$$

The rational part of the plant is $\boldsymbol{G_O}(s) = \boldsymbol{G_N}(s)\boldsymbol{G_{MP}}(s)$. *The element of* $\boldsymbol{G_O}(s)$ *has a time delay.*

When the numerator or denominator of the element in $\boldsymbol{G_{MP}}(s)$ involves multiple time delays, the design problem becomes very involved. In this case, it is recommended to use a rational approximation to reduce the order of $\boldsymbol{G_O}(s)$ or $\boldsymbol{G_{MP}}(s)$. The rational approximation makes the design simpler and easier without loss of too much precision. The higher the order of the approximate numerator or denominator, the better the precision. The designer has to tradeoff between the complexity and the precision.

12.2 Quasi-H_∞ Controller Design

The design of a MIMO control system is similar to that of a SISO control system. First, a desired closed-loop transfer function matrix $\boldsymbol{T}(s)$ is constructed based on the factorization developed in the last section. Utilizing the $\boldsymbol{T}(s)$, the controller $\boldsymbol{Q}(s)$ is then derived. This is a "no-weight" design. The designer is not required to choose weighting functions in the procedure.

Assume that the controller is designed for step inputs. Factorize the plant according to Definition 12.1.1 and Definition 12.1.2: $\boldsymbol{G}(s) = \boldsymbol{G_D}(s)\boldsymbol{G_N}(s)\boldsymbol{G_{MP}}(s)$. If the plant is stable, along the same lines as that in the SISO design, the desired closed-loop transfer function matrix can be chosen as

$$\boldsymbol{T}(s) = \boldsymbol{T_{opt}}(s)\boldsymbol{J}(s), \tag{12.2.1}$$

where

$$\boldsymbol{T_{opt}}(s) = \boldsymbol{G_D}(s)\boldsymbol{G_N}(s), \tag{12.2.2}$$

$$\boldsymbol{J}(s) = \begin{bmatrix} J_1(s) & \cdots & 0 \\ \vdots & \ddots & \vdots \\ 0 & \cdots & J_n(s) \end{bmatrix}, \tag{12.2.3}$$

and

$$J_i(s) = \frac{1}{(\lambda_i s + 1)^{n_i}}, i = 1, 2, ..., n. \tag{12.2.4}$$

Here $\lambda_i (i = 1, 2, ..., n)$ are the performance degrees.

Let $\boldsymbol{T_{opt}}(s) = \boldsymbol{G}(s)\boldsymbol{Q_{opt}}(s)$. Then

$$\boldsymbol{Q_{opt}}(s) = \boldsymbol{G}^{-1}(s)\boldsymbol{T_{opt}}(s) = \boldsymbol{G_{MP}}^{-1}(s). \tag{12.2.5}$$

Introduce the notion of relative degree, that is, the degree of a transfer function's denominator polynomial minus the degree of its numerator polynomial. Denote the smallest relative degree of all elements in the ith column of $\boldsymbol{Q_{opt}}(s)$ as α_i. Then, $n_i = -\alpha_i$ for improper columns (that is, at least one element of the column is improper) and $n_i = 1$ for the others.

Example 12.2.1. *This example is given to illustrate the definition of α_i. Assume that $\boldsymbol{Q_{opt}}(s)$ is described by*

$$\boldsymbol{Q_{opt}}(s) = \begin{bmatrix} * & \frac{(s+1)^2}{s+2} & * \\ * & (s+1)^3 & * \\ * & \frac{s+2}{s+1} & * \end{bmatrix}.$$

The relative degrees of the elements in the second column of $\boldsymbol{Q_{opt}}(s)$ are -1, -3, and 0, respectively. Then $\alpha_2 = -3$.

With the desired closed-loop transfer function matrix $\boldsymbol{T}(s)$ and the factorization of $\boldsymbol{G}(s)$, the controller is obtained as follows:

$$\boldsymbol{Q}(s) = \boldsymbol{G}^{-1}(s)\boldsymbol{T}(s) = \boldsymbol{G_{MP}}^{-1}(s)\boldsymbol{J}(s). \tag{12.2.6}$$

The controller can be implemented in the IMC structure, or in the unity feedback loop:

$$C(s) = Q(s)[I - G(s)Q(s)]^{-1}. \tag{12.2.7}$$

When the plant is unstable, the desired closed-loop transfer function matrix can be chosen as

$$T(s) = T_{opt}(s)J(s), \tag{12.2.8}$$

where $T_{opt}(s)$ and $J(s)$ keep the same form as that for stable plants, but

$$J_i(s) = \frac{N_{xi}(s)}{(\lambda_i s + 1)^{n_i}}, i = 1, 2, ..., n. \tag{12.2.9}$$

Here $N_{xi}(s)(i = 1, 2, ..., n)$ are polynomials with all roots in the LHP, and $N_{xi}(0) = 1$. $n_i = \deg\{N_{xi}(s)\} - \alpha_i$ for improper columns and $n_i = \deg\{N_{xi}(s)\} + 1$ for the others.

Assume that $G(s)$ has r_p unstable poles; the multiplicity of the unstable pole p_j ($\mathrm{Re}(p_j) \geq 0, j = 1, 2, ..., r_p$) is l_j; l_{ij} is the largest multiplicity of p_j in the ith row of $G(s)$; the ith elements of $G_D(s)$ and $G_N(s)$ are $G_{Di}(s)$ and $G_{Ni}(s)$, respectively. Then $N_{xi}(s)$ is determined by

$$\lim_{s \to p_j} \frac{d^k}{ds^k}[I - G_{Di}(s)G_{Ni}(s)J_i(s)] = 0, \tag{12.2.10}$$
$$i = 1, 2, ..., n; j = 1, 2, ..., r_p; k = 0, 1, ..., l_{ij} - 1$$

with $\deg\{N_{xi}(s)\} = \sum_{j=1}^{r_p} l_{ij}$.

For unstable plants, the controller must be implemented in the unity feedback loop:

$$C(s) = Q(s)[I - G(s)Q(s)]^{-1}. \tag{12.2.11}$$

Furthermore, it is required that all RHP zero-pole cancellations in $[I - G(s)Q(s)]G(s)$ are removed. Recall the discussion in Section 8.1. This can be achieved by employing a rational approximation. It will be shown in the next section that this condition, together with (12.2.10), guarantees the internal stability of the closed-loop system.

It should be pointed out that the condition

$$\lim_{s \to p_j} \frac{d^k}{ds^k}\det[I - G(s)Q(s)] = 0, \tag{12.2.12}$$
$$j = 1, 2, ..., r_p; k = 0, 1, ..., l_{ij} - 1$$

is not sufficient for internal stability. By comparing (12.2.10) with (12.2.12), it can be seen that more closed RHP zeros have to be introduced in $I - G(s)Q(s)$ to guarantee the internal stability. This is the price for decoupling.

Example 12.2.2. *Consider the plant described by the following transfer function matrix:*

$$G(s) = \begin{bmatrix} \frac{1}{s+3} & \frac{1}{s-2} \\ \frac{2}{s+3} & \frac{s-1}{s-2} \end{bmatrix}.$$

The plant is NMP. It has one pole at $s = -3$, one pole at $s = 2$, and one zero at $s = 3$. Assume that the controller is

$$Q(s) = \begin{bmatrix} \frac{-(s+3)(s-1)}{3(s+1)^2} & \frac{(s+3)(13s+1)}{3(s+1)^2} \\ \frac{2(s-2)}{3(s+1)^2} & \frac{-(s-2)(13s+1)}{3(s+1)^2} \end{bmatrix}.$$

We have

$$I - G(s)Q(s) = \begin{bmatrix} \frac{s(s+7/3)}{(s+1)^2} & 0 \\ 0 & \frac{16s(s-2)}{3(s+1)^2} \end{bmatrix}.$$

It has one zero at $s = 2$. The RHP zero number of $I - G(s)Q(s)$ is the same as the RHP pole number of the plant. No more closed RHP zeros are introduced in $I - G(s)Q(s)$ by the controller. Since

$$[I - G(s)Q(s)]G(s) = \begin{bmatrix} \frac{s(s+7/3)}{(s+3)(s+1)^2} & \frac{s(s+7/3)}{(s-2)(s+1)^2} \\ \frac{32s(s-2)}{3(s+3)(s+1)^2} & \frac{16s(s-1)}{3(s+1)^2} \end{bmatrix},$$

$[I - G(s)Q(s)]G(s)$ is not stable, the closed-loop system is not internally stable.

Now consider another controller described by

$$Q(s) = \begin{bmatrix} \frac{-(s+3)(s-1)(40s+1)}{3(s+1)^3} & \frac{(s+3)(13s+1)}{3(s+1)^2} \\ \frac{2(s-2)(40s+1)}{3(s+1)^3} & \frac{-(s-2)(13s+1)}{3(s+1)^2} \end{bmatrix}.$$

One readily obtains

$$I - G(s)Q(s) = \begin{bmatrix} \frac{s(s-2)(s+55/3)}{(s+1)^3} & 0 \\ 0 & \frac{16s(s-2)}{3(s+1)^2} \end{bmatrix}.$$

It can be seen that one more closed RHP zero is introduced at $s = 2$ by the controller. Since

$$[I - G(s)Q(s)]G(s) = \begin{bmatrix} \frac{s(s-2)(s+55/3)}{(s+3)(s+1)^3} & \frac{s(s+55/3)}{(s+1)^3} \\ \frac{32s(s-2)}{3(s+3)(s+1)^2} & \frac{16s}{3(s+1)^2} \end{bmatrix},$$

$[I - G(s)Q(s)]G(s)$ is stable.

Now, a question of interest is when more closed RHP zeros need to be introduced in $I - G(s)Q(s)$ for internal stability. This has to be done if the closed RHP poles and their multiplicities in the elements of at least one row of $G(s)$ are not the same.

Example 12.2.3. *Consider the following plant:*

$$G(s) = \begin{bmatrix} \frac{1}{s+3} & \frac{2}{s+3} \\ \frac{1}{s-2} & \frac{s-1}{s-2} \end{bmatrix}.$$

The plant has one pole at $s = -3$, one pole at $s = 2$, and one zero at $s = 3$. The multiplicities of the closed RHP poles in every row are the same. If the following controller is taken:

$$Q(s) = \begin{bmatrix} \frac{(s+3)(s-1)}{-3(s+1)^2} & \frac{2(s-2)(13s+1)}{3(s+1)^2} \\ \frac{s+3}{3(s+1)^2} & -\frac{(s-2)(13s+1)}{3(s+1)^2} \end{bmatrix},$$

the closed-loop response is decoupled. It is evident that both

$$I - G(s)Q(s) = \begin{bmatrix} \frac{s(s+7/3)}{(s+1)^2} & 0 \\ 0 & \frac{16s(s-2)}{3(s+1)^2} \end{bmatrix}$$

and

$$[I - G(s)Q(s)]G(s) = \begin{bmatrix} \frac{s(s+7/3)}{(s+3)(s+1)^2} & \frac{2s(s+7/3)}{(s+3)(s+1)^2} \\ \frac{16s}{3(s+1)^2} & \frac{16s(s-1)}{3(s+1)^2} \end{bmatrix}$$

are stable. There is only one closed RHP zero in $I - G(s)Q(s)$.

The quasi-H_∞ controller can also be designed with the following procedure:

1. If the plant does not contain any time delay (that is, $G_D(s) = I$), turn to 3.
2. If the plant contains a time delay, take the rational part $G_O(s)$ as the nominal plant.
3. If $G_O(s)$ has no zeros in the RHP (that is, $G_N(s) = I$), take its inverse as $Q_{opt}(s)$ and turn to 5.
4. If $G_O(s)$ has zeros in the RHP, remove the factor that contains the zeros (that is, $G_N(s)$) and take the inverse of the remainder as $Q_{opt}(s)$.
5. Introduce a filter to $Q_{opt}(s)$, compute $C(s)$ and remove the RHP zero-pole cancellations in $C(s)$.

12.3 Analysis for Quasi-H_∞ Control Systems

Consider the nominal stability first. According to the discussion in Section 10.4, the closed-loop system is internally stable if and only if all elements in the following matrix are stable:

$$\boldsymbol{H}(s) = \begin{bmatrix} \boldsymbol{G}(s)\boldsymbol{Q}(s) & [\boldsymbol{I} - \boldsymbol{G}(s)\boldsymbol{Q}(s)]\boldsymbol{G}(s) \\ \boldsymbol{Q}(s) & -\boldsymbol{Q}(s)\boldsymbol{G}(s) \end{bmatrix}. \tag{12.3.1}$$

The following theorems provide necessary and sufficient conditions for the internal stability of the system designed in the last section.

Theorem 12.3.1. *The unity feedback control system is internally stable for the plant with time delay if and only if*

1. $\boldsymbol{Q}(s)$ *is stable.*
2. $[\boldsymbol{I} - \boldsymbol{G}(s)\boldsymbol{Q}(s)]\boldsymbol{G}(s)$ *is stable.*

Proof. Necessity is obvious. Consider sufficiency.

Assume that $[\boldsymbol{I} - \boldsymbol{G}(s)\boldsymbol{Q}(s)]\boldsymbol{G}(s)$ is stable but $\boldsymbol{G}(s)\boldsymbol{Q}(s)$ is not stable. As $\boldsymbol{Q}(s)$ is stable, the unstable pole of $\boldsymbol{G}(s)\boldsymbol{Q}(s)$ should be at the unstable poles of $\boldsymbol{G}(s)$. This implies that $\boldsymbol{I} - \boldsymbol{G}(s)\boldsymbol{Q}(s)$ is unbounded at the unstable poles of $\boldsymbol{G}(s)$, which contradicts the assumption. Hence, if $[\boldsymbol{I} - \boldsymbol{G}(s)\boldsymbol{Q}(s)]\boldsymbol{G}(s)$ is stable, $\boldsymbol{G}(s)\boldsymbol{Q}(s)$ is stable.

If $[\boldsymbol{I} - \boldsymbol{G}(s)\boldsymbol{Q}(s)]\boldsymbol{G}(s)$ is stable, $\boldsymbol{Q}(s)\boldsymbol{G}(s)$ is also stable, since

$$[\boldsymbol{I} - \boldsymbol{G}(s)\boldsymbol{Q}(s)]\boldsymbol{G}(s) = \boldsymbol{G}(s)[\boldsymbol{I} - \boldsymbol{Q}(s)\boldsymbol{G}(s)].$$

□

Theorem 12.3.2. *Assume that*

1. $\lim_{s\to p_j} \frac{d^k}{ds^k}[I - G_{Di}(s)G_{Ni}(s)J_i(s)] = 0, i = 1, 2, ..., n; j = 1, 2, ..., r_p; k = 0, 1, ..., l_{ij} - 1.$
2. *All RHP zero-pole cancellations in* $[I - \boldsymbol{G}(s)\boldsymbol{Q}(s)]\boldsymbol{G}(s)$ *are removed.*

Then, the unity feedback control system with a quasi-H_∞ *controller is internally stable.*

Proof. It is sufficient to prove that the two conditions in Theorem 12.3.1 hold.

It is evident that $\boldsymbol{Q}(s) = \boldsymbol{G_{MP}}^{-1}(s)\boldsymbol{J}(s)$ is stable, since $\boldsymbol{G_{MP}}(s)$ is MP and $\boldsymbol{J}(s)$ is stable in quasi-H_∞ control.

Now consider the stability of $[\boldsymbol{I} - \boldsymbol{G}(s)\boldsymbol{Q}(s)]\boldsymbol{G}(s)$. Since

$$\lim_{s\to p_j} \frac{d^k}{ds^k}[I - G_{Di}(s)G_{Ni}(s)J_i(s)] = 0,$$
$$i = 1, 2, ..., n; j = 1, 2, ..., r_p; k = 0, 1, ..., l_{ij} - 1,$$

all unstable poles of $\boldsymbol{G}(s)$ are cancelled by $\boldsymbol{I} - \boldsymbol{G}(s)\boldsymbol{Q}(s)$ when all RHP zero-pole cancellations in $[I - \boldsymbol{G}(s)\boldsymbol{Q}(s)]\boldsymbol{G}(s)$ are removed. Therefore, $[\boldsymbol{I} - \boldsymbol{G}(s)\boldsymbol{Q}(s)]\boldsymbol{G}(s)$ must be stable. □

When the nominal performance is considered, it is required that the system should have a zero steady-state error. The quasi-H_∞ controller is designed for step inputs. In view of the discussion in Section 10.4, the system should be of Type 1 for tracking step inputs asymptotically. In other words, the closed-loop transfer function matrix should satisfy the following condition:

$$\lim_{s\to 0} \boldsymbol{T}(s) = \boldsymbol{I}. \tag{12.3.2}$$

Evidently, the quasi-H_∞ controller satisfies the condition.

When the requirement on robustness is proposed, the robustness of the closed-loop system can be tested by the rigorous criteria introduced in Chapter 10. As an alternative, the engineering tuning method for SISO systems can be directly extended to MIMO systems; that is, increase the performance degrees monotonically until the required response is obtained. The advantage of this tuning method is that it is simple and can be used to tune both nominal performance and robustness.

One may find that some time delays or closed RHP zeros influence other channels while some others not. To distinguish them, two definitions are given.

Definition 12.3.1. *A time delay is canonical if at least one element of $\boldsymbol{G_D}(s)$ contains a time delay, provided that the greatest common time delay of all elements of $\boldsymbol{G_D}(s)$ has been removed.*

Definition 12.3.2. *A RHP zero is canonical if at least one element of $\boldsymbol{G_N}(s)$ has this zero, provided that the greatest common factor of all elements of $\boldsymbol{G_N}(s)$ has been removed.*

A canonical time delay or a canonical RHP zero does not spread its influence over all channels, whereas a non-canonical time delay or a non-canonical RHP zero affects all channels.

Example 12.3.1. *Consider the plant with the following transfer function matrix:*

$$\boldsymbol{G}(s) = \frac{1}{(s+3)(s-1)} \begin{bmatrix} s-2 & 2(s-2) \\ 1 & s-1 \end{bmatrix}.$$

The plant has a zero at $s = 2$ and a zero at $s = 3$. Since

$$\boldsymbol{G_N}(s) = \begin{bmatrix} \frac{-(s-2)}{s+2} & 0 \\ 0 & 1 \end{bmatrix} \frac{-(s-3)}{s+3},$$

the zero at s=2 is canonical and the zero at s=3 is non-canonical.

Example 12.3.2. *This example illustrates the effect of non-canonical zeros. Consider the plant in Example 12.2.2:*

$$\boldsymbol{G}(s) = \begin{bmatrix} \frac{1}{s+3} & \frac{1}{s-2} \\ \frac{2}{s+3} & \frac{s-1}{s-2} \end{bmatrix}.$$

If the controller is

$$\boldsymbol{Q}(s) = \begin{bmatrix} \frac{-(s+3)(s-1)(40s+1)}{3(s+1)^3} & \frac{(s+3)(13s+1)}{3(s+1)^2} \\ \frac{2(s-2)(40s+1)}{3(s+1)^3} & \frac{-(s-2)(13s+1)}{3(s+1)^2} \end{bmatrix},$$

the closed-loop transfer function matrix is

$$\boldsymbol{T}(s) = \begin{bmatrix} \frac{40s+1}{(s+1)^3} & 0 \\ 0 & \frac{13s+1}{(s+1)^2} \end{bmatrix} \left(-\frac{s}{3}+1\right).$$

The zero at $s = 3$ affects all channels.

12.4 Increasing Time Delays for Performance Improvement

Two classes of methods have been proposed in literature to enhance the closed-loop performance:

1. Developing control schemes that have better ability for time delay compensation.
2. Modifying the plant to reduce the effect of NMP factors.

The discussion in Section 12.2 falls into the first class, while the second class will be discussed in this section.

A fact in MIMO systems is that decreasing or increasing time delays may result in improved performance. Usually, it is impossible to decrease time delays in a plant, but in many cases it is possible to increase them. For example, the time delay can be increased by simply increasing the length of the pipe that connects process units. This section considers the strategy for enhancing performance by increasing time delays.

Let $\theta_{si} (i = 1, 2, ..., n)$ be the smallest time delay of the ith row of $\boldsymbol{G}(s)$, that is, $\theta_{si} = \min_j \theta_{ij}, j = 1, 2, ..., n$. Then the minimum time needed for any input to affect the output i is θ_{si}.

Theorem 12.4.1. *The element with the largest prediction in each column of $\boldsymbol{G}^{-1}(s)$ is on the diagonal if and only if the rows and columns of $\boldsymbol{G}(s)$ are rearranged so that the smallest time delay of $\boldsymbol{G}(s)$ in each row is on the diagonal. If this is true, the largest prediction is $e^{\theta_{si}}$.*

Proof. Prove sufficiency first. Assume that $\boldsymbol{G}(s)$ has been rearranged so that the smallest time delay in each row is on the diagonal, that is, $\theta_{si} = \theta_{jj}, j = 1, 2, ..., n$. It is known that

$$\boldsymbol{G}^{-1}(s) = \frac{\text{adj}[\boldsymbol{G}(s)]}{\det[\boldsymbol{G}(s)]}.$$

The largest time delay that can be separated from $\det[\boldsymbol{G}(s)]$ is

$$\theta_{det} = \theta_{11} + \theta_{22} + ... + \theta_{nn}.$$

Consider the jth column of $\mathrm{adj}[\boldsymbol{G}(s)]$. Compare the maximum time delays that can be separated from the diagonal element (denoted by θ_{ajj}) and non-diagonal element (denoted by $\theta_{aij}, i \neq j$):

$$\begin{aligned}\theta_{ajj} &= \theta_{11} + ... + \theta_{(i-1)(i-1)} + \theta_{ii} + \theta_{(i+1)(i+1)} + \\ &\quad ... + \theta_{(j-1)(j-1)} + \theta_{(j+1)(j+1)} + ... + \theta_{nn}. \\ \theta_{aij} &= \theta_{11} + ... + \theta_{(i-1)(i-1)} + \theta_{ij} + \theta_{(i+1)(i+1)} + \\ &\quad ... + \theta_{(j-1)(j-1)} + \theta_{(j+1)(j+1)} + ... + \theta_{nn}.\end{aligned}$$

Since

$$\theta_{ii} \leq \theta_{ij}, \forall i, j = 1, 2, ..., n,$$

the result is

$$\theta_{ajj} \leq \theta_{aij}.$$

Subtracting the two sides of the inequality from θ_{det} yields

$$\theta_{det} - \theta_{ajj} \geq \theta_{det} - \theta_{aij}.$$

$\theta_{det} - \theta_{ajj} = \theta_{jj} = \theta_{si}$ is the prediction of the diagonal element in the jth column of $\boldsymbol{G}^{-1}(s)$, while $(\theta_{det} - \theta_{aij})$ is the prediction of the non-diagonal element in the jth column of $\boldsymbol{G}^{-1}(s)$.

The sufficiency proof is reversible. □

It may not be feasible to make the smallest time delay in each row on the diagonal by rearranging rows and columns of a plant. The idea here is to increase individual delay θ_{ij} by the minimum amount, so that the minimum time delays occur on the diagonal. For simple cases, this work can be done manually. In the general case, optimization techniques may have to be used. The key is to formulate this problem mathematically.

Let $b_{ij}(i, j = 1, 2, ..., n)$ be a continuous variable that represents the modified time delay. The binary variable y_{ij} is introduced, which takes values of $0 - 1$ and is associated with the element b_{ij}. When $y_{ij} = 1$, the corresponding element has the smallest time delay. The minimum time delay needed to improve the closed-loop response is given by the solution to the following optimization problem:

$$\mathrm{Min} \sum_{i} \sum_{j} b_{ij}, \tag{12.4.1}$$

subject to

$$\sum_{i} y_{ij} = 1, j = 1, 2, ..., n, \tag{12.4.2}$$

$$\sum_j y_{ij} = 1, i = 1, 2, ..., n, \quad (12.4.3)$$

$$\sum_j (y_{ij} b_{ij}) - b_{ij} \leq 0, i, j = 1, 2, ..., n, \quad (12.4.4)$$

$$\theta_{ij} \leq b_{ij}, i, j = 1, 2, ..., n, \quad (12.4.5)$$

$$b_{ij} \leq \max_j \theta_{ij}, i, j = 1, 2, ..., n, \quad (12.4.6)$$

$$y_{ij} = 0, 1, i, j = 1, 2, ..., n. \quad (12.4.7)$$

Constraints (12.4.2) and (12.4.3) imply that only one element is picked up as the element with the smallest time delay. Constraint (12.4.4) states that the selected element has the smallest time delay indeed. The first term in its left-hand side picks up one element from each row, and compares it with other elements in this row (that is, the second term in its left-hand side). Finally, Constraints (12.4.5) and (12.4.6) provide the lower and upper bounds for the continuous variable b_{ij}.

This mathematical formulation is a mixed-integer nonlinear programming problem, since the constraint (12.4.4) involves the bilinearity of the form $y_{ij}b_{ij}$. The constraint (12.4.4) is not convex. This implies that it is hard to obtain the global optimal solution.

The problem can be overcome by converting the nonlinearity programming into a linear programming. The bilinearity of the form $y_{ij}b_{ij}$, in which y_{ij} is an integer and b_{ij} is a continuous variable, can be substituted by a continuous variable h_{ij} with four additional constraints:

$$h_{ij} = y_{ij} b_{ij}, \quad (12.4.8)$$

$$b_{ij} - U(1 - y_{ij}) \leq h_{ij}, \quad (12.4.9)$$

$$L y_{ij} \geq h_{ij}, \quad (12.4.10)$$

$$b_{ij} - L(1 - y_{ij}) \leq h_{ij}, \quad (12.4.11)$$

$$U y_{ij} \geq h_{ij}, \quad (12.4.12)$$

$$i, j = 1, 2, ..., n,$$

where the scalars L and U satisfy

$$L \leq b_{ij} \leq U, i, j = 1, 2, ..., n. \quad (12.4.13)$$

In the above formulation, one can take $L = 0$ and $U = \max_{ij}(\theta_{ij})$.

Utilizing the conversion, the formulation becomes

$$\text{Min} \sum_{\text{i}} \sum_j b_{ij}, \quad (12.4.14)$$

subject to

$$\sum_i y_{ij} = 1, j = 1, 2, ..., n, \quad (12.4.15)$$

$$\sum_j y_{ij} = 1, i = 1, 2, ..., n, \tag{12.4.16}$$

$$\sum_j h_{ij} - b_{ij} \leq 0, i, j = 1, 2, ..., n, \tag{12.4.17}$$

$$b_{ij} - (\max_{ij} \theta_{ij})(1 - y_{ij}) \leq h_{ij}, i, j = 1, 2, ..., n, \tag{12.4.18}$$

$$0 \leq h_{ij}, i, j = 1, 2, ..., n, \tag{12.4.19}$$

$$\theta_{ij} \leq b_{ij}, i, j = 1, 2, ..., n, \tag{12.4.20}$$

$$b_{ij} \leq \max_j \theta_{ij}, i, j = 1, 2, ..., n, \tag{12.4.21}$$

$$y_{ij} = 0, 1, i, j = 1, 2, ..., n. \tag{12.4.22}$$

This is a mixed-integer linear programming problem. It can be solved easily with the help of computer software.

The mathematical tool adopted in this section is thoroughly different from those introduced in the preceding chapters. For detailed discussion, please refer to the monographs addressing this topic.

12.5 A Design Example for Quasi-H_∞ Control

In this section, an example is given to illustrate the design procedure of the quasi-H_∞ controller. It will be shown that the performance is significantly improved by employing the optimization technique introduced in the last section.

Example 12.5.1. *Consider the model of a pilot scale ethanol and water distillation column:*

$$G(s) = \begin{bmatrix} \frac{0.66e^{-6s}}{6.7s+1} & \frac{-0.005e^{-s}}{9.1s+1} \\ \frac{-34.7e^{-9.2s}}{8.1s+1} & \frac{0.87(11.6s+1)e^{-s}}{(3.9s+1)(18.8s+1)} \end{bmatrix}.$$

In this system the outputs are

- *Overhead ethanol mole fraction.*
- *Bottom composition temperature (°C).*

The inputs are

- *Reflux flow rate (gpm).*
- *Reboiler steam pressure (psig).*

The inverse of the plant is

$$G^{-1}(s) = \frac{adj[G(s)]}{det[G(s)]},$$

where

$$adj[G(s)] = \begin{bmatrix} \frac{0.87(11.6s+1)e^{-s}}{(3.9s+1)(18.8s+1)} & \frac{0.005e^{-s}}{9.1s+1} \\ \frac{34.7e^{-9.2s}}{8.1s+1} & \frac{0.66e^{-6s}}{6.7s+1} \end{bmatrix},$$

$$\det[G(s)] = \left[\frac{0.66 * 0.87(11.6s+1)}{(6.7s+1)(3.9s+1)(18.8s+1)} - \frac{34.7 * 0.005e^{-3.2s}}{(8.1s+1)(9.1s+1)}\right] e^{-7s}.$$

$G(s)$ is factorized into

$$G(s) = G_D(s)G_O(s).$$

Both maximum predictions of the first and second columns of $G^{-1}(s)$ are 6. According to Definition 12.1.1, $G_D(s)$ is given by

$$G_D(s) = \begin{bmatrix} e^{-6s} & 0 \\ 0 & e^{-6s} \end{bmatrix}.$$

$G_O(s) = G_D{}^{-1}(s)G(s)$ has no RHP zeros, which implies that $G_N(s) = I$ and $G_{MP}(s) = G_O(s)$. In light of (12.2.5), the quasi-H_∞ controller is

$$Q_{opt}(s) = \frac{\begin{bmatrix} \frac{0.87(11.6s+1)}{(3.9s+1)(18.8s+1)} & \frac{0.005}{9.1s+1} \\ \frac{34.7e^{-8.2s}}{8.1s+1} & \frac{0.66e^{-5s}}{6.7s+1} \end{bmatrix}}{\frac{0.66*0.87(11.6s+1)}{(6.7s+1)(3.9s+1)(18.8s+1)} - \frac{34.7*0.005e^{-3.2s}}{(8.1s+1)(9.1s+1)}}.$$

This controller is exact, but too complex. Model reduction techniques are used here to simplify the design result. Suppose that the relative degree of the approximate denominator is chosen as 2. The following result can be obtained by applying fitting techniques:

$$\frac{0.66 * 0.87(11.6s+1)}{(6.7s+1)(3.9s+1)(18.8s+1)} - \frac{34.7 * 0.005e^{-3.2s}}{(8.1s+1)(9.1s+1)} \approx \frac{0.4007}{3s^2+16s+1}.$$

With this approximate denominator, we have

$$Q_{opt}(s) = \begin{bmatrix} \frac{0.87(11.6s+1)}{(3.9s+1)(18.8s+1)} & \frac{0.005}{9.1s+1} \\ \frac{34.7e^{-8.2s}}{8.1s+1} & \frac{0.66e^{-5s}}{6.7s+1} \end{bmatrix} \frac{3s^2+16s+1}{0.4007}.$$

Since the plant is stable, the filter can be chosen as

$$J(s) = \begin{bmatrix} \frac{1}{\lambda_1 s+1} & 0 \\ 0 & \frac{1}{\lambda_2 s+1} \end{bmatrix}.$$

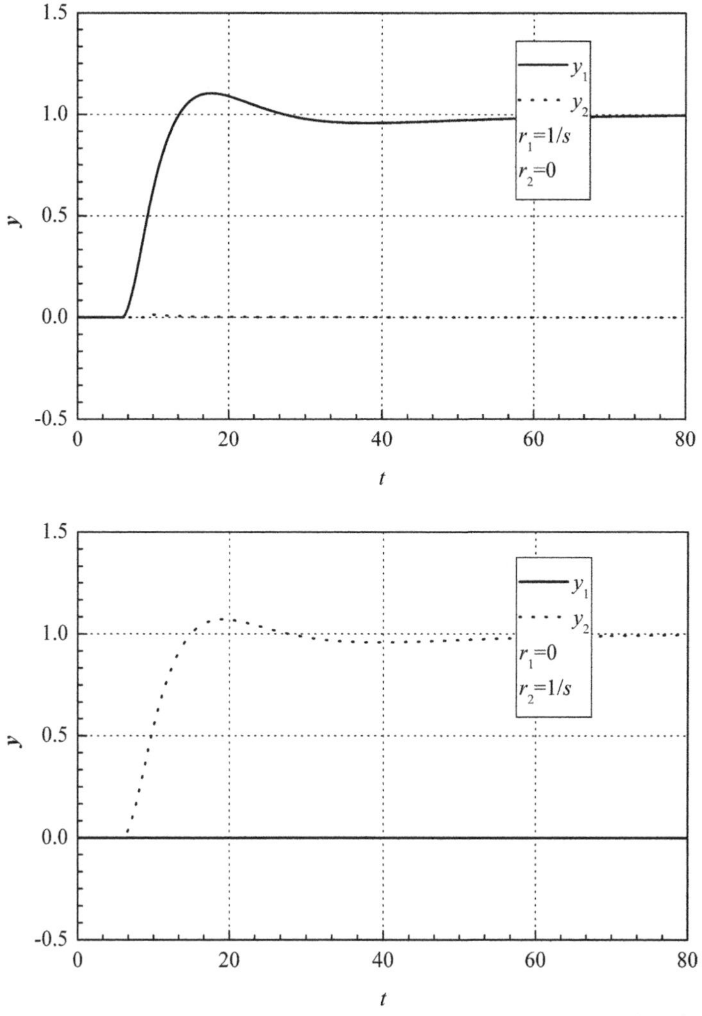

FIGURE 12.5.1
Closed-loop responses with $\lambda_1 = 3.2$ and $\lambda_2 = 4$.

The suboptimal controller is

$$Q(s) = \begin{bmatrix} \frac{0.87(11.6s+1)}{(3.9s+1)(18.8s+1)(\lambda_1 s+1)} & \frac{0.005}{(9.1s+1)(\lambda_2 s+1)} \\ \frac{34.7e^{-8.2s}}{(8.1s+1)(\lambda_1 s+1)} & \frac{0.66e^{-5s}}{(6.7s+1)(\lambda_2 s+1)} \end{bmatrix} \frac{3s^2+16s+1}{0.4007}.$$

If there is not any requirement on the closed-loop response, λ_1 and λ_2 can be selected freely. Suppose the design requirement on the closed-loop response is 10% overshoot for each channel. With the engineering tuning method, it is easy to obtain that $\lambda_1 = 3.2$ and $\lambda_2 = 4$. The closed-loop responses are shown in Figure 12.5.1.

In the last section, an optimization method was introduced to improve the performance. Using this method, one gets

$$\text{Min} \sum_{i} \sum_{j} b_{ij} = 22.2,$$

$$b_{11} = 6, b_{12} = 6, b_{21} = 9.2, b_{22} = 1,$$

$$y_{11} = 1, y_{12} = 0, y_{21} = 0, y_{22} = 1.$$

This implies that the closed-loop performance could be improved by increasing θ_{12} in $G(s)$ to 6 while keeping other time delays the same. In this case,

$$\boldsymbol{G_D}(s) = \begin{bmatrix} e^{-6s} & 0 \\ 0 & e^{-s} \end{bmatrix},$$

and the controller is

$$Q(s) = \begin{bmatrix} \frac{0.87(11.6s+1)}{(3.9s+1)(18.8s+1)(\lambda_1 s+1)} & \frac{0.005}{(9.1s+1)(\lambda_2 s+1)} \\ \frac{34.7e^{-8.2s}}{(8.1s+1)(\lambda_1 s+1)} & \frac{0.66}{(6.7s+1)(\lambda_2 s+1)} \end{bmatrix} \frac{3s^2+16s+1}{0.4007}.$$

For the sake of comparison, the same controller parameters are used as those in the system with unchanged time delays. It can be seen from Figure 12.5.2 that the response of the second output is significantly improved.

12.6 Multivariable PID Controller Design

The PID controller is widely used in practice. Design techniques of the SISO PID controller have been well developed. However, the design problem of multivariable PID controllers remains a subject of study, since the MIMO case is much more intricate than the SISO one.

Section 12.2 introduced a simple design method for multivariable controllers, of which one main feature is that the controller is analytical. Once

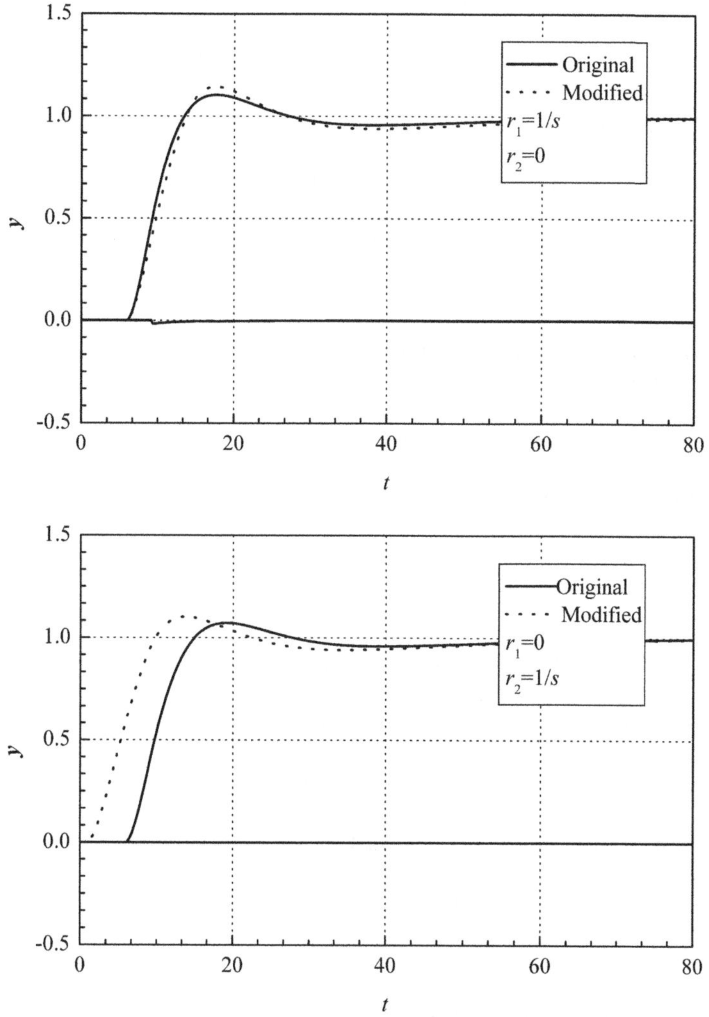

FIGURE 12.5.2
Performance improvement by increasing the time delay.

an analytical result is obtained, one can directly use the design procedures in Section 5.5 and Section 5.6 to reduce every element of the multivariable controller $\boldsymbol{C}(s)$ to a SISO PID controller.

When the desired PID controllers have the same form (that is, all controllers are in the form of a PI controller, or all controllers in the form of a PID controller), the design procedure can be expressed by matrix and vector notations. As an example, the design procedure for Maclaurin PID controllers is provided here.

Consider the quasi-H_∞ design method. First, the plant is factorized as

$$\boldsymbol{G}(s) = \boldsymbol{G_D}(s)\boldsymbol{G_N}(s)\boldsymbol{G_{MP}}(s). \tag{12.6.1}$$

The desired closed-loop transfer function matrix is chosen as

$$\boldsymbol{T}(s) = \boldsymbol{G_D}(s)\boldsymbol{G_N}(s)\boldsymbol{J}(s). \tag{12.6.2}$$

With this closed-loop transfer function matrix, the obtained unity feedback loop controller can be expressed as

$$\boldsymbol{C}(s) = \boldsymbol{G_{MP}}^{-1}(s)\boldsymbol{J}(s)[\boldsymbol{I} - \boldsymbol{G_D}(s)\boldsymbol{G_N}(s)\boldsymbol{J}(s)]^{-1}.$$

Rewrite the controller as

$$\boldsymbol{C}(s) = s^{-1}\boldsymbol{f}(s), \tag{12.6.3}$$

where

$$\boldsymbol{f}(s) = s\boldsymbol{G_{MP}}^{-1}(s)\boldsymbol{J}(s)[\boldsymbol{I} - \boldsymbol{G_D}(s)\boldsymbol{G_N}(s)\boldsymbol{J}(s)]^{-1}. \tag{12.6.4}$$

Expand the controller in a Maclaurin series:

$$\boldsymbol{C}(s) = s^{-1}\left[\boldsymbol{f}(0) + \boldsymbol{f}'(0)s + \boldsymbol{f}''(0)s^2/2! + \boldsymbol{f}^{(3)}(0)s^3/3! + ...\right]. \tag{12.6.5}$$

The first three terms form a standard PID controller:

$$\boldsymbol{C}(s) = \boldsymbol{K_C} + \boldsymbol{K_I}s^{-1} + \boldsymbol{K_D}s, \tag{12.6.6}$$

whose parameters are

$$\boldsymbol{K_C} = \boldsymbol{f}'(0), \boldsymbol{K_I} = \boldsymbol{f}(0), \boldsymbol{K_D} = \boldsymbol{f}''(0)/2. \tag{12.6.7}$$

To simplify the representation, let

$$\begin{aligned} \boldsymbol{N}(s) &= s^{-1}[\boldsymbol{I} - \boldsymbol{G_D}(s)\boldsymbol{G_N}(s)\boldsymbol{J}(s)]\boldsymbol{J}^{-1}(s) \\ &= s^{-1}[\boldsymbol{J}^{-1}(s) - \boldsymbol{G_D}(s)\boldsymbol{G_N}(s)]. \end{aligned} \tag{12.6.8}$$

Then we have

$$\boldsymbol{f}(s) = \boldsymbol{G_{MP}}^{-1}(s)\boldsymbol{N}^{-1}(s). \tag{12.6.9}$$

The values of $\boldsymbol{f}(s)$ and its derivatives at the origin are

$$\begin{aligned}
\boldsymbol{f}(0) &= \boldsymbol{G_{MP}}^{-1}(0)\boldsymbol{N}^{-1}(0),\\
\boldsymbol{f}'(0) &= [\boldsymbol{G_{MP}}^{-1}(0)]'\boldsymbol{N}^{-1}(0) + \boldsymbol{G}_{MP}^{-1}(0)[\boldsymbol{N}^{-1}(0)]',\\
\boldsymbol{f}''(0) &= [\boldsymbol{G_{MP}}^{-1}(0)]''\boldsymbol{N}^{-1}(0) +\\
&\qquad 2[\boldsymbol{G_{MP}}^{-1}(0)]'[\boldsymbol{N}^{-1}(0)]' + \boldsymbol{G_{MP}}^{-1}(0)[\boldsymbol{N}^{-1}(0)]'',
\end{aligned}$$

where

$$\begin{aligned}
\boldsymbol{N}(0) &= \left[\boldsymbol{J}^{-1}(0)\right]' - [\boldsymbol{G_N}^{-1}(0)\boldsymbol{G_D}^{-1}(0)]',\\
[\boldsymbol{N}(0)]' &= \{[\boldsymbol{J}^{-1}(0)]'' - [\boldsymbol{G_N}^{-1}(0)\boldsymbol{G_D}^{-1}(0)]''\}/2!,\\
[\boldsymbol{N}(0)]'' &= \{[\boldsymbol{J}^{-1}(0)]^{(3)} - [\boldsymbol{G_N}^{-1}(0)\boldsymbol{G_D}^{-1}(0)]^{(3)}\}/3!,\\
\left[\boldsymbol{N}^{-1}(0)\right]' &= -\boldsymbol{N}^{-1}(0)[\boldsymbol{N}(0)]'\boldsymbol{N}^{-1}(0),\\
\left[\boldsymbol{N}^{-1}(0)\right]'' &= -[\boldsymbol{N}^{-1}(0)]'[\boldsymbol{N}(0)]'\boldsymbol{N}^{-1}(0) -\\
&\qquad \boldsymbol{N}^{-1}(0)[\boldsymbol{N}(0)]''\boldsymbol{N}^{-1}(0) -\\
&\qquad \boldsymbol{N}^{-1}(0)[\boldsymbol{N}(0)]'[\boldsymbol{N}^{-1}(0)]',\\
\left[\boldsymbol{G}^{-1}(0)\right]' &= -\boldsymbol{G}_{MP}^{-1}(0)[\boldsymbol{G}_{MP}(0)]'\boldsymbol{G_{MP}}^{-1}(0),\\
\left[\boldsymbol{G}^{-1}(0)\right]'' &= -[\boldsymbol{G_{MP}}^{-1}(0)]'[\boldsymbol{G}_{MP}(0)]'\boldsymbol{G_{MP}}^{-1}(0) -\\
&\qquad \boldsymbol{G_{MP}}^{-1}(0)[\boldsymbol{G}_{MP}(0)]''\boldsymbol{G_{MP}}^{-1}(0) -\\
&\qquad \boldsymbol{G_{MP}}^{-1}(0)[\boldsymbol{G}_{MP}(0)]'[\boldsymbol{G_{MP}}^{-1}(0)]'.
\end{aligned}$$

Example 12.6.1. *Consider a binary distillation column for separating a mixture of methanol and water (the feed) into a bottom product (mostly water) and a methanol distillate. Schematically, the distillation process functions as follows (Figure 12.6.1):*

- *Steam flows into the reboiler and vaporizes the bottom liquid. This vapor is reinjected into the column and mixes with the feed.*
- *Methanol, being more volatile than water, tends to concentrate in the vapor moving upward. Meanwhile, water tends to flow downward and accumulate as the bottom liquid.*
- *The vapor exiting at the top of the column is condensed by a flow of cooling water. Part of this condensed vapor is extracted as the distillate, and the rest of the condensate (the reflux) is sent back to the column.*
- *Part of the bottom liquid is collected as the bottom product (waste).*

In this application, the objective is to control the amount of the bottom and top methanol by manipulating the steam flow rate and the reflux flow rate, respectively. Since a change in either steam flow rate or reflux flow rate upsets

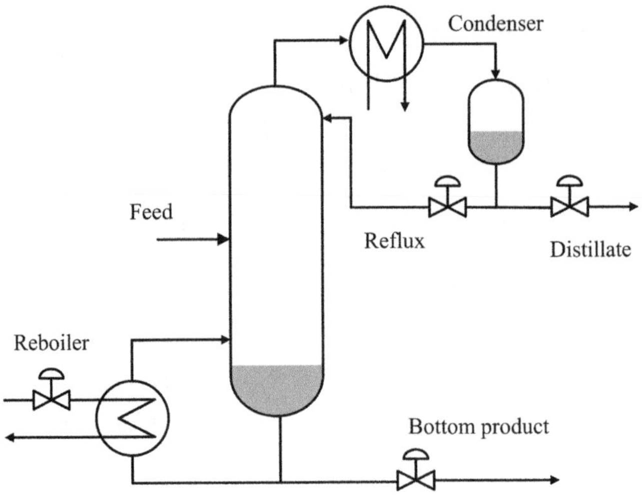

FIGURE 12.6.1
A binary distillation column.

both methanols, we have an interacting system. The model of the distillation column is described by

$$\boxed{G(s) = \begin{bmatrix} \frac{12.8e^{-s}}{16.7s+1} & \frac{-18.9e^{-3s}}{21s+1} \\ \frac{6.6e^{-7s}}{10.9s+1} & \frac{-19.4e^{-3s}}{14.4s+1} \end{bmatrix}}.$$

The inverse of the plant is

$$G^{-1}(s) = \frac{adj[G(s)]}{det[G(s)]},$$

where

$$\begin{aligned} adj[G(s)] &= \begin{bmatrix} \frac{-19.4e^{-3s}}{14.4s+1} & \frac{18.9e^{-3s}}{21s+1} \\ \frac{-6.6e^{-7s}}{10.9s+1} & \frac{12.8e^{-s}}{16.7s+1} \end{bmatrix}, \\ \det[G(s)] &= \left(\frac{-248.3}{240.5s^2+31.1s+1} - \frac{-124.7e^{-6s}}{228.9s^2+31.9s+1} \right) e^{-4s}. \end{aligned}$$

$G(s)$ is factorized into

$$G(s) = G_D(s)G_O(s).$$

Since the largest predictions of the first and the second columns of $G^{-1}(s)$ are 1 and 3 respectively, $G_D(s)$ is given by

$$G_D(s) = \begin{bmatrix} e^{-1s} & 0 \\ 0 & e^{-3s} \end{bmatrix}.$$

$\boldsymbol{G_O}(s)$ is MP, thus $\boldsymbol{G_N}(s) = \boldsymbol{I}$. The plant is stable. According to (12.2.1) the desired closed-loop transfer function matrix is

$$\boldsymbol{T}(s) = \begin{bmatrix} \dfrac{e^{-s}}{\lambda_1 s + 1} & 0 \\ 0 & \dfrac{e^{-3s}}{\lambda_2 s + 1} \end{bmatrix}.$$

Then

$$\boldsymbol{Q}(s) = \frac{\begin{bmatrix} \frac{-19.4}{(14.4s+1)(\lambda_1 s+1)} & \frac{18.9e^{-2s}}{(21s+1)(\lambda_2 s+1)} \\ \frac{-6.6e^{-4s}}{(10.9s+1)(\lambda_1 s+1)} & \frac{12.8}{(16.7s+1)(\lambda_2 s+1)} \end{bmatrix}}{\frac{124.7e^{-6s}}{228.9s^2+31.9s+1} - \frac{248.3}{240.5s^2+31.1s+1}},$$

and the unity feedback loop controller is

$$\boldsymbol{C}(s) = \frac{\begin{bmatrix} \frac{-19.4}{(14.4s+1)(\lambda_1 s+1-e^{-s})} & \frac{18.9e^{-2s}}{(21s+1)(\lambda_2 s+1-e^{-3s})} \\ \frac{-6.6e^{-4s}}{(10.9s+1)(\lambda_1 s+1-e^{-s})} & \frac{12.8}{(16.7s+1)(\lambda_2 s+1-e^{-3s})} \end{bmatrix}}{\frac{124.7e^{-6s}}{228.9s^2+31.9s+1} - \frac{248.3}{240.5s^2+31.1s+1}}.$$

This is a rigorously analytical result.

Suppose a multivariable PI controller is desired. Take $\lambda_1 = 4.5$ and $\lambda_2 = 4$. The controller parameters given by (12.6.7) are

$$\boldsymbol{K_C} = \begin{bmatrix} 0.2833 & -0.0411 \\ 0.0915 & -0.1210 \end{bmatrix}, \quad \boldsymbol{K_I} = \begin{bmatrix} 0.0285 & -0.0219 \\ 0.0097 & -0.0148 \end{bmatrix}.$$

If there is no constraint on the structure of PID controller, one can choose the PID controller with the best achievable performance (see Section 5.6). Nevertheless, the resulting controller is unstable. To overcome this problem, the controllers in the first column of $\boldsymbol{C}(s)$ are chosen in the form of a PI controller, while the controllers in the second column are PID controllers with the best achievable performance. Take $\lambda_1 = 3.8$ and $\lambda_2 = 3.5$. The multivariable PID controller is

$$\boldsymbol{C}(s) = \begin{bmatrix} \frac{0.3251s+0.0327}{s} & -\frac{3.0901s^2+0.8804s+0.0235}{s(35.4901s+1)} \\ \frac{1.0506s+0.1112}{10s} & -\frac{8.6640s^2+1.1570s+0.0159}{s(64.3898s+1)} \end{bmatrix}.$$

The closed-loop responses are shown in Figure 12.6.2. Because a rational approximation is used, the interaction among different channels cannot be thoroughly decoupled in the systems with PI or PID controllers.

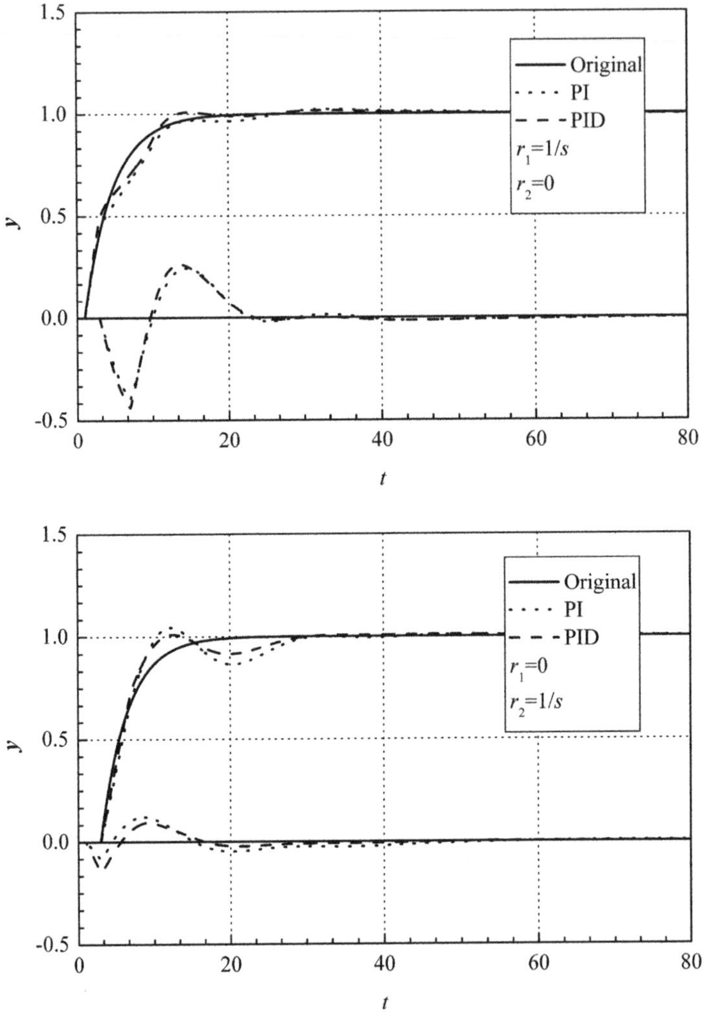

FIGURE 12.6.2
Closed-loop responses of multivariable PID controllers.

12.7 Summary

The design idea of the quasi-H_∞ control is constructing a desired closed-loop transfer function matrix first, and then deriving the controller based on this transfer function matrix. Similar ideas were also used by many other design methods, for example, the pole assignment method, Dahlin controller, and model predictive control. The central problem of such methods is how to define the desired closed-loop transfer function matrix.

The quasi-H_∞ control chooses the following transfer function matrix as the desired closed-loop transfer function matrix:

$$\boldsymbol{T}(s) = \boldsymbol{G_D}(s)\boldsymbol{G_N}(s)\boldsymbol{J}(s).$$

The resulting unity feedback loop controller is

$$\boldsymbol{C}(s) = \boldsymbol{G}_{\boldsymbol{MP}}^{-1}(s)\boldsymbol{J}(s)[\boldsymbol{I} - \boldsymbol{T}(s)]^{-1}.$$

$\boldsymbol{G_D}(s), \boldsymbol{G_N}(s)$, and $\boldsymbol{G_{MP}}(s)$ are the transfer function matrices defined in the factorization for MIMO plants. The stability and performance of the closed-loop system is directly analyzed in this chapter with developed techniques. It has been proved that the closed-loop transfer function matrix and the factorization can guarantee the internal stability and asymptotical tracking. Similar to SISO systems, the closed-loop responses can also be tuned quantitatively.

The MIMO control system design is complicated. The obtained closed-loop performance depends on not only the optimal design method, but also the distribution of time delays. It is shown in this chapter that, by arranging or adjusting the element time delays appropriately, significant improvement on performance can be obtained.

Another important problem discussed in this chapter is the design problem of multivariable PID controllers. Based on the discussion for SISO controller design, a general design method is proposed for multivariable PID controllers and suboptimal performance is obtained.

Exercises

1. Consider the following plant:

$$\boldsymbol{G}(s) = \begin{bmatrix} \frac{1/2}{s+1} & \frac{2}{s+2} \\ \frac{-1}{s+1} & \frac{2e^{-s}}{s+1} \end{bmatrix}.$$

The plant has a time delay in one of its elements. Is this plant NMP?

2. A plant is described by the following model:

$$\boldsymbol{G}(s) = \frac{1}{(s+3)(s-1)} \begin{bmatrix} s-2 & 2(s-2) \\ 1 & s-1 \end{bmatrix}.$$

Design a quasi-H_∞ controller.

3. Suppose that a controller is in the form of

$$C(s) = \frac{c(s)}{a(s) + b(s)e^{-\theta s}},$$

where $a(s), b(s)$ and $c(s)$ are polynomials with roots in the LHP. Such a controller can be exactly implemented by means of a block with an inner feedback loop. Give the implementation.

4. Assume that the system input is a ramp. What a condition should the desired closed-loop transfer function matrix satisfy for asymptotical tracking?

5. Following the idea in Chapter 9, design a 2 DOF quasi-H_∞ controller.

Notes and References

The factorization of the stable MIMO plant was discussed in many papers, for example, Holt and Morari (1985), Jerome and Ray (1986), and Wang et al. (2002).

Section 12.4 is based on Psarris and Floudas (1990) (Psarris P. and C. A. Floudas, Improving dynamic operability in MIMO systems with time delays, *Chemical Engineering Science*, 1990, 45(12), 3505–3524. ©Elsevier). Theorem 12.4.1 belongs to Holt and Morari (1985). Detailed discussion on the optimization theory can be found in Floudas (1995).

The plant in Example 12.5.1 was given by Ogunnaike et al. (1983).

The idea of designing a MIMO PID controller with Maclaurin series was first proposed by Dong and Brosilow (1997). The plant in Example 12.6.1 was given by Wood and Berry (1973). This plant is very famous. It has been used in many papers for performance comparison.

For the MIMO system decoupling control, one can also refer to Wang (2003) and Goodwin et al. (2001, Chapter 26).

13

H_2 Decoupling Control

CONTENTS

H_∞ control and H_2 control are two prevailing design methods. In Section 3.4, it was seen that the necessary and sufficient condition for robust performance analysis could be obtained when the system performance was specified in terms of ∞-norm. For controller design, however, 2-norm is mathematically more convenient to treat.

In the SISO design, it could be seen that the methodologies of quasi-H_∞ control and H_2 control were thoroughly different. In the quasi-H_∞ control, the controller was designed by constructing a desired closed-loop transfer function, while in the H_2 control the controller was derived by solving an optimization problem. In this chapter, the SISO H_2 design will be extended to the decoupling control system. First, all stabilizing decoupling controllers are parameterized. Then, diagonal factorizations are defined for MIMO plants. Finally, based on the parameterization and factorizations, the controller and decoupler are analytically derived in one step.

13.1 Controller Parameterization for MIMO Systems

Consider the control system consisting of an $n \times n$ plant $\boldsymbol{G}(s)$ and an $n \times n$ controller $\boldsymbol{C}(s)$.

$$\boldsymbol{C}(s) = \boldsymbol{Q}(s)[\boldsymbol{I} - \boldsymbol{G}(s)\boldsymbol{Q}(s)]^{-1}, \tag{13.1.1}$$

where $\boldsymbol{Q}(s)$ is the IMC controller.

In the design method of H_2 decoupling control in this chapter, the plant can be proper, have a time delay, have poles on the imaginary axis, or have poles and zeros in the open RHP.

It is assumed that

1. There is no unstable hidden mode in $\boldsymbol{G}(s)$.
2. $\boldsymbol{G}(s)$ is of full normal rank.

These assumptions are the same as those in the quasi-H_∞ decoupling control. Normally, the finite zeros on the imaginary axis is not considered, because they may cause the problem of internal instability or poor performance.

From Theorem 12.3.1 it is known that the closed-loop system is internally stable if and only if

1. $\boldsymbol{Q}(s)$ is stable.
2. $[\boldsymbol{I} - \boldsymbol{G}(s)\boldsymbol{Q}(s)]\boldsymbol{G}(s)$ is stable.

Theorem 13.1.1. *Assume that $\boldsymbol{G}(s)$ is a plant with time delay. The unity feedback control system is internally stable if and only if*

1. $\boldsymbol{Q}(s)$ is stable.

2. $\boldsymbol{I} - \boldsymbol{G}(s)\boldsymbol{Q}(s)$ has zeros wherever $\boldsymbol{G}(s)$ has unstable poles.

3. All RHP zero-pole cancellations in $[I - \boldsymbol{G}(s)\boldsymbol{Q}(s)]\boldsymbol{G}(s)$ are removed.

Proof. It is enough to prove the equivalency between the stability of $[\boldsymbol{I} - \boldsymbol{G}(s)\boldsymbol{Q}(s)]\boldsymbol{G}(s)$ and the second and third conditions.

Assume that $[\boldsymbol{I} - \boldsymbol{G}(s)\boldsymbol{Q}(s)]\boldsymbol{G}(s)$ is stable. Evidently, $\boldsymbol{I} - \boldsymbol{G}(s)\boldsymbol{Q}(s)$ must have zeros wherever $\boldsymbol{G}(s)$ has unstable poles and all RHP zero-pole cancellations in $[I - \boldsymbol{G}(s)\boldsymbol{Q}(s)]\boldsymbol{G}(s)$ are removed. Otherwise, $[\boldsymbol{I} - \boldsymbol{G}(s)\boldsymbol{Q}(s)]\boldsymbol{G}(s)$ will have unstable poles, which contradicts the assumption.

Now, assume that $\boldsymbol{I} - \boldsymbol{G}(s)\boldsymbol{Q}(s)$ has zeros wherever $\boldsymbol{G}(s)$ has unstable poles and all RHP zero-pole cancellations in $[I - \boldsymbol{G}(s)\boldsymbol{Q}(s)]\boldsymbol{G}(s)$ are removed. As the only unstable poles of $[\boldsymbol{I} - \boldsymbol{G}(s)\boldsymbol{Q}(s)]\boldsymbol{G}(s)$ are those of $\boldsymbol{G}(s)$, the above assumptions imply that all unstable poles in $[\boldsymbol{I} - \boldsymbol{G}(s)\boldsymbol{Q}(s)]\boldsymbol{G}(s)$ are removed by the zeros of $\boldsymbol{I} - \boldsymbol{G}(s)\boldsymbol{Q}(s)$. □

In multivariable systems, there exists such a possibility that $\boldsymbol{I} - \boldsymbol{G}(s)\boldsymbol{Q}(s)$ has zeros wherever $\boldsymbol{G}(s)$ has unstable poles, but there are irremovable RHP zero-pole cancellations in $[\boldsymbol{I} - \boldsymbol{G}(s)\boldsymbol{Q}(s)]\boldsymbol{G}(s)$. Only when there is no irremovable RHP zero-pole cancellation in $[\boldsymbol{I} - \boldsymbol{G}(s)\boldsymbol{Q}(s)]\boldsymbol{G}(s)$, can the stability of $[\boldsymbol{I} - \boldsymbol{G}(s)\boldsymbol{Q}(s)]\boldsymbol{G}(s)$ be guaranteed.

Example 13.1.1. *This example illustrates the irremovable RHP zero-pole cancellation in $[\boldsymbol{I} - \boldsymbol{G}(s)\boldsymbol{Q}(s)]\boldsymbol{G}(s)$. The plant is described by the following transfer function matrix:*

$$\boldsymbol{G}(s) = \begin{bmatrix} \frac{1}{s+3} & \frac{1}{s-2} \\ \frac{2}{s+3} & \frac{s-1}{s-2} \end{bmatrix}.$$

It has two poles at $s = -3$ and $s = 2$, and one zero at $s = 3$. Assume that the controller is

$$\boldsymbol{Q}(s) = \begin{bmatrix} \frac{-(s-1)}{s+1} & \frac{22s+1}{(s+1)^2} \\ \frac{2(s-2)}{(s+3)(s+1)} & \frac{-(s-2)(22s+1)}{(s+3)(s+1)^2} \end{bmatrix}.$$

$\boldsymbol{Q}(s)$ is stable. Since

$$\boldsymbol{I} - \boldsymbol{G}(s)\boldsymbol{Q}(s) = \begin{bmatrix} \frac{s(s+5)}{(s+3)(s+1)} & 0 \\ 0 & \frac{s(s+29)(s-2)}{(s+3)(s+1)^2} \end{bmatrix},$$

$\boldsymbol{I} - \boldsymbol{G}(s)\boldsymbol{Q}(s)$ has zeros wherever $\boldsymbol{G}(s)$ has unstable poles. Nevertheless,

$$[\boldsymbol{I} - \boldsymbol{G}(s)\boldsymbol{Q}(s)]\boldsymbol{G}(s) = \begin{bmatrix} \frac{s(s+5)}{(s+3)^2(s+1)} & \frac{s(s+5)}{(s-2)(s+3)(s+1)} \\ \frac{2s(s+29)(s-2)}{(s+3)^2(s+1)^2} & \frac{s(s+29)(s-1)}{(s+3)(s+1)^2} \end{bmatrix}$$

is not stable. It has a RHP pole at $s = 2$ and a RHP zero at $s = 2$. The RHP zero-pole cancellation cannot be removed.

Let the multiplicity of the unstable pole p_j ($\text{Re}(p_j) \geq 0, j = 1, 2, ..., r_p$) be l_j, and let l_{ij} be the largest multiplicity of p_j in the ith row of $\boldsymbol{G}(s)$.

Theorem 13.1.2. *All controllers that make the unity feedback control system internally stable can be parameterized as*

$$\boldsymbol{C}(s) = \boldsymbol{Q}(s)[\boldsymbol{I} - \boldsymbol{G}(s)\boldsymbol{Q}(s)]^{-1},$$

where $\boldsymbol{Q}(s)$ is any stable proper transfer function matrix that satisfies

$$\lim_{s \to p_j} \frac{d^k}{ds^k} \det[\boldsymbol{I} - \boldsymbol{G}(s)\boldsymbol{Q}(s)] = 0, j = 1, 2, ..., r_p; k = 0, 1, ..., l_j - 1,$$

and all RHP zero-pole cancellations in $[I - \boldsymbol{G}(s)\boldsymbol{Q}(s)]\boldsymbol{G}(s)$ are removed.

Proof. To guarantee the internal stability of the closed-loop system, first, $\boldsymbol{Q}(s)$ should be stable. This implies that $\boldsymbol{Q}(s)$ should be proper.

Next, $[\boldsymbol{I} - \boldsymbol{G}(s)\boldsymbol{Q}(s)]\boldsymbol{G}(s)$ should be stable. This implies that $\boldsymbol{I} - \boldsymbol{G}(s)\boldsymbol{Q}(s)$ has to cancel all closed RHP poles of $\boldsymbol{G}(s)$. To achieve this, $\boldsymbol{Q}(s)$ must satisfy that

$$\lim_{s \to p_j} \frac{d^k}{ds^k} \det[\boldsymbol{I} - \boldsymbol{G}(s)\boldsymbol{Q}(s)] = 0, j = 1, 2, ..., r_p; k = 0, 1, ..., l_j - 1.$$

This condition cannot guarantee the stability of $[\boldsymbol{I} - \boldsymbol{G}(s)\boldsymbol{Q}(s)]\boldsymbol{G}(s)$ unless all RHP zero-pole cancellations in $[I - \boldsymbol{G}(s)\boldsymbol{Q}(s)]\boldsymbol{G}(s)$ are removed. □

Corollary 13.1.3. *Assume that $\boldsymbol{G}(s)$ is a stable plant. All controllers that make the unity feedback control system internally stable can be parameterized as*

$$\boldsymbol{C}(s) = \boldsymbol{Q}(s)[\boldsymbol{I} - \boldsymbol{G}(s)\boldsymbol{Q}(s)]^{-1},$$

where $\boldsymbol{Q}(s)$ is any stable proper transfer function matrix.

It is noted that, for stable plants, the new paramerization is identical to Youla parameterization.

When the system performance is considered, it is always desirable that the system should possess the asymptotic tracking property. If the system inputs are steps, the closed-loop transfer function matrix $\boldsymbol{T}(s) = \boldsymbol{G}(s)\boldsymbol{Q}(s)$ should satisfy the following condition for the asymptotic tracking property:

$$\lim_{s \to 0}[\boldsymbol{I} - \boldsymbol{T}(s)] = \boldsymbol{0}. \tag{13.1.2}$$

Theorem 13.1.4. *All controllers that make the unity feedback control system internally stable and possess the asymptotic tracking property for step inputs can be parameterized as*

$$\boldsymbol{C}(s) = \boldsymbol{Q}(s)[\boldsymbol{I} - \boldsymbol{G}(s)\boldsymbol{Q}(s)]^{-1},$$

where

$$\boldsymbol{Q}(s) = \boldsymbol{G}^{-1}(0)[\boldsymbol{I} + s\boldsymbol{Q}_1(s)],$$

$\boldsymbol{Q}_1(s)$ is any stable transfer function matrix that makes $\boldsymbol{Q}(s)$ proper and satisfies

$$\lim_{s \to p_j} \frac{d^k}{ds^k} \det[\boldsymbol{I} - \boldsymbol{G}(s)\boldsymbol{G}^{-1}(0) - s\boldsymbol{G}(s)\boldsymbol{G}^{-1}(0)\boldsymbol{Q}_1(s)] = 0,$$
$$j = 1, 2, ..., r_p; k = 0, 1, ..., l_j - 1,$$

and all RHP zero-pole cancellations in $[I - \boldsymbol{G}(s)\boldsymbol{Q}(s)]\boldsymbol{G}(s)$ are removed.

Proof. If

$$\lim_{s \to 0}[\boldsymbol{I} - \boldsymbol{G}(s)\boldsymbol{Q}(s)] = \boldsymbol{0},$$

or equivalently,

$$\boldsymbol{Q}(0) = \boldsymbol{G}^{-1}(0),$$

the closed-loop system possesses the asymptotic tracking property. All transfer function matrices that satisfy the condition can be written as

$$\boldsymbol{Q}(s) = \boldsymbol{G}^{-1}(0)[\boldsymbol{I} + s\boldsymbol{Q}_1(s)].$$

To guarantee the internal stability of the closed-loop system, $\boldsymbol{Q}(s)$ should be stable. This implies that $\boldsymbol{Q}(s)$ is proper. Furthermore, $\boldsymbol{I}-\boldsymbol{G}(s)\boldsymbol{Q}(s)$ has to cancel all closed RHP poles of $\boldsymbol{G}(s)$. To achieve this, $\boldsymbol{Q}(s)$ must satisfy that

$$\lim_{s\to p_j}\frac{d^k}{ds^k}\det[\boldsymbol{I}-\boldsymbol{G}(s)\boldsymbol{G}^{-1}(0)-s\boldsymbol{G}(s)\boldsymbol{G}^{-1}(0)\boldsymbol{Q}_1(s)]=0,$$
$$j=1,2,...,r_p;k=0,1,...,l_j-1.$$

The condition cannot guarantee the stability of $[\boldsymbol{I}-\boldsymbol{G}(s)\boldsymbol{Q}(s)]\boldsymbol{G}(s)$ unless all RHP zero-pole cancellations in $[I-\boldsymbol{G}(s)\boldsymbol{Q}(s)]\boldsymbol{G}(s)$ are removed. □

It is easy to obtain the closed-loop transfer function matrix:

$$\boldsymbol{T}(s)=\boldsymbol{G}(s)\boldsymbol{Q}(s)=\boldsymbol{G}(s)\boldsymbol{G}^{-1}(0)[\boldsymbol{I}+s\boldsymbol{Q}_1(s)]. \tag{13.1.3}$$

In practice, it may be required that the closed-loop response is decoupled, that is, the closed-loop transfer function matrix $\boldsymbol{T}(s)$ is diagonal. Let the ith element of $\boldsymbol{T}(s)$ be $T_i(s)$.

Corollary 13.1.5. *Assume that the closed-loop response is decoupled. All controllers that make the unity feedback control system internally stable and possess the asymptotic tracking property for step inputs can be parameterized as*

$$\boldsymbol{C}(s)=\boldsymbol{Q}(s)[\boldsymbol{I}-\boldsymbol{G}(s)\boldsymbol{Q}(s)]^{-1},$$

where

$$\boldsymbol{Q}(s)=\boldsymbol{G}^{-1}(0)[\boldsymbol{I}+s\boldsymbol{Q}_1(s)],$$

$\boldsymbol{Q}_1(s)$ *is any stable transfer function matrix that makes* $\boldsymbol{Q}(s)$ *proper and* $\boldsymbol{T}(s)$ *diagonal, and satisfies*

$$\lim_{s\to p_j}\frac{d^k}{ds^k}[1-T_i(s)]=0,$$
$$i=1,2,...,n;j=1,2,...,r_p;k=0,1,...,l_{ij}-1,$$

and all RHP zero-pole cancellations in $[I-\boldsymbol{G}(s)\boldsymbol{Q}(s)]\boldsymbol{G}(s)$ *are removed.*

The well-known Youla parameterization is not used here because of several reasons:

1. It cannot be directly used for plants with time delay.
2. It needs a coprime factorization, which cannot be obtained with analytical methods.
3. It does not directly relate to the IMC controller $\boldsymbol{Q}(s)$.
4. It cannot be directly used for decoupling control.

13.2 Diagonal Factorization for H_2 Control

In this section, a diagonal factorization will be defined for H_2 decoupling control. The factorization is similar to, but not exactly the same as that for the quasi-H_∞ decoupling control. The introduction follows the original developing procedure of the factorization, which explains why such a factorization is constructed.

Assume that the plant is expressed as

$$\boldsymbol{G}(s) = \begin{bmatrix} G_{11}(s)e^{-\theta_{11}s} & \cdots & G_{1n}(s)e^{-\theta_{1n}s} \\ \vdots & \ddots & \vdots \\ G_{n1}(s)e^{-\theta_{n1}s} & \cdots & G_{nn}(s)e^{-\theta_{nn}s} \end{bmatrix}, \tag{13.2.1}$$

where $G_{ij}(s)(i, j = 1, 2, ..., n)$ are scalar rational transfer functions and $\theta_{ij} \geq 0$ are time delays.

Recall that, in the decentralized control, only those diagonal elements of the plant are treated; the non-diagonal elements are regarded as uncertainty. Along the same lines, it seems that the plant can be factorized into the following form:

$$\boldsymbol{G}(s) = \boldsymbol{G_D}(s)\boldsymbol{G_O}(s),$$

where

$$\boldsymbol{G_D}(s) = \begin{bmatrix} e^{-\theta_{11}s} & \cdots & 0 \\ \vdots & \ddots & \vdots \\ 0 & \cdots & e^{-\theta_{nn}s} \end{bmatrix}, \tag{13.2.2}$$

$$\boldsymbol{G_O}(s) = \begin{bmatrix} G_{11}(s) & \cdots & G_{1n}(s)e^{-(\theta_{1n}-\theta_{11})s} \\ \vdots & \ddots & \vdots \\ G_{n1}(s)e^{-(\theta_{n1}-\theta_{nn})s} & \cdots & G_{nn}(s) \end{bmatrix} \tag{13.2.3}$$

Unfortunately, such a factorization is not feasible for controller design, because there may be predictions in the resulting controller.

To find a feasible factorization, let us review the SISO design first. Given a plant $G(s)$, the ideal controller is

$$Q_{opt}(s) = G^{-1}(s). \tag{13.2.4}$$

In this case, the closed-loop transfer function is 1. The performance is evidently optimal. However, such a controller is not physically realizable when $G(s)$ has a time delay. An alternative is to factorize the plant into two parts: $G(s) = G_D(s)G_O(s)$, where $G_D(s)$ is the time delay part and $G_O(s)$ is the delay-free part. If $G_O(s)$ is NMP, a further factorizing should be made: $G_O(s) =$

$G_N(s)G_{MP}(s)$, where $G_N(s)$ is the all-pass part and $G_{MP}(s)$ is the MP part. The optimal controller can be taken as

$$Q_{opt}(s) = G^{-1}(s)G_D(s)G_N(s) = G_{MP}^{-1}(s). \tag{13.2.5}$$

This controller is proved to be the optimal realizable inverse of $G^{-1}(s)$. When $Q_{opt}(s)$ is improper, a filter will be introduced to make it proper.

Now consider the MIMO case. The factorization of the time delay part is defined first. Let the inverse of the plant be

$$\boldsymbol{G}^{-1}(s) = \begin{bmatrix} G^{11}(s)e^{-\theta^{11}s} & \cdots & G^{1n}(s)e^{-\theta^{1n}s} \\ \cdots & \ddots & \cdots \\ G^{n1}(s)e^{-\theta^{n1}s} & \cdots & G^{nn}(s)e^{-\theta^{nn}s} \end{bmatrix}, \tag{13.2.6}$$

where $G^{ji}(s)e^{-\theta^{ji}s}(j, i = 1, 2, ..., n)$ are the elements of $\boldsymbol{G}^{-1}(s)$; θ^{ji} is the maximum time delay that can be separated from each element.

It is conjectured that the H_2 optimal decoupling controller can be designed in a similar way to the SISO controller. To carry out such a design, the first step is to factorize the plant $\boldsymbol{G}(s)$ into two parts: $\boldsymbol{G}(s) = \boldsymbol{G_D}(s)\boldsymbol{G_O}(s)$, where $\boldsymbol{G_D}(s)$ is the time delay part. $\boldsymbol{G_D}(s)$ is diagonal and should be chosen so that

1. It counteracts the predictions in $\boldsymbol{G}^{-1}(s)$, that is, $\boldsymbol{G_O}^{-1}(s) = \boldsymbol{G}^{-1}(s)\boldsymbol{G_D}(s)$ does not involve predictions.
2. No additional time delays are introduced in $\boldsymbol{G_O}^{-1}(s)$.

Because the elements of $\boldsymbol{G_D}(s)$ are time delays, no RHP zeros and poles are cancelled in forming $\boldsymbol{G_O}(s) = \boldsymbol{G_D}^{-1}(s)\boldsymbol{G}(s)$. If $\boldsymbol{G_O}(s)$ is MP, the optimal controller can be taken as

$$\boldsymbol{Q_{opt}}(s) = \boldsymbol{G_O}^{-1}(s). \tag{13.2.7}$$

Definition 13.2.1. *Let $\theta_{li}(i = 1, 2, ..., n)$ be the largest prediction of the ith column of $\boldsymbol{G}^{-1}(s)$, that is, $\theta_{li} = \max_j \theta^{ji}, j = 1, 2, ..., n$. The H_2 diagonal factorization for time delay is defined as*

$$\boldsymbol{G_D}(s) = \begin{bmatrix} e^{-\theta_{l1}s} & \cdots & 0 \\ \vdots & \ddots & \vdots \\ 0 & \cdots & e^{-\theta_{ln}s} \end{bmatrix}.$$

In particular, for rational plants $\boldsymbol{G_D}(s) = \boldsymbol{I}$.

The definition is the same as that in the quasi-H_∞ decoupling control. This is because any factorization with time delays shorter than the ones in this factorization will not thoroughly counteract the predictions in $\boldsymbol{G}^{-1}(s)$; only an irrealizable controller can be obtained.

It is easy to verify that $\boldsymbol{G_D}^H(j\omega)\boldsymbol{G_D}(j\omega) = \boldsymbol{I}$ for all ω and $\boldsymbol{G_D}(0) = \boldsymbol{I}$.

The property $\boldsymbol{G_D}^H(j\omega)\boldsymbol{G_D}(j\omega) = \boldsymbol{I}$ is the generalization of the concept of all-pass.

Next, the factorization for the RHP zero part is defined. $\boldsymbol{G_O}(s)$ may be NMP. In this case, $\boldsymbol{G_O}(s)$ has to be factorized into two parts: $\boldsymbol{G_O}(s) = \boldsymbol{G_N}(s)\boldsymbol{G_{MP}}(s)$, where $\boldsymbol{G_N}(s)$ is a diagonal matrix satisfying the following requirement:

1. It is all-pass and $\boldsymbol{G_A}(s) = \boldsymbol{I}$.
2. It counteracts the RHP poles in $\boldsymbol{G_O}^{-1}(s)$, so that $\boldsymbol{G_{MP}}^{-1}(s) = \boldsymbol{G_O}^{-1}(s)\boldsymbol{G_N}(s)$ is stable.
3. No additional RHP zeros are introduced in $\boldsymbol{G_{MP}}^{-1}(s)$.

The $\boldsymbol{G_{MP}}(s)$ obtained in such a way is the MP part of $\boldsymbol{G_O}(s)$. The optimal controller can be taken as

$$\boldsymbol{Q_{opt}}(s) = \boldsymbol{G_{MP}}^{-1}(s). \tag{13.2.8}$$

Assume that $z_j(\mathrm{Re}(z_j) > 0, j = 1, 2, ..., r_z)$ are the unstable poles of $\boldsymbol{G_O}^{-1}(s)$.

Definition 13.2.2. *Let $k_{ij}(i = 1, 2, ..., n)$ be the largest multiplicity of the unstable pole $z_j(j = 1, 2, ..., r_z)$ in the ith column of $\boldsymbol{G_O}^{-1}(s)$. The H_2 diagonal factorization for closed RHP zeros is*

$$\boldsymbol{G_N}(s) = \begin{bmatrix} \prod_{j=1}^{r_z} \left(\frac{-s+z_j}{s+\bar{z}_j}\right)^{k_{1j}} & \cdots & 0 \\ \vdots & \ddots & \vdots \\ 0 & \cdots & \prod_{j=1}^{r_z} \left(\frac{-s+z_j}{s+\bar{z}_j}\right)^{k_{nj}} \end{bmatrix}.$$

In particular, for MP plants $\boldsymbol{G_N}(s) = \boldsymbol{I}$.

It can be verified that $\boldsymbol{G_N}^H(j\omega)\boldsymbol{G_N}(j\omega) = \boldsymbol{I}$ for all ω and $\boldsymbol{G_N}(0) = \boldsymbol{I}$. Finally, the diagonal all-pass factorization is defined. Let

$$\boldsymbol{G_A}(s) = \boldsymbol{G_D}(s)\boldsymbol{G_N}(s).$$

The diagonal all-pass factorization of $\boldsymbol{G}(s)$ can be expressed as

$$\boldsymbol{G}(s) = \boldsymbol{G_A}(s)\boldsymbol{G_{MP}}(s). \tag{13.2.9}$$

The factorization is unique and has the following features:

1. $\boldsymbol{G_A}(s)$ is diagonal.
2. $\boldsymbol{G_A}(s)$ is stable and all-pass.
3. $\boldsymbol{G_{MP}}(s)$ is MP.

Although the factorization defined in this section is similar to the inner-outer factorization, it has two features different from the inner-outer factorization:

1. The new factorization has a diagonal all-pass part.
2. It is defined for both stable and unstable plants.

13.3 H_2 Optimal Decoupling Control

The subject of this section is to derive the H_2 optimal decoupling controller with the controller parameterization in Section 13.1 and the diagonal factorization in Section 13.2 (Figure 13.3.1). It will be shown that the conjecture in the last section about the optimal controller is correct. The diagonal factorization results in the optimal solution indeed.

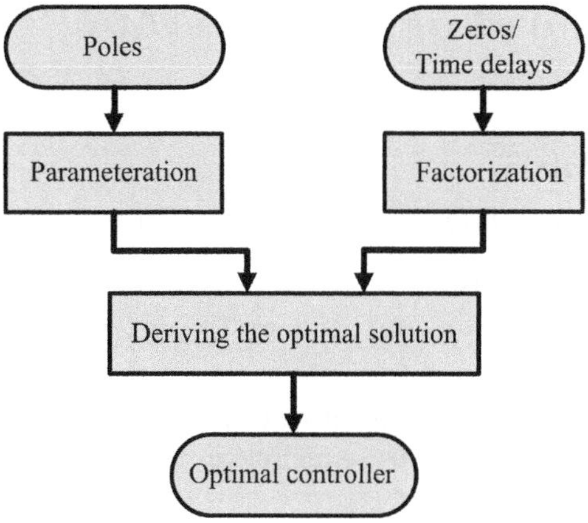

FIGURE 13.3.1
Design procedure for the H_2 decoupling controller.

As discussed in Section 10.4, the H_2 optimal control problem can be expressed as

$$\min \|\boldsymbol{S}(s)\boldsymbol{W}(s)\|_2, \tag{13.3.1}$$

where $\boldsymbol{W}(s) = \boldsymbol{I}/s$ is the weighting function and

$$\boldsymbol{S}(s) = \boldsymbol{I} - \boldsymbol{G}(s)\boldsymbol{Q}(s).$$

For decoupling control, $\boldsymbol{S}(s)$ should be diagonal. Since a fixed weighting function is adopted, the designer is not required to select a weighting function.

The design problem here can be considered as an optimization problem over the class of all stabilizing controllers with asymptotic tracking property. To carry out the design, some preliminaries are needed.

Let H_2 be the set of all strictly proper stable functions, $\mathrm{H}_2^\perp$ be the set of all strictly proper transfer function matrices without poles in the open LHP. Then $\mathrm{L}_2 := \mathrm{H}_2 + \mathrm{H}_2^\perp$ denotes the set of all strictly proper transfer function matrices without poles on the imaginary axis. Given a $\boldsymbol{T}(s)$ in L_2, it can be uniquely expressed as

$$\boldsymbol{T}(s) = \boldsymbol{T_1}(s) + \boldsymbol{T_2}(s), \tag{13.3.2}$$

where $\boldsymbol{T_1}(s) \in \mathrm{H}_2$ and $\boldsymbol{T_2}(s) \in \mathrm{H}_2^\perp$.

Let the superscript * denote the conjugate transpose of a system: $\boldsymbol{T}^*(s) = \boldsymbol{T}^T(-s)$. Readers should not confuse it with the complex conjugate transpose, which is defined for a complex matrix: $\boldsymbol{T}^H(j\omega) = \bar{\boldsymbol{T}}^T(j\omega)$. For complex matrices, $\boldsymbol{T}^*(j\omega) = \boldsymbol{T}^H(j\omega)$.

Lemma 13.3.1. *If* $\boldsymbol{T_1}(s) \in H_2$ *and* $\boldsymbol{T_2}(s) \in H_2^\perp$, *then*

$$\|\boldsymbol{T_1}(s) + \boldsymbol{T_2}(s)\|_2^2 = \|\boldsymbol{T_1}(s)\|_2^2 + \|\boldsymbol{T_2}(s)\|_2^2.$$

Proof.

$$\begin{aligned}&\|\boldsymbol{T_1}(s) + \boldsymbol{T_2}(s)\|_2^2\\ &= \frac{1}{2\pi}\int \mathrm{Trace}\{[\boldsymbol{T_1}(j\omega) + \boldsymbol{T_2}(j\omega)]^H[\boldsymbol{T_1}(j\omega) + \boldsymbol{T_2}(j\omega)]\}d\omega\\ &= \|\boldsymbol{T_1}(s)\|_2^2 + \|\boldsymbol{T_2}(s)\|_2^2 +\\ &\quad \frac{1}{2\pi}\int \mathrm{Trace}[\boldsymbol{T_1}^H(j\omega)\boldsymbol{T_2}(j\omega) + \boldsymbol{T_2}^H(j\omega)\boldsymbol{T_1}(j\omega)]d\omega.\end{aligned}$$

Consider the last term. Convert it into a contour integral by closing the imaginary axis with an infinite-radius semicircle in the LHP:

$$\begin{aligned}&\frac{1}{2\pi}\int \mathrm{Trace}[\boldsymbol{T_1}^H(j\omega)\boldsymbol{T_2}(j\omega) + \boldsymbol{T_2}^H(j\omega)\boldsymbol{T_1}(j\omega)]d\omega\\ &= \frac{1}{2\pi j}\oint \mathrm{Trace}[\boldsymbol{T_1}^*(s)\boldsymbol{T_2}(s) + \boldsymbol{T_2}^*(s)\boldsymbol{T_1}(s)]ds.\end{aligned}$$

According to Lemma 3.1.3, if a function has no poles in a bounded open set, then its integral on a closed contour in the set equals zero. Therefore, the right-hand side of the above equation equals zero. □

Now consider the design problem. Using the controller parameterization developed in Corollary 13.1.5, we have

$$\begin{aligned}&\|\boldsymbol{S}(s)\boldsymbol{W}(s)\|_2^2\\ &= \left\|s^{-1}[\boldsymbol{I} - \boldsymbol{G}(s)\boldsymbol{Q}(s)]\right\|_2^2\\ &= \left\|s^{-1}[\boldsymbol{I} - \boldsymbol{G}(s)\boldsymbol{G}^{-1}(0) - s\boldsymbol{G}(s)\boldsymbol{G}^{-1}(0)\boldsymbol{Q_1}(s)]\right\|_2^2.\end{aligned} \tag{13.3.3}$$

Theorem 13.3.2. *Assume that the plant with time delay can be uniquely factorized into two parts according to Definition 13.2.1 and Definition 13.2.2:*

$$\boldsymbol{G}(s) = \boldsymbol{G_A}(s)\boldsymbol{G_{MP}}(s),$$

where $\boldsymbol{G_A}(s) = \boldsymbol{G_D}(s)\boldsymbol{G_N}$. *Then the optimal solution to the H_2 decoupling control problem is*

$$\boldsymbol{Q_{opt}}(s) = \boldsymbol{G_{MP}}^{-1}(s).$$

Proof. An important property of the all-pass function is that it does not affect the value of 2-norm, that is,

$$\|\boldsymbol{G_A}(s)\boldsymbol{G_1}(s)\|_2 = \|\boldsymbol{G_1}(s)\|_2,$$

where $\boldsymbol{G_1}(s)$ is a transfer function matrix without poles on the imaginary axis. With this property and the diagonal factorization given in the last section, the H_2 performance index can be written as

$$\begin{aligned}&\|\boldsymbol{S}(s)\boldsymbol{W}(s)\|_2^2\\ &= \left\|s^{-1}[\boldsymbol{G_A}^{-1}(s) - \boldsymbol{I}] + s^{-1}\{\boldsymbol{I} - \boldsymbol{G_{MP}}(s)\boldsymbol{G}^{-1}(0)[\boldsymbol{I} + s\boldsymbol{Q_1}(s)]\}\right\|_2^2.\end{aligned}$$

Since $\boldsymbol{G_A}(0) = \boldsymbol{I}$ and $\boldsymbol{G_{MP}}(0)\boldsymbol{G}^{-1}(0) = \boldsymbol{I}$, s must be a factor of $\boldsymbol{G_A}^{-1}(s) - \boldsymbol{I}$ and $\boldsymbol{I} - \boldsymbol{G_{MP}}(s)\boldsymbol{G}^{-1}(0)[\boldsymbol{I} + s\boldsymbol{Q_1}(s)]$. $s^{-1}[\boldsymbol{G_A}^{-1}(s) - \boldsymbol{I}]$ is strictly proper. $s^{-1}\{\boldsymbol{I} - \boldsymbol{G_{MP}}(s)\boldsymbol{G}^{-1}(0)[\boldsymbol{I} + s\boldsymbol{Q_1}(s)]\}$ is also strictly proper if $\boldsymbol{Q}(s) = \boldsymbol{G}^{-1}(0)[\boldsymbol{I} + s\boldsymbol{Q_1}(s)]$ is proper.

On the other hand, in view of the definition of $\boldsymbol{G_A}(s)$, $s^{-1}[\boldsymbol{G_A}^{-1}(s) - \boldsymbol{I}]$ is analytic in the open LHP. $s^{-1}\{\boldsymbol{I} - \boldsymbol{G_{MP}}(s)\boldsymbol{G}^{-1}(0)[\boldsymbol{I} + s\boldsymbol{Q_1}(s)]\}$ is stable. To see this, let us consider the following equality:

$$\begin{aligned}&\boldsymbol{G_A}(s)\{\boldsymbol{I} - \boldsymbol{G_{MP}}(s)\boldsymbol{G}^{-1}(0)[\boldsymbol{I} + s\boldsymbol{Q_1}(s)]\}\\ =\ &[\boldsymbol{G_A}(s) - \boldsymbol{I}] + \{\boldsymbol{I} - \boldsymbol{G}(s)\boldsymbol{G}^{-1}(0)[\boldsymbol{I} + s\boldsymbol{Q_1}(s)]\}.\end{aligned}$$

As $\boldsymbol{G_A}(s)$ is stable, the first term in the right-hand side is stable. By Corollary 13.1.5, the second term is stable, too.

Applying Lemma 13.3.1, we have

$$\begin{aligned}&\|\boldsymbol{S}(s)\boldsymbol{W}(s)\|_2^2\\ &= \left\|s^{-1}[\boldsymbol{G_A}^{-1}(s) - \boldsymbol{I}]\right\|_2^2 + \left\|s^{-1}\{\boldsymbol{I} - \boldsymbol{G_{MP}}(s)\boldsymbol{G}^{-1}(0)[\boldsymbol{I} + s\boldsymbol{Q_1}(s)]\}\right\|_2^2.\end{aligned}$$

Minimizing the right-hand side of the equation yields

$$\boldsymbol{Q_{1opt}}(s) = s^{-1}\boldsymbol{G}(0)\boldsymbol{G_{MP}}^{-1}(s)[\boldsymbol{I} - \boldsymbol{G_{MP}}(s)\boldsymbol{G}^{-1}(0)].$$

Hence, the optimal solution is

$$\boldsymbol{Q_{opt}}(s) = \boldsymbol{G_{MP}}^{-1}(s).$$

□

It should be emphasized that the proof procedure for Theorem 13.3.2 is general enough to allow general weighting functions, although only the prespecified weighting function is considered in this section.

It can be seen from the deriving procedure that it is the parameterization proposed in Section 13.1 and the factorization defined in Section 13.2 that make it possible to obtain an analytical solution. With the optimal analytical solution, the optimal performance is readily obtained as follows:

$$\min \|\boldsymbol{S}(s)\boldsymbol{W}(s)\|_2 = \left\|s^{-1}[\boldsymbol{G_A}^{-1}(s) - \boldsymbol{I}]\right\|_2. \tag{13.3.4}$$

No matter what method is used, this is the best decoupling performance that can be reached, provided the performance index is specified as minimizing the 2-norm of the weighted sensitivity function.

Sometimes, the decoupled response can also reach the optimal performance in a general sense, rather than the diagonal optimal performance. When the time delays in each row of $\boldsymbol{G}(s)$ are the same, and all RHP zeros have the same multiplicities in each row of $\boldsymbol{G}(s)$, the diagonal optimal solution is identical to the optimal solution. Evidently, for MP plants, the results of the optimal control and the decoupling optimal control are the same.

13.4 Analysis for H_2 Decoupling Control Systems

The design procedure of a MIMO H_2 controller is similar to that of a SISO H_2 controller. After the optimal controller is obtained, the next step is to introduce a filter: $\boldsymbol{Q}(s) = \boldsymbol{Q_{opt}}(s)\boldsymbol{J}(s)$. The filter has two main functions:

1. The optimal controller $\boldsymbol{Q_{opt}}(s)$ is usually improper. The filter is introduced to make it proper.
2. The filter is used to tune the shape of the closed-loop response and to satisfy the performance and robustness requirements, which are achieved with weights in many other methods.

The filter should satisfy the following requirements:

1. The closed-loop system is internally stable.
2. The controller $\boldsymbol{Q}(s)$ is proper.
3. The asymptotic tracking can be achieved.

Since the decoupled closed-loop response is required, $\boldsymbol{J}(s)$ should be chosen as a diagonal matrix:

$$\boldsymbol{J}(s) = \begin{bmatrix} J_1(s) & \cdots & 0 \\ \vdots & \ddots & \vdots \\ 0 & \cdots & J_n(s) \end{bmatrix}, \tag{13.4.1}$$

with

$$J_i(s) = \frac{N_{xi}(s)}{(\lambda_i s + 1)^{n_i}}, i = 1, 2, ..., n. \tag{13.4.2}$$

Here $\lambda_i (i = 1, 2, ..., n)$ are performance degrees. $N_{xi}(s)$ are polynomials with all roots in the LHP.

It is not difficult to satisfy the first requirement on the filter. Let the ith element of $\boldsymbol{G_A}(s)$ be $G_{Ai}(s)(i = 1, 2, ..., n)$. By Corollary 13.1.5, the closed-loop system is internally stable if

$$\lim_{s \to p_j} \frac{d^k}{ds^k}[1 - G_{Ai}(s)J_i(s)] = 0,$$
$$i = 1, 2, ..., n; j = 1, 2, ..., r_p; k = 0, 1, ..., l_{ij} - 1. \tag{13.4.3}$$

Here $\deg\{N_{xi}(s)\} = \sum_{j=1}^{r_p} l_{ij} (i = 1, 2, ..., n)$.

Assume that the smallest relative degree of all elements in the ith column of $\boldsymbol{Q_{opt}}(s)$ is α_i. The second requirement can be satisfied by choosing $n_i = \deg\{N_{xi}(s)\} - \alpha_i$ for improper columns and $n_i = \deg\{N_{xi}(s)\} + 1$ for the others.

To track the input asymptotically, the filter should satisfy

$$\boldsymbol{J}(0) = \boldsymbol{I}. \tag{13.4.4}$$

This implies that the third requirement can be satisfied by choosing $N_{xi}(0) = 1$.

Each element of $\boldsymbol{J}(s)$ has an adjustable performance degree λ_i, which should be determined by design requirements. As the closed-loop response is decoupled, each channel can be independently tuned. The ith element of the closed-loop transfer function matrix is

$$T_i(s) = \frac{N_{xi}(s)}{(\lambda_i s + 1)^{n_i}} \prod_{j=1}^{r_z} \left(\frac{-s + z_j}{s + \bar{z}_j} \right)^{k_{ij}} e^{-\theta_{ii} s}, i = 1, 2, ..., n. \tag{13.4.5}$$

Since the performance degree is related to the nominal closed-loop response monotonically, the system can be tuned conveniently for quantitative responses. When there exists uncertainty, every channel is mainly affected by the performance degree of this channel. The tuning procedure is similar.

For stable plants, the controller can be implemented in the IMC structure. An important advantage of such an implementation is that the decoupling property can be thoroughly preserved even when a rational approximation is used for $\boldsymbol{Q}(s)$. For unstable plants, the controller must be implemented in the unity feedback loop.

Control systems with unstable plants usually exhibit excessive overshoots. This problem can be well solved by employing a 2 DOF structure (Figure

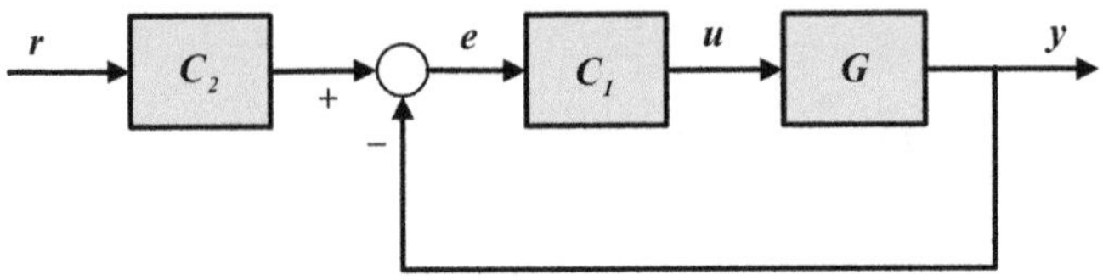

FIGURE 13.4.1
A 2 DOF MIMO system.

13.4.1), which can isolate the disturbance response from the reference response and thus make a better response possible.

It is easy to design 2 DOF controllers in the framework of this section. The controller of the disturbance loop is exactly the controller of the unity feedback loop:

$$\boldsymbol{C_1}(s) = \boldsymbol{Q}(s)[\boldsymbol{I} - \boldsymbol{G}(s)\boldsymbol{Q}(s)]^{-1} \tag{13.4.6}$$

with $\boldsymbol{Q}(s) = \boldsymbol{Q_{opt}}(s)\boldsymbol{J_1}(s)$. The closed-loop transfer function matrix of the corresponding unity feedback loop is

$$\boldsymbol{T}(s) = \boldsymbol{G}(s)\boldsymbol{Q}(s) = \boldsymbol{G_D}(s)\boldsymbol{G_N}(s)\boldsymbol{J_1}(s). \tag{13.4.7}$$

Regard $\boldsymbol{T}(s)$ as a new plant and $\boldsymbol{C_2}(s)$ as the IMC controller in the next step. The following optimal controller for the reference loop can be obtained:

$$\boldsymbol{C_{2opt}}(s) = \boldsymbol{J_1}^{-1}(s)\boldsymbol{G_N}^{-1}(s). \tag{13.4.8}$$

Introduce a diagonal filter $\boldsymbol{J_2}(s)$ to the optimal controller to obtain a physically realizable controller:

$$\boldsymbol{C_2}(s) = \boldsymbol{C_{2opt}}(s)\boldsymbol{J_2}(s). \tag{13.4.9}$$

$\boldsymbol{J_2}(s)$ has a similar structure to $\boldsymbol{J_1}(s)$. It is noticed that the response from $\boldsymbol{r}(s)$ to $\boldsymbol{y}(s)$ can be tuned only by $\boldsymbol{J_2}(s)$.

13.5 Design Examples for H_2 Decoupling Control

The design procedure for H_2 decoupling controllers can be formulated as follows:

1. If the plant does not contain any time delay (that is, $\boldsymbol{G_D}(s) = \boldsymbol{I}$), turn to 3.

2. If the plant contains a time delay, take the rational part $\boldsymbol{G_O}(s)$ as the nominal plant.
3. If $\boldsymbol{G_O}(s)$ has no zeros in the RHP (that is, $\boldsymbol{G_N}(s) = \boldsymbol{I}$), take its inverse as $\boldsymbol{Q_{opt}}(s)$ and turn to 5.
4. If $\boldsymbol{G_O}(s)$ has zeros in the RHP, construct an all-pass transfer function matrix with the factor that contains these zeros (that is, $\boldsymbol{G_N}(s)$) and then remove the all-pass transfer function matrix. Take the inverse of the remainder as $\boldsymbol{Q_{opt}}(s)$.
5. Introduce a filter $\boldsymbol{J}(s)$ to $\boldsymbol{Q_{opt}}(s)$, compute $\boldsymbol{C}(s)$ and remove the RHP zero-pole cancellations in $\boldsymbol{C}(s)$.

In this section, two examples are provided to help readers understand the design procedure of H_2 decoupling control. In the first example, a constructed unstable plant is used, since it is not easy to find an appropriate real unstable plant for the design problem illustrated. The plant in the second example is a real one.

Example 13.5.1. *The plant is described by the following transfer function matrix:*

$$\boxed{\boldsymbol{G}(s) = \frac{1}{(s+3)(s-1)} \begin{bmatrix} s-2 & 2(s-2) \\ 1 & s-1 \end{bmatrix}},$$

which is NMP and unstable. The plant has two two-multiplicity poles at $s = -3$ and $s = 1$ respectively, and two zeros at $s = 2$ and $s = 3$ respectively.

According to Definition 13.2.1, $\boldsymbol{G}(s)$ is factorized into

$$\boldsymbol{G}(s) = \boldsymbol{G_D}(s)\boldsymbol{G_O}(s).$$

Since $\boldsymbol{G}(s)$ is rational, $\boldsymbol{G_D}(s) = \boldsymbol{I}$. The inverse of $\boldsymbol{G_O}(s)$ is

$$\boldsymbol{G_O}^{-1}(s) = \frac{\begin{bmatrix} s-1 & -2(s-2) \\ -1 & s-2 \end{bmatrix}}{\frac{(s-2)(s-3)}{(s-1)(s+3)}}.$$

It has unstable poles. Hence, $\boldsymbol{G_O}(s)$ has to be factorized as follows:

$$\boldsymbol{G_O}(s) = \boldsymbol{G_N}(s)\boldsymbol{G_{MP}}(s).$$

In view of Definition 13.2.2, we have

$$\begin{aligned} \boldsymbol{G_N}(s) &= \begin{bmatrix} \frac{-(s-2)}{s+2} & 0 \\ 0 & 1 \end{bmatrix} \frac{-(s-3)}{s+3}, \\ \boldsymbol{G_{MP}}(s) &= \frac{-1}{(s-3)(s-1)} \begin{bmatrix} -(s+2) & -2(s+2) \\ 1 & s-1 \end{bmatrix}. \end{aligned}$$

By Theorem 13.3.2, the optimal controller is

$$Q_{opt}(s) = G_{MP}{}^{-1}(s) = \frac{\begin{bmatrix} s-1 & 2(s+2) \\ -1 & -(s+2) \end{bmatrix}}{\frac{s+2}{s-1}}.$$

Since both of the largest multiplicities of the RHP poles in the first and the second row of $G(s)$ are 1, the following filter is chosen:

$$J(s) = \begin{bmatrix} \frac{\beta_1 s+1}{(\lambda_1 s+1)^2} & 0 \\ 0 & \frac{\beta_2 s+1}{(\lambda_2 s+1)^2} \end{bmatrix}.$$

The suboptimal controller is

$$Q(s) = Q_{opt}(s)J(s) = \begin{bmatrix} \frac{(s-1)(\beta_1 s+1)}{(\lambda_1 s+1)^2} & \frac{2(s+2)(\beta_2 s+1)}{(\lambda_2 s+1)^2} \\ \frac{-(\beta_1 s+1)}{(\lambda_1 s+1)^2} & \frac{-(s+2)(\beta_2 s+1)}{(\lambda_2 s+1)^2} \end{bmatrix} \frac{s-1}{s+2}.$$

Simple computation gives

$$\begin{aligned} S(s) &= I - G(s)Q(s) \\ &= \begin{bmatrix} 1 - \frac{(s-2)(s-3)(\beta_1 s+1)}{(s+2)(s+3)(\lambda_1 s+1)^2} & 0 \\ 0 & 1 - \frac{-(s-3)(\beta_2 s+1)}{(s+3)(\lambda_2 s+1)^2} \end{bmatrix}. \end{aligned}$$

Based on (13.4.3), it is readily obtained that

$$\beta_1 = 6(\lambda_1 + 1)^2 - 1, \beta_2 = 2(\lambda_2 + 1)^2 - 1.$$

The unity feedback controller is

$$C(s) = \begin{bmatrix} \frac{(s-1)(s+3)\{[6(\lambda_1+1)^2-1]s+1\}}{s[\lambda_1^2 s^2-(5+10\lambda_1)s+60\lambda_1+36\lambda_1^2+20]} & \frac{2(s+3)\{[2(\lambda_2+1)^2-1]s+1\}}{s(\lambda_2^2 s+6\lambda_2^2+6\lambda_2+1)} \\ \frac{-(s+3)\{[6(\lambda_1+1)^2-1]s+1\}}{s[\lambda_1^2 s^2-(5+10\lambda_1)s+60\lambda_1+36\lambda_1^2+20]} & \frac{-(s+3)\{[2(\lambda_2+1)^2-1]s+1\}}{s(\lambda_2^2 s+6\lambda_2^2+6\lambda_2+1)} \end{bmatrix}.$$

It is observed that the RHP zero-pole cancellations in $C(s)$ have been removed. If the performance degrees are taken as $\lambda_1 = \lambda_2 = 1$, the controller is

$$\boxed{C(s) = \begin{bmatrix} \frac{23s^3+47s^2-67s-3}{s(s^2-15s+116)} & \frac{14s^2+44s+6}{s(s+13)} \\ \frac{-(23s^2+70s+3)}{s(s^2-15s+116)} & \frac{-(7s^2+22s+3)}{s(s+13)} \end{bmatrix}}.$$

The closed-loop responses are shown in Figure 13.5.1.

Because the plant is unstable, there are large overshoots in the reference responses. If a 2 DOF structure is adopted, the controller of the disturbance loop is

$$\boxed{C_1(s) = C(s).}$$

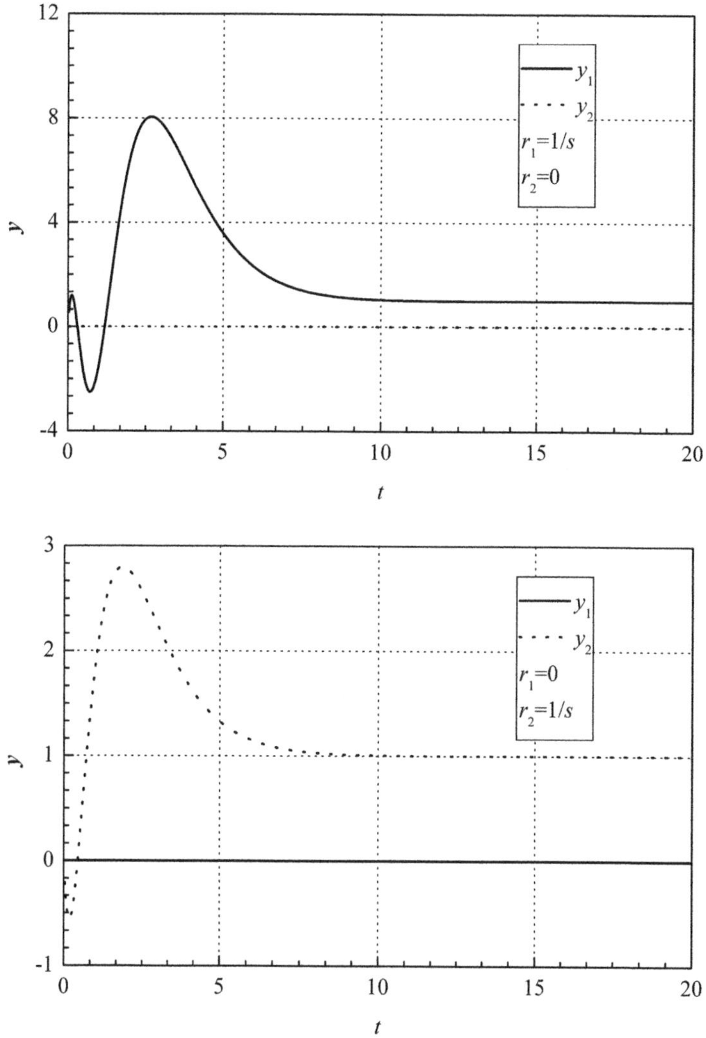

FIGURE 13.5.1
Closed-loop responses with $\lambda_1 = \lambda_2 = 1$.

The controller of the reference loop can be obtained with (13.4.9):

$$C_2(s) = \begin{bmatrix} \frac{(\lambda_1 s+1)^2}{(\beta_1 s+1)(\lambda_1' s+1)} & 0 \\ 0 & \frac{(\lambda_2 s+1)^2}{(\beta_2 s+1)(\lambda_2' s+1)} \end{bmatrix}.$$

For 10% undershoot in each reference loop, $\lambda_1' = 1.4$ *and* $\lambda_2' = 0.8$. *The controller is*

$$\boxed{C_2(s) = \begin{bmatrix} \frac{s^2+2s+1}{32.2s^2+24.4s+1} & 0 \\ 0 & \frac{s^2+2s+1}{5.6s^2+7.8s+1} \end{bmatrix}}.$$

The closed-loop responses are shown in Figure 13.5.2. Now there are no overshoots in the reference responses.

Example 13.5.2. *Consider a heavy oil fractionator (Figure 13.5.3). The linearized model is*

$$\boxed{G(s) = \begin{bmatrix} \frac{4.05e^{-27s}}{27s+1} & \frac{1.77e^{-28s}}{60s+1} \\ \frac{5.39e^{-18s}}{50s+1} & \frac{5.72e^{-14s}}{60s+1} \end{bmatrix}},$$

in which the time constants and time delays are expressed in minutes. The main objective is to maintain process outputs $y_1(t)$ *and* $y_2(t)$ *at the specification* 0.0 ± 0.005 *in the steady state, while at the same time the process inputs* $u_1(t)$ *and* $u_2(t)$ *are subject to the saturation* ± 0.5 *with respect to every unit step reference. The anti-windup scheme based on the modified control structure is not preferred here, because that will significantly increase the complexity of this MIMO control problem. The design method introduced in this chapter can reach the goal without modifying the control structure.*

The inverse of the plant is

$$G^{-1}(s) = \frac{adj[G(s)]}{det[G(s)]},$$

where

$$\begin{aligned} adj[G(s)] &= \begin{bmatrix} \frac{5.72e^{-14s}}{60s+1} & -\frac{1.77e^{-28s}}{60s+1} \\ -\frac{5.39e^{-18s}}{50s+1} & \frac{4.05e^{-27s}}{27s+1} \end{bmatrix}, \\ \det[G(s)] &= \left(\frac{4.05}{27s+1} \frac{5.72}{60s+1} - \frac{1.77}{60s+1} \frac{5.39e^{-5s}}{50s+1} \right) e^{-41s}. \end{aligned}$$

It can be verified that

$$G_D(s) = \begin{bmatrix} e^{-27s} & 0 \\ 0 & e^{-14s} \end{bmatrix},$$

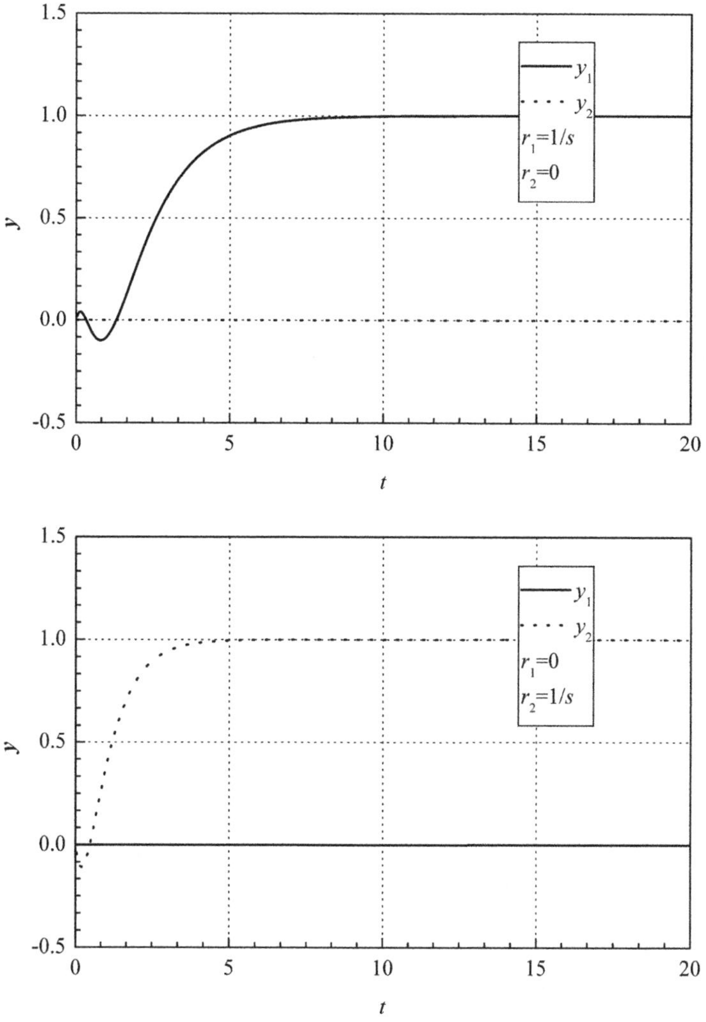

FIGURE 13.5.2
Closed-loop responses with $\lambda_1' = 1.4$ and $\lambda_2' = 0.8$.

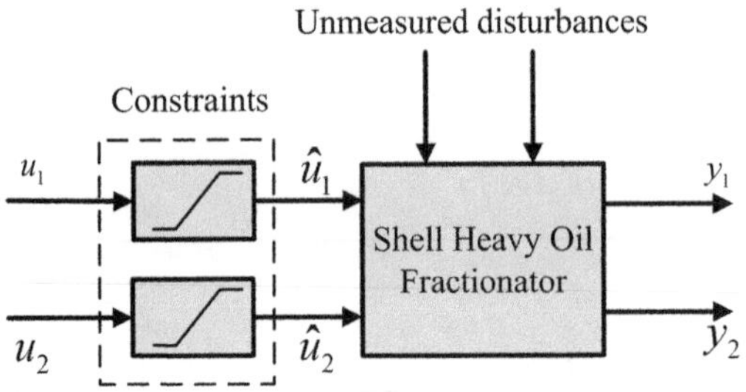

FIGURE 13.5.3
Shell heavy oil fractionator.

$$\boldsymbol{G_N}(s) = \boldsymbol{I}.$$

Therefore,

$$\begin{aligned}\boldsymbol{G_O}(s) &= \boldsymbol{G_D}^{-1}(s)\boldsymbol{G}(s) = \begin{bmatrix} \frac{4.05}{27s+1} & \frac{1.77e^{-s}}{60s+1} \\ \frac{5.39e^{-4s}}{50s+1} & \frac{5.72}{60s+1} \end{bmatrix}, \\ \boldsymbol{G_{MP}}(s) &= \boldsymbol{G_N}^{-1}(s)\boldsymbol{G_O}(s) = \boldsymbol{G_O}(s).\end{aligned}$$

A little algebra gives

$$\begin{aligned}\boldsymbol{Q_{opt}}(s) = \boldsymbol{G_{MP}}^{-1}(s) &= \begin{bmatrix} \frac{4.05}{27s+1} & \frac{1.77e^{-s}}{60s+1} \\ \frac{5.39e^{-4s}}{50s+1} & \frac{5.72}{60s+1} \end{bmatrix}^{-1} \\ &= \frac{\begin{bmatrix} \frac{5.72}{60s+1} & -\frac{1.77e^{-s}}{60s+1} \\ -\frac{5.39e^{-4s}}{50s+1} & \frac{4.05}{27s+1} \end{bmatrix}}{\frac{4.05}{27s+1}\frac{5.72}{60s+1} - \frac{1.77}{60s+1}\frac{5.39e^{-5s}}{50s+1}}.\end{aligned}$$

This rigorously analytical controller is of infinite dimension. With the help of fitting techniques, one obtains

$$\frac{4.05}{27s+1}\frac{5.72}{60s+1} - \frac{1.77}{60s+1}\frac{5.39e^{-5s}}{50s+1} \approx \frac{13.6257}{1193.2s^2+67.4s+1}.$$

For step inputs, choose the following filter:

$$\boldsymbol{J}(s) = \begin{bmatrix} \frac{1}{\lambda_1 s+1} & 0 \\ 0 & \frac{1}{\lambda_2 s+1} \end{bmatrix}.$$

The controller is

$$\boxed{\boldsymbol{Q}(s) = (1193.2s^2+67.4s+1)\begin{bmatrix} \frac{0.4198}{(60s+1)(\lambda_1 s+1)} & -\frac{0.1299e^{-s}}{(60s+1)(\lambda_2 s+1)} \\ -\frac{0.3956e^{-4s}}{(50s+1)(\lambda_1 s+1)} & \frac{0.2972}{(27s+1)(\lambda_2 s+1)} \end{bmatrix}.}$$

Increase the performance degrees from small to large. It is found that $\lambda_1 = 19$ and $\lambda_2 = 26$ can provide the required responses. The closed-loop responses are shown in Figure 13.5.4 and the manipulated variable responses are shown in Figure 13.5.5. Since the IMC structure is used, the closed-loop responses are thoroughly decoupled even for the approximate controller.

13.6 Summary

In this chapter, a new controller parameterization is presented. The advantages of this parameterization are as follows:

1. It does not require a coprime factorization.
2. It can be used for the decoupling control of systems with multiple time delays.

A diagonal factorization is then proposed, which is different from the inner-outer factorization. The optimal controller is obtained analytically based on the new parameterization and the diagonal factorization.

The first step in the H_2 decoupling design is factorizing the plant into two parts:

$$\boldsymbol{G}(s) = \boldsymbol{G_A}(s)\boldsymbol{G_{MP}}(s).$$

The second step is computing the optimal controller:

$$\boldsymbol{Q_{opt}}(s) = \boldsymbol{G_{MP}^{-1}}(s).$$

The third step is introducing a filter and adjusting the closed-loop response quantitatively:

$$\boldsymbol{Q}(s) = \boldsymbol{Q_{opt}}(s)\boldsymbol{J}(s).$$

In the design, the parameterization is only used to prove the optimal solution. When the obtained optimal solution is employed, those complex procedures, such as computing the controller parameterization and deriving the optimal controller, are not needed anymore.

Compared to the classical decoupling design, the notable feature of the H_2 decoupling design is that the controller is suboptimal and applicable to general plants. The stability and performance of the closed-loop system can be analyzed directly, and the performance limit is clearly known.

The common features of the SISO design methods in this book are that

1. The designer is not required to choose a weighting function.
2. The result is analytical.

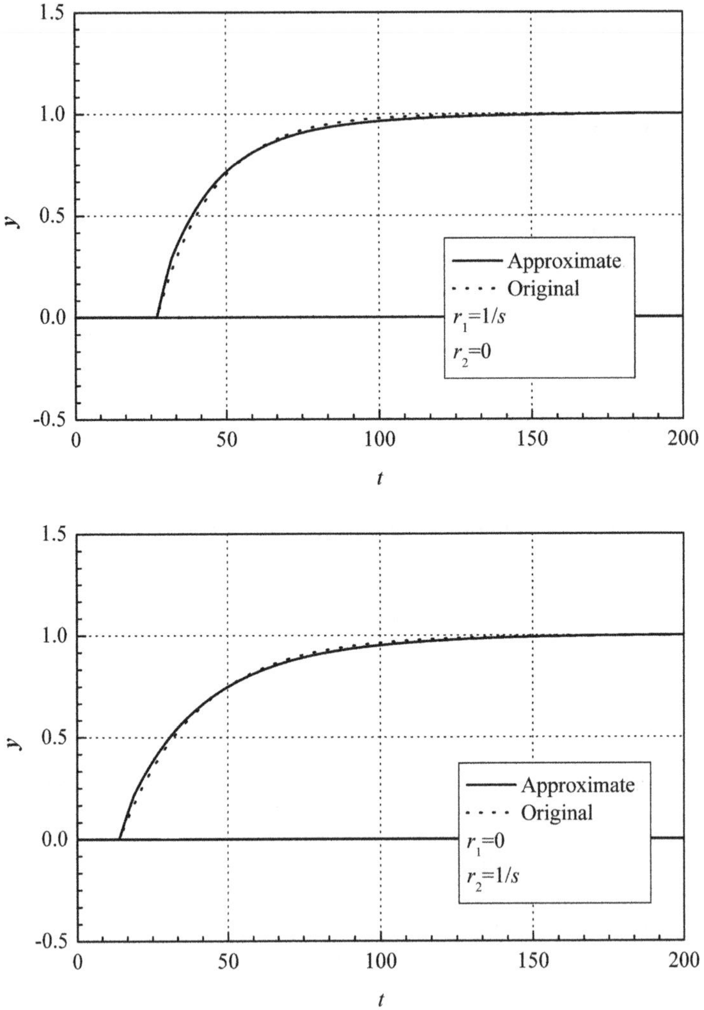

FIGURE 13.5.4
Closed-loop responses with $\lambda_1 = 19$ and $\lambda_2 = 26$.

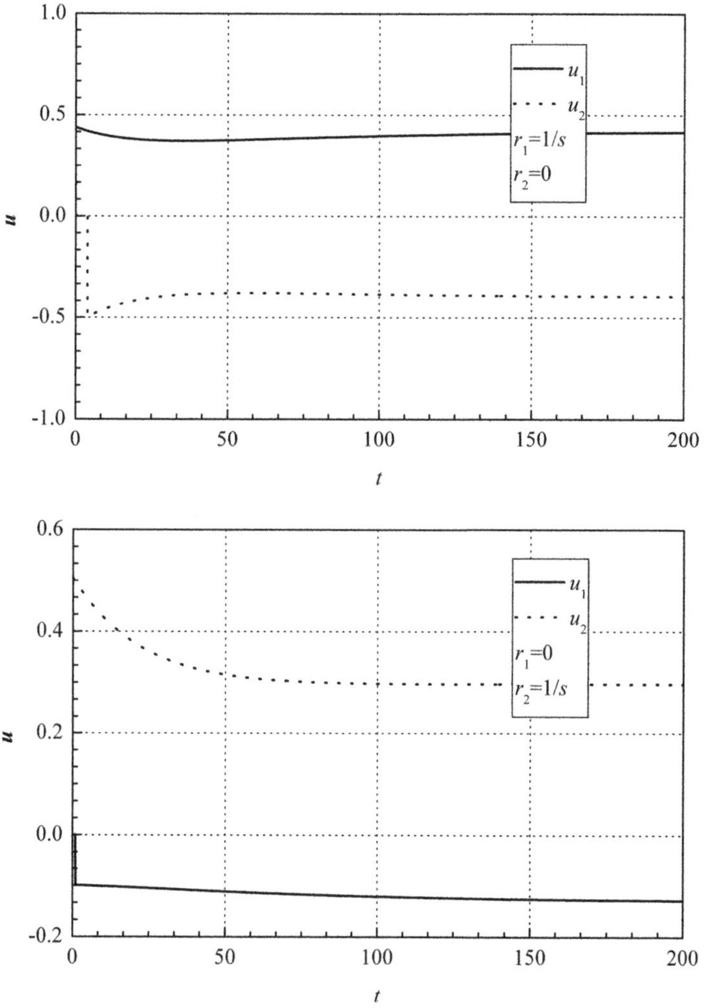

FIGURE 13.5.5
Responses of manipulated variables for $\lambda_1 = 19$ and $\lambda_2 = 26$.

3. The closed-loop response can be quantitatively adjusted.

The three features are preserved in both H_∞ and H_2 decoupling control.

Exercises

1. Assume that the plant $\boldsymbol{G}(s)$ has only simple RHP zeros $z_j(j = 1, 2, ..., r_z)$. Prove that there exists a stable H_2 controller $\boldsymbol{Q}(s)$ such that the complementary sensitivity function matrix, $\boldsymbol{G}(s)\boldsymbol{Q}(s)$, is equal to the desired closed-loop transfer function matrix $\boldsymbol{T}(s)$ if and only if $\boldsymbol{T}(s)$satisfies

$$\boldsymbol{v_j}\boldsymbol{T}(z_j) = 0$$

for all RHP zeros of $\boldsymbol{G}(s)$. Here $\boldsymbol{v_j}(s)$ is a nonzero row vector satisfying $\boldsymbol{v_j}\boldsymbol{G}(z_j) = 0$.

2. If the time delays in each column are the same, the plant can be factorized as follows:

$$\boldsymbol{G}(s) = \boldsymbol{G_O}(s)\boldsymbol{G_D}(s),$$

where $\boldsymbol{G_O}(s)$ is the rational part and $\boldsymbol{G_D}(s)$ is the diagonal time delay part. Is it possible to design an H_2 controller based on this factorization?

3. Assume that $\boldsymbol{G}(s)$ has a zero at $z_1, \mathrm{Re}(z_1) \geq 0$. $\boldsymbol{G}^{-1}(s)$ can be expressed in the form of a partial fraction expansion:

$$\boldsymbol{G}^{-1}(s) = \frac{\boldsymbol{G_1}}{s - z_1} + \boldsymbol{G_0}(s),$$

where $\boldsymbol{G_1}$ is the matrix of residue and $\boldsymbol{G_0}(s)$ is a term without poles at z_1. Consider the following plant:

$$\boldsymbol{G}(s) = \frac{1}{s+1}\begin{bmatrix} 1 & 1 \\ 1+2s & 2 \end{bmatrix}.$$

Find the partial fraction expansion of $\boldsymbol{G}(s)^{-1}$.

4. Explain why $\boldsymbol{G_N}(0) = \boldsymbol{I}$ when z_j in Definition 13.2.2 is a complex number.

5. Consider the following plant

$$\boldsymbol{G}(s) = \frac{1}{(s+2)(s-3)}\begin{bmatrix} (s-2)e^{-4s} & (s-2)e^{-s} \\ e^{-2}e^{-3s} & e^{-2s} \end{bmatrix}.$$

Compute the H_2 optimal decoupling controller $\boldsymbol{Q_{opt}}(s)$.

6. Suppose $\boldsymbol{G}(\infty) = \boldsymbol{I}$. Prove that $\boldsymbol{G}(s)$ has the same number of finite zeros and finite poles.

Notes and References

Section 13.1 is based on Zhang et al. (2006) (Zhang W. D., F. Allgöwer and T. Liu. Controller parameterization for SISO and MIMO plants with time delay, *System Control Letters*, 2006, 55(10), 794–802. ©Elsevier). Foias et al. (1996) discussed the application of Youla parameterization in systems with time delay.

Sections 13.2–13.5 follows Zhang et al. (2006) closely (Zhang W. D., C. Lin, and L. L. Ou. Algebraic solution to H_2 control problems - Part II: the multivariable decoupling case, *Ind. Eng. Chem. Res.*, 2006, 45(21), 7163–7176. ©ACS). The design of the filter is similar to that in Morari and Zafiriou (1989), but the parameterization and factorizations are thoroughly different. The factorization adopted in Morari and Zafiriou (1989) is the inner-outer factorization. Some related discussions can also be found in Wang (2003).

The plant in Example 13.5.2 is drawn from the 2×3 plant in the famous Shell problem. The Shell problem is an extremely difficult problem which includes many possibly conflicting requirements. It has appealed to many researcher's interest (Prett et al., 1990).

A 2 DOF MIMO control system was discussed in Liu et al. (2007). An extension of the H_2 decoupling control to nonsquare plants was given in Zhang and Lin (2006).

Exercise 1 is extracted from Morari and Zafiriou (1989, Section 13.2).

Exercise 3 is adapted from Morari and Zafiriou (1989, Section 13.2).

Exercise 5 is based on Zhang et al. (2006).

14

Multivariable H_2 Optimal Control

CONTENTS

In the last two chapters, the decoupling control problem was discussed in detail. In the two decoupling control schemes, the MIMO system was decomposed into a series of independent SISO loops. Optimal design procedures were developed and analytical controllers were obtained. The common feature of the two methods is that the closed-loop response is decoupled and thus the response of each channel is determined by only one parameter. This is important for applications, since when a channel is tuned, it is usually desirable that other channels not be affected. Furthermore, this feature makes the tuning of decoupling control systems be simplified significantly.

Nevertheless, the optimal control in general sense (the closed-loop response may be non-decoupled) is theoretically more important than the decoupling optimal control. Compared with the decoupling optimality, the general optimality implies that the minimum error is reached. In this chapter, the design procedure for H_2 optimal control is introduced. Again the steps of solving this problem are factorizing the plant, analytically deriving the optimal controller with the controller parameterization and plant factorization, and then developing a quantitative tuning procedure. Since the closed-loop response may not be decoupled, the tuning procedure is more complicated than that for decoupled systems.

14.1 Factorization for Simple RHP Zeros

In the preceding chapters, the plant was permitted to have a time delay. However, in this chapter the plant is restricted to be rational. The factorization, which is the core of H_2 optimal design, is still an open question for systems with time delay.

Let $\boldsymbol{G}(s)$ denote an $n \times n$ plant and $\boldsymbol{C}(s)$ an $n \times n$ controller. It is assumed in this chapter that

1. There is no unstable hidden mode in $\boldsymbol{G}(s)$.
2. $\boldsymbol{G}(s)$ is of full normal rank.
3. $\boldsymbol{G}(s)$ has no finite zeros on the imaginary axis.

The assumptions are the same as those for the H_2 decoupling control.

The factorization used in this chapter is an extension of the inner-outer factorization. The inner-outer factorization plays a pivotal role in system and control theory. Its goal is to compute the factorization of a stable nonsingular transfer function matrix $\boldsymbol{G}(s)$ as the product of a proper stable inner factor $\boldsymbol{G_A}(s)$ and a stable MP outer factor $\boldsymbol{G_{MP}}(s)$, that is,

$$\boldsymbol{G}(s) = \boldsymbol{G_A}(s)\boldsymbol{G_{MP}}(s). \tag{14.1.1}$$

A matrix $\boldsymbol{G_A}(s)$ is said to be inner if it is stable and satisfies $\boldsymbol{G_A}^*(s)\boldsymbol{G_A}(s) = \boldsymbol{I}$; A matrix $\boldsymbol{G_{MP}}(s)$ is said to be outer if it is stable and has no zeros in the RHP. $\boldsymbol{G_A}(s)$ is a full matrix in general. It is an all-pass function, since $\boldsymbol{G_A}^*(j\omega)\boldsymbol{G_A}(j\omega) = \boldsymbol{I}$ for all ω. $\boldsymbol{G_{MP}}(s)$ may have imaginary zeros according to the original definition of inner-outer factorization.

The inner-outer factorization is not unique. For example, multiplying $\boldsymbol{G_A}(s)$ and $\boldsymbol{G_{MP}}(s)$ by an orthogonal matrix does not change the poles and zeros. The obtained is still an inner-outer factorization. To obtain a unique inner-outer factorization, a constraint must be imposed on the inner factor or the outer factor. For example, let $\boldsymbol{G_A}(\infty) = I$.

The original inner-outer factorization is only defined for stable plants. Aimed at the design problem of general linear systems, the extended inner-outer factorization is defined in this chapter, which is applicable to both stable and unstable plants.

Definition 14.1.1. *$\boldsymbol{G_A}(s)$ and $\boldsymbol{G_{MP}}(s)$ are said to be the extended inner-outer factorization of $\boldsymbol{G}(s)$ if*

1. $\boldsymbol{G_A}^*(s)\boldsymbol{G_A}(s) = \boldsymbol{I}$.
2. *$\boldsymbol{G_A}(s)$ and $\boldsymbol{G_{MP}}^{-1}(s)$ are stable.*
3. *$\boldsymbol{G_{MP}}(s)$ has the same closed RHP poles as $\boldsymbol{G}(s)$.*

In the new definition, Condition 1 is from the original definition. It implies that $\boldsymbol{G_A}(s)$ has the same number of zeros and poles. Conditions 2 and 3 are from the design requirement of H_2 optimal controllers. In Condition 2 it is required that $\boldsymbol{G_A}(s)$ must be stable even if the plant is unstable, while $\boldsymbol{G_{MP}}(s)$ is not necessarily stable. Since $\boldsymbol{G_{MP}}(s)$ is required to be MP, $\boldsymbol{G_{MP}}^{-1}(s)$ must be stable. If Condition 3 is not satisfied, the plant has to be factorized for the second time, so that a MP part with the same closed RHP poles as $\boldsymbol{G}(s)$ is obtained. Such a factorization is needed in deriving the optimal solution. In particular, to obtain the optimal controller analytically, an analytical solution is expected for the extended inner-outer factorization.

The extended inner-outer factorization is a control-oriented definition. One can regard the original inner-outer factorization as a special case of the new factorization except for the imaginary axis case. It is noticed that, in the new definition, the outer factor does not have imaginary zero anymore. Otherwise, an internally unstable system may be obtained in the design.

For clarity of presentation, the factorization for plants with simple open RHP zeros is introduced first. Then the result is extended to plants with multiple open RHP zeros.

Assume that $z_j (j = 1, 2, ..., r_z)$ are the simple zeros of $\boldsymbol{G}(s)$.

Definition 14.1.2. *The nonzero* $1 \times n$ *vector* $\boldsymbol{v_j}$ *satisfying* $\boldsymbol{v_j G}(z_j) = \boldsymbol{0}$ *is called the direction of the zero* z_j.

In the definition, the zero is not necessarily a RHP zero. The vector $\boldsymbol{v_j^T}$ is the eigenvector of $\boldsymbol{G}^T(z_j)$ associated with the eigenvalue zero. It is called the zero direction since for a system input of the form $\boldsymbol{c}e^{z_j t}$ (where $\boldsymbol{c}$ is an arbitrary complex vector, $t \geq 0$), the output caused by the input in the direction of $\boldsymbol{v_j}$ is identically zero, given appropriate initial conditions.

Now let $\boldsymbol{G_A}(s)$ be an $n \times n$ transfer function matrix, $\det[\boldsymbol{G_A}(s)]$ is not identically zero (or equivalently, $\boldsymbol{G_A}(s)$ is of full normal rank), and $\det[\boldsymbol{G_A}(s)]$ does not have any pole at z_j.

Lemma 14.1.1. $\boldsymbol{G_A}(s)$ *has a simple zero* z_j *with zero direction* $\boldsymbol{v_j}$ *if and only if* $\boldsymbol{G_A}^{-1}(s)$ *has an expression of the form*

$$\boldsymbol{G_A}^{-1}(s) = (-s + z_j)^{-1}\boldsymbol{\alpha_j v_j} + \boldsymbol{G_0}(s), \tag{14.1.2}$$

where $\boldsymbol{G_0}(s)$ *is a term without poles at* z_j *and* $\boldsymbol{\alpha_j}$ *is a nonzero* $n \times 1$ *vector.*

Proof. Suppose that $\boldsymbol{G_A}^{-1}(s)$ has an expression of the form

$$\boldsymbol{G_A}^{-1}(s) = (-s + z_j)^{-1}\boldsymbol{\alpha_j v_j} + \boldsymbol{G_0}(s),$$

where $\boldsymbol{G_0}(s)$ is a term without poles at z_j.

As $\boldsymbol{\alpha_j}$ is a nonzero column vector and $\boldsymbol{v_j}$ is a nonzero row vector, $\boldsymbol{G_A}(s)$ has at least one zero at z_j.

Multiply both sides of (14.1.2) on the right by $(-s + z_j)\boldsymbol{G_A}(s)$. We have

$$(-s + z_j)\boldsymbol{I} = [\boldsymbol{\alpha_j v_j} + (-s + z_j)\boldsymbol{G_0}(s)]\boldsymbol{G_A}(s).$$

Taking determinants yields

$$(-s+z_j)^n = \det[\boldsymbol{\alpha_j v_j} + (-s+z_j)\boldsymbol{G_O}(s)]\det[\boldsymbol{G_A}(s)].$$

Since the rows of $\boldsymbol{\alpha_j v_j}$ are the multiples of $\boldsymbol{v_j}$, $\boldsymbol{\alpha_j v_j}$ is of rank 1. From Theorem 10.1.1, it is known that $\det[\boldsymbol{\alpha_j v_j} + (-s+z_j)\boldsymbol{G_O}(s)]$ has a zero at z_j with multiplicity at least $n-1$. It is deduced that $\det[\boldsymbol{G_A}(s)]$ has a zero at z_j with multiplicity at most 1. Therefore, z_j is a simple zero of $\boldsymbol{G_A}(s)$.

Conversely, suppose that $\boldsymbol{G_A}(s)$ has a simple zero z_j with the zero direction $\boldsymbol{v_j}$. Since $\det[\boldsymbol{G_A}(s)]$ is not identically zero, $\boldsymbol{G_A}(s)$ is invertible and $\boldsymbol{G_A}^{-1}(s)$ has a simple pole at z_j. Apply the Laurent expression for rational transfer function matrices. $\boldsymbol{G_A}^{-1}(s)$ can be written in the following form:

$$\begin{aligned}\boldsymbol{G_A}^{-1}(s) &= \sum_{i=-1}^{\infty}(-s+z_j)^i \boldsymbol{R_{i+2}} \\ &= (-s+z_j)^{-1}\sum_{i=-1}^{\infty}(-s+z_j)^{i+1}\boldsymbol{R_{i+2}}.\end{aligned}$$

Here $\boldsymbol{R_{i+2}}(i=-1,0,...,\infty)$ are $n\times n$ constant matrices. As $\boldsymbol{G_A}(s)$ has no poles at z_j, $\boldsymbol{R_1}\neq \boldsymbol{0}$.

Choose a $1\times n$ vector $\boldsymbol{\beta}$ satisfying $\boldsymbol{\beta R_1}\neq\boldsymbol{0}$. We have

$$\boldsymbol{\beta G_A}^{-1}(s) = (-s+z_j)^{-1}\sum_{i=-1}^{\infty}\boldsymbol{\beta}(-s+z_j)^{i+1}\boldsymbol{R_{i+2}}.$$

Rewrite the equation as

$$(-s+z_j)\boldsymbol{\beta} = \left[\sum_{i=-1}^{\infty}\boldsymbol{\beta}(-s+z_j)^{i+1}\boldsymbol{R_{i+2}}\right]\boldsymbol{G_A}(s).$$

In particular, at $s=z_j$ we have

$$\boldsymbol{\beta R_1 G_A}(z_j) = \boldsymbol{0}.$$

The definition of zero direction gives

$$\boldsymbol{v_j G_A}(z_j) = \boldsymbol{0}.$$

It is evident that $\boldsymbol{\beta R_1}$ is a multiple of $\boldsymbol{v_j}$ for any nonzero row vector $\boldsymbol{\beta}$. This happens if and only if there exists a nonzero column vector $\boldsymbol{\alpha_j}$ such that $\boldsymbol{R_1} = \boldsymbol{\alpha_j v_j}$. □

In the theorem, $\boldsymbol{\alpha_j v_j}$ is in fact the residue of $\boldsymbol{G_A}^{-1}(s)$ at z_j.

Lemma 14.1.2. *If an $n\times n$ stable transfer function matrix $\boldsymbol{G_A}(s)$ satisfies*

1. $\boldsymbol{G_A}^*(s)\boldsymbol{G_A}(s) = \boldsymbol{I}$;

2. $z_j(Re(z_j) > 0; j = 1, 2, ..., r_z)$ are the only open RHP zeros of $\boldsymbol{G_A}(s)$ with zero directions v_j,

then $\boldsymbol{G_A}(s)$ is the inner factor of $\boldsymbol{G}(s)$ and $\boldsymbol{G_{MP}}(s) = \boldsymbol{G_A}^{-1}(s)\boldsymbol{G}(s)$ is the outer factor.

Proof. It is enough to prove that $\boldsymbol{G_{MP}}(s)$ is MP and has the same closed RHP poles as $\boldsymbol{G}(s)$.

Since $z_j(j = 1, 2, ..., r_z)$ are the only open RHP zeros of $\boldsymbol{G_A}(s)$, the only open RHP poles of $\boldsymbol{G_A}^{-1}(s)$ are z_j. Recall Lemma 14.1.1. If z_j is the simple zero of $\boldsymbol{G_A}(s)$, $\boldsymbol{G_A}^{-1}(s)$ has an expression of the form

$$\boldsymbol{G_A}^{-1}(s) = (-s + z_j)^{-1}\boldsymbol{\alpha_j v_j} + \boldsymbol{G_0}(s),$$

where $\boldsymbol{\alpha_j}$ is a nonzero column vector and $\boldsymbol{G_0}(s)$ is a term without poles at z_j.

Multiply both sides on the right by $\boldsymbol{G}(s)$. We have

$$\boldsymbol{G_A}^{-1}(s)\boldsymbol{G}(s) = (-s + z_j)^{-1}\boldsymbol{\alpha_j v_j G}(s) + \boldsymbol{G_0}(s)\boldsymbol{G}(s).$$

The second term in the right-hand side is analytic at z_j for $j = 1, 2, ..., r_z$, since both $\boldsymbol{G_0}(s)$ and $\boldsymbol{G}(s)$ do not have poles at z_j. In the first term, $\boldsymbol{v_j G}(z_j) = \boldsymbol{0}$ for $j = 1, 2, ..., r_z$. Hence, $\boldsymbol{G_A}^{-1}(s)\boldsymbol{G}(s)$ does not have a pole at z_j for $j = 1, 2, ..., r_z$; that is, all open RHP poles of $\boldsymbol{G_A}^{-1}(s)$ are cancelled by those open RHP zeros of $\boldsymbol{G}(s)$. $\boldsymbol{G_{MP}}(s) = \boldsymbol{G_A}^{-1}(s)\boldsymbol{G}(s)$ has the same closed RHP poles as $\boldsymbol{G}(s)$.

Furthermore, those open RHP zeros of $\boldsymbol{G}(s)$ are the only possible open RHP zeros of $\boldsymbol{G_{MP}}(s)$. All of these zeros are cancelled. Therefore, $\boldsymbol{G_{MP}}(s)$ is MP. □

The following theorem gives the main result of this section.

Theorem 14.1.3. *The $n \times n$ transfer function matrix $\boldsymbol{G_A}(s)$ of the form*

$$\boldsymbol{G_A}(s) = \boldsymbol{I} - \boldsymbol{B}^*(s\boldsymbol{I} + \bar{\boldsymbol{A}})^{-1}\boldsymbol{F}^{-1}\boldsymbol{B} \tag{14.1.3}$$

is inner. Here

$$\boldsymbol{A} = \begin{bmatrix} z_1 & \cdots & 0 \\ \vdots & \ddots & \vdots \\ 0 & \cdots & z_{r_z} \end{bmatrix}, \boldsymbol{B} = \begin{bmatrix} \boldsymbol{v_1} \\ \vdots \\ \boldsymbol{v_{r_z}} \end{bmatrix},$$

$$\boldsymbol{F} = [f_{ij}], f_{ij} = \frac{\boldsymbol{v_i v_j}^*}{\bar{z}_j + z_i}; i, j = 1, 2, ..., r_z.$$

Proof. First, it is shown that $\boldsymbol{G_A}^*(s)\boldsymbol{G_A}(s) = \boldsymbol{I}$. $z_j(j = 1, 2, ..., r_z)$ are the simple zeros of $\boldsymbol{G_A}(s)$. With this fact, it can be proved that

$$\boldsymbol{G_A}^*(s) = \boldsymbol{I} - \boldsymbol{B}^*(\boldsymbol{F}^*)^{-1}(-s\boldsymbol{I} + \boldsymbol{A}^T)^{-1}\boldsymbol{B}. \tag{14.1.4}$$

Since $\boldsymbol{F} = \boldsymbol{F}^*$,

$$
\begin{aligned}
& \boldsymbol{G_A}^*(s)\boldsymbol{G_A}(s) \\
= \; & [\boldsymbol{I} - \boldsymbol{B}^*\boldsymbol{F}^{-1}(-s\boldsymbol{I} + \boldsymbol{A}^T)^{-1}\boldsymbol{B}][\boldsymbol{I} - \boldsymbol{B}^*(s\boldsymbol{I} + \bar{\boldsymbol{A}})^{-1}\boldsymbol{F}^{-1}\boldsymbol{B}] \\
= \; & \boldsymbol{I} - \boldsymbol{B}^*\boldsymbol{F}^{-1}(-s\boldsymbol{I} + \boldsymbol{A}^T)^{-1}\boldsymbol{B} - \boldsymbol{B}^*(s\boldsymbol{I} + \bar{\boldsymbol{A}})^{-1}\boldsymbol{F}^{-1}\boldsymbol{B} + \\
& \boldsymbol{B}^*\boldsymbol{F}^{-1}(-s\boldsymbol{I} + \boldsymbol{A}^T)^{-1}\boldsymbol{B}\boldsymbol{B}^*(s\boldsymbol{I} + \bar{\boldsymbol{A}})^{-1}\boldsymbol{F}^{-1}\boldsymbol{B} \\
= \; & \boldsymbol{I} - \boldsymbol{B}^*\boldsymbol{F}^{-1}(-s\boldsymbol{I} + \boldsymbol{A}^T)^{-1}\{\boldsymbol{F}(s\boldsymbol{I} + \bar{\boldsymbol{A}}) + \\
& (-s\boldsymbol{I} + \boldsymbol{A}^T)\boldsymbol{F} - \boldsymbol{B}\boldsymbol{B}^*\}(s\boldsymbol{I} + \bar{\boldsymbol{A}})^{-1}\boldsymbol{F}^{-1}\boldsymbol{B} \\
= \; & \boldsymbol{I} - \boldsymbol{B}^*\boldsymbol{F}^{-1}(-s\boldsymbol{I} + \boldsymbol{A}^T)^{-1}(\boldsymbol{F}\bar{\boldsymbol{A}} + \boldsymbol{A}^T\boldsymbol{F} - \boldsymbol{B}\boldsymbol{B}^*)(s\boldsymbol{I} + \bar{\boldsymbol{A}})^{-1}\boldsymbol{F}^{-1}\boldsymbol{B}.
\end{aligned}
$$

It is easy to verify that $\boldsymbol{F}$ satisfies the following Lyapunov equation:

$$\boldsymbol{F}\bar{\boldsymbol{A}} + \boldsymbol{A}^T\boldsymbol{F} = \boldsymbol{B}\boldsymbol{B}^*.$$

One readily obtains $\boldsymbol{G_A}^*(s)\boldsymbol{G_A}(s) = \boldsymbol{I}$.

Next, examine the zero of $\boldsymbol{G_A}(s)$. Since

$$\boldsymbol{G_A}^{-1}(s) = \boldsymbol{G_A}^*(s) = \boldsymbol{I} - \boldsymbol{B}^*\boldsymbol{F}^{-1}(-s\boldsymbol{I} + \boldsymbol{A}^T)^{-1}\boldsymbol{B},$$

$z_j (j = 1, 2, ..., r_z)$ are the only open RHP zeros of $\boldsymbol{G_A}(s)$. By Lemma 14.1.2 it is known that $\boldsymbol{G_A}(s)$ is inner. □

Evidently, $\boldsymbol{G_A}(s)$ is unique, since $\boldsymbol{G_A}(\infty) = \boldsymbol{I}$.

14.2 Construction Procedure of Factorization

One may be interested in how the inner matrix in Theorem 14.1.3 was conceived. The procedure will be explained in detail in this section. An important step is constructing a transfer function matrix with the desired zero-pole distribution and, at the same time, without the RHP zero-pole cancellation. This is very easy for scalar plants, while for multivariable plants the constructing procedure is an involved problem, since the zero of the plant may not be the zero of its elements.

In the last chapter, the inner-outer factorization problem for decoupling control was studied. It was shown that the inner matrix could be obtained by analyzing the inverse of a plant, since all zeros of the plant would emerge in its inverse matrix. This is an important idea that inspires the solving of the factorization problem in this chapter. Furthermore, it is found in interpolation study that the zero-pole distribution of a transfer function matrix is closely related to the partial fraction expansion of its inverse matrix. This fact will be used to compute the factorization.

To obtain a unique inner-outer factorization, let $\boldsymbol{G_A}(\infty) = I$. Assume that

$z_j (j = 1, 2, ..., r_z)$ are the simple zeros of $\boldsymbol{G_A}(s)$ with the zero directions $\boldsymbol{v_j}$. By Lemma 14.1.1, $\boldsymbol{G_A}^{-1}(s)$ has the following partial fraction expansion:

$$\boldsymbol{G_A}^{-1}(s) = \boldsymbol{I} + \sum_{j=1}^{r_z} (-s + z_j)^{-1} \boldsymbol{\alpha_j v_j}, \tag{14.2.1}$$

where α_j are unknown nonzero $n \times 1$ vectors. Based on this expression, the form of $\boldsymbol{G_A}(s)$ is given in the following lemma.

Lemma 14.2.1. *If*

$$\boldsymbol{G_A}^{-1}(s) = \boldsymbol{I} + \sum_{j=1}^{r_z} (-s + z_j)^{-1} \boldsymbol{\alpha_j v_j},$$

then

$$\boldsymbol{G_A}(s) = \boldsymbol{I} + \boldsymbol{C_v}[s\boldsymbol{I} - (\boldsymbol{A_v} + \boldsymbol{B_v C_v})]^{-1} \boldsymbol{B_v}, \tag{14.2.2}$$

where

$$\boldsymbol{A_v} = \begin{bmatrix} z_1 & \cdots & 0 \\ \vdots & \ddots & \vdots \\ 0 & \cdots & z_{r_z} \end{bmatrix},$$

$$\boldsymbol{B_v} = [\boldsymbol{v_1}^T, \boldsymbol{v_2}^T, \cdots, \boldsymbol{v_{r_z}}^T]^T, \boldsymbol{C_v} = [\boldsymbol{\alpha_1}, \boldsymbol{\alpha_2}, \cdots, \boldsymbol{\alpha_{r_z}}].$$

Proof. This can be shown by the following multiplication:

$$\begin{aligned}
&\left[\boldsymbol{I} - \sum_{j=1}^{r_z} (s - z_j)^{-1} \boldsymbol{\alpha_j v_j}\right] [\boldsymbol{I} + \boldsymbol{C_v}[s\boldsymbol{I} - (\boldsymbol{A_v} + \boldsymbol{B_v C_v})]^{-1} \boldsymbol{B_v}] \\
= \;& \boldsymbol{I} - \sum_{j=1}^{r_z} (s - z_j)^{-1} \boldsymbol{\alpha_j v_j} + \boldsymbol{C_v}[s\boldsymbol{I} - (\boldsymbol{A_v} + \boldsymbol{B_v C_v})]^{-1} \boldsymbol{B_v} - \\
& \left[\sum_{j=1}^{r_z} (s - z_j)^{-1} \boldsymbol{\alpha_j v_j}\right] \boldsymbol{C_v}[s\boldsymbol{I} - (\boldsymbol{A_v} + \boldsymbol{B_v C_v})]^{-1} \boldsymbol{B_v} \\
= \;& \boldsymbol{I} - \boldsymbol{C_v}(s\boldsymbol{I} - \boldsymbol{A_v})^{-1} \boldsymbol{B_v} + \boldsymbol{C_v}[s\boldsymbol{I} - (\boldsymbol{A_v} + \boldsymbol{B_v C_v})]^{-1} \boldsymbol{B_v} - \\
& \boldsymbol{C_v}(s\boldsymbol{I} - \boldsymbol{A_v})^{-1} \boldsymbol{B_v C_v}[s\boldsymbol{I} - (\boldsymbol{A_v} + \boldsymbol{B_v C_v})]^{-1} \boldsymbol{B_v} \\
= \;& \boldsymbol{I} - \boldsymbol{C_v}(s\boldsymbol{I} - \boldsymbol{A_v})^{-1} \{[s\boldsymbol{I} - (\boldsymbol{A_v} + \boldsymbol{B_v C_v})] - (s\boldsymbol{I} - \boldsymbol{A_v}) + \boldsymbol{B_v C_v}\} \\
& [s\boldsymbol{I} - (\boldsymbol{A_v} + \boldsymbol{B_v C_v})]^{-1} \boldsymbol{B_v} \\
= \;& \boldsymbol{I}.
\end{aligned}$$

□

Since α_j are unknown, more conditions are needed to compute $\boldsymbol{G_A}(s)$.

Consider the transfer function matrix with the same number of zeros and poles, since the inner factor possesses the feature. Assume that $p_j (j = 1, 2, ..., r_z)$ are simple zeros of $\boldsymbol{G_A}^{-1}(s)$ with the zero directions $\boldsymbol{w_j}$, and $z_j \neq p_j$. Then p_j are simple poles of $\boldsymbol{G_A}(s)$.

On one hand, from Lemma 14.1.1 $\boldsymbol{G_A}(s)$ can be expressed as

$$\boldsymbol{G_A}(s) = \boldsymbol{I} + \sum_{j=1}^{r_z} (s - p_j)^{-1} \boldsymbol{w_j} \boldsymbol{\beta_j}, \tag{14.2.3}$$

where β_j are unknown nonzero $1 \times n$ vectors. The expression can be rewritten as

$$\boldsymbol{G_A}(s) = \boldsymbol{I} + \boldsymbol{C_w}(s\boldsymbol{I} - \boldsymbol{A_w})^{-1}\boldsymbol{B_w}, \tag{14.2.4}$$

where

$$\boldsymbol{A_w} = \begin{bmatrix} p_1 & \cdots & 0 \\ \vdots & \ddots & \vdots \\ 0 & \cdots & p_{r_z} \end{bmatrix},$$

$$\boldsymbol{B_w} = [\boldsymbol{\beta_1}^T, \boldsymbol{\beta_2}^T, \cdots, \boldsymbol{\beta_{r_z}}^T]^T, \boldsymbol{C_w} = [\boldsymbol{w_1}, \boldsymbol{w_2}, \cdots, \boldsymbol{w_{r_z}}].$$

On the other hand, Lemma 14.2.1 shows that the poles of $\boldsymbol{G_A}(s)$ are the eigenvalues of the matrix $\boldsymbol{A_v} + \boldsymbol{B_v C_v}$. In other words, $p_j (j = 1, 2, ..., r_z)$ are the eigenvalues of $\boldsymbol{A_v} + \boldsymbol{B_v C_v}$. This implies that $\boldsymbol{A_v} + \boldsymbol{B_v C_v}$ is similar to $\boldsymbol{A_w}$, or equivalently, there exists an invertible matrix $\boldsymbol{F}$ such that

$$\boldsymbol{A_v} + \boldsymbol{B_v C_v} = \boldsymbol{F A_w F}^{-1}. \tag{14.2.5}$$

Substitute this expression into (14.2.2). One obtains that

$$\begin{aligned} \boldsymbol{G_A}(s) &= \boldsymbol{I} + \boldsymbol{C_v}[s\boldsymbol{I} - (\boldsymbol{A_v} + \boldsymbol{B_v C_v})]^{-1}\boldsymbol{B_v} \\ &= \boldsymbol{I} + \boldsymbol{C_v}(s\boldsymbol{I} - \boldsymbol{F A_w F}^{-1})^{-1}\boldsymbol{B_v} \\ &= \boldsymbol{I} + \boldsymbol{C_v F}(s\boldsymbol{I} - \boldsymbol{A_w})^{-1}\boldsymbol{F}^{-1}\boldsymbol{B_v}. \end{aligned} \tag{14.2.6}$$

(14.2.4) and (14.2.6) are identical when $\boldsymbol{C_w} = \boldsymbol{C_v F}$ and $\boldsymbol{B_w} = \boldsymbol{F}^{-1}\boldsymbol{B_v}$. If $\boldsymbol{F}$ is known, $\boldsymbol{G_A}(s)$ can readily be derived by (14.2.4) or (14.2.6).

Consider the computation of $\boldsymbol{F}$. Substituting the expressions of $\boldsymbol{C_w}$ into (14.2.5) yields the following Lyapunov equation:

$$\boldsymbol{F A_w} - \boldsymbol{A_v F} = \boldsymbol{B_v C_w}. \tag{14.2.7}$$

Let $f_{ij} (i, j = 1, 2, ..., r_z)$ be the (i, j)th element of $\boldsymbol{F}$. Evidently, the above matrix equation is equivalent to the following scalar equations:

$$f_{ij} p_j - z_i f_{ij} = \boldsymbol{v_i w_j}. \tag{14.2.8}$$

The unique solution to this equation is

$$f_{ij} = \frac{\boldsymbol{v_i w_j}}{p_j - z_i}. \tag{14.2.9}$$

The matrix $\boldsymbol{F}$ is now obtained.

When $\boldsymbol{B_v}$ and $\boldsymbol{F}$ are known, $\boldsymbol{B_w} = \boldsymbol{F}^{-1}\boldsymbol{B_v}$, or equivalently,

$$\begin{bmatrix} \boldsymbol{\beta_1} \\ \vdots \\ \boldsymbol{\beta_{r_z}} \end{bmatrix} = \begin{bmatrix} f_{11} & \cdots & f_{1r_z} \\ \vdots & \ddots & \vdots \\ f_{r_z1} & \cdots & f_{r_zr_z} \end{bmatrix}^{-1} \begin{bmatrix} \boldsymbol{v_1} \\ \vdots \\ \boldsymbol{v_{r_z}} \end{bmatrix}. \tag{14.2.10}$$

The computing procedure for $\boldsymbol{G_A}(s)$ is summarized as the following theorem.

Theorem 14.2.2. *Assume that $z_j (j = 1, 2, ..., r_z)$ are the simple zeros of $\boldsymbol{G_A}(s)$ with the zero directions $\boldsymbol{v_j}$, p_j are simple zeros of $\boldsymbol{G_A}^{-1}(s)$ with the zero directions $\boldsymbol{w_j}$, $z_j \neq p_j$, and $\boldsymbol{G_A}(\infty) = \boldsymbol{I}$. Then $\boldsymbol{G_A}(s)$ can be written as*

$$\boldsymbol{G_A}(s) = \boldsymbol{I} + \sum_{j=1}^{r_z} (s - p_j)^{-1} \boldsymbol{w_j \beta_j},$$

where

$$\begin{aligned} \begin{bmatrix} \boldsymbol{\beta_1} \\ \vdots \\ \boldsymbol{\beta_{r_z}} \end{bmatrix} &= \boldsymbol{F}^{-1} \begin{bmatrix} \boldsymbol{v_1} \\ \vdots \\ \boldsymbol{v_{r_z}} \end{bmatrix}, \\ \boldsymbol{F} = [f_{ij}], f_{ij} &= \frac{\boldsymbol{v_i w_j}}{p_j - z_i}; i, j = 1, 2, ..., r_z. \end{aligned}$$

Proof. Follows the foregoing analysis procedure. □

When the zero of $\boldsymbol{G_A}(s)$ is known, the zero direction can be computed by definition. The obtained zero direction is not unique. If v_j is the zero direction of $\boldsymbol{G_A}(s)$, kv_j (k is a nonzero constant) is evidently the zero direction of $\boldsymbol{G_A}(s)$. However, in view of (14.2.9) and (14.2.10), this does not affect the ultimate result of Theorem 14.2.2.

Consider a special case: $z_j(\text{Re}(z_j) > 0; j = 1, 2, ..., r_z)$ are the simple zeros and the only open RHP zeros of $\boldsymbol{G_A}(s)$ with zero directions $\boldsymbol{v_j}$, while $p_j = -\bar{z}_j$ are the simple zeros of $\boldsymbol{G_A}^{-1}(s)$ with zero directions $\boldsymbol{w_j} = -\boldsymbol{v_j}^*$. The following lemma shows that this is the case encountered in the extended inner-outer factorization problem.

Lemma 14.2.3. *Assume that $\boldsymbol{G_A}(s)$ is the inner factor of $\boldsymbol{G}(s)$. If z_j are the simple open RHP zeros of $\boldsymbol{G_A}(s)$, then $-\bar{z}_j$ are the simple poles of $\boldsymbol{G_A}(s)$.*

Proof. $\det[\boldsymbol{G_A}^*(s)] = \det[\boldsymbol{G_A}^T(-s)] = \det[\boldsymbol{G_A}(-s)]$. It is easy to know that $-z_j$ are the simple zeros of $\boldsymbol{G_A}^*(s)$. The coefficients in the zero polynomial of $\boldsymbol{G_A}^*(s)$ are real numbers. This implies that $-\bar{z}_j$ are the simple zeros of $\boldsymbol{G_A}^*(s)$, too.

Since $\boldsymbol{G_A}^*(s)\boldsymbol{G_A}(s) = \boldsymbol{I}$, $\boldsymbol{G_A}^{-1}(s) = \boldsymbol{G_A}^*(s)$. $-\bar{z}_j$ are simple zeros of $\boldsymbol{G_A}^{-1}(s)$ and thus simple poles of $\boldsymbol{G_A}(s)$. □

As introduced, $\boldsymbol{C_v} = \boldsymbol{C_w}\boldsymbol{F}^{-1}$ and $\boldsymbol{w_j} = -\boldsymbol{v_j}^*$. We have $\boldsymbol{C_w} = -\boldsymbol{B_v}^*$ and $\boldsymbol{C_v} = -\boldsymbol{B_v}^*(\boldsymbol{F}^*)^{-1}$. Substituting these into (14.2.1) gives

$$\begin{aligned}\boldsymbol{G_A}^{-1}(s) &= \boldsymbol{I} + \boldsymbol{C_v}(-s\boldsymbol{I} + \boldsymbol{A_v})^{-1}\boldsymbol{B_v} \\ &= \boldsymbol{I} - \boldsymbol{B_v}^*(\boldsymbol{F}^*)^{-1}(-s\boldsymbol{I} + \boldsymbol{A_v})^{-1}\boldsymbol{B_v}.\end{aligned}$$

Since $\boldsymbol{G_A}^*(s) = \boldsymbol{G_A}^{-1}(s)$, $\boldsymbol{A_v} = \boldsymbol{A}$, and $\boldsymbol{B_v} = \boldsymbol{B}$, $\boldsymbol{G_A}^*(s)$ can be expressed in the form of (14.1.4).

Theorem 14.2.4. *Assume that $z_j(j = 1, 2, ..., r_z)$ are the simple zeros of $\boldsymbol{G_A}(s)$ with the zero directions $\boldsymbol{v_j}$, $-\bar{z}_j$ are simple zeros of $\boldsymbol{G_A}^{-1}(s)$ with the zero directions $-\boldsymbol{v_j}^*$, and $\boldsymbol{G_A}(\infty) = \boldsymbol{I}$. Then the following transfer function matrix is inner:*

$$\boldsymbol{G_A}(s) = \boldsymbol{I} + \sum_{j=1}^{r_z} (s + \bar{z}_j)^{-1}\boldsymbol{v_j}^*\boldsymbol{\beta_j},$$

where

$$\begin{aligned}\begin{bmatrix} \boldsymbol{\beta_1} \\ \vdots \\ \boldsymbol{\beta_{r_z}} \end{bmatrix} &= \boldsymbol{F}^{-1}\begin{bmatrix} \boldsymbol{v_1} \\ \vdots \\ \boldsymbol{v_{r_z}} \end{bmatrix}, \\ \boldsymbol{F} = [f_{ij}], f_{ij} &= \frac{\boldsymbol{v_i}\boldsymbol{v_j}^*}{\bar{z}_j + z_i}; i, j = 1, 2, ..., r_z\end{aligned}$$

Let

$$\boldsymbol{A} = \begin{bmatrix} z_1 & \cdots & 0 \\ \vdots & \ddots & \vdots \\ 0 & \cdots & z_{r_z} \end{bmatrix}, \boldsymbol{B} = \begin{bmatrix} \boldsymbol{v_1} \\ \vdots \\ \boldsymbol{v_{r_z}} \end{bmatrix}. \tag{14.2.11}$$

Writing $\boldsymbol{G_A}(s)$ in the matrix form yields the expression in Theorem 14.1.3.

The constructing procedure for $\boldsymbol{G_A}(s)$ is illustrated in Figure 14.2.1.

14.3 Factorization for Multiple RHP Zeros

In this section, the factorization for simple zeros will be extended to the case of multiple zeros. More precisely, the parameters $\boldsymbol{A}$, $\boldsymbol{B}$, and $\boldsymbol{F}$ in Theorem 14.1.3 will be derived for plants with multiple RHP zeros.

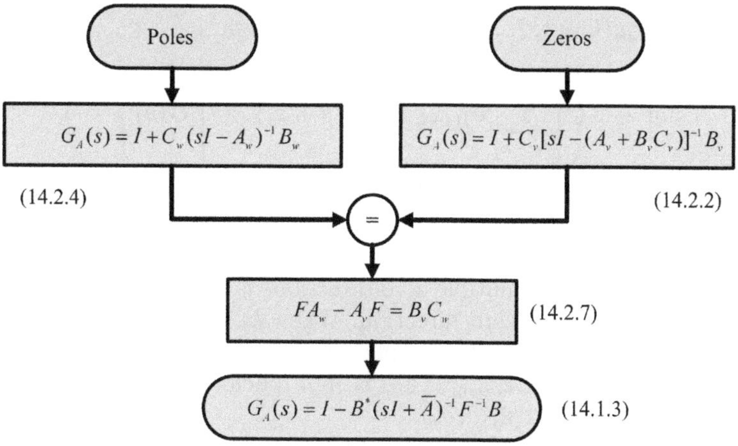

FIGURE 14.2.1
Constructing procedure for $\boldsymbol{G_A}(s)$.

In the factorization of the plant with simple RHP zeros, it is seen that the construction of $\boldsymbol{A}$ depends on the open RHP zero, while the construction of $\boldsymbol{B}$ and the computation of $\boldsymbol{F}$ depend on the zero direction (Figure 14.3.1). The zero direction is only defined for simple zero. Hence, the first step in this section is defining the zero direction for multiple zeros. The second step is constructing $\boldsymbol{A}$ and $\boldsymbol{B}$, and computing $\boldsymbol{F}$ based on the definition.

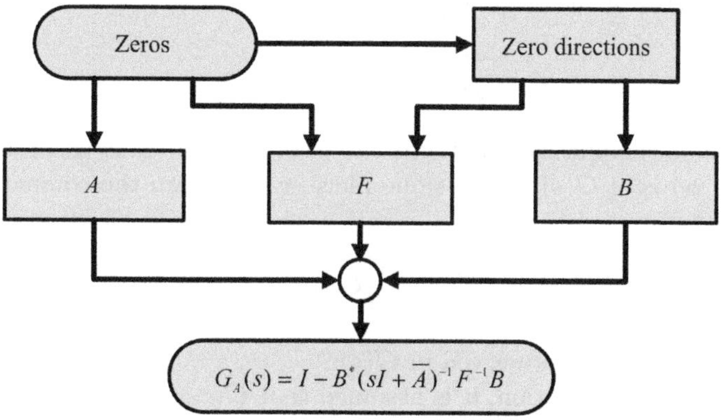

FIGURE 14.3.1
Computation of the inner factor.

Definition 14.3.1. *Assume that z_j is a k_j multiplicity zero of $\mathbf{G}(s)$. The*

$1 \times n$ vectors $\boldsymbol{v_{jk}}(k = 1, 2, ..., k_j; \boldsymbol{v_{j1}} \neq \boldsymbol{0})$ satisfying

$$\lim_{s \to z_j} \frac{d^l}{ds^l} \left\{ \left[\sum_{i=-k_j}^{-1} \boldsymbol{v_{j(i+k_j+1)}}(-s + z_j)^{i+k_j} \right] \boldsymbol{G}(s) \right\} = 0,$$

$$l = 0, 1, ..., k_j - 1.$$

are called the zero directions of z_j.

The definition of the multiple zero directions is a natural extension of the original definition of the zero direction. When $k_j = 1$, $\boldsymbol{v}_{j1}\boldsymbol{G}(z_j) = \boldsymbol{0}$. This definition reduces to the one for simple zeros.

Assume that $z_j (j = 1, 2, ..., r_z)$ are k_j multiplicity RHP zeros of $\boldsymbol{G}(s)$ with zero directions $\boldsymbol{v_{jk}}(k = 1, 2, ..., k_j)$.

A special case is that some z_j are the common zero of all elements in $\boldsymbol{G}(s)$. In this case, the common zero in $\boldsymbol{G}(s)$ should be separated before the extended inner-outer factorization is carried out. Otherwise, $\boldsymbol{G}(z_j) = \boldsymbol{0}$; the corresponding zero direction can be any nonzero vector. This can be achieved by removing the following factor from $\boldsymbol{G}(s)$:

$$\frac{-s/\bar{z}_j + 1}{s/z_j + 1}, \tag{14.3.1}$$

and then factorize the remainder of $\boldsymbol{G}(s)$:

$$\frac{s/z_j + 1}{-s/\bar{z}_j + 1} \boldsymbol{G}(s) = \boldsymbol{G_A}(s) \boldsymbol{G_{MP}}(s). \tag{14.3.2}$$

The inner factor of the original plant $\boldsymbol{G}(s)$ is

$$\frac{-s/\bar{z}_j + 1}{s/z_j + 1} \boldsymbol{G_A}(s). \tag{14.3.3}$$

It should be emphasized that only those common zeros are separated. Some z_j are the zeros of $\boldsymbol{G}(s)$ at the same place, rather than the common zero of all elements in

$$\frac{s/z_j + 1}{-s/\bar{z}_j + 1} \boldsymbol{G}(s).$$

These zeros should be preserved in $\boldsymbol{G}(s)$.

To simplify presentation, it is assumed that $\boldsymbol{G}(z_j) \neq \boldsymbol{0}$.

Let $\boldsymbol{G_A}(s)$ be an $n \times n$ transfer function matrix; $\det[\boldsymbol{G_A}(s)]$ is not identically zero and $\boldsymbol{G_A}(s)$ has no poles at z_j. It will be seen that $\boldsymbol{G_A}^{-1}(s)$ has a wonderful form in its expansion.

Lemma 14.3.1. *$\boldsymbol{G_A}(s)$ has a k_j multiplicity zero z_j with zero directions $\boldsymbol{v_{jk}}(k = 1, 2, ..., k_j)$ if and only if $\boldsymbol{G_A}^{-1}(s)$ can be expressed as*

$$\boldsymbol{G_A}^{-1}(s) = \tag{14.3.4}$$

$$(-s+z_j)^{-k_j}\boldsymbol{\alpha}_{j1}\boldsymbol{v}_{j1}+$$
$$(-s+z_j)^{-k_j+1}[\boldsymbol{\alpha}_{j1}\boldsymbol{v}_{j2}+\boldsymbol{\alpha}_{j2}\boldsymbol{v}_{j1}]+...+$$
$$(-s+z_j)^{-1}[\boldsymbol{\alpha}_{j1}\boldsymbol{v}_{jk_j}+\boldsymbol{\alpha}_{j2}\boldsymbol{v}_{j(k_j-1)}+...+\boldsymbol{\alpha}_{jk_j}\boldsymbol{v}_{j1}]+\boldsymbol{G}_0(s),$$

where $\boldsymbol{\alpha}_{j1}$ is a nonzero column vector and $\boldsymbol{G}_0(s)$ is a term without poles at z_j.

Proof. First, suppose that $\boldsymbol{G_A}^{-1}(s)$ can be expressed in the form of (14.3.4).

From the definition of zero directions it is known that $\boldsymbol{\alpha}_{j1}\boldsymbol{v}_{j1}\neq\boldsymbol{0}$. This implies that $\boldsymbol{G_A}(s)$ has at least k_j zeros at z_j.

Multiply both sides of (14.3.4) on the right by $(-s+z_j)^{k_j}\boldsymbol{G_A}(s)$. We have

$$\begin{aligned}&(-s+z_j)^{k_j}\boldsymbol{I}\\ =\;&\{\boldsymbol{\alpha}_{j1}\boldsymbol{v}_{j1}+\\ &(-s+z_j)[\boldsymbol{\alpha}_{j1}\boldsymbol{v}_{j2}+...]+...+\\ &(-s+z_j)^{k_j-1}[\boldsymbol{\alpha}_{j1}\boldsymbol{v}_{jk_j}+...]+(-s+z_j)^{k_j}\boldsymbol{G}_0(s)\}\boldsymbol{G_A}(s).\end{aligned}$$

Taking the determinants of both sides yields

$$\begin{aligned}&(-s+z_j)^{k_j\times n}\\ =\;&\det\{\boldsymbol{\alpha}_{j1}\boldsymbol{v}_{j1}+\\ &(-s+z_j)[\boldsymbol{\alpha}_{j1}\boldsymbol{v}_{j2}+...]+...+\\ &(-s+z_j)^{k_j-1}[\boldsymbol{\alpha}_{j1}\boldsymbol{v}_{jk_j}+...]+(-s+z_j)^{k_j}\boldsymbol{G}_0(s)\}\det[\boldsymbol{G_A}(s)].\end{aligned}$$

Since all of the lth ($l=0,1,...,k_j-1$) derivatives of

$$\begin{aligned}&\boldsymbol{\alpha}_{j1}\boldsymbol{v}_{j1}+\\ &(-s+z_j)[\boldsymbol{\alpha}_{j1}\boldsymbol{v}_{j2}+...]+...+\\ &(-s+z_j)^{k_j-1}[\boldsymbol{\alpha}_{j1}\boldsymbol{v}_{jk_j}+...]+(-s+z_j)^{k_j}\boldsymbol{G}_0(s)\end{aligned}$$

at z_j are of rank 1, by Theorem 10.1.1,

$$\begin{aligned}&\det\{\boldsymbol{\alpha}_{j1}\boldsymbol{v}_{j1}+\\ &(-s+z_j)[\boldsymbol{\alpha}_{j1}\boldsymbol{v}_{j2}+...]+...+\\ &(-s+z_j)^{k_j-1}[\boldsymbol{\alpha}_{j1}\boldsymbol{v}_{jk_j}+...]+(-s+z_j)^{k_j}\boldsymbol{G}_0(s)\}\end{aligned}$$

has at least $k_j\times(n-1)$ zeros at z_j. It is deduced that $\det[\boldsymbol{G_A}(s)]$ has at most k_j zeros at z_j.

Therefore, $\det[\boldsymbol{G_A}(s)]$ has a zero at z_j with multiplicity k_j.

Conversely, suppose that $\boldsymbol{G_A}(s)$ has a k_j multiplicity zero z_j with zero direction $\boldsymbol{v}_{jk}(k=1,2,...,k_j)$. Since $\det[\boldsymbol{G_A}(s)]$ is not identically zero, $\boldsymbol{G_A}(s)$ is invertible. $\boldsymbol{G_A}^{-1}(s)$ has k_j multiplicity poles at z_j. It has the following Laurent expression:

$$\boldsymbol{G_A}^{-1}(s)=\sum_{i=-k_j}^{\infty}(-s+z_j)^i\boldsymbol{R}_{j(i+k_j+1)}.$$

Here $\boldsymbol{R}_{j(i+k_j+1)}(i = -k_j, -k_j+1, ..., \infty)$ are $n \times n$ constant matrices and $\boldsymbol{R}_{j1} \neq \boldsymbol{0}$.

Choose a nonzero $1 \times n$ vector $\boldsymbol{\beta}$ such that $\boldsymbol{\beta}\boldsymbol{R}_{j1} \neq \boldsymbol{0}$. We have

$$\boldsymbol{\beta}\boldsymbol{G_A}^{-1}(s) = (-s+z_j)^{-k_j} \sum_{i=-k_j}^{\infty} \boldsymbol{\beta}(-s+z_j)^{i+k_j} \boldsymbol{R}_{j(i+k_j+1)}.$$

Rewrite the equation as

$$(-s+z_j)^{k_j}\boldsymbol{\beta} = \left[\sum_{i=-k_j}^{\infty} \boldsymbol{\beta}(-s+z_j)^{i+k_j} \boldsymbol{R}_{j(i+k_j+1)} \right] \boldsymbol{G_A}(s).$$

Compute the lth $(l = 0, 1, ..., k_j - 1)$ derivatives of the two sides at z_j:

$$\lim_{s \to z_j} \frac{d^l}{ds^l} \left\{ \left[\sum_{i=-k_j}^{\infty} \boldsymbol{\beta}(-s+z_j)^{i+k_j} \boldsymbol{R}_{j(i+k_j+1)} \right] \boldsymbol{G_A}(s) \right\} = 0,$$
$$l = 0, 1, ..., k_j - 1,$$

which can be reduced to

$$\lim_{s \to z_j} \frac{d^l}{ds^l} \left\{ \left[\sum_{i=-k_j}^{-1} \boldsymbol{\beta}(-s+z_j)^{i+k_j} \boldsymbol{R}_{j(i+k_j+1)} \right] \boldsymbol{G_A}(s) \right\} = 0,$$
$$l = 0, 1, ..., k_j - 1.$$

Compare the results with the definition of multiple zero directions. It is trivial to prove that $\boldsymbol{G_A}^{-1}(s)$ can be expressed in the form of (14.3.4). □

The following two examples are given to help readers understand Lemma 14.3.1.

Example 14.3.1. *Consider the following plant:*

$$\boldsymbol{G}(s) = \begin{bmatrix} \frac{(s-1)^2}{(s+1)^2} & \frac{(s-1)^2}{(s+1)^2} \\ \frac{-1}{s+1} & \frac{s-2}{s+1} \end{bmatrix}.$$

The plant has one three-multiplicity RHP zero at $s = 1$. It can be verified that the zero directions are

$$\boldsymbol{v}_{11} = [1\ 0], \boldsymbol{v}_{12} = [1\ 0], \boldsymbol{v}_{13} = [0\ 1/2].$$

$\boldsymbol{G}^{-1}(s)$ *can be expressed as*

$$\begin{aligned} \boldsymbol{G}^{-1}(s) \quad = \quad & (-s+1)^{-3}\boldsymbol{\alpha}_{11}\boldsymbol{v}_{11} + \\ & (-s+1)^{-2}[\boldsymbol{\alpha}_{11}\boldsymbol{v}_{12} + \boldsymbol{\alpha}_{12}\boldsymbol{v}_{11}] + \end{aligned}$$

$$(-s+1)^{-1}[\boldsymbol{\alpha_{11}v_{13}}+\boldsymbol{\alpha_{12}v_{12}}+\boldsymbol{\alpha_{13}v_{11}}]+\boldsymbol{G_0}(s),$$

where

$$\boldsymbol{\alpha_{11}}=\begin{bmatrix}4\\-4\end{bmatrix},\boldsymbol{\alpha_{12}}=\begin{bmatrix}-4\\8\end{bmatrix},\boldsymbol{\alpha_{13}}=\begin{bmatrix}1\\-9\end{bmatrix},\boldsymbol{G_0}(s)=\begin{bmatrix}1&-1\\0&1\end{bmatrix}.$$

Example 14.3.2. *Consider the following plant:*

$$\boldsymbol{G}(s)=\begin{bmatrix}\frac{5s^2-2s-3}{5(s+1)^2}&\frac{-4(s-1)}{5(s+1)^2}\\\frac{-4}{5(s+1)}&\frac{5s-3}{5(s+1)}\end{bmatrix}.$$

The plant has one two-multiplicity RHP zero at $s=1$. It can be verified that the zero directions are

$$\boldsymbol{v_{11}}=[1\ 0],\boldsymbol{v_{12}}=[1\ -1].$$

$\boldsymbol{G}^{-1}(s)$ *can be expressed as*

$$\begin{aligned}\boldsymbol{G}^{-1}(s)&=(-s+1)^{-2}\boldsymbol{\alpha_{11}v_{11}}+\\&\quad(-s+1)^{-1}[\boldsymbol{\alpha_{11}v_{12}}+\boldsymbol{\alpha_{12}v_{11}}]+\boldsymbol{G_0}(s),\end{aligned}$$

where

$$\boldsymbol{\alpha_{11}}=\begin{bmatrix}0.8\\1.6\end{bmatrix},\boldsymbol{\alpha_{12}}=\begin{bmatrix}-3.2\\-2.4\end{bmatrix},\boldsymbol{G_0}(s)=\boldsymbol{I}.$$

Lemma 14.3.2. *If an $n\times n$ stable transfer function matrix $\boldsymbol{G_A}(s)$ satisfies*

1. $\boldsymbol{G_A}^*(s)\boldsymbol{G_A}(s)=\boldsymbol{I}$;
2. z_j *with multiplicity $k_j(j=1,2,...,r_z)$ are the only open RHP zeros of $\boldsymbol{G_A}(s)$ with zero directions $\boldsymbol{v_{jk}}(k=1,2,...,k_j)$,*

then $\boldsymbol{G_A}(s)$ is the inner factor of $\boldsymbol{G}(s)$ and $\boldsymbol{G_{MP}}(s)=\boldsymbol{G_A}^{-1}(s)\boldsymbol{G}(s)$ is the outer factor of $\boldsymbol{G}(s)$.

Proof. It is enough to prove that $\boldsymbol{G_{MP}}(s)$ is MP and has the same closed RHP poles as $\boldsymbol{G}(s)$.

Since $z_j(j=1,2,...,r_z)$ are the only open RHP zeros of $\boldsymbol{G_A}(s)$, the only open RHP poles of $\boldsymbol{G_A}^{-1}(s)$ are z_j. Recall Lemma 14.3.1. If z_j is a k_j multiplicity zero of $\boldsymbol{G_A}(s)$, $\boldsymbol{G_A}^{-1}(s)$ is in the form of (14.3.4). Multiply both sides on the right by $\boldsymbol{G}(s)$:

$$\begin{aligned}\boldsymbol{G_A}^{-1}(s)\boldsymbol{G}(s)&=(-s+z_j)^{-k_j}\boldsymbol{\alpha_{j1}v_{j1}G}(s)+\\&\quad(-s+z_j)^{-k_j+1}[\boldsymbol{\alpha_{j1}v_{j2}}+...]\boldsymbol{G}(s)+...+\\&\quad(-s+z_j)^{-1}[\boldsymbol{\alpha_{j1}v_{jk_j}}+...]\boldsymbol{G}(s)+...+\boldsymbol{G_0}(s)\boldsymbol{G}(s).\end{aligned}$$

To obtain the Laurent expression of $\boldsymbol{G_A}^{-1}(s)\boldsymbol{G}(s)$ at z_j, all coefficients of the terms with $(-s+z_j)^{-i}(i=1,2,...,k_j)$ have to be computed. This can be achieved by multiplying both sides by $(-s+z_j)^{-k_j}$, and then computing the lth $(l=0,1,...,k_j-1)$ derivatives at z_j. The computing procedure is similar to that in Lemma 14.3.1. With the definition of zero directions, it can be found that all coefficients are zero for $j=1,2,...,r_z$. In other words, all open RHP poles of $\boldsymbol{G_A}^{-1}(s)$ are cancelled by those open RHP zeros of $\boldsymbol{G}(s)$. The implication of this fact is twofold:

1. $\boldsymbol{G_A}^{-1}(s)$ does not introduce any closed RHP poles to $\boldsymbol{G_A}^{-1}(s)\boldsymbol{G}(s)$. $\boldsymbol{G_{MP}}(s)$ has the same closed RHP poles as $\boldsymbol{G}(s)$.
2. All open RHP zeros of $\boldsymbol{G}(s)$ are removed. $\boldsymbol{G_{MP}}(s)$ does not have any open RHP zero.

This completes the proof. □

Theorem 14.3.3. *The matrix* $\boldsymbol{G_A}(s)$ *of the form*

$$\boldsymbol{G_A}(s) = \boldsymbol{I} - \boldsymbol{B}^*(s\boldsymbol{I}+\bar{\boldsymbol{A}})^{-1}\boldsymbol{F}^{-1}\boldsymbol{B}$$

is inner. Here

$$\boldsymbol{A} = \begin{bmatrix} \boldsymbol{A}_1 & \cdots & 0 \\ \vdots & \ddots & \vdots \\ 0 & \cdots & \boldsymbol{A}_{r_z} \end{bmatrix}, \boldsymbol{A_j} = \begin{bmatrix} z_j & -1 & & \\ & z_j & \ddots & \\ & & \ddots & -1 \\ & & & z_j \end{bmatrix},$$

$$\boldsymbol{B} = \begin{bmatrix} \boldsymbol{B}_1 \\ \vdots \\ \boldsymbol{B}_{r_z} \end{bmatrix}, \boldsymbol{B_j} = \begin{bmatrix} \boldsymbol{v}_{j1} \\ \vdots \\ \boldsymbol{v}_{jk_j} \end{bmatrix},$$

$$j = 1,2,...,r_z.$$

$\boldsymbol{F}$ *is the solution to the following Lyapunov equation:*

$$\boldsymbol{F}\bar{\boldsymbol{A}} + \boldsymbol{A}^T\boldsymbol{F} = \boldsymbol{B}\boldsymbol{B}^*.$$

Proof. First, it is shown that $\boldsymbol{G_A}^*(s)\boldsymbol{G_A}(s) = \boldsymbol{I}$. With a similar procedure to that for simple zero, it can be proved that

$$\boldsymbol{G_A}^*(s) = \boldsymbol{I} - \boldsymbol{B}^*\boldsymbol{F}^{-1}(-s\boldsymbol{I}+\boldsymbol{A}^T)^{-1}\boldsymbol{B}.$$

Since $\boldsymbol{F}=\boldsymbol{F}^*$,

$$\begin{aligned} & \boldsymbol{G_A}^*(s)\boldsymbol{G_A}(s) \\ = \ & [\boldsymbol{I} - \boldsymbol{B}^*(s\boldsymbol{I}+\bar{\boldsymbol{A}})^{-1}\boldsymbol{F}^{-1}\boldsymbol{B}]^*[\boldsymbol{I} - \boldsymbol{B}^*(s\boldsymbol{I}+\bar{\boldsymbol{A}})^{-1}\boldsymbol{F}^{-1}\boldsymbol{B}] \\ = \ & [\boldsymbol{I} - \boldsymbol{B}^*\boldsymbol{F}^{-1}(-s\boldsymbol{I}+\boldsymbol{A}^T)^{-1}\boldsymbol{B}][\boldsymbol{I} - \boldsymbol{B}^*(s\boldsymbol{I}+\bar{\boldsymbol{A}})^{-1}\boldsymbol{F}^{-1}\boldsymbol{B}] \\ = \ & \boldsymbol{I} - \boldsymbol{B}^*\boldsymbol{F}^{-1}(-s\boldsymbol{I}+\boldsymbol{A}^T)^{-1}\boldsymbol{B} - \boldsymbol{B}^*(s\boldsymbol{I}+\bar{\boldsymbol{A}})^{-1}\boldsymbol{F}^{-1}\boldsymbol{B} + \end{aligned}$$

$$
\begin{aligned}
& \boldsymbol{B}^*\boldsymbol{F}^{-1}(-s\boldsymbol{I}+\boldsymbol{A}^T)^{-1}\boldsymbol{B}\boldsymbol{B}^*(s\boldsymbol{I}+\bar{\boldsymbol{A}})^{-1}\boldsymbol{F}^{-1}\boldsymbol{B} \\
= \;& \boldsymbol{I}-\boldsymbol{B}^*\boldsymbol{F}^{-1}(-s\boldsymbol{I}+\boldsymbol{A}^T)^{-1}\{\boldsymbol{F}(s\boldsymbol{I}+\bar{\boldsymbol{A}})+(-s\boldsymbol{I}+\boldsymbol{A}^T)\boldsymbol{F}- \\
& \boldsymbol{B}\boldsymbol{B}^*\}(s\boldsymbol{I}+\bar{\boldsymbol{A}})^{-1}\boldsymbol{F}^{-1}\boldsymbol{B} \\
= \;& \boldsymbol{I}-\boldsymbol{B}^*\boldsymbol{F}^{-1}(-s\boldsymbol{I}+\boldsymbol{A}^T)^{-1}(\boldsymbol{F}\bar{\boldsymbol{A}}+\boldsymbol{A}^T\boldsymbol{F}-\boldsymbol{B}\boldsymbol{B}^*)(s\boldsymbol{I}+\bar{\boldsymbol{A}})^{-1}\boldsymbol{F}^{-1}\boldsymbol{B}.
\end{aligned}
$$

It is known that $\boldsymbol{F}$ satisfies the following Lyapunov equation:

$$
\boldsymbol{F}\bar{\boldsymbol{A}}+\boldsymbol{A}^T\boldsymbol{F}=\boldsymbol{B}\boldsymbol{B}^*.
$$

One readily obtains that $\boldsymbol{G_A}^*(s)\boldsymbol{G_A}(s)=\boldsymbol{I}$.

Next, examine the zero of $\boldsymbol{G_A}(s)$. Since

$$
\boldsymbol{G_A}^{-1}(s)=\boldsymbol{G_A}^*(s)=\boldsymbol{I}-\boldsymbol{B}^*\boldsymbol{F}^{-1}(-s\boldsymbol{I}+\boldsymbol{A}^T)^{-1}\boldsymbol{B},
$$

$z_j(j=1,2,...,r_z)$ are the only open RHP zeros of $\boldsymbol{G_A}(s)$.

By Lemma 14.3.2 it is concluded that $\boldsymbol{G_A}(s)$ is inner. □

It is easy to verify that Theorem 14.1.3 is a special case of Theorem 14.3.3.

14.4 Analysis and Computation

In the last three sections, an analytical solution to the extended inner-outer factorization was developed. Provided that the plant zeros are given, the factorization can be computed with a formula in closed form.

Normally, the plant zeros are known for the sake of analyzing the stability or estimating the performance. The computation of zero is analytical for low-order plants, but not analytical for high-order plants. It is equivalent to computing the roots of an equation in a single unknown. If the order of the equation is more than 4, there is no analytical formula for computation.

When the plant zeros are known, the computation complexity of the factorization depends on the multiplicity of the zero in the RHP. As it was seen in the preceding sections, to compute the inner matrix, $\boldsymbol{A}$, $\boldsymbol{B}$, and $\boldsymbol{F}$ must be obtained first:

1. $\boldsymbol{A}$ is from construction. It is exactly known.
2. $\boldsymbol{B}$ is also from construction, but the zero directions have to be computed first. Once the zero directions are known, $\boldsymbol{B}$ is exactly known.
3. $\boldsymbol{F}$ has to be computed on the basis of the zero directions.

The zero directions can be computed by the following formulas:

$$
\boldsymbol{v}_{j1}\boldsymbol{G}(z_j) \;=\; 0,
$$

$$\begin{aligned} \boldsymbol{v}_{j2}\boldsymbol{G}(z_j) &= \boldsymbol{v}_{j1}\frac{d}{ds}\boldsymbol{G}(z_j), \\ &\ldots, \\ \boldsymbol{v}_{jk_j}\boldsymbol{G}(z_j) &= \sum_{i=1}^{k_j-1}(-1)^{k_j-i+1}\frac{\boldsymbol{v}_{ji}}{(k_j-i)!}\frac{d^{k_j-i}}{ds^{k_j-i}}\boldsymbol{G}(z_j), \\ j &= 1,2,...,r_z. \end{aligned} \tag{14.4.1}$$

These formulas can be directly derived from the definition of multiple zero directions.

If the plant has only simple open RHP zeros, the zero directions can be obtained by

$$\boldsymbol{v}_{j1}\boldsymbol{G}(z_j) = \boldsymbol{0}, j = 1,2,...,r_z.$$

Because $\boldsymbol{G}(s)$ loses rank at z_j and $\boldsymbol{v}_{j1}$ can be any nonzero vector satisfying (14.4.1), the computation of $\boldsymbol{v}_{j1}$ is very simple. For example, if $\boldsymbol{G}(z_j) = [0\ 0; 1\ 2]$, one can simply take $\boldsymbol{v}_{j1} = [1\ 0]$. With (14.4.1), the computation of $\boldsymbol{v}_{j2}$ is easy when $\boldsymbol{v}_{j1}$ is known. As to the computation of $\boldsymbol{v}_{jk} (k > 2)$, the complexity is mainly from the computation of the derivative of $\boldsymbol{G}(s)$ at z_j.

Example 14.4.1. *Consider the following plant:*

$$\boldsymbol{G}(s) = \frac{1}{s+1}\begin{bmatrix} s-1 & s-1 \\ -1 & s-2 \end{bmatrix}.$$

The plant has a RHP zero with multiplicity 2 at $s = 1$ (that is, r_z=1, $z_1 = 1$, and k_j=2) and an LHP pole with multiplicity 2 at $s = -1$. Since

$$\boldsymbol{G}(1) = \begin{bmatrix} 0 & 0 \\ -1/2 & -1/2 \end{bmatrix},$$

one can take

$$\boldsymbol{v}_{11} = [1\ 0].$$

Furthermore,

$$\frac{d}{ds}\boldsymbol{G}(s) = \begin{bmatrix} \frac{2}{(s+1)^2} & \frac{2}{(s+1)^2} \\ \frac{1}{(s+1)^2} & \frac{3}{(s+1)^2} \end{bmatrix}.$$

Then

$$\frac{d}{ds}\boldsymbol{G}(1) = \begin{bmatrix} 1/2 & 1/2 \\ 1/4 & 3/4 \end{bmatrix}.$$

According to (14.4.1), $\boldsymbol{v}_{12}\boldsymbol{G}(1) = \boldsymbol{v}_{11}\frac{d}{ds}\boldsymbol{G}(1)$; that is,

$$\boldsymbol{v}_{12}\begin{bmatrix} 0 & 0 \\ -1/2 & -1/2 \end{bmatrix} = [1/2\ 1/2].$$

A simple choice is

$$\boldsymbol{v_{12}} = [1 \;\; -1].$$

Now it is shown that, when the zero directions are exactly known, $\boldsymbol{F}$ can be analytically computed. This is the key to obtain the analytical solution of the extended inner-outer factorization. $\boldsymbol{F}$ is the solution to the following Lyapunov equation:

$$\boldsymbol{F}\bar{\boldsymbol{A}} + \boldsymbol{A}^T\boldsymbol{F} = \boldsymbol{B}\boldsymbol{B}^*.$$

Express $\boldsymbol{F}$ as a block matrix:

$$\boldsymbol{F} = [\boldsymbol{F_{ij}}], \boldsymbol{F_{ij}} = \left[f_{xy}^{ij}\right], i, j = 1, 2, ..., r_z; x, y = 1, 2, ..., k_j. \quad (14.4.2)$$

Let $f_{x0}^{ij} = f_{0y}^{ij} = 0$ for all i, j, x, y. $\boldsymbol{F}$ can be directly computed with the following formula:

$$f_{xy}^{ij} = \frac{\boldsymbol{v_{ix}}\boldsymbol{v_{jy}}^*}{\bar{z}_j + z_i} + \frac{f_{(x-1)y}^{ij} + f_{x(y-1)}^{ij}}{\bar{z}_j + z_i}. \quad (14.4.3)$$

In particular, when the plant has only simple RHP zeros, we have

$$\boldsymbol{F} = \left[f_{11}^{ij}\right], f_{11}^{ij} = \frac{\boldsymbol{v_{i1}}\boldsymbol{v_{j1}^*}}{\bar{z}_j + z_i}. \quad (14.4.4)$$

This is the case discussed in Section 14.1. It is seen that the computation of $\boldsymbol{F}$ is exact and elegant, without any numerical error.

Example 14.4.2. *Consider the plant in Example 14.4.1. It is known that* $r_z = 1$, $z_1 = 1$, $k_j = 2$, *and the zero directions are*

$$\boldsymbol{v_{11}} = [1 \;\; 0], \boldsymbol{v_{12}} = [1 \;\; -1].$$

Then

$$\boldsymbol{F} = [\boldsymbol{F_{11}}], \boldsymbol{F_{11}} = \begin{bmatrix} f_{11}^{11} & f_{12}^{11} \\ f_{21}^{11} & f_{22}^{11} \end{bmatrix},$$

where

$$\begin{aligned} f_{11}^{11} &= \frac{\boldsymbol{v_{11}}\boldsymbol{v_{11}}^*}{z_1 + z_1} + \frac{f_{01}^{11} + f_{10}^{11}}{z_1 + z_1} = \frac{1}{2} + 0 = \frac{1}{2}, \\ f_{12}^{11} &= \frac{\boldsymbol{v_{11}}\boldsymbol{v_{12}}^*}{z_1 + z_1} + \frac{f_{01}^{11} + f_{11}^{11}}{z_1 + z_1} = \frac{1}{2} + \frac{1}{4} = \frac{3}{4}, \\ f_{21}^{11} &= \frac{\boldsymbol{v_{12}}\boldsymbol{v_{11}}^*}{z_1 + z_1} + \frac{f_{11}^{11} + f_{10}^{11}}{z_1 + z_1} = \frac{1}{2} + \frac{1}{4} = \frac{3}{4}, \\ f_{22}^{11} &= \frac{\boldsymbol{v_{12}}\boldsymbol{v_{12}}^*}{z_1 + z_1} + \frac{f_{12}^{11} + f_{21}^{11}}{z_1 + z_1} = 1 + \frac{3}{4} = \frac{7}{4}. \end{aligned}$$

The deriving procedure of the factorization formula is complicated. However, the result is simple, because it is analytical. The computation is summarized as follows.

Given an $n \times n$ transfer function matrix $\boldsymbol{G}(s)$, its k_j multiplicity open RHP zeros $z_j (j = 1, 2, ..., r_z)$, and zero directions $\boldsymbol{v_{jk}} (k = 1, 2, ..., k_j)$, the inner matrix can be exactly computed through the following steps:

1. $\boldsymbol{A} = \begin{bmatrix} \boldsymbol{A_1} & \cdots & 0 \\ \vdots & \ddots & \vdots \\ 0 & \cdots & \boldsymbol{A_{r_z}} \end{bmatrix}, \boldsymbol{A_j} = \begin{bmatrix} z_j & -1 & & \\ & z_j & \ddots & \\ & & \ddots & -1 \\ & & & z_j \end{bmatrix}.$

2. $\boldsymbol{B} = \begin{bmatrix} \boldsymbol{B_1} \\ \vdots \\ \boldsymbol{B_{r_z}} \end{bmatrix}, \boldsymbol{B_j} = \begin{bmatrix} \boldsymbol{v_{j1}} \\ \vdots \\ \boldsymbol{v_{jk_j}} \end{bmatrix}.$

3. $\boldsymbol{F} = [\boldsymbol{F_{ij}}], \boldsymbol{F_{ij}} = \left[f_{xy}^{ij}\right], f_{xy}^{ij} = \frac{\boldsymbol{v_{ix}}\boldsymbol{v_{jy}}^*}{\bar{z}_j + z_i} + \frac{f_{(x-1)y}^{ij} + f_{x(y-1)}^{ij}}{\bar{z}_j + z_i}$, and $f_{x0}^{ij} = f_{0y}^{ij} = 0$.

4. $\boldsymbol{G_A}(s) = \boldsymbol{I} - \boldsymbol{B}^*(s\boldsymbol{I} + \bar{\boldsymbol{A}})^{-1}\boldsymbol{F}^{-1}\boldsymbol{B}$.

Two typical examples are given here to illustrate the use of these formulas.

Example 14.4.3. *Consider the plant in Example 14.4.1:*

$$\boldsymbol{G}(s) = \frac{1}{s+1}\begin{bmatrix} s-1 & s-1 \\ -1 & s-2 \end{bmatrix}.$$

It is known from Example 14.4.1 and Example 14.4.2 that

$$\boldsymbol{v_{11}} = [1\ 0], \boldsymbol{v_{12}} = [1\ -1], \qquad \boldsymbol{F} = \begin{bmatrix} 1/2 & 3/4 \\ 3/4 & 7/4 \end{bmatrix},$$

$$\boldsymbol{A} = \begin{bmatrix} 1 & -1 \\ 0 & 1 \end{bmatrix}, \qquad \boldsymbol{B} = \begin{bmatrix} 1 & 0 \\ 1 & -1 \end{bmatrix}.$$

Then

$$\begin{aligned} & \boldsymbol{G_A}(s) \\ = & \ \boldsymbol{I} - \boldsymbol{B}^*(s\boldsymbol{I} + \bar{\boldsymbol{A}})^{-1}\boldsymbol{F}^{-1}\boldsymbol{B} \\ = & \ \boldsymbol{I} - \begin{bmatrix} 1 & 1 \\ 0 & -1 \end{bmatrix}\left(s\boldsymbol{I} + \begin{bmatrix} 1 & -1 \\ 0 & 1 \end{bmatrix}\right)^{-1}\begin{bmatrix} 1/2 & 3/4 \\ 3/4 & 7/4 \end{bmatrix}^{-1}\begin{bmatrix} 1 & 0 \\ 1 & -1 \end{bmatrix} \\ = & \begin{bmatrix} \frac{5s^2-2s-3}{5(s+1)^2} & \frac{-4(s-1)}{5(s+1)^2} \\ \frac{-4}{5(s+1)} & \frac{5s-3}{5(s+1)} \end{bmatrix}, \end{aligned}$$

and

$$\boldsymbol{G_{MP}}(s) = \begin{bmatrix} \frac{5s+7}{5(s+1)} & \frac{5s+11}{5(s+1)} \\ \frac{-1}{5(s+1)} & \frac{5s+2}{5(s+1)} \end{bmatrix}.$$

Example 14.4.4. *Consider the plant in Example 14.3.1:*

$$G(s) = \begin{bmatrix} \frac{(s-1)^2}{(s+1)^2} & \frac{(s-1)^2}{(s+1)^2} \\ \frac{-1}{s+1} & \frac{s-2}{s+1} \end{bmatrix}.$$

The plant has a RHP zero with multiplicity 3 at $s = 1$ (that is, $r_z = 1$, $z_1 = 1$, and $k_j = 3$) and an LHP pole with multiplicity 3 at $s = -1$. The zero directions can be obtained based on (14.4.1):

$$\boldsymbol{v_{11}} = [1\ 0], \boldsymbol{v_{12}} = [1\ 0], \boldsymbol{v_{13}} = [0\ 1/2].$$

As the open RHP zeros and zero directions are known, it is readily obtained that

$$\boldsymbol{A} = \begin{bmatrix} 1 & -1 & 0 \\ 0 & 1 & -1 \\ 0 & 0 & 1 \end{bmatrix}, \boldsymbol{B} = \begin{bmatrix} 1 & 0 \\ 1 & 0 \\ 0 & 1/2 \end{bmatrix}.$$

$\boldsymbol{F}$ *can be computed based on (14.4.2) and (14.4.3):*

$$\boldsymbol{F} = \begin{bmatrix} 1/2 & 3/4 & 3/8 \\ 3/4 & 5/4 & 13/16 \\ 3/8 & 13/16 & 15/16 \end{bmatrix}.$$

Consequently,

$$\begin{aligned} \boldsymbol{G_A}(s) &= \boldsymbol{I} - \boldsymbol{B}^*(s\boldsymbol{I} + \bar{\boldsymbol{A}})^{-1}\boldsymbol{F}^{-1}\boldsymbol{B} \\ &= \boldsymbol{I} - \begin{bmatrix} 1 & 1 & 0 \\ 0 & 0 & 1/2 \end{bmatrix} \left(s\boldsymbol{I} + \begin{bmatrix} 1 & -1 & 0 \\ 0 & 1 & -1 \\ 0 & 0 & 1 \end{bmatrix} \right)^{-1} \\ &\quad \begin{bmatrix} 1/2 & 3/4 & 3/8 \\ 3/4 & 5/4 & 13/16 \\ 3/8 & 13/16 & 15/16 \end{bmatrix}^{-1} \begin{bmatrix} 1 & 0 \\ 1 & 0 \\ 0 & 1/2 \end{bmatrix} \\ &= \begin{bmatrix} \frac{5s^3-7s^2-s+3}{5(s+1)^3} & \frac{-4(s-1)^2}{5(s+1)^3} \\ \frac{-4}{5(s+1)} & \frac{5s-3}{5(s+1)} \end{bmatrix}, \end{aligned}$$

and

$$\boldsymbol{G_{MP}}(s) = \begin{bmatrix} \frac{5s+7}{5(s+1)} & \frac{5s+11}{5(s+1)} \\ \frac{-1}{5(s+1)} & \frac{5s+2}{5(s+1)} \end{bmatrix}.$$

Since the computation is analytical, the result is exact.

14.5 Solution to the H_2 Optimal Control Problem

In this section, the parameterization in Section 13.1 and the extended inner-outer factorization in the last section will be used to derive H_2 optimal controller analytically.

The idea of H_2 optimal control is finding a controller that stabilizes the system and minimizes ISE. As indicated in Chapter 10, the following H_2 performance index is the focus of attention:

$$\min \|\boldsymbol{S}(s)\boldsymbol{W}(s)\|_2, \tag{14.5.1}$$

where $\boldsymbol{W}(s) = \boldsymbol{I}/s$ is the weighting function and

$$\boldsymbol{S}(s) = \boldsymbol{I} - \boldsymbol{G}(s)\boldsymbol{Q}(s).$$

Here $\boldsymbol{Q}(s)$ is the IMC controller. When $\boldsymbol{Q}(s)$ is known, the unity feedback loop controller can be obtained as follows:

$$\boldsymbol{C}(s) = \boldsymbol{Q}(s)[\boldsymbol{I} - \boldsymbol{G}(s)\boldsymbol{Q}(s)]^{-1}. \tag{14.5.2}$$

Lemma 14.5.1. *Assume that $\boldsymbol{G_A}(s)$ is the inner factor of $\boldsymbol{G}(s)$. $\boldsymbol{G_A}^{-1}(s) - \boldsymbol{G_A}^{-1}(0)$ has only unstable poles.*

Proof. It has been known that

$$\boldsymbol{G_A}^{-1}(s) = \boldsymbol{I} - \boldsymbol{B}^*\boldsymbol{F}^{-1}(-s\boldsymbol{I} + \boldsymbol{A}^T)^{-1}\boldsymbol{B}.$$

$\boldsymbol{G_A}(0)$ is a constant matrix. It does not affect the pole distribution. From the expression of $\boldsymbol{G_A}^{-1}(s)$, it is known that $\boldsymbol{G_A}^{-1}(s) - \boldsymbol{G_A}^{-1}(0)$ has only unstable poles. □

In Theorem 13.1.4, all stabilizing controllers with asymptotic tracking property are parameterized. Substituting the parameterization into the H_2 optimization problem, we have:

$$\begin{aligned} & \left\|s^{-1}\boldsymbol{S}(s)\right\|_2^2 \\ = & \left\|s^{-1}[\boldsymbol{I} - \boldsymbol{G}(s)\boldsymbol{Q}(s)]\right\|_2^2 \\ = & \left\|s^{-1}\{\boldsymbol{I} - \boldsymbol{G}(s)\boldsymbol{G}^{-1}(0)[\boldsymbol{I} + s\boldsymbol{Q}_1(s)]\}\right\|_2^2. \end{aligned} \tag{14.5.3}$$

Theorem 14.5.2. *Assume that the plant can be factorized into two parts:*

$$\boldsymbol{G}(s) = \boldsymbol{G_A}(s)\boldsymbol{G_{MP}}(s),$$

where $\boldsymbol{G_A}(s)$ is the inner factor given by Theorem 14.3.3 and $\boldsymbol{G_{MP}}^{-1}(s)$ is the corresponding outer factor. Then the unique optimal solution of the H_2 control problem is

$$\boldsymbol{Q_{opt}}(s) = \boldsymbol{G_{MP}}^{-1}(s)\boldsymbol{G_A}^{-1}(0).$$

Proof. Since $\boldsymbol{G_A}^*(s)\boldsymbol{G_A}(s) = I$, we have

$$
\begin{aligned}
&\left\|s^{-1}\boldsymbol{S}(s)\right\|_2^2 \\
= &\left\| \boldsymbol{G_A}(s)s^{-1}\{\boldsymbol{G_A}^{-1}(s) - \boldsymbol{G_{MP}}(s)\boldsymbol{G}^{-1}(0)[\boldsymbol{I} + s\boldsymbol{Q_1}(s)]\} \right\|_2^2 \\
= &\left\| s^{-1}\{\boldsymbol{G_A}^{-1}(s) - \boldsymbol{G_{MP}}(s)\boldsymbol{G}^{-1}(0)[\boldsymbol{I} + s\boldsymbol{Q_1}(s)]\} \right\|_2^2 \\
= &\left\| \begin{matrix} s^{-1}[\boldsymbol{G_A}^{-1}(s) - \boldsymbol{G_A}^{-1}(0)] + \\ s^{-1}\{\boldsymbol{G_A}^{-1}(0) - \boldsymbol{G_{MP}}(s)\boldsymbol{G}^{-1}(0)[\boldsymbol{I} + s\boldsymbol{Q_1}(s)]\} \end{matrix} \right\|_2^2 .
\end{aligned}
$$

s is a factor of

$$\boldsymbol{G_A}^{-1}(s) - \boldsymbol{G_A}^{-1}(0).$$

Since $\boldsymbol{G_{MP}}(0)\boldsymbol{G}^{-1}(0) = \boldsymbol{G_A}^{-1}(0)$, s must be a factor of

$$\boldsymbol{G_A}^{-1}(0) - \boldsymbol{G_{MP}}(s)\boldsymbol{G}^{-1}(0)[\boldsymbol{I} + s\boldsymbol{Q_1}(s)].$$

It is evident that

$$s^{-1}[\boldsymbol{G_A}^{-1}(s) - \boldsymbol{G_A}^{-1}(0)]$$

is strictly proper.

$$s^{-1}\{\boldsymbol{G_A}^{-1}(0) - \boldsymbol{G_{MP}}(s)\boldsymbol{G}^{-1}(0)[\boldsymbol{I} + s\boldsymbol{Q_1}(s)]\}$$

is also strictly proper if $\boldsymbol{Q}(s) = \boldsymbol{G}^{-1}(0)[\boldsymbol{I} + s\boldsymbol{Q_1}(s)]$ is proper.

On the other hand, from Lemma 14.5.1 it is known that

$$s^{-1}[\boldsymbol{G_A}^{-1}(s) - \boldsymbol{G_A}^{-1}(0)]$$

has only unstable poles.

$$s^{-1}\{\boldsymbol{G_A}^{-1}(0) - \boldsymbol{G_{MP}}(s)\boldsymbol{G}^{-1}(0)[\boldsymbol{I} + s\boldsymbol{Q_1}(s)]\}$$

is stable. To see this, let us consider the following equality:

$$
\begin{aligned}
&\boldsymbol{G_A}(s)\{\boldsymbol{G_A}^{-1}(0) - \boldsymbol{G_{MP}}(s)\boldsymbol{G}^{-1}(0)[\boldsymbol{I} + s\boldsymbol{Q_1}(s)]\} \\
= &[\boldsymbol{G_A}(s)\boldsymbol{G_A}^{-1}(0) - \boldsymbol{I}] + \{\boldsymbol{I} - \boldsymbol{G}(s)\boldsymbol{G}^{-1}(0)[\boldsymbol{I} + s\boldsymbol{Q_1}(s)]\}.
\end{aligned}
$$

As $\boldsymbol{G_A}(s)$ is stable, the first term in the right-hand side is stable. By Theorem 13.1.4, the second term is stable, too.

To find the optimal controller, the constrained search will be replaced with an unconstrained one in the design procedure. Temporarily relax the constraint on $\boldsymbol{Q}(s)$. We have

$$
\begin{aligned}
&\left\|s^{-1}\boldsymbol{S}(s)\right\|_2^2 \\
= &\left\|s^{-1}[\boldsymbol{G_A}^{-1}(s) - \boldsymbol{G_A}^{-1}(0)]\right\|_2^2 +
\end{aligned}
$$

$$\left\|s^{-1}\{\boldsymbol{G_A}^{-1}(0) - \boldsymbol{G_{MP}}(s)\boldsymbol{G}^{-1}(0)[\boldsymbol{I} + s\boldsymbol{Q}_1(s)]\}\right\|_2^2.$$

Minimizing the right-hand side of the equation, we obtain that

$$\boldsymbol{Q_{1opt}}(s) = s^{-1}[\boldsymbol{G}(0)\boldsymbol{G_{MP}}^{-1}(s)\boldsymbol{G_A}^{-1}(0) - \boldsymbol{I}].$$

A little algebra yields

$$\boldsymbol{Q_{opt}}(s) = \boldsymbol{G_{MP}}^{-1}(s)\boldsymbol{G_A}^{-1}(0).$$

□

It can be seen that the controller order is directly related to the plant order. Since the optimal solution is obtained, the achievable performance can be directly estimated.

Corollary 14.5.3. *The optimal performance of the H_2 controller is*

$$\left\|s^{-1}\boldsymbol{B}^*\boldsymbol{F}^{-1}[(\boldsymbol{A}^T)^{-1} - (-s\boldsymbol{I} + \boldsymbol{A}^T)^{-1}]\boldsymbol{B}\right\|_2. \tag{14.5.4}$$

Proof.

$$\begin{aligned} & \boldsymbol{G_A}^{-1}(s) - \boldsymbol{G_A}^{-1}(0) \\ = & [\boldsymbol{I} - \boldsymbol{B}^*\boldsymbol{F}^{-1}(-s\boldsymbol{I} + \boldsymbol{A}^T)^{-1}\boldsymbol{B}] - [\boldsymbol{I} - \boldsymbol{B}^*\boldsymbol{F}^{-1}(\boldsymbol{A}^T)^{-1}\boldsymbol{B}] \\ = & \boldsymbol{B}^*\boldsymbol{F}^{-1}(\boldsymbol{A}^T)^{-1}\boldsymbol{B} - \boldsymbol{B}^*\boldsymbol{F}^{-1}(-s\boldsymbol{I} + \boldsymbol{A}^T)^{-1}\boldsymbol{B} \\ = & \boldsymbol{B}^*\boldsymbol{F}^{-1}[(\boldsymbol{A}^T)^{-1} - (-s\boldsymbol{I} + \boldsymbol{A}^T)^{-1}]\boldsymbol{B}. \end{aligned}$$

Therefore,

$$\min \left\|s^{-1}\boldsymbol{S}(s)\right\|_2 = \left\|s^{-1}\boldsymbol{B}^*\boldsymbol{F}^{-1}[(\boldsymbol{A}^T)^{-1} - (-s\boldsymbol{I} + \boldsymbol{A}^T)^{-1}]\boldsymbol{B}\right\|_2.$$

□

Based on the discussion in Section 13.3, we have the following conclusions:

1. When all RHP zeros have the same multiplicities in each row of $\boldsymbol{G}(s)$, the above optimal performance is identical to that of the H_2 decoupling controller.
2. When the time delays in each row of $\boldsymbol{G}(s)$ are the same, the design procedure given in this section can be directly extended to the control of plants with time delay.

14.6 Filter Design

The optimal controller $\boldsymbol{Q_{opt}}(s)$ is usually improper. To implement the controller, a filter $\boldsymbol{J}(s)$ must be introduced. When the model is exactly known, the optimal solution can be arbitrarily approached by choosing an appropriate filter while the internal stability is kept. However, the optimal solution can never be reached, because the optimal controller is physically irrealizable.

In general, the filter should satisfy the following requirements:

1. $\boldsymbol{Q}(s) = \boldsymbol{Q_{opt}}(s)\boldsymbol{J}(s)$ is proper.
2. The closed-loop system is internally stable.
3. The asymptotic tracking can be reached.

For clarity of presentation, it is assumed that the system inputs are steps. Depending on different plants, the filter is chosen in different ways.

Stable plants For stable plants, the filter can be chosen as a diagonal one:

$$\boldsymbol{J}(s) = \begin{bmatrix} J_1(s) & \cdots & 0 \\ \vdots & \ddots & \vdots \\ 0 & \cdots & J_n(s) \end{bmatrix}, \tag{14.6.1}$$

with

$$J_i(s) = \frac{1}{(\lambda_i s + 1)^{n_i}}, i = 1, 2, ..., n, \tag{14.6.2}$$

where $\lambda_i (i = 1, 2, ..., n)$ are performance degrees.

The first condition is easy to satisfy. Assume that the smallest relative degree in any element of the ith column of $\boldsymbol{Q_{opt}}(s)$ is α_i. To satisfy the first condition, one can take $n_i = -\alpha_i$ for improper columns and $n_i = 1$ for the others.

The second condition and the third condition have already been satisfied. Since $\boldsymbol{J}(s)$ is stable, the closed-loop system must be internally stable. $\boldsymbol{J}(0) = \boldsymbol{I}$. Hence,

$$\begin{aligned} & \lim_{s \to 0} \det[\boldsymbol{I} - \boldsymbol{G}(s)\boldsymbol{Q}(s)] \\ = & \lim_{s \to 0} \det[\boldsymbol{I} - \boldsymbol{G_A}(s)\boldsymbol{G_A}^{-1}(0)\boldsymbol{J}(s)] \\ = & \ 0. \end{aligned}$$

The tuning method for quantitative performance and robustness is similar to that in the H_2 decoupling control. As different loops are coupled, the tuning procedure is more complex than that for the decoupling control.

Unstable MP plants For unstable MP plants, the filter can also be chosen as a diagonal one:

$$\boldsymbol{J}(s) = \begin{bmatrix} J_1(s) & \cdots & 0 \\ \vdots & \ddots & \vdots \\ 0 & \cdots & J_n(s) \end{bmatrix}, \tag{14.6.3}$$

with

$$J_i(s) = \frac{N_{xi}(s)}{(\lambda_i s + 1)^{n_i}}, i = 1, 2, ..., n, \tag{14.6.4}$$

where $\lambda_i (i = 1, 2, ..., n)$ are performance degrees, $N_{xi}(s)$ are polynomials with all roots in the LHP and $N_{xi}(0) = 1$. Suppose l_{ij} is the largest multiplicity of the unstable pole $p_j (j = 1, 2, ..., r_p)$ in the ith row of $\boldsymbol{G}(s)$. $\deg\{N_{xi}(s)\} = \sum_{j=1}^{r_p} l_{ij}$.

Assume that the smallest relative degree in any element of the ith column of $\boldsymbol{Q_{opt}}(s)$ is α_i. To satisfy the first condition, one can take $n_i = \deg\{N_{xi}(s)\} - \alpha_i$ for the improper columns and $n_i = \deg\{N_{xi}(s)\} + 1$ for the others.

For MP plants, $\boldsymbol{G}(s)\boldsymbol{Q_{opt}}(s) = \boldsymbol{I}$. The closed-loop response is decoupled. To make the closed-loop system internally stable, the ith element of the filter should satisfy

$$\lim_{s \to p_j} \frac{d^k}{ds^k}[1 - J_i(s)] = 0, \tag{14.6.5}$$

$$i = 1, 2, ..., n; j = 1, 2, ..., r_p; k = 0, 1, ..., l_{ij} - 1.$$

Since $\boldsymbol{J}(0) = \boldsymbol{I}$, the third condition has already been satisfied.

Unstable NMP plants For unstable NMP plants, a more complex structure may be necessary for the filter. The first condition is easy to satisfy. As it is known, an improper transfer function implies that the degree of its numerator is greater than that of its denominator. To make it proper, a pole-zero excess should be introduced by utilizing the filter. This is not difficult.

The second condition is normally not easy to satisfy, because $\boldsymbol{J}(s)$ is determined by

$$\lim_{s \to p_j} \frac{d^k}{ds^k} \det[\boldsymbol{I} - \boldsymbol{G_A}(s)\boldsymbol{G_A}^{-1}(0)\boldsymbol{J}(s)] = 0, \tag{14.6.6}$$

$$j = 1, 2, ..., r_p; k = 0, 1, ..., l_j - 1.$$

To solve the problem, let $\boldsymbol{J}(s) = \boldsymbol{J_F}(s)\boldsymbol{J_D}(s)$, where the subscripts F and D denote "full matrix" and "diagonal matrix," respectively,

$$\boldsymbol{J_F}(s) = \boldsymbol{G_A}(0)\boldsymbol{G_A}^{-1}(s), \tag{14.6.7}$$

$$\boldsymbol{J_D}(s) = \begin{bmatrix} J_1(s) & \cdots & 0 \\ \vdots & \ddots & \vdots \\ 0 & \cdots & J_n(s) \end{bmatrix}, \tag{14.6.8}$$

with

$$J_i(s) = \prod_{j=1}^{r_z} (-s/z_j + 1)^{k_{ij}} \frac{N_{xi}(s)}{(\lambda_i s + 1)^{n_i}}, i = 1, 2, ..., n, \tag{14.6.9}$$

where $\lambda_i (i = 1, 2, ..., n)$ are performance degrees, $N_{xi}(s)$ are polynomials with all roots in the LHP, and k_{ij} is the largest multiplicity of z_j in the ith column of $\boldsymbol{J_F}(s)$. As $\boldsymbol{J_D}(s)$ removes all unstable poles of $\boldsymbol{J_F}(s)$, $\boldsymbol{J}(s)$ is stable. Suppose l_{ij} is the largest multiplicity of the unstable pole $p_j (j = 1, 2, ..., r_p)$ in the ith row of $\boldsymbol{G}(s)$. $\deg\{N_{xi}(s)\} = \sum_{j=1}^{r_p} l_{ij}$. The second condition reduces to

$$\lim_{s \to p_j} \frac{d^k}{ds^k} [1 - J_i(s)] = 0, \tag{14.6.10}$$
$$i = 1, 2, ..., n; j = 1, 2, ..., r_p; k = 0, 1, ..., l_{ij} - 1.$$

The third condition can be satisfied by choosing $N_{xi}(0) = 1$. The order of $\boldsymbol{J_D}(s)$ should be chosen so that $\boldsymbol{Q}(s)$ is proper.

As $\boldsymbol{T}(s) = \boldsymbol{G}(s)\boldsymbol{Q}(s) = \boldsymbol{J_D}(s)$, the obtained response is decoupled. As a matter of fact, the response is identical to that in Section 12.2.

Compared to the design with the weighting functions $\boldsymbol{W_{p1}}(s)$ and $\boldsymbol{W_{p2}}(s)$, the introduction of a filter simplifies the design task. The designer is not required to choose weighting functions by trial and error.

Now let us see how to simplify the choosing of weighting functions when a filter is used.

Consider the weighting function $\boldsymbol{W_{p1}}(s)$ first. In Section 10.4, it is assumed that the inputs are unit steps, and the controller is designed only for the prespecified weighting function $\boldsymbol{W_{p1}}(s) = s^{-1}\boldsymbol{I}$ and the following performance index:

$$\min \|\boldsymbol{W_{p2}}(s)\boldsymbol{S}(s)\boldsymbol{W_{p1}}(s)\|_2^2 . \tag{14.6.11}$$

However, the system inputs may be complicated. They may be steps with lags (for example, $\boldsymbol{r}(s) = \boldsymbol{I}/s/(s + 1)$) or the ramp (that is, $\boldsymbol{r}(s) = \boldsymbol{I}/s/s$). If the weight function $\boldsymbol{W}_{p1}(s)$ is chosen to equal the input, the design procedure will be complex. In this case, one can choose the weighting function as $s^{-1}\boldsymbol{I}$ and adopt the following simple design procedure:

1. Design the controller for unit steps.

2. Choose an appropriate filter $\boldsymbol{J}(s)$ to satisfy the constraints imposed by the asymptotic tracking property.

This design procedure can be used for both the optimal control and the decoupling control.

Now consider the weighting function $\boldsymbol{W}_{p2}(s)$. As it is known, $\boldsymbol{W}_{p2}(s)$ is used to weight errors over different frequency ranges. Although the optimal controller can be derived for a general weighting function $\boldsymbol{W}_{p2}(s)$, the procedure and the obtained controller will be complicated. The optimal solution for the general $\boldsymbol{W}_{p2}(s)$ is

$$\begin{aligned} \boldsymbol{Q_{opt}}(s) \;=\; & \boldsymbol{G_{MP}}^{-1}(s)[\boldsymbol{W_{p2}}(s)\boldsymbol{G_A}(s)]_{\boldsymbol{MP}}^{-1} \\ & \{s\{[\boldsymbol{W_{p2}}(s)\boldsymbol{G_A}(s)]_{\boldsymbol{MP}}\boldsymbol{G_A}^{-1}(s)/s - \\ & [\boldsymbol{W_{p2}}(0)\boldsymbol{G_A}(0)]_{\boldsymbol{MP}}\boldsymbol{G_A}^{-1}(0)/s\}_* + \\ & [\boldsymbol{W_{p2}}(0)\boldsymbol{G_A}(0)]_{\boldsymbol{MP}}\boldsymbol{G_A}^{-1}(0)\}, \end{aligned} \tag{14.6.12}$$

where $\boldsymbol{W_{p2}}(s)\boldsymbol{G_A}(s) = [\boldsymbol{W_{p2}}(s)\boldsymbol{G_A}(s)]_{\boldsymbol{A}}[\boldsymbol{W_{p2}}(s)\boldsymbol{G_A}(s)]_{\boldsymbol{MP}}$, $[\boldsymbol{W_{p2}}(s)\boldsymbol{G_A}(s)]_{\boldsymbol{A}}$ and $[\boldsymbol{W_{p2}}(s)\boldsymbol{G_A}(s)]_{\boldsymbol{MP}}$ denote the all-pass and MP parts of $\boldsymbol{W_{p2}}(s)\boldsymbol{G_A}(s)$, respectively. $[\boldsymbol{W_{p2}}(0)\boldsymbol{G_A}(0)]_{\boldsymbol{MP}}$ denotes the value of $[\boldsymbol{W_{p2}}(s)\boldsymbol{G_A}(s)]_{\boldsymbol{MP}}$ at $s = 0$. $\{\cdot\}_*$ denotes that, after a partial fraction expansion of the function, all terms involving the poles z_j are removed. This is why the controller is designed only for a simple weighting function $\boldsymbol{W}_{p2}(s) = \boldsymbol{I}$. With the help of a filter, the errors can be weighted in an easy way; that is, the weighting is achieved by tuning.

14.7 Examples for H$_2$ Optimal Controller Design

The purpose of this section is to illustrate the H$_2$ optimal design procedure. The design procedure is summarized as follows:

1. Factorize the plant: $\boldsymbol{G}(s) = \boldsymbol{G_A}(s)\boldsymbol{G_{MP}}(s)$, where $\boldsymbol{G_A}(s) = \boldsymbol{I} - \boldsymbol{B}^*(s\boldsymbol{I} + \bar{\boldsymbol{A}})^{-1}\boldsymbol{F}^{-1}\boldsymbol{B}$.
2. Compute the optimal controller: $\boldsymbol{Q_{opt}}(s) = \boldsymbol{G_{MP}}^{-1}(s)\boldsymbol{G_A}^{-1}(0)$.
3. Introduce a filter to the optimal controller: $\boldsymbol{Q}(s) = \boldsymbol{Q_{opt}}(s)\boldsymbol{J}(s)$. The unity feedback controller is $\boldsymbol{C}(s) = \boldsymbol{Q}(s)\left[\boldsymbol{I} - \boldsymbol{G}(s)\boldsymbol{Q}(s)\right]^{-1}$.

Three examples are provided in this section. In the first example, the controller is analytically designed and tuned for the required quantitative undershoot. The second example is given to illustrate the quantitative tuning for weighting errors of different channels. In the third example, a real plant with frequency domain design requirements is considered. It is shown how the quantitative requirements can be easily met with the design method introduced in this chapter.

Example 14.7.1. *Consider the following plant:*

$$\boxed{\boldsymbol{G}(s)=\frac{1}{(s+1)^3}\begin{bmatrix}(s-1)^2 & (s-1)^2\\ (s-1)(s-2) & 2(s-1)(s-2)\end{bmatrix}.}$$

The plant has three NMP zeros at $s=1$, one NMP zero at $s=2$, and 6 stable poles at $s=-1$. One zero at $s=1$ is the common zero of all elements of $\boldsymbol{G}(s)$. As introduced in Section 14.3, the first step is separating the common zero by removing the following factor:

$$\frac{-s+1}{s+1}.$$

The next step is factorizing the remainder of $\boldsymbol{G}(s)$. Let

$$\boldsymbol{G_r}(s)=\frac{s+1}{-s+1}\boldsymbol{G}(s)=\frac{-1}{(s+1)^2}\begin{bmatrix}s-1 & s-1\\ s-2 & 2(s-2)\end{bmatrix}.$$

Since

$$\boldsymbol{A}=\begin{bmatrix}1 & 0\\ 0 & 2\end{bmatrix},\boldsymbol{B}=\begin{bmatrix}1 & 0\\ 0 & 1\end{bmatrix},\boldsymbol{F}=\begin{bmatrix}1/2 & 0\\ 0 & 1/4\end{bmatrix},$$

the inner factor of $\boldsymbol{G_r}(s)$ is

$$\boldsymbol{G_A}(s)=\begin{bmatrix}\frac{s-1}{s+1} & 0\\ 0 & \frac{s-2}{s+2}\end{bmatrix}.$$

The inner factor of the original plant $\boldsymbol{G}(s)$ is

$$\frac{-s+1}{s+1}\boldsymbol{G_A}(s)=\frac{-s+1}{s+1}\begin{bmatrix}\frac{s-1}{s+1} & 0\\ 0 & \frac{s-2}{s+2}\end{bmatrix}.$$

Therefore

$$\boldsymbol{G_{MP}}(s)=\frac{s+1}{-s+1}\boldsymbol{G_A}^{-1}(s)\boldsymbol{G}(s)=\frac{-1}{(s+1)^2}\begin{bmatrix}s+1 & s+1\\ s+2 & 2(s+2)\end{bmatrix}.$$

It is easy to verify that, for this special plant, H_2 optimal control and H_2 decoupling control result in the same factorization.

By Theorem 14.5.2, the optimal controller is

$$\boldsymbol{Q_{opt}}(s)=\boldsymbol{G_{MP}}^{-1}(s)\boldsymbol{G_A}^{-1}(0)=\frac{s+1}{s+2}\begin{bmatrix}2(s+2) & -(s+1)\\ -(s+2) & s+1\end{bmatrix}.$$

The plant is stable. For step inputs, choose

$$\boldsymbol{J}(s)=\begin{bmatrix}\frac{1}{\lambda_1 s+1} & 0\\ 0 & \frac{1}{\lambda_2 s+1}\end{bmatrix}.$$

The suboptimal controller is

$$\boldsymbol{Q}(s) = \boldsymbol{Q_{opt}}(s)\boldsymbol{J}(s) = \frac{s+1}{s+2}\begin{bmatrix} \frac{2(s+2)}{\lambda_1 s+1} & \frac{-(s+1)}{\lambda_2 s+1} \\ \frac{-(s+2)}{\lambda_1 s+1} & \frac{s+1}{\lambda_2 s+1} \end{bmatrix}.$$

The sensitivity function is

$$\begin{aligned} \boldsymbol{S}(s) &= \boldsymbol{I} - \boldsymbol{G}(s)\boldsymbol{Q}(s) \\ &= \begin{bmatrix} 1 - \frac{(s-1)^2}{(s+1)^2(\lambda_1 s+1)} & 0 \\ 0 & 1 - \frac{(s-1)(s-2)}{(s+1)(s+2)(\lambda_2 s+1)} \end{bmatrix}. \end{aligned}$$

The unity feedback loop controller is

$$\boxed{\boldsymbol{C}(s) = \begin{bmatrix} \frac{2(s+1)^3}{s(\lambda_1 s^2+2\lambda_1 s+\lambda_1+4)} & \frac{-(s+1)^3}{s(\lambda_2 s^2+3\lambda_2 s+2\lambda_2+6)} \\ \frac{-(s+1)^3}{s(\lambda_1 s^2+2\lambda_1 s+\lambda_1+4)} & \frac{(s+1)^3}{s(\lambda_2 s^2+3\lambda_2 s+2\lambda_2+6)} \end{bmatrix}.}$$

This controller is exact. There is no numerical error.

The performance degrees are determined by the desired closed-loop responses, such as the overshoot, amplitudes of coupled responses, shape of $\boldsymbol{S}(s)$*, and so on. Suppose the design specification is a 20% undershoot with the shortest rise time for both loops. One can take* $\lambda_1 = 1.25$ *and* $\lambda_2 = 1.05$*. The closed-loop responses are shown in Figure 14.7.1.*

Example 14.7.2. *Consider the plant in Example 14.4.1:*

$$\boxed{\boldsymbol{G}(s) = \frac{1}{s+1}\begin{bmatrix} s-1 & s-1 \\ -1 & s-2 \end{bmatrix}.}$$

It has been obtained in Example 14.4.3 that

$$\boldsymbol{G_A}(s) = \begin{bmatrix} \frac{5s^2-2s-3}{5(s+1)^2} & \frac{-4(s-1)}{5(s+1)^2} \\ \frac{-4}{5(s+1)} & \frac{5s-3}{5(s+1)} \end{bmatrix}$$

and

$$\boldsymbol{G_{MP}}(s) = \begin{bmatrix} \frac{5s+7}{5(s+1)} & \frac{5s+11}{5(s+1)} \\ \frac{-1}{5(s+1)} & \frac{5s+2}{5(s+1)} \end{bmatrix}.$$

Hence, the optimal controller is

$$\boldsymbol{Q_{opt}}(s) = \boldsymbol{G_{MP}}^{-1}(s)\boldsymbol{G_A}^{-1}(0) = \frac{1}{5(s+1)}\begin{bmatrix} -(7s+10) & -(s-5) \\ 4s+5 & -(3s+5) \end{bmatrix}.$$

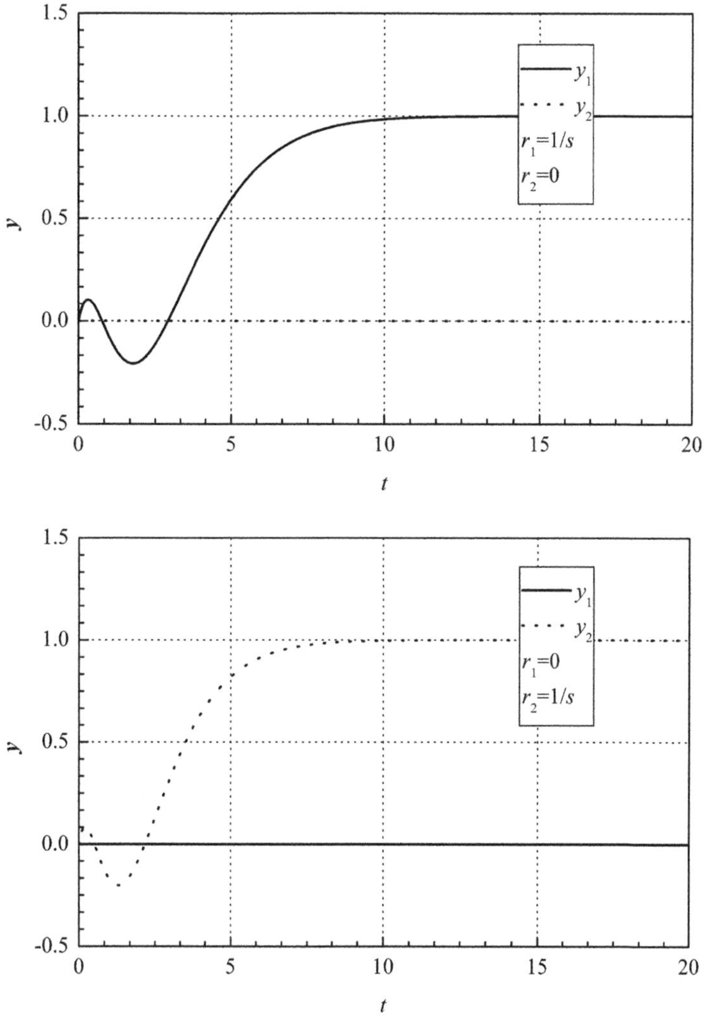

FIGURE 14.7.1
Responses of the system with $\lambda_1 = 1.25$ and $\lambda_2 = 1.05$.

Introduce the following filter for step inputs:

$$\boldsymbol{J}(s) = \begin{bmatrix} \frac{1}{\lambda_1 s+1} & 0 \\ 0 & \frac{1}{\lambda_2 s+1} \end{bmatrix}.$$

The suboptimal controller is

$$\boldsymbol{Q}(s) = \boldsymbol{Q}_{\boldsymbol{opt}}(s)\boldsymbol{J}(s) = \frac{1}{5(s+1)} \begin{bmatrix} \frac{-(7s+10)}{\lambda_1 s+1} & \frac{-(s-5)}{\lambda_2 s+1} \\ \frac{4s+5}{\lambda_1 s+1} & \frac{-(3s+5)}{\lambda_2 s+1} \end{bmatrix}.$$

The sensitivity function is

$$\begin{aligned} \boldsymbol{S}(s) &= \boldsymbol{I} - \boldsymbol{G}(s)\boldsymbol{Q}(s) \\ &= \begin{bmatrix} 1 - \frac{5s^2-2s-3}{5(s+1)^2(\lambda_1 s+1)} & \frac{4(s-1)}{5(s+1)^2} \\ \frac{4}{5(s+1)(\lambda_1 s+1)} & 1 - \frac{5s-3}{5(s+1)(\lambda_2 s+1)(\lambda_1 s+1)} \end{bmatrix}. \end{aligned}$$

The unity feedback loop controller is

$$\boldsymbol{C}(s) = \boldsymbol{Q}(s)[\boldsymbol{I} - \boldsymbol{G}(s)\boldsymbol{Q}(s)].$$

Suppose that the design specification is $y_2(t) < 0.3$ *for* $r_1(s) = 1/s$ *and* $r_2(s) = 0$, $y_1(t) < 0.3$ *for* $r_1(s) = 0$ *and* $r_2(s) = 1/s$. *One can take* $\lambda_1 = 1$ *and* $\lambda_2 = 0.5$. *The controller is*

$$\boxed{\boldsymbol{C}(s) = \frac{1}{s(5s^2+34s+65)} \begin{bmatrix} -(7s^2+41s+34) & -2(s^2-8s-9) \\ 4s^2+17s+17 & -2(3s^2+16s+13) \end{bmatrix}.}$$

The closed-loop responses are shown in Figure 14.7.2.

Example 14.7.3. *The longitudinal dynamics of an aircraft trimmed at 25 000ft and 0.9 Mach is unstable and has two RHP phugoid modes (Figure 14.7.3). The linear model can be expressed in the form of*

$$\boxed{\boldsymbol{G}(s) = \frac{\begin{bmatrix} n_{11}(s) & n_{12}(s) \\ n_{21}(s) & n_{22}(s) \end{bmatrix}}{d(s)}},$$

where

$$\begin{aligned} n_{11}(s) &= -5.1240s^4 - 1099.4s^3 - 28390s^2 - 568.48s + 24.076, \\ n_{12}(s) &= -948.12s^3 - 30325s^2 - 56482s - 1215.3, \\ n_{21}(s) &= -0.14896s^4 + 655.67s^3 + 19817s^2 + 385.44s - 61.970, \\ n_{22}(s) &= 671.88s^3 + 21446s^2 + 38716s + 916.45, \\ d(s) &= s^6 + 64.554s^5 + 1167.0s^4 + 3728.6s^3 - \\ &\quad 5495.4s^2 + 1102.0s + 708.10. \end{aligned}$$

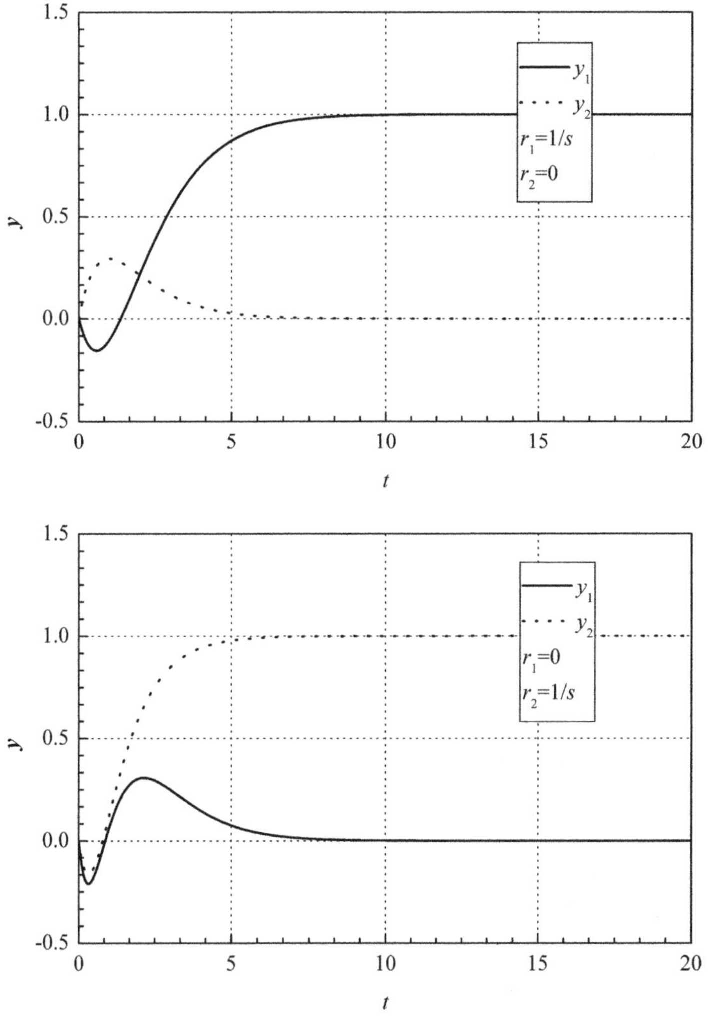

FIGURE 14.7.2
Responses of the system with $\lambda_1 = 1$ and $\lambda_2 = 0.5$.

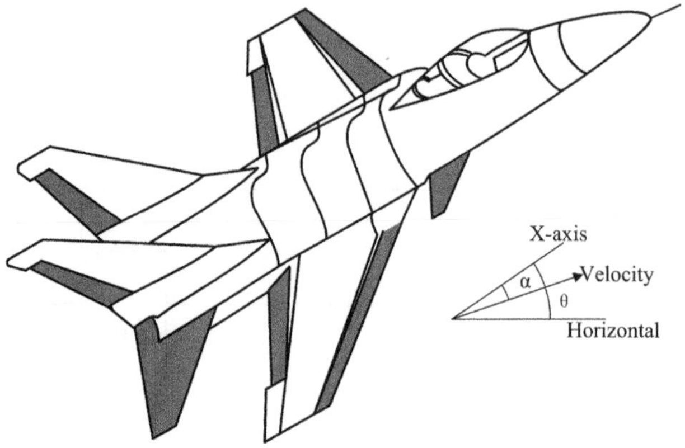

FIGURE 14.7.3
Aircraft and vertical plane geometry.

The control variables are the angles of two flaps and the system outputs are the angle of attack (α) and the attitude angle (θ). The singular value design specification is as follows:

1. *Robustness specification: −40 dB/decade attenuation and at least −20 dB at 100 rad/sec.*

2. *Performance specification: Minimizing the sensitivity function as much as possible.*

The plant is unstable and MP. There are two unstable poles at $s = 0.6898+0.2488i$ and $s = 0.6898-0.2488i$. The optimal solution is

$$\boldsymbol{Q}_{opt}(s) = \boldsymbol{G}^{-1}(s).$$

Since the largest relative degree of the first column is −2, the largest relative degree of the second column is −3, and the plant has two unstable poles, the following filter is chosen:

$$\boldsymbol{J}(s) = \begin{bmatrix} \frac{\beta_{12}s^2+\beta_{11}s+1}{(\lambda_1 s+1)^4} & 0 \\ 0 & \frac{\beta_{22}s^2+\beta_{21}s+1}{(\lambda_2 s+1)^5} \end{bmatrix}.$$

With the following constraints:

$$\lim_{s\to 0.6898+0.2488i}[1-J_i(s)] = 0, \lim_{s\to 0.6898-0.2488i}[1-J_i(s)] = 0, i = 1, 2,$$

we have

$$\beta_{12} = -10.2624\lambda_1 + 6\lambda_1^2 - 0.5377\lambda_1^4 +$$

$$
\begin{aligned}
& 1.3796(7.4387\lambda_1 - 4\lambda_1^3 + 1.3796\lambda_1^4), \\
\beta_{11} &= 0.5377(7.4387\lambda_1 - 4\lambda^3 + 1.3796\lambda_1^4), \\
\beta_{22} &= -12.828\lambda_2 + 10\lambda_2^2 - 2.6887\lambda_2^4 + 0.7419\lambda_2^5 + \\
& 1.3796(9.2984\lambda_2 - 10\lambda_2^3 + 6.898\lambda_2^4 - 1.3656\lambda_2^5), \\
\beta_{21} &= 0.53773(9.2984\lambda_2 - 10\lambda_2^3 + 6.898\lambda_2^4 - 1.3656\lambda_2^5),
\end{aligned}
$$

Therefore, the closed-loop transfer function matrix is

$$
\boxed{\boldsymbol{T}(s) = \begin{bmatrix} \frac{\beta_{12}s^2+\beta_{11}s+1}{(\lambda_1 s+1)^4} & 0 \\ 0 & \frac{\beta_{22}s^2+\beta_{21}s+1}{(\lambda_2 s+1)^5} \end{bmatrix},}
$$

and the sensitivity function matrix is

$$
\boldsymbol{S}(s) = \boldsymbol{I} - \begin{bmatrix} \frac{\beta_{12}s^2+\beta_{11}s+1}{(\lambda_1 s+1)^4} & 0 \\ 0 & \frac{\beta_{22}s^2+\beta_{21}s+1}{(\lambda_2 s+1)^5} \end{bmatrix}.
$$

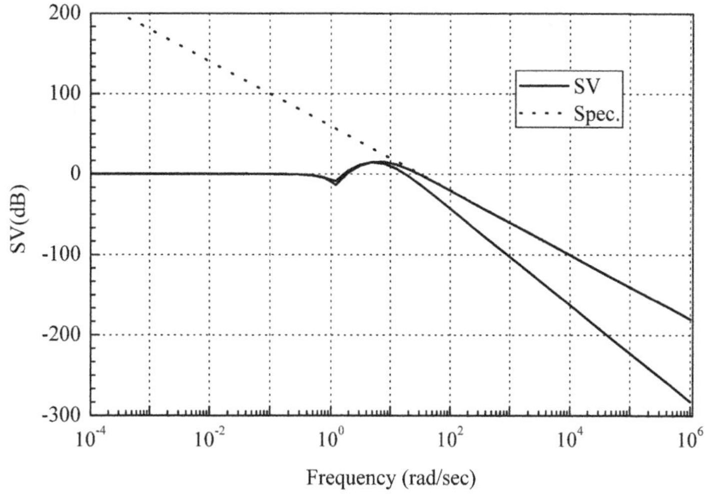

FIGURE 14.7.4
Response of the system with $\lambda_1 = \lambda_2 = 0.16$.

It is seen that the closed-loop response is thoroughly decoupled. Since both of the relative degrees of the two loops in the system are greater than 2, the singular value satisfies the specification of -40 *dB/decade roll-off. For simplicity, let the two performance degrees be the same. Increase the performance degrees until* -20 *dB at 100 rad/sec is reached. The performance degrees are* $\lambda_1 = \lambda_2 = 0.16$. *The frequency domain responses of the closed-loop system are shown in Figure 14.7.4.*

Once the critical $\boldsymbol{T}(s)$ is determined, $\boldsymbol{S}(s)$ is determined at the same time owing to the constraint $\boldsymbol{T}(s) + \boldsymbol{S}(s) = \boldsymbol{I}$.

14.8 Summary

In this chapter, a design procedure is developed for the H_2 optimal controller. By utilizing the parameterization and the extended inner-outer factorization, the unique optimal solution is analytically derived.

Given the unity feedback control loop with a plant $\boldsymbol{G}(s)$, the goal of H_2 optimal control is to design a controller $\boldsymbol{C}(s)$ such that the closed-loop system is internally stable and minimizes the quadratic cost function. The optimal solution to this problem is obtained as follows:

$$\boldsymbol{Q_{opt}}(s) = \boldsymbol{G_{MP}}^{-1}(s)\boldsymbol{G_A}^{-1}(0),$$

where $\boldsymbol{G_A}(s)$ and $\boldsymbol{G_{MP}}(s)$ are the all-pass part and the MP part of $\boldsymbol{G}(s)$, respectively.

$$\begin{aligned} \boldsymbol{G_A}(s) &= \boldsymbol{I} - \boldsymbol{B}^*(s\boldsymbol{I} + \bar{\boldsymbol{A}})^{-1}\boldsymbol{F}^{-1}\boldsymbol{B}. \\ \boldsymbol{G_{MP}}(s) &= \boldsymbol{G_A}^{-1}(s)\boldsymbol{G}(s) \end{aligned}$$

The optimal performance is

$$\left\| s^{-1}\boldsymbol{B}^*\boldsymbol{F}^{-1}[(\boldsymbol{A}^T)^{-1} - (-s\boldsymbol{I} + \boldsymbol{A}^T)^{-1}]\boldsymbol{B} \right\|_2 .$$

An important insight provided by the result is that, even for MIMO plants, the optimal solution can be obtained with only input-output information.

It is seen in this design that the internal stability is guaranteed by controller parameterization, no weighting function needs to be chosen, and the optimal controller is analytically derived. The analytical solution to the H_2 optimal control problem owes to the analytical solution to the extended inner-outer factorization, of which the key is to obtain the analytical solution to a Lyapunov equation. The quantitative tuning of the obtained multivariable controller follows the same way as that in the decoupling control system.

A question unsolved in this chapter is the factorization of the MIMO plant with time delay. This remains a challenging problem. It is conjectured that a rigorous solution does not exist for general plants.

Exercises

1. The zero direction $\boldsymbol{v_j}$ of z_j may be obtained from an SVD of $\boldsymbol{G}(z_j) = \boldsymbol{U\Sigma V}^*$. $\boldsymbol{v_j}^T$ is the last column of $\boldsymbol{U}$. Given the plant

$$\boldsymbol{G}(s) = \frac{1}{s+2}\begin{bmatrix} s-1 & 4 \\ 4.5 & 2(s-1) \end{bmatrix},$$

compute the zero direction of this plant with SVD.

2. Compute the inner factor of the following unstable plant:

$$\boldsymbol{G}(s) = \frac{1}{s-3}\begin{bmatrix} s-1 & s-1 \\ s-2 & 2(s-2) \end{bmatrix}.$$

3. A nonsquare matrix does not have an inverse or a determinant. A partial replacement for the inverse is provided by the Moore-Penrose pseudo-inverse. Assume that $\boldsymbol{G}(s)$ has more inputs than outputs. The Moore-Penrose pseudo-inverse of $\boldsymbol{G}(s)$ is called the right inverse:

$$\boldsymbol{G}^+(s) = \boldsymbol{G}^*(s)[\boldsymbol{G}(s)\boldsymbol{G}^*(s)]^{-1}.$$

Consider the following plant:

$$\boldsymbol{G}(s) = \begin{bmatrix} \frac{s-1}{s+1} & \frac{s-2}{s+2} \end{bmatrix}.$$

(a) Is this plant MP?

(b) Is the right inverse of this plant stable?

4. Assume that the performance index is

$$\min \|\boldsymbol{W_{p2}}(s)\boldsymbol{S}(s)/s\|_2^2.$$

Prove the optimal solution is

$$\begin{aligned}\boldsymbol{Q_{opt}}(s) \quad = \quad & \boldsymbol{G_{MP}}^{-1}(s)[\boldsymbol{W_{p2}}(s)\boldsymbol{G_A}(s)]_{\boldsymbol{MP}}^{-1} \\ & \{s\{[\boldsymbol{W_{p2}}(s)\boldsymbol{G_A}(s)]_{\boldsymbol{MP}}\boldsymbol{G_A}^{-1}(s)/s - \\ & [\boldsymbol{W_{p2}}(0)\boldsymbol{G_A}(0)]_{\boldsymbol{MP}}\boldsymbol{G_A}^{-1}(0)/s\}_* + \\ & [\boldsymbol{W_{p2}}(0)\boldsymbol{G_A}(0)]_{\boldsymbol{MP}}\boldsymbol{G_A}^{-1}(0)\}.\end{aligned}$$

5. Consider the following plant

$$\boldsymbol{G}(s) = \frac{1}{75s+1}\begin{bmatrix} 0.878 & -0.864 \\ 1.082 & -1.096 \end{bmatrix}\begin{bmatrix} k_{11}e^{-\theta_{11}s} & 0 \\ 0 & k_{22}e^{-\theta_{22}s} \end{bmatrix},$$

where $k_{ii} \in [0.8, 1.2]$ and $\theta_{ii} \in [0, 1.0]$, $i = 1, 2$. Physically, this model corresponds to a high-purity distillation column. The aim is to design a controller that meets the following quantitative robust stability and robust performance specifications:

(a) Closed-loop stability.

(b) For a unit step reference in channel 1 at $t = 0$, the plant outputs y_1 (tracking) and y_2 (interaction) should satisfy:

i. $y_1(t) \geq 0.9$ for all $t > 30$min;
ii. $y_1(t) \leq 1.1$ for all t;
iii. $0.99 \leq y_1(\infty) \leq 1.01$;
iv. $y_2(t) \leq 0.5$ for all t;
v. $-0.01 \leq y_2(\infty) \leq 0.01$.

The corresponding requirements hold for a unit step demand in channel 2.

(c) To avoid the controllers with unrealistic gains and bandwidths, the transfer function matrix between output disturbances and plant inputs be gain limited to about 50 dB and the unity gain cross over frequency of the largest singular value should be below 150 rad/min.

Design a controller.

6. For Unstable NMP plants, choose a filter such that the closed-loop response is identical to that in Section 13.4.

7*. A state feedback system with an observer can be converted into an output feedback system. Let $\boldsymbol{G}(s) = \boldsymbol{C}(s\boldsymbol{I} - \boldsymbol{A})^{-1}\boldsymbol{B}$ be the transfer function matrix of the plant, where $\boldsymbol{A}, \boldsymbol{B}$, and $\boldsymbol{C}$ are matrices of appropriate dimensions. $\boldsymbol{K}$ is the feedback gain. The reduced-order observer can be expressed as

$$\begin{aligned}\dot{\boldsymbol{z}} &= \boldsymbol{F}\boldsymbol{z} + \boldsymbol{G}\boldsymbol{y} + \boldsymbol{H}\boldsymbol{u}, \\ \hat{\boldsymbol{x}} &= \boldsymbol{Q}_1\boldsymbol{y} + \boldsymbol{Q}_2\boldsymbol{z},\end{aligned}$$

where $\boldsymbol{F}, \boldsymbol{G}, \boldsymbol{H}, \boldsymbol{Q}_1$, and $\boldsymbol{Q}_2$ are matrices of appropriate dimensions (Figure E14.1). The system is equivalent to the one in Figure E14.2. Derive the expression of $\boldsymbol{G}_1(s)$ and $\boldsymbol{G}_2(s)$.

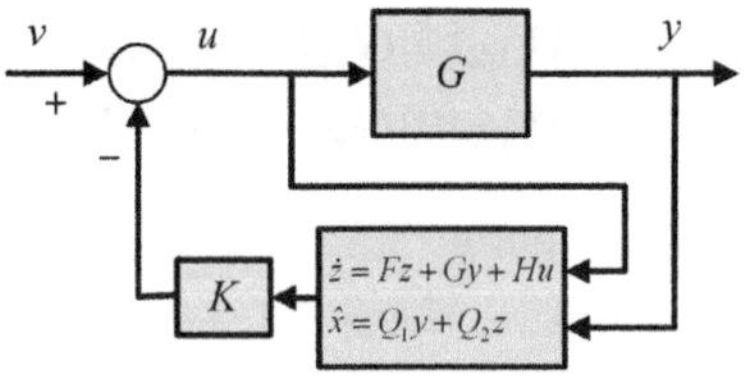

FIGURE E14.1
State feedback system with an observer.

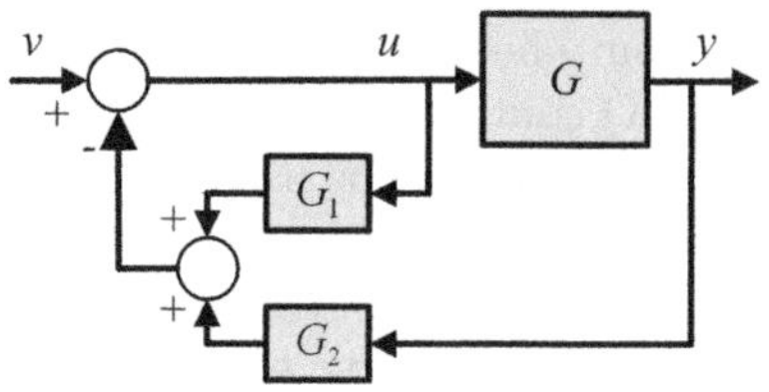

FIGURE E14.2
The equivalent output feedback system.

Notes and References

The basis of this chapter is Zhang et al. (2011) (Zhang W. D., S. W. Gao, and D. Y. Gu, No-weight design of H_2 controllers for square plants, *IET Control Theory and Applications*, 2011, 5(6), 785–794. ©IET). This work was inspired by Morari and Zafiriou (1989, Chapter 12).

The zero direction for a simple zero was first formulated by McFarlane and Karcanias (1976).

The inner-outer factorization problem discussed in Sections 14.1–14.3 is closely related to another problem called the spectral factorization (that is, to find the spectral factor of a matrix-valued spectral density). One can find the algorithms for the original inner-outer factorization in, for example, Morari and Zafiriou (1989) and Oara and Varga (2000). The famous control software MATLAB® provides the command iofr/iofc for the original inner-outer factorization.

The result in Sections 14.1–14.3 can be regarded as a special case of the Nevanlinna-Pick interpolation problem, which was discussed by Ball et al. (1990) in detail. The dual problem of the original inner-outer factorization was studied by Shaked (1989).

Kucera (2007) discussed a transfer function solution to the H_2 control problem.

The plant in Example 14.7.3 is from *The User Guide for Robust Control Toolbox* (Mathworks, 2001, p. 63), where the plant is used to illustrate the design procedures of H_2 control and H_∞ control.

The plant in Exercise 1 is from Skogestad and Postlethwaite (2005, Section 4.5).

Exercise 3 gives an important feature of the nonsquare plant.

The plant in Exercise 5 was given by Skogestad and Morari (1988). The plant and the design specification were reformulated by Limebeer and then published in the IEEE CDC as a Benchmark Problem (Limebeer, 1991). Although the plant is simple, the design problem is difficult. Zhang et al. (2012) gave a simple solution.

Exercise 7 is adapted from Zheng (1990, p. 215). The observer-based feedback is very important in LQ control, because a full state feedback is usually hard to come by, while the observer can be used to "infer" the missing state information from a few output measurements. Nevertheless, the observer-based feedback causes two new problems: the system structure is complicated and the system performance becomes sensitive to uncertainty.

Bibliography

Alcantara, S., W. D. Zhang, C. Pedreta, R. Vilanova, and S. Skogestad (2011). IMC-like analytical H_∞ design with S/PS mixed sensitivity consideration: Utility in PID tuning guidance. *J. of Process Control 21*(6), 976–985.

Astrom, K. J. (1970). *Introduction to Stochastic Control Theory.* London: Academic Press.

Astrom, K. J. and T. Hagglund (2005). *Advanced PID Control.* NC: ISA.

Astrom, K. J., C. C. Hang, and B. C. Lim (1994). A new Smith predictor for controlling a process with an integrator and long dead time. *IEEE Trans. Auto. Control 39*(2), 343–345.

Astrom, K. J., C. C. Hang, P. Persson, and W. Ho (1992). Towards intelligent PID control. *Automatica 28*(1), 1–9.

Atay, F. M. (2010). *Complex Time-Delay Systems: Theory and Applications.* Berlin: Springer.

Ball, J. A., I. Gohberg, and L. Rodman (1990). *Interpolation of Rational Matrix Functions.* Berlin: Birkhauser Verlag.

Bhattacharyya, S. P., A. Datta, and L. H. Keel (2009). *Linear Control Theory: Structure, Robustness, and Optimization.* NY: CRC Press.

Bittar, A. and R. M. Sales (1994). H2 and H infinity control for maglev vehicles. *IEEE Control Systems Magazine 18*(4), 18–25.

Boyd, S., L. Ghaoui, E. Feron, and V. Balakrishnan (1994). *Linear Matrix Inequivalities in System and Control Theory.* PA: SIAM.

Bristol, E. H. (1966). On a new measure of interaction for multivariable process control. *IEEE Trans. Auto. Control 11*(1), 133–134.

Brosilow, C. and B. Joseph (2002). *Techniques of Model-Based Control.* Prentice Hall International Series in the Physical and Chemical Engineering Sciences. NY: Prentice Hall PTR.

Brosilow, C. B. and M. Tong (1978). The structure and dynamics of inferential control system. *AIChE J. 24*(3), 492–499.

Camacho, E. F. and C. Bordons (1999). *Model Predictive Control.* London: Springer.

Chien, I. and P. S. Fruehauf (1990). Consider IMC tuning to improve controller performance. *Chemical Engineering Progress 86*(10), 33–41.

Chien, I., S. C. Peng, and J. H. Liu (2002). Simple control method for integrating processes with long deadtime. *J. of Process Control 12*(3), 391–404.

Cohen, G. H. and G. A. Coon (1953). Theoretical considerations of retarded control. *Trans. ASME 75*(7), 827–834.

Coulibaly, E., S. Maiti, and C. Brosilow (1990). Internal model predictive control. In *AIChE Annual Meeting*, Miami, FL.

Culter, C. R. and B. L. Ramaker (1979). Dynamic matrix control-a computer control algorithm. In *AIChE 86th National Meeting*, Houston,TX.

Dahlin, E. B. (1968). Designing and tuning digital controllers. *Instr. and Contr. Systms 41*(7), 77–83.

D'Azzo, J. J. and C. H. Houpis (1988). *Linear Control System Analysis and Design.* NY: McGraw Hill.

de la Barra S., B. A. L. (1994). On undershoot in SISO systems. *IEEE Trans. Auto. Control 39*(3), 578–58.

Deshpande, P. B. and R. H. Ash (1981). *Elements of Computer Process Control.* North Carolina: Instrument Society of American, Research Triangle Park.

Dong, J. and C. B. Brosilow (1997). Design of robust multivariable PID controller via imc. In *Proc. Amer. Contr. Conf.*, Albuquerque, NM, pp. 3380–3384.

Dorato, P. (2000). *Analytical Feedback System Design.* CA: Brooks/Cole.

Dorato, P. R., L. Fortuna, and G. Muscato (1992). *Robust Control for Unstructured Perturbations–An Introduction.* Lecture Notes in Control and Information Sciences. NY: Springer.

Dorf, R. C. and R. H. Bishop (2001). *Modern Control Systems* (9th ed.). NJ: Prentice Hall.

Doyle, J. C. (1982). Analysis of control systems with structured uncertainty. *IEE Proc. - D 129*(6), 242–250.

Doyle, J. C. (1983). Synthesis of robust controllers and filters. In *Proc. IEEE Conference on Decision and Cotrol*, San Antonio, TX, pp. 109–114.

Doyle, J. C., B. A. Francis, and A. R. Tannenbaum (1992). *Feedback Control Theory.* NY: Macmillan Publishing Company.

Edwards, C. and I. Postlethwaite (1998). Antiwindup and bumpless transfer scheme. *Automatica 34*(2), 199–210.

Enns, D. H., H. Ozbay, and A. Tannenbaum (1992). Abstract model and controller design for an unstable aircraft. *J. of Guidance Control and Dynamics 15*(2), 498–508.

Floudas, C. A. (1995). *Nonlinear and Mixed-Integer Optimization: Fundamentals and Applications.* Oxford, UK: Oxford University Press.

Foias, C., H. Ozbay, and A. Tannenbaum (1996). *Robust Control of Infinite Dimensional Systems.* Lecture Notes in Control and Information Sciences. London: Springer.

Francis, B. A. (1987). *A course in H_∞ Control Theory.* Lecture Notes in Control and Information Sciences. London: Springer.

Frei, C. W., E. Bullinger, A. Gentilini, A. H. Glattfelder, T. J. Sieber, and A. Zbinden (2000). Artifact-tolerant controllers for automatic drug delivery in anesthesia. *Crit. Rev. Biomed. Eng. 28*(1–2), 187–192.

Garcia, C. E. and M. Morari (1982). Internal model control−1:a unifying review and some new results. *Ind. Eng. Chem. Proc. Des. & Dev. 21*(2), 308–323.

Glaria, J. J. and G. C. Goodwin (1994). A parameterization for the class of all stabilizing controllers for linear minimum phase plants. *IEEE Trans. Auto. Control 39*(2), 433–435.

Golten, J. and A. Verwer (1991). *Control System Design and Simulation.* NY: McGraw Hill.

Goodwin, G. C., S. F. Graebe, and M. E. Salgado (2001). *Control System Design.* Beijing, China: Tsinghua Universtiy Press.

Gu, D. Y., Y. C. Tang, L. L. Ou, and W. D. Zhang (2005). Relationship between two typical modeling methods in process control and its application in relay feedback autotuning. *Dynamics of Continous Discrete and Impulsive Systems-Series A-Mathmatical Analysis 3*(4), 1650–1657.

Gu, K. Q., V. L. Kharitonov, and J. Chen (2003). *Stability of Time-Delay Systems.* Berlin: Springer.

Hang, C. C., K. J. Astrom, and W. K. Ho (1991). Refinements of the Ziegler–Nichols tuning formula. *IEE Proc. -D 138*(2), 111–118.

Holt, B. R. and M. Morari (1985). Design of resilient processing plants V–the effect of deadtime on dynamic resilience. *Chemical Engineering Science 40*(7), 1229–1237.

Howell, J. R. (1996). Comment regarding "on undershoot in SISO systems." *IEEE Trans. Auto. Control. 41*(12), 1845–1846.

Iinoya, K. and R. J. Altpeter (1962). Inverse response in process control. *Ind. Eng. Chem. 54*(7), 39–43.

Jerome, N. F. and W. H. Ray (1986). High-performance multivariable control strategies for systems having time delays. *AIChE Journal 32*(6), 914–931.

Jin, Y. H. (1993). *Process Control (in Chinese)*. Beijing, China: Tsinghua University Press.

Kaya, I. (2001). Improving performance using cascade control and a Smith predictor. *ISA Trans. 40*(3), 223–234.

Kharitonov, V. L. (1978). Asymptotic stability of an equilibrium position of a family of systems of linear differential equations (in Russian). *Differentialnye Uraveniya 14*(11), 2086–2088.

Kucera, V. (1979). *Discrete Linear Control: the Polynomial Equation Approach*. NY: Wiley.

Kucera, V. (2007). The H_2 control problem: a general transfer-function solution. *Int. J. of Control 80*(5), 800–815.

Kuo, B. C. (2003). *Automatic Control Systems* (9th ed.). NY: John Wiley & Sons.

Kwak, H. J., S. W. Sung, I. Lee, and J. Y. Park (1999). Modified Smith predictor with a new structure for unstable processes. *Ind. Eng. Chem. Res. 38*(2), 405–411.

Kwak, H. J., S. W. Sung, and I. B. Lee (2001). Modified Smith predictors for integrating processes: comparisons and proposition. *Ind. Eng. Chem. Res. 40*(6), 1500–1506.

Kwok, W. W. and D. E. Davison (2007). Implementation of stabilizing control laws - how many controller blocks are needed for a universally good implementation? *IEEE Control Systems Magazine 27*(1), 55–60.

Laub, A. J. and B. C. Moore (1978). Calculation of transmission zeros using QZ techniques. *Automatica 14*(6), 557–566.

Laughlin, D. L., D. E. Rivera, and M. Morari (1987). Smith predictor design for robust performance. *Int. J. Control 46*(2), 477–504.

Lee, Y., J. Lee, and S. Park (2000). PID controller tuning for integrating and unstable processes with time delay. *Chem. Eng. Sci. 55*(17), 3481–3493.

Lee, Y., M. Lee, S. Park, and C. Brisilow (1998). PID controller tuning for desired closed loop responses for SISO systems. *AIChE J. 44*(1), 106–115.

Lewin, D. R. and C. Scali (1988). Feedforward control in the presence of uncertainty. *Ind. Eng. Chem. Res. 27*(12), 2323–2331.

Limebeer, D. J. N. (1991). The specification and purpose of a controller design case study. In *Proc. IEEE Conference on Decision and Control*, Brighton, UK, pp. 1579–1580.

Lin, S. K. and C. J. Fang (1997). Nonovershooting and monotone nondecreasing step responses of a third order SISO system. *IEEE Trans. Auto. Control 42*(9), 1299–1303.

Linkens, D. A. (1993). Anaesthesia simulators for the design of supervisory rule-based control in the operating theatre. *Computing and Control Engineering Journal 4*(2), 55–62.

Lipschutz, S. (1968). *Linear Algebra.* NY: McGraw Hill.

Liu, C. H. (1983). *General decoupling theory of multivariable process control systems.* Lecture Notes in Control and Information Sciences. Berlin: Springer.

Liu, F., J. W. Zhang, and Q. Z. Qu (2002). 4WS control system design based on QFT (in Chinese). *Automotive Engineering 24*(1), 68–72.

Liu, T., X. He, D. Y. Gu, and W. D. Zhang (2004). Analytical decoupling control design for dynamic plants with time delay and double integrators. *IEE Proc. -D 151*(6), 745–753.

Liu, T., W. D. Zhang, and F. R. Gao (2007). Analytical two-degrees-of-freedom (2-DOF) decoupling control scheme for multiple-input-multiple-output (MIMO) processes with time delays. *Ind. Eng. Chem. Res. 46*(20), 6546–6557.

Luyben, W. L. (1986). Simple method for tuning SISO controllers in multivariable systems. *Ind. Eng. Chem. Proc. Des. & Dev. 25*(3), 654–660.

Luyben, W. L. (2001). Effect of derivative algorithm and tuning selection on the PID control of dead time processes. *Ind. Eng. Chem. Res. 40*(16), 3605–3611.

Luyben, W. L. and M. Melcic (1978). Consider reactor control lags. *Hydrocarbon Processing* (3), 115–117.

Maciejowski, J. M. (1989). *Multivariable Feedback Design.* UK: Addison Wesley.

Mackenroth, U. (2010). *Robust Control Systems: Theory and Case Studies.* Berlin: Springer.

Maffezzoni, C., N. Schiavoni, and G. Ferretti (1990). Robust design of cascade control. *IEEE Control Systems Magzine 10*(1), 21–25.

Marshall, J. E. (1979). *Control of Time Delay System.* UK: Peter Peregrinus.

Mathworks, T. (2001). *The User's Guide for Robust Control Toolbox (version 2.08, release 12.1).* The MathWorks.

Mayr, O. (1970). *The Origins of Feedback Control.* Cambridge, MA: MIT Press.

McFarlane, A. G. J. and N. Karcanias (1976). Poles and zeros of linear multivariable systems: A survey of the algebraic, geometric and complex variable theory. *Int. J. Control 24*(1), 33–74.

Morari, M. and E. Zafiriou (1989). *Robust Process Control.* NJ: Prentice Hall, Englewood Cliffs.

Newton, G. C., L. A. Gould, and J. F. Kraiser (1957). *Analytical Design of Feedback Controls.* NY: Wiley.

Normey-Rico, J. E. and E. F. Camacho (2002). A unified approach to design dead–time compensators for stable and integrative processes with dead–time. *IEEE Trans. Auto. Control 47*(2), 299–305.

Oara, C. and A. Varga (2000). Computation of general inner-outer and spectral factorizations. *IEEE Trans. Auto. Control 45*(12), 2307–2325.

O'Dwyer, A. (2006). *Handbook of PI and PID Controller Tuning Rules.* London: Imperial College Press.

Ogata, K. (2002). *Modern Control Engineering.* Prentice Hall.

Ogunnaike, B. A., J. P. Lemaire, M. Morari, and W. H. Ray (1983). Advanced multivariable control of a pilot-scale distillation column. *AIChE J. 29*(4), 632–640.

Ou, L. L., W. D. Zhang, and L. Yu (2009). Stabilizing low-order controllers for LTI systems with time delay. *IEEE Trans. Auto. Control 54*(6), 774–787.

Paraskevopoulos, P., G. Pasgianos, and K. Arvanitis (2006). PID-type controller tuning for unstable first order plus dead time processes based on gain and phase margin specifications. *IEEE Trans. Control Sys. Tech. 14*(5), 926–936.

Postlethwaite, I. and A. G. J. MacFarlane (1979). *A Complex Variable Approach to the Analysis of Linear Multivariable Feedback Systems.* Berlin: Springer.

Powell, J. D., N. P. Fekete, and C. F. Chang (1998). Observer based air–fuel ratio control. *IEEE Control Systems Magzine 18*(5), 72–83.

Prett, D. M., C. E. Garcia, and M. Morari (1990). *The Second Shell Process Control Workshop: Solutions to the Shell Standard Control Problem.* London: Butterworth-Heinemann.

Psarris, P. and C. A. Floudas (1990). Improving dynamic operability in MIMO systems with time delays. *Chemical Engineering Science 45*(12), 3505–3524.

Qiu, L. and K. M. Zhou (2009). *Introduction to Feedback Control.* NJ: Prentice Hall.

Richalet, J. A., A. Rault, J. L. Testud, and J. Papon (1978). Model predictive heuristic control applications to an industry process. *Automatica 14*(5), 413–428.

Rivera, D. E., M. Morari, and S. Skogestad (1986). Internal model control–4: PID controller design. *Ind. Eng. Chem. Proc. Des. & Dev. 25*(1), 252–265.

Saff, E. B. and R. S. Varga (Eds.) (1977). *Pade and Rational Approximation.* London: Academic Press.

Semino, D. and A. Brambilla (1996). An efficient structure for parallel cascade control. *Ind. Eng. Chem. Res. 35*(6), 1845–1852.

Shaked, U. (1989). An explicit expression for the minimum-phase image of transfer function matrices. *IEEE Trans. Auto. Control 34*(12), 1290–1293.

Shamash, Y. (1975). Model reduction using Routh stability criterion and the Pade approximation. *Int. J. Control 21*(3), 475–484.

Shinskey, F. G. (2002). Process control diagnostics. *ISA 2002 www.isa.org*, Technical Papers.

Silva, G. J., A. Datta, and S. P. Bhattachcharyya (2002). New results on the synthesis of PID controllers. *IEEE Trans. Auto. Control 47*(2), 241–252.

Singh, A. and D. H. Mcewan (1975). The control of a process having appreciable transport lag – a laboratory case study. *IEEE Trans. Ind. Elec. and Contr. Instr. 22*(3), 396–401.

Skogestad, S. and M. Morari (1988). LV control of a high-purity distillation column. *Chem. Eng. Sci. 43*(1), 33–48.

Skogestad, S. and I. Postlethwaite (2005). *Multivariable Feedback Control: Analysis and Design* (2th ed.). West Sussex, England: Wiley-Interscience.

Stahl, H. and P. Hippe (1987). Design of pole placing controllers for stable and unstable systems with pure time delay. *Int. J. Control 45*(6), 2173–2182.

Stefani, R. T., B. Shahian, C. J. S. Jr., and G. H. Hostetter (2002). *Design of Feedback Control Systems.* NY: Oxford University Press.

Stephanopoulos, G. (1984). *Chemical Process Control: An Introduction to Theory and Practice* (4th ed.). NJ: Prentice Hall.

Sun, Y., P. W. Nelson, and A. G. Ulsoy (2010). *Time-delay Systems: Analysis and Control Using the Lambert W Function.* NJ: World Scientific Publishing Company.

Sun, Y. X. (1993). *Control Paper Making Processes (in Chinese).* Hangzhou, China: Zhangjiang University Press.

Thirunavukkarasu, I., V. I. George, G. S. Kumar, and A. Ramakalyan (2009). Robust stability and performance analysis of unstable process with dead time using mu synthesis. *ARPN Journal of Engineering and Applied Sciences 4*(2), 1–4.

Vidyasagar, M. (1985). *Control System Synthesis: A Factorization Approach.* MA: MIT Press.

Visioli, A. (2006). *Practical PID Control.* London: Springer.

Waller, K. V. T. and C. G. Nygardas (1975). On inverse response in process control. *Ind. Eng. Chem. 14*(3), 221–223.

Wang, Q., Y. Zhang, and M.-S. Chiu (2002). Decoupling internal model control for multivariable systems with multiple time delays. *Chemical Engineering Science 57*(1), 115–124.

Wang, Q. G. (2003). *Decoupling Control.* Lecture Notes in Control and Information Sciences. Berlin: Springer.

Wang, Y. C. and X. Z. Ren (1986). *Design Samples of Industrial Control Systems (in Chinese).* Beijing, China: Science Press.

Wang, Y. G. and H. H. Shao (2000). Optimal tuning for PI controller. *Automatica 36*(1), 147–152.

Wood, R. K. and M. W. Berry (1973). Terminal composition control of a binary distillation column. *Chem. Eng. Sci. 28*(9), 1707–1717.

Wu, M., Y. He, and J. H. She (2010). *Stability Analysis and Robust Control of Time-Delay Systems.* Beijing, China: Science Press.

Xiang, C., Q. Wang, X. Lu, L. Nguyen, and T. Lee (2007). Stabilization of second-order unstable delay processes by simple controllers. *J. of Process Control 17*(8), 675–682.

Youla, D. C., H. A. Jabr, and J. J. Bongiorno (1976). Modern Wiener Hopf design of optimal controllers – Part II: The multivariable case. *IEEE Trans. Auto. Control 21*(1), 319–338.

Zaccarian, L. and A. R. Teel (2002). A common framework for anti-windup, bumpless transfer and reliable designs. *Automatica 38*(10), 1735–1744.

Zames, G. (1981). Feedback and optimal sensitivity: model reference transformations, multiplicative seminorms and approximate inverse. *IEEE Trans. Auto. Control 26*(2), 301–320.

Zames, G. and B. A. Francis (1983). Feedback, minimax sensitivity, and optimal robustness. *IEEE Trans. Auto. Control 28*(5), 585–601.

Zhang, W., S. H. Chen, and W. D. Zhang (2012). Two-degree-of-freedom controller design for an ill-conditioned process using h2 decoupling control. *Ind. Eng. Chem. Res. 51*(45), 1472–1478.

Zhang, W. D. (1996). *Robust Control of Systems with Time Delay (in Chinese).* Ph. D. thesis, Zhejiang University, Hangzhou.

Zhang, W. D. (1998). *Analytical Design for Process Control (in Chinese).* Post-Doctoral Research Report. Shanghai Jiaotong University.

Zhang, W. D. (2006). Optimal design of the RZN PID controller for stable and unstable processes with time delay. *Ind. Eng. Chem. Res. 45*(4), 1408–1419.

Zhang, W. D., F. Allgower, and T. Liu (2006). Controller parameterization for SISO and MIMO plants with time delay. *System Control Letters 55*(10), 794–802.

Zhang, W. D., S. W. Gao, and D. Y. Gu (2011). No-weight design of H2 controllers for square plants. *IET Control Theory and Applications 5*(6), 785–794.

Zhang, W. D., D. Y. Gu, F. Li, Q. Z. Zhang, and L. Wang (2005). Design of PID controllers with best achievable performance for MIMO systems (in Chinese). China Patent 200510112231.2.

Zhang, W. D., D. Y. Gu, W. Wang, and X. M. Xu (2004). Quantitative performance design of a modified Smith predictor for unstable processes with time delay. *Ind. Eng. Chem. Res. 43*(1), 56–62.

Zhang, W. D., D. Y. Gu, and X. M. Xu (2002). Dual parameterization for linear nonminimum phase plants. *IEE Proc. -D 149*(6), 494–496.

Zhang, W. D. and C. Lin (2006). Multivariable Smith predictors design for nonsquare plants. *IEEE Trans. Control Systems Tech. 14*(6), 1145–1149.

Zhang, W. D., C. Lin, and L. L. Ou (2006). Algebraic solution to H2 control problems – Part II: The multivariable decoupling case. *Ind. Eng. Chem. Res. 45*(21), 7163–7176.

Zhang, W. D., L. L. Ou, and D. Y. Gu (2006). Algebraic solution to H2 control problems– Part I: The scalar case. *Ind. Eng. Chem. Res. 45*(21), 7151–7162.

Zhang, W. D. and Y. X. Sun (1995). Universal basis weight and moisture control model for paper machines (in Chinese). *Zhongguo Zaozhi Xuebao/Transactions of China Pulp and Paper 10*(s1), 84–92.

Zhang, W. D. and Y. X. Sun (1996a). H2 suboptimal control for uncertain time delay system (in Chinese). *Kongzhi Lilun yu Yingyong/Control Theory and Applications 13*(4), 495–499.

Zhang, W. D. and Y. X. Sun (1996b). Modified Smith predictor for the integrator/time delay processes. *Ind. Eng. Chem. Res. 35*(8), 2769–2772.

Zhang, W. D. and Y. X. Sun (1997). Optimal design of Smith predictor using Taylor series (in Chinese). *Yiqi Yibiao Xuebao/Chinese Journal of Scientific Instrument 18*(2), 213–216,224.

Zhang, W. D., Y. X. Sun, and X. M. Xu (1998a). New two degree-of-freedom Smith predictor for processes with time delay. *Automatica 34*(10), 1279–1282.

Zhang, W. D., Y. X. Sun, and X. M. Xu (1998b). Robust digital controller design of process with dead time – some new results. *IEE Proc. -D 145*(2), 159–164.

Zhang, W. D., Y. X. Sun, and X. M. Xu (2001). Modeling of fourdrinier and cylinder machines. *Development in Chem. Eng. and Mineral Processing 9*(1), 69–76.

Zhang, W. D., H. Wang, and X. M. Xu (2002). Analytical formulas for near-H infinity control of linear systems with time delay. In *Proc. Amer. Contr. Conf.*, Ankarage, A.K., pp. 2233–2238.

Zhang, W. D., Y. G. Xi, G. K. Yang, and X. M. Xu (2002). Design PID controllers for desired time domain and frequency domain response. *ISA Trans. 41*(4), 511–520.

Zhang, W. D. and X. M. Xu (2000). Counterexamples for sufficient and necessary condition of internal stability. *IEE Proc. -D 147*(3), 371–372.

Zhang, W. D. and X. M. Xu (2002a). H infinity PID controller design for runaway processes with time delay. *ISA Trans. 41*(3), 317–322.

Zhang, W. D. and X. M. Xu (2002b). Quantitative performance design for H_2 controllers of unstable processes with time delay (in Chinese). *Yiqi Yibiao Xuebao/ Chinese J. of Scientific Instrument 23*(3), 221–225.

Zhang, W. D. and X. M. Xu (2003). Comparison of several well-known controllers used in process control. *ISA Trans. 42*(2), 317–325.

Zhang, W. D., X. M. Xu, and Y. X. Sun (1999). Quantitative control of integrating processes with time delay. *Automatica 35*(4), 719–723.

Zhang, W. D., X. M. Xu, and Y. X. Sun (2000). Quantitative performance design for inverse response processes. *Ind. Eng. Chem. Res. 39*(6), 2056–2061.

Zheng, A., M. V. Kothare, and M. Morari (1994). Anti-windup design for internal model control. *Int. J. Control 60*(5), 1015–1024.

Zheng, D. Z. (1990). *Linear System Theory (in Chinese).* Beijing: Tsinghua University Press.

Zhou, K. M. and J. C. Doyle (1997). *Essentials Of Robust Control.* NJ: Prentice Hall.

Zhou, K. M., J. C. Doyle, and K. G. Glover (1996). *Robust and optimal control.* NJ: Prentice Hall.

Zhu, L. (2005). *The Measurement and Control Strategy for Molten Steel Level of Strip Casting (in Chinese).* Ph. D. thesis, Shanghai.

Ziegler, J. G. and N. B. Nichols (1942). Optimum settings for automatic controllers. *Trans. ASME 64*(11), 759–768.

Index

Printed and bound by CPI Group (UK) Ltd, Croydon, CR0 4YY
18/10/2024
01776261-0016